HVAC
Level One

Trainee Guide

Pearson

Boston   Columbus   Indianapolis   New York   San Francisco   Amsterdam
Cape Town Dubai London Madrid Milan Munich Paris Montreal Toronto Delhi
Mexico City Sao Paulo Sydney Hong Kong Seoul Singapore Taipei Tokyo

NCCER

President: Don Whyte
Vice President: Steve Greene
Chief Operations Officer: Katrina Kersch
HVAC Project Manager: Chris Wilson
Senior Development Manager: Mark Thomas

Senior Production Manager: Tim Davis
Quality Assurance Coordinator: Karyn Payne
Desktop Publishing Coordinator: James McKay
Permissions Specialists: Kelly Sadler
Production Specialist: Kelly Sadler
Editors: Graham Hack, Adrienne Payne

Writing and development services provided by Topaz Publications, Liverpool, NY

Lead Writer/Project Manager: Troy Staton
Desktop Publisher: Joanne Hart
Art Director: Alison Richmond

Permissions Editor: Andrea LaBarge
Writers: Troy Staton, Thomas Burke, Terry Egolf

Pearson

Director of Alliance/Partnership Management: Andrew Taylor
Editorial Assistant: Collin Lamothe
Program Manager: Alexandrina B. Wolf
Assistant Content Producer: Alma Dabral
Digital Content Producer: Jose Carchi
Director of Marketing: Leigh Ann Simms

Senior Marketing Manager: Brian Hoehl
Composition: NCCER
Printer/Binder: LSC Communications
Cover Printer: LSC Communications
Text Fonts: Palatino and Univers

Credits and acknowledgments for content borrowed from other sources and reproduced, with permission, in this textbook appear at the end of each module.

26 2023

Paper Bound:	ISBN-13:	978-0-13-518509-4
	ISBN-10:	0-13-518509-2
Case Bound:	ISBN-13:	978-0-13-489811-7
	ISBN-10:	0-13-489811-7

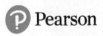

To the Trainee

Heating, ventilating, and air-conditioning (HVAC) systems technicians install and repair systems that regulate the temperature, humidity, and the total air quality in residential, commercial, industrial, and other buildings. Refrigeration systems, which are used to transport or store food, medicine, and other perishable items, are also a major part of the HVAC technician's job. Hydronics (water-based heating systems), solar panels, and other specialized heating, cooling, and refrigeration systems are also important facets of the HVAC industry.

As a technician, you must be able to install, maintain, and troubleshoot problems throughout the entire system. You must know how to follow drawings or other specifications to maintain any system. You will have to understand air movement, temperatures, and pressures. You may also need basic working knowledge of other crafts such as sheet metal, welding, basic pipefitting, and electrical practices.

Nearly all buildings and homes in the United States alone use forms of heating, cooling, and ventilation to maintain comfort. The increasing advancement of HVAC technology causes employers to recognize the importance of continuous education and keeping up to speed with the latest equipment and skills. Hence, technical school training or apprenticeship programs often provide an advantage and a higher qualification for employment.

NCCER's HVAC program has been designed by highly qualified subject matter experts. The four levels present an apprentice approach to the HVAC field, including theoretical and practical skills essential to your success as an HVAC technician. The US Department of Labor projects faster than average job growth in the HVAC industry.

We wish you the best as you begin an exciting and promising career. This newly revised HVAC curriculum will help you enter the workforce with the knowledge and skills needed to perform productively in either the residential or commercial market.

New with *HVAC Level One*

NCCER is proud to release the newest edition of *HVAC Level One* with updates to the curriculum that will engage you and give you the best training possible.

The module "Basic Electricity" now covers OSHA regulations for lockout and tagging of equipment. It also offers additional training on three-phase power. The "Introduction to Heating" module offers new coverage of dual-fuel systems and geothermal heat pumps. In "Introduction to Cooling," there is addi-

tional training on vane compressors as well as a look at the industry's refrigerant landscape. The "Air Distribution Systems" module offers information about both natural air movement and forced air, including new diagrams showing the relationship between static, velocity, and total pressures.

We wish you success as you progress through this training program. If you have any comments on how NCCER might improve upon this textbook, please complete the User Update form located at the back of each module and send it to us. We will always consider and respond to input from our customers.

We invite you to visit the NCCER website at **www.nccer.org** for information on the latest product releases and training, as well as online versions of the *Cornerstone* magazine and Pearson's NCCER product catalog.

Your feedback is welcome. You may email your comments to **curriculum@nccer.org** or send general comments and inquiries to **info@nccer.org**.

NCCER Standardized Curricula

NCCER is a not-for-profit 501(c)(3) education foundation established in 1996 by the world's largest and most progressive construction companies and national construction associations. It was founded to address the severe workforce shortage facing the industry and to develop a standardized training process and curricula. Today, NCCER is supported by hundreds of leading construction and maintenance companies, manufacturers, and national associations. The NCCER Standardized Curricula was developed by NCCER in partnership with Pearson, the world's largest educational publisher.

Some features of the NCCER Standardized Curricula are as follows:

- An industry-proven record of success
- Curricula developed by the industry, for the industry
- National standardization providing portability of learned job skills and educational credits
- Compliance with the Office of Apprenticeship requirements for related classroom training (*CFR 29:29*)
- Well-illustrated, up-to-date, and practical information

NCCER also maintains the NCCER Registry, which provides transcripts, certificates, and wallet cards to individuals who have successfully completed a level of training within a craft in NCCER's Curricula. *Training programs must be delivered by an NCCER Accredited Training Sponsor in order to receive these credentials.*

Special Features

In an effort to provide a comprehensive and user-friendly training resource, this curriculum showcases several informative features. Whether you are a visual or hands-on learner, these features are intended to enhance your knowledge of the construction industry as you progress in your training. Some of the features you may find in the curriculum are explained below.

Introduction

This introductory page, found at the beginning of each module, lists the module Objectives, Performance Tasks, and Trade Terms. The Objectives list the knowledge you will acquire after successfully completing the module. The Performance Tasks give you an opportunity to apply your knowledge to real-world tasks. The Trade Terms are industry-specific vocabulary that you will learn as you study this module.

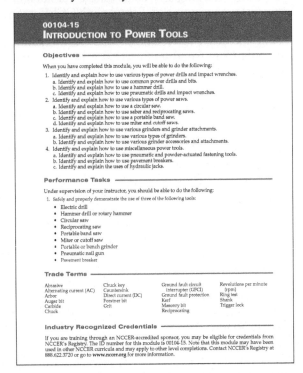

Figures and Tables

Photographs, drawings, diagrams, and tables are used throughout each module to illustrate important concepts and provide clarity for complex instructions. Text references to figures and tables are emphasized with *italic* type.

Notes, Cautions, and Warnings

Safety features are set off from the main text in highlighted boxes and categorized according to the potential danger involved. Notes simply provide additional information. Cautions flag a hazardous issue that could cause damage to materials or equipment. Warnings stress a potentially dangerous situation that could result in injury or death to workers.

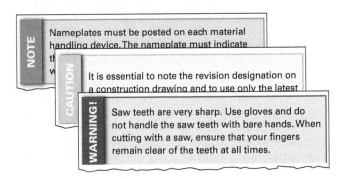

Trade Features

Trade features present technical tips and professional practices based on real-life scenarios similar to those you might encounter on the job site.

Bowline Trivia

Some people use this saying to help them remember how to tie a bowline: "The rabbit comes out of his hole, around a tree, and back into the hole."

Case History

Case History features emphasize the importance of safety by citing examples of the costly (and often devastating) consequences of ignoring best practices or OSHA regulations.

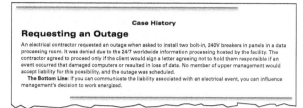

Case History

Requesting an Outage

An electrical contractor requested an outage when asked to install two bolt-in, 240V breakers in panels in a data processing room. It was denied due to the 24/7 worldwide information processing hosted by the facility. The contractor agreed to proceed only if the client would sign a letter agreeing not to hold them responsible if an event occurred that damaged computers or resulted in loss of data. No member of upper management would accept liability for this possibility, and the outage was scheduled.

The Bottom Line: If you can communicate the liability associated with an electrical event, you can influence management's decision to work energized.

Going Green

Going Green features present steps being taken within the construction industry to protect the environment and save energy, emphasizing choices that can be made on the job to preserve the health of the planet.

GOING GREEN

Reducing Your Carbon Footprint

Many companies are taking part in the paperless movement. They reduce their environmental impact by reducing the amount of paper they use. Using email helps to reduce the amount of paper used,

Did You Know

Did You Know features introduce historical tidbits or interesting and sometimes surprising facts about the trade.

Did You Know?

Safety First

Safety training is required for all activities. Never operate tools, machinery, or equipment without prior training. Always refer to the manufacturer's instructions.

Step-by-Step Instructions

Step-by-step instructions are used throughout to guide you through technical procedures and tasks from start to finish. These steps show you how to perform a task safely and efficiently.

Perform the following steps to erect this system area scaffold:

Step 1 Gather and inspect all scaffold equipment for the scaffold arrangement.

Step 2 Place appropriate mudsills in their approximate locations.

Step 3 Attach the screw jacks to the mudsills.

Trade Terms

Each module presents a list of Trade Terms that are discussed within the text and defined in the Glossary at the end of the module. These terms are presented in the text with bold, blue type upon their first occurrence. To make searches for key information easier, a comprehensive Glossary of Trade Terms from all modules is located at the back of this book.

During a rigging operation, the load being lifted or moved must be connected to the apparatus, such as a crane, that will provide the power for movement. The connector—the link between the load and the apparatus—is often a sling made of synthetic, chain, or wire rope materials. This section focuses on three types of slings:

Section Review

Each section of the module wraps up with a list of Additional Resources for further study and Section Review questions designed to test your knowledge of the Objectives for that section.

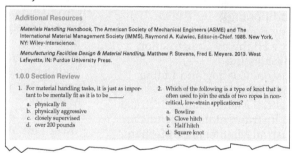

Additional Resources

Materials Handling Handbook, The American Society of Mechanical Engineers (ASME) and The International Material Management Society (IMMS), Raymond A. Kulwiec, Editor-in-Chief. 1985. New York, NY: Wiley-Interscience.

Manufacturing Facilities Design & Material Handling, Matthew P. Stevens, Fred E. Meyers. 2013. West Lafayette, IN: Purdue University Press.

1.0.0 Section Review

1. For material handling tasks, it is just as important to be mentally fit as it is to be _____.
   a. physically fit
   b. physically aggressive
   c. closely supervised
   d. over 200 pounds

2. Which of the following is a type of knot that is often used to join the ends of two ropes in non-critical, low-strain applications?
   a. Bowline
   b. Clove hitch
   c. Half hitch
   d. Square knot

Review Questions

The end-of-module Review Questions can be used to measure and reinforce your knowledge of the module's content.

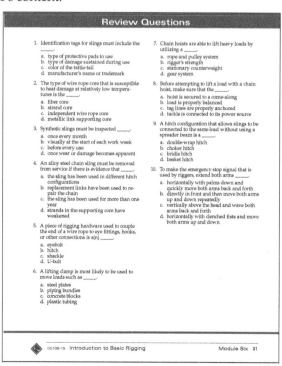

Review Questions

1. Identification tags for slings must include the _____.
   a. type of protective pads to use
   b. type of damage sustained during use
   c. color of the tattle-tail
   d. manufacturer's name or trademark

2. The type of wire rope core that is susceptible to heat damage at relatively low temperatures is the _____.
   a. fiber core
   b. strand core
   c. independent wire rope core
   d. metallic link supporting core

3. Synthetic slings must be inspected _____.
   a. once every month
   b. visually at the start of each work week
   c. before every use
   d. once wear or damage becomes apparent

4. An alloy steel chain sling must be removed from service if there is evidence that _____.
   a. the sling has been used in different hitch configurations
   b. replacement links have been used to repair the chain
   c. the sling has been used for more than one year
   d. strands in the supporting core have weakened

5. A piece of rigging hardware used to couple the end of a wire rope to eye fittings, hooks, or other connections is a(n) _____.
   a. eyebolt
   b. hitch
   c. shackle
   d. U-bolt

6. A lifting clamp is most likely to be used to move loads such as _____.
   a. steel plates
   b. piping bundles
   c. concrete blocks
   d. plastic tubing

7. Chain hoists are able to lift heavy loads by utilizing a _____.
   a. rope and pulley system
   b. rigger's strength
   c. stationary counterweight
   d. gear system

8. Before attempting to lift a load with a chain hoist, make sure that the _____.
   a. hoist is secured to a come-along
   b. load is properly balanced
   c. tag lines are properly anchored
   d. tackle is connected to its power source

9. A hitch configuration that allows slings to be connected to the same load without using a spreader beam is a _____.
   a. double-wrap hitch
   b. choker hitch
   c. bridle hitch
   d. basket hitch

10. To make the emergency stop signal that is used by riggers, extend both arms _____.
    a. horizontally with palms down and quickly move both arms back and forth
    b. directly in front and then move both arms up and down repeatedly
    c. vertically above the head and wave both arms back and forth
    d. horizontally with clenched fists and move both arms up and down

NCCER Standardized Curricula

NCCER's training programs comprise more than 80 construction, maintenance, pipeline, and utility areas and include skills assessments, safety training, and management education.

Boilermaking
Cabinetmaking
Carpentry
Concrete Finishing
Construction Craft Laborer
Construction Technology
Core Curriculum: Introductory
 Craft Skills
Drywall
Electrical
Electronic Systems Technician
Heating, Ventilating, and Air
 Conditioning
Heavy Equipment Operations
Heavy Highway Construction
Hydroblasting
Industrial Coating and Lining
 Application Specialist
Industrial Maintenance Electrical
 and Instrumentation Technician
Industrial Maintenance Mechanic
Instrumentation
Ironworking
Manufactured Construction
 Technology
Masonry
Mechanical Insulating
Millwright
Mobile Crane Operations
Painting
Painting, Industrial
Pipefitting
Pipelayer
Plumbing
Reinforcing Ironwork
Rigging
Scaffolding
Sheet Metal
Signal Person
Site Layout
Sprinkler Fitting
Tower Crane Operator
Welding

Maritime

Maritime Industry Fundamentals
Maritime Pipefitting
Maritime Structural Fitter

Green/Sustainable Construction

Building Auditor
Fundamentals of Weatherization
Introduction to Weatherization
Sustainable Construction
 Supervisor
Weatherization Crew Chief
Weatherization Technician
Your Role in the Green
 Environment

Energy

Alternative Energy
Introduction to the Power Industry
Introduction to Solar Photovoltaics
Power Generation Maintenance
 Electrician
Power Generation I&C
 Maintenance Technician
Power Generation Maintenance
 Mechanic
Power Line Worker
Power Line Worker: Distribution
Power Line Worker: Substation
Power Line Worker: Transmission
Solar Photovoltaic Systems Installer
Wind Energy
Wind Turbine Maintenance
 Technician

Pipeline

Abnormal Operating Conditions,
 Control Center
Abnormal Operating Conditions,
 Field and Gas
Corrosion Control
Electrical and Instrumentation
Field and Control Center
 Operations
Introduction to the Pipeline
 Industry
Maintenance
Mechanical

Safety

Field Safety
Safety Orientation
Safety Technology

Supplemental Titles

Applied Construction Math
Tools for Success

Management

Construction Workforce
 Development Professional
Fundamentals of Crew Leadership
Mentoring for Craft Professionals
Project Management
Project Supervision

Spanish Titles

Acabado de concreto: nivel uno
 (*Concrete Finishing Level One*)
Aislamiento: nivel uno
 (*Insulating Level One*)
Albañilería: nivel uno
 (*Masonry Level One*)
Andamios (*Scaffolding*)
Carpintería: Formas para
 carpintería, nivel tres
 (*Carpentry: Carpentry Forms, Level
 Three*)
Currículo básico: habilidades
 introductorias del oficio
 (*Core Curriculum: Introductory Craft
 Skills*)
Electricidad: nivel uno
 (*Electrical Level One*)
Herrería: nivel uno
 (*Ironworking Level One*)
Herrería de refuerzo: nivel uno
 (*Reinforcing Ironwork Level One*)
Instalación de rociadores: nivel uno
 (*Sprinkler Fitting Level One*)
Instalación de tuberías: nivel uno
 (*Pipefitting Level One*)
Instrumentación: nivel uno, nivel
 dos, nivel tres, nivel cuatro
 (*Instrumentation Levels One through
 Four*)
Orientación de seguridad
 (*Safety Orientation*)
Paneles de yeso: nivel uno
 (*Drywall Level One*)
Seguridad de campo
 (*Field Safety*)

Acknowledgments

This curriculum was revised as a result of the farsightedness and leadership of the following sponsors:

Builders Association of North Central Florida
CareerSafety Center
Center for Employment Training
Duke Energy
Fort Scott Community College
Hubbard Construction

Industrial Management and Training Institute, Inc.
Lee Company
Lincoln Tech
Santa Fe College
Windham School District

This curriculum would not exist were it not for the dedication and unselfish energy of those volunteers who served on the Authoring Team. A sincere thanks is extended to the following:

Corey Driggs
Art Grant
Joseph Pietrzak
Norman Sparks

John Stronkowski
Tony Vazquez
Ted Watts

A sincere thanks is also extended to the dedication and assistance provided by the following technical advisors:

Senobio Aguilera   Lenny Joseph   Chris Sterrett

NCCER Partners

American Council for Construction Education
American Fire Sprinkler Association
Associated Builders and Contractors, Inc.
Associated General Contractors of America
Association for Career and Technical Education
Association for Skilled and Technical Sciences
Construction Industry Institute
Construction Users Roundtable
Design Build Institute of America
GSSC – Gulf States Shipbuilders Consortium
ISN
Manufacturing Institute
Mason Contractors Association of America
Merit Contractors Association of Canada
NACE International
National Association of Women in Construction
National Insulation Association
National Technical Honor Society
National Utility Contractors Association
NAWIC Education Foundation
North American Crane Bureau
North American Technician Excellence
Pearson

Prov
SkillsUSA®
Steel Erectors Association of America
U.S. Army Corps of Engineers
University of Florida, M. E. Rinker Sr., School of Construction Management
Women Construction Owners & Executives, USA

NCCER Business Partners

Contents

Module One
Introduction to HVAC

Covers the basic principles of heating, ventilating, and air conditioning, career opportunities in HVAC, and how apprenticeship programs are constructed. Basic safety principles, as well as trade licensure and EPA guidelines, are also introduced. (Module ID 03101; 7.5 Hours)

Module Two
Trade Mathematics

Explains how to solve HVACR trade-related problems involving the measurement of lines, area, volume, weights, angles, pressure, vacuum, and temperature. Also includes a review of scientific notation, powers, roots, and basic algebra and geometry. (Module ID 03102; 10 Hours)

Module Three
Basic Electricity

Introduces the concept of power generation and distribution, common electrical components, AC and DC circuits, and electrical safety as it relates to the HVAC field. Introduces reading and interpreting wiring diagrams. (Module ID 03106; 12.5 Hours)

Module Four
Introduction to Heating

Covers the fundamentals of heating systems and the combustion process. The different types and designs of gas furnaces and their components, as well as basic procedures for their installation and service, are also provided. (Module ID 03108; 15 Hours)

Module Five
Introduction to Cooling

Explains the fundamental operating concepts of the refrigeration cycle and identifies both primary and secondary components found in typical HVACR systems. Common refrigerants are introduced as well. Describes the principles of heat transfer and the essential pressure-temperature relationships of refrigerants. Basic control concepts for simple systems are also introduced. (Module ID 03107; 30 Hours)

Module Six
Introduction to Air Distribution Systems

Describes the factors related to air movement and its measurement in common air distribution systems. The required mechanical equipment and materials used to create air distribution systems are also present-ed. Basic system design principles for both hot and cold climates are introduced. (Module ID 03109; 15 Hours)

Module Seven
Basic Copper and Plastic Piping Practices

Explains how to identify types of copper tubing and fittings used in the HVACR industry and how they are mechanically joined. The identification and application of various types of plastic piping, along with their common assembly and installation practices, are also presented. (Module ID 03103; 10 Hours)

Module Eight
Soldering and Brazing

Introduces the equipment, techniques, and materials used to safely join copper tubing through both soldering and brazing. The required PPE, preparation, and work processes are covered in detail. The procedures for brazing copper to dissimilar materials are also provided. (Module ID 03104; 10 Hours)

Module Nine

Basic Carbon Steel Piping Practices

Explains how to identify various carbon steel piping materials and fittings. The joining and installation of threaded and grooved carbon steel piping systems is covered, with detailed coverage of threading and grooving techniques included. (Module ID 03105; 10 Hours)

Glossary

Index

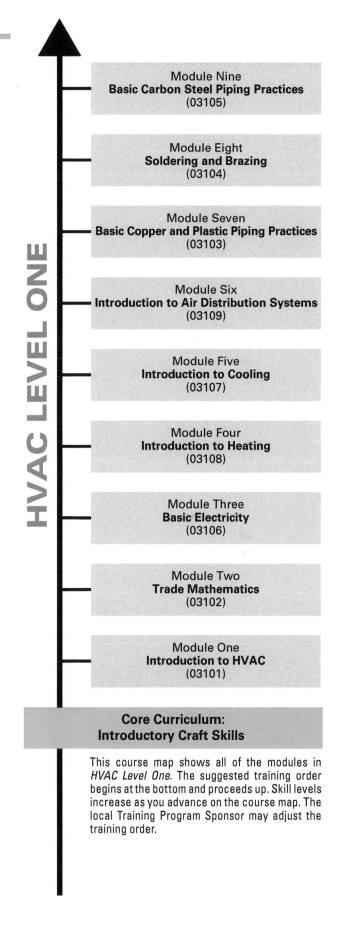

HVAC LEVEL ONE

Module Nine
Basic Carbon Steel Piping Practices
(03105)

Module Eight
Soldering and Brazing
(03104)

Module Seven
Basic Copper and Plastic Piping Practices
(03103)

Module Six
Introduction to Air Distribution Systems
(03109)

Module Five
Introduction to Cooling
(03107)

Module Four
Introduction to Heating
(03108)

Module Three
Basic Electricity
(03106)

Module Two
Trade Mathematics
(03102)

Module One
Introduction to HVAC
(03101)

**Core Curriculum:
Introductory Craft Skills**

This course map shows all of the modules in *HVAC Level One*. The suggested training order begins at the bottom and proceeds up. Skill levels increase as you advance on the course map. The local Training Program Sponsor may adjust the training order.

Introduction to HVAC

OVERVIEW

Virtually all of the millions of homes and businesses in the United States have a heating system. A large percentage of homes and businesses have comfort cooling systems as well. In addition, there are many thousands of stores and restaurants that use refrigeration equipment. Workers trained in the HVAC industry have the opportunity to install systems in new construction, service equipment in existing construction, and replace aging systems.

Module 03101

Trainees with successful module completions may be eligible for credentialing through the NCCER Registry. To learn more, go to **www.nccer.org** or contact us at **1.888.622.3720**. Our website has information on the latest product releases and training, as well as online versions of our *Cornerstone* magazine and Pearson's product catalog.

Your feedback is welcome. You may email your comments to **curriculum@nccer.org**, send general comments and inquiries to **info@nccer.org**, or fill in the User Update form at the back of this module.

This information is general in nature and intended for training purposes only. Actual performance of activities described in this manual requires compliance with all applicable operating, service, maintenance, and safety procedures under the direction of qualified personnel. References in this manual to patented or proprietary devices do not constitute a recommendation of their use.

Objectives

When you have completed this module, you will be able to do the following:

1. Explain the basic principles of heating, ventilation, air conditioning, and refrigeration.
   a. Explain the principles of heating.
   b. Explain the principles of ventilation.
   c. Explain the principles of air conditioning.
   d. Explain the principles of refrigeration.
2. Describe the principles that guide HVACR installation and service techniques.
   a. Identify common safety principles and organizations.
   b. Describe the importance of LEED construction and energy management.
   c. Describe trade licensing and certification requirements.
   d. Identify important codes and permits.
3. Identify career paths available in the HVACR trade.
   a. Identify the responsibilities and characteristics needed to be a successful HVACR technician.
   b. Identify residential, commercial, and industrial career opportunities.
   c. Describe opportunities provided by equipment manufacturers.

Performance Tasks

This is a knowledge-based module; there are no performance tasks.

Trade Terms

Bond
Carbon monoxide (CO)
Chiller
Chlorofluorocarbon (CFC) refrigerant
Compressor
Condenser
Easement
Evaporator
Expansion device
Heat pump
Heat transfer
Hydrochlorofluorocarbon (HCFC) refrigerant

Hydronic
International Building Code (IBC)
Mechanical refrigeration
Mechanical refrigeration cycle
Noxious
On the job learning (OJL)
Reclaimed
Recovery
Recycling
Sustainable construction
Toxic

Industry Recognized Credentials

If you are training through an NCCER-accredited sponsor, you may be eligible for credentials from NCCER's Registry. The ID number for this module is 03101. Note that this module may have been used in other NCCER curricula and may apply to other level completions. Contact NCCER's Registry at 888.622.3720 or go to **www.nccer.org** for more information.

Contents

1.0.0 HVACR Principles.. 1

   1.1.0 Heating 1

   1.2.0 Ventilation 4

   1.3.0 Air Conditioning 5

   1.4.0 Refrigeration 7

2.0.0 Guiding Principles for HVACR Service Technicians 10

   2.1.0 Safety 10

   2.2.0 Energy Conservation and LEED Construction Principles 11

      2.2.1 LEED Program.. 12

   2.3.0 Licensing and Certification... 12

      2.3.1 EPA Certification 13

   2.4.0 Codes and Permits 15

3.0.0 Careers in HVACR 18

   3.1.0 Employee Responsibilities and Characteristics 18

      3.1.1 Professionalism 18

      3.1.2 Honesty 18

      3.1.3 Loyalty 18

      3.1.4 Willingness to Learn... 19

      3.1.5 Willingness to Take Responsibility 19

      3.1.6 Willingness to Cooperate 19

      3.1.7 Tardiness and Absenteeism.. 19

   3.2.0 Career Opportunities 20

      3.2.1 Residential/Light Commercial 20

      3.2.2 Commercial/Industrial.. 20

      3.2.3 HVACR Training... 21

   3.3.0 Opportunities in Manufacturing 24

Figures and Tables

Figure 1 Forced-air heating 2
Figure 2 Hot water heating 3
Figure 3 Humidifier and electronic air cleaner in a residential
heating system 5
Figure 4 Basic refrigeration cycle for an air conditioner 6
Figure 5 Heat pump heating cycle 7
Figure 6 Retail refrigeration equipment 8
Figure 7 Comfort cooling and refrigeration temperature ranges 8
Figure 8 Heat recovery ventilator (HRV) 11
Figure 9 Duct joints sealed with mastic 12
Figure 10 Refrigerant cylinders 15
Figure 11 Refrigerant recovery equipment 15
Figure 12 Career opportunities in the HVACR trade 21

Table 1 Common Refrigerant Environmental Impact Data 14

1.0.0 HVACR PRINCIPLES

Objective

Explain the basic principles of heating, ventilation, air conditioning, and refrigeration.

  a. Explain the principles of heating.
  b. Explain the principles of ventilation.
  c. Explain the principles of air conditioning.
  d. Explain the principles of refrigeration.

Trade Terms

Compressor: In a refrigeration system, the mechanical device that converts low-pressure, low-temperature refrigerant gas into high-temperature, high-pressure refrigerant gas.

Condenser: A heat exchanger that transfers heat from the refrigerant flowing inside it to the air or water flowing over it.

Evaporator: A heat exchanger that transfers heat from the air flowing over it to the cooler refrigerant flowing through it.

Expansion device: Also known as the liquid metering device or metering device. Provides a pressure drop that converts the high-temperature, high-pressure liquid refrigerant from the condenser into the low-temperature, low-pressure liquid refrigerant entering the evaporator.

Heat pump: A comfort air conditioner that is able to produce heat by reversing the mechanical refrigeration cycle.

Heat transfer: The transfer of heat from a warmer substance to a cooler substance.

Hydronic: A heating or air conditioning system that uses water as a heat transfer medium.

Mechanical refrigeration: The use of machinery to provide cooling.

Mechanical refrigeration cycle: The process by which a circulating refrigerant absorbs heat from one location and transfers it to another location.

Noxious: Harmful to health.

Toxic: Poisonous.

Today, the Heating, Ventilating, Air Conditioning, and Refrigeration (HVACR) industry provides the means to control the temperature, humidity, and even the cleanliness of the air in our homes, schools, offices, and factories. Comfort air conditioning and product/process refrigeration are based on the same principles, so there are many common elements in the training required for the two areas.

1.1.0 Heating

Early humans burned fuel as a source of heat. That hasn't changed; what's different between then and now is the way it's done. It is no longer necessary to huddle around a wood fire to keep warm. Instead, a central heating source such as a furnace or boiler does the job using the heat transfer principle; that is, heat is created in one place and carried to another place by means of air or water.

For example, in a common household furnace, fuel oil, natural gas, or propane gas is burned to create heat, which warms metal plates known as heat exchangers (*Figure 1*). Air from living spaces is circulated over the heat exchangers and returned to the living spaces as heated air. This type of system is known as a forced-air system. It is the most common type of central heating system used in the United States.

Water is also used as a heat exchange medium. The water is heated in a boiler (*Figure 2*), then pumped through pipes to heat exchangers, where the heat it contains is transferred to the surrounding air. The heat exchangers are usually baseboard heating elements located in the space to be heated. Systems that use water or steam as a heat exchange medium are known as hydronic heating systems. This type of system is more common in the Northeast and Midwest than in other parts of the United States.

Natural gas and fuel oil are, by far, the most widely used heating fuels. Of the two, natural gas is far more common. Oil heat is more common in the Northeast than in other parts of the country. Propane gas is used in many parts of the country in place of oil or natural gas. Propane fuels are primarily used in rural areas of the country where natural gas pipelines are not readily available. Propane is stored in tanks on the property where the equipment is located.

Electricity is also used as a heat source. In an electric heating system, electricity flows through coils of heavy wire, causing the coils to become hot. Air from the conditioned space is passed over the coils and the heat from the coils is transferred to the air. Because electricity can be expensive, total electric resistance heating is no longer common in cold climates. Total electric heat is more likely to be used in warm climates where heat is seldom required. Heat pumps are used in areas with moderate climates, such as the Southeast-

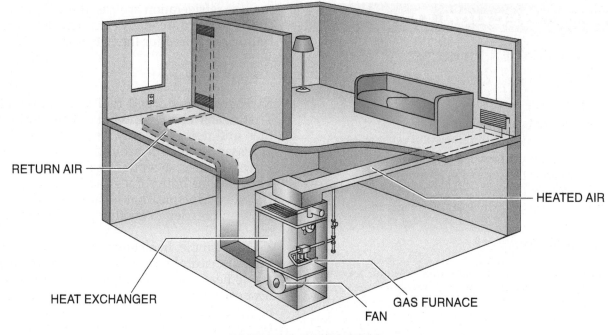

RETURN AIR

HEATED AIR

HEAT EXCHANGER

GAS FURNACE

FAN

BASEMENT INSTALLATION

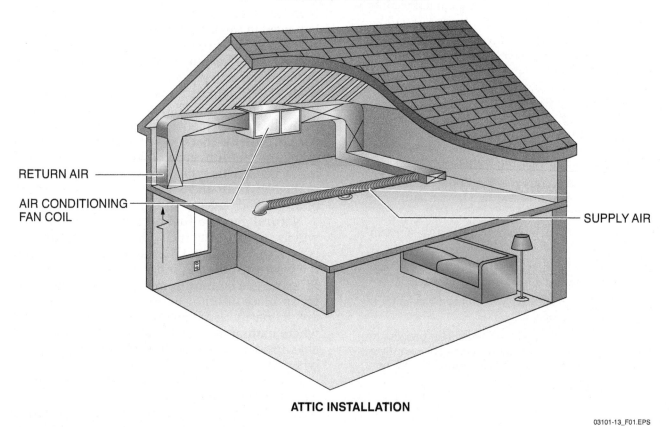

RETURN AIR

AIR CONDITIONING
FAN COIL

SUPPLY AIR

ATTIC INSTALLATION

03101-13_F01.EPS

Figure 1 Forced-air heating.

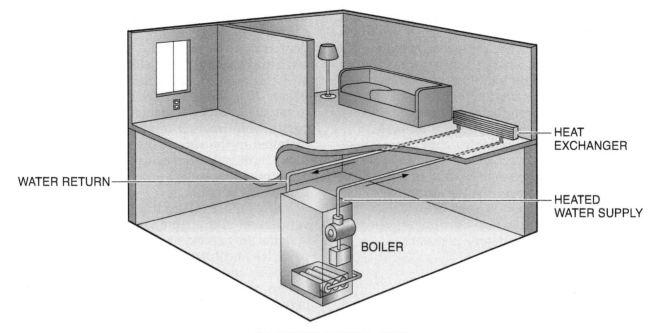

BASEMENT INSTALLATION

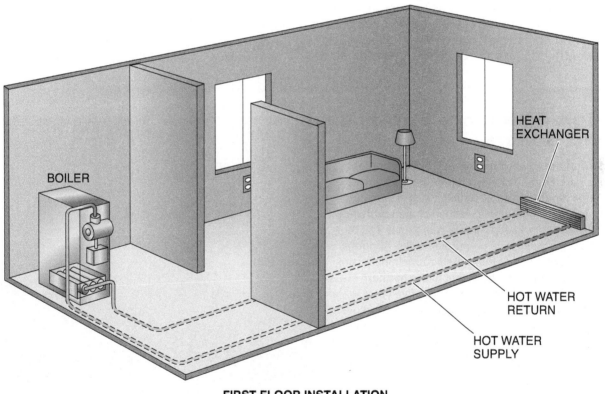

FIRST-FLOOR INSTALLATION

03101-13_F02.EPS

Figure 2 Hot water heating.

ern United States, and in most cases are preferred over total electric heat. Heat pumps are often augmented with electric resistance heating coils for those instances where the heat pump cannot efficiently provide the required level of heat. Although heat pumps are another form of electric heating, they can produce heat more effectively by using the mechanical refrigeration cycle.

In some areas of the country, a main or supplementary heating system may be fueled by wood or coal. Due to increases in the cost of fossil fuels such as oil and gas over the last several decades, and the fact that wood is a relatively cheap and renewable resource, wood-burning stoves and furnaces have remained quite popular.

1.2.0 Ventilation

Ventilation is the introduction of fresh air into a closed space in order to control air quality. Fresh air entering a building provides oxygen. The air in our homes, schools, and offices contains dust, pollen, and molds, as well as vapors and odors from a variety of sources. Relatively simple air circulation and filtration methods, including natural ventilation, are used to help keep the air in these environments clean and fresh. Among the common methods of improving the air in residential applications is the addition of humidifiers, ultraviolet light, and electronic air cleaners to air-handling systems such as forced-air furnaces (*Figure 3*). Humidifiers are used to add moisture to the air in order to overcome dryness that occurs in cold-weather months. They are typically installed in the supply duct of a furnace. Ultraviolet light is used to control bacteria, fungi, and other microorganisms in evaporator coils, drain pans, and ductwork.

Many industrial environments require special ventilation and air-management systems. Such systems are needed to eliminate noxious or toxic particles and fumes that may be created by the processes and materials used at the facility.

The energy-efficient buildings constructed between the mid-1970s and early 1990s actually contributed to indoor air quality problems. Building construction became so tight that the interior was robbed of the fresh air that would normally infiltrate around windows and doors and tiny openings in the structure. In the 1990s, indoor air quality became a major concern and the term *sick building* was coined to represent structures

High-Efficiency Furnaces and Boilers

Many of the newer furnaces and boilers available today are high-efficiency units with efficiency ratings ranging from 90 percent to 96.6 percent for gas. This compares to ratings of approximately 70 percent for old-style gas-burning units and 65 percent for older-style oil-burning units. Although high-efficiency equipment costs more initially, the energy savings for a customer in the northern states can provide a payback in as little as five years when compared with less efficient equipment. Moreover, high-efficiency units used for replacement purposes are often eligible for energy saving cash incentives from local utilities. The furnace shown is known as a multipoise furnace, which means it can be used in any airflow configuration: upflow, downflow, and to either side. At one time, it was necessary to order a furnace specifically for the airflow configuration.

03101-13_SA01.EPS

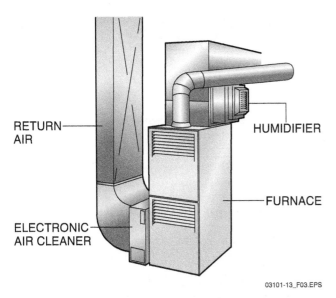

RETURN AIR

HUMIDIFIER

FURNACE

ELECTRONIC AIR CLEANER

03101-13_F03.EPS

Figure 3 Humidifier and electronic air cleaner in a residential heating system.

in which the air being breathed by the occupants lacked sufficient fresh air and contained excessive amounts of dust, germs, molds, and other pollutants. The recognition of this problem has led to the much wider use of electronic air cleaners, fresh air ventilation, duct cleaning, and other measures aimed at improving the indoor air we breathe. The problem has also created more opportunities in the HVACR industry to manufacture, sell, install, and service equipment designed to improve indoor air quality.

The US government has strict regulations governing indoor air quality (IAQ) in both commercial and industrial environments, and the release of toxic materials to the outside air. Where noxious or toxic fumes may be present, the indoor air must be constantly replaced with fresh air. Fans and other ventilating devices are normally used for this purpose. Special filtering devices may also be required. These devices not only protect the health of building occupants, but also prevent the release of toxic materials to the outdoor air.

1.3.0 Air Conditioning

The most common method of cooling indoor spaces is based on what is known as mechanical refrigeration. This method, which came into use early in the twentieth century, is based on a principle known as the mechanical refrigeration cycle (*Figure 4*).

Simply stated, the refrigeration cycle relies on the ability of chemical refrigerants to absorb heat. If a cold refrigerant flows through a warm space, it absorbs heat from the space. Having given up heat to the refrigerant, the space becomes cooler. The colder the refrigerant, the more heat it will absorb, and the cooler the space will become. If the super-hot refrigerant flows to a cooler location, the outdoors for example, the refrigerant gives up the heat it absorbed from the indoors and becomes cool again.

A mechanical refrigeration system is a sealed system operating under pressure. The main elements of a mechanical refrigeration system are the:

- Compressor – A compressor provides the force that circulates the refrigerant and creates the pressure differential necessary for the refrigeration cycle to work. A special refrigerant compressor is used.
- Evaporator – An evaporator is a heat exchanger where the heat in the warm indoor air is transferred to the cold refrigerant.
- Condenser – A condenser is also a heat exchanger. In the condenser, the heat absorbed by the refrigerant is transferred to relatively cooler outdoor air. Even though it may be hot outdoors, the temperature of the refrigerant flowing through the condenser will be higher.
- Expansion device – The relatively simple expansion device is essential to the refrigeration cycle. It provides a pressure drop that lowers the pressure and boiling point of the refrigerant as it enters the evaporator. This allows the refrigerant to become a cold liquid/gas mixture and absorb heat in the evaporator.

The Mechanical Refrigeration Cycle

Many people think that an air conditioner adds cool air to an indoor space. In reality, the basic principle of air conditioning and the mechanical refrigeration cycle is that heat is extracted from the indoor air and transferred to another location (the outdoors) by the refrigerant that flows through the cycle.

A special type of air conditioner known as a heat pump is widely used in moderate climates to provide both cooling and heating. Heat pumps are extremely efficient; however, they are most effective in climates where the temperature generally does not fall below 25°F or 30°F. In colder areas, heat pumps can be combined with a gas- or oil-fired, forced-air furnace. In such arrangements, the furnace automatically takes over heating duties when the outdoor temperature falls below the efficient range of the heat pump. Like high-efficiency furnaces and boilers, heat pumps are often eligible for energy saving cash incentives from local power companies.

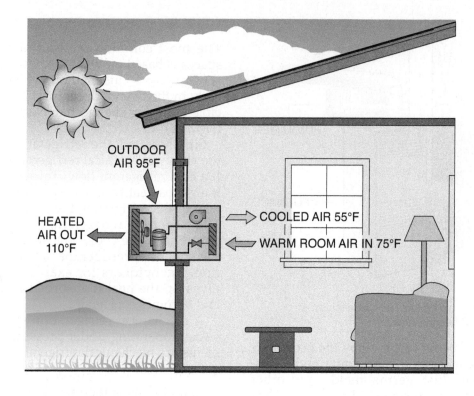

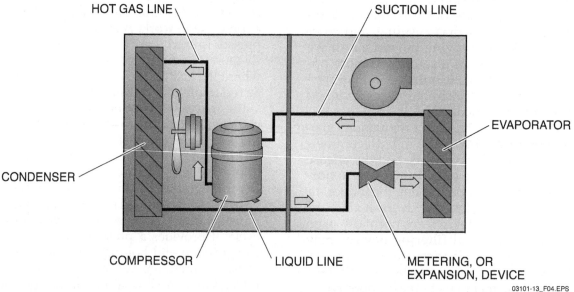

OUTDOOR
AIR 95°F

HEATED
AIR OUT
110°F

COOLED AIR 55°F

WARM ROOM AIR IN 75°F

HOT GAS LINE

SUCTION LINE

EVAPORATOR

CONDENSER

COMPRESSOR

LIQUID LINE

METERING, OR
EXPANSION, DEVICE

03101-13_F04.EPS

Figure 4 Basic refrigeration cycle for an air conditioner.

Absolute Zero

Absolute zero represents the theoretical point where there is a total absence of heat. It is roughly expressed as −459.67°F or −273.15°C. On another temperature scale that is based on absolute zero, known as the Kelvin scale, it is expressed as 0 K; no degree symbol is used. The Kelvin scale is incrementally the same as the Celsius scale, with 273 K (0°C) being the freezing point of water.

Although the Kelvin scale is by far the most popular in the scientific community, there is yet another scale that also begins at absolute zero. The Rankine scale is a companion to the Fahrenheit scale. Like the Kelvin and Celsius scales, the Rankine and Fahrenheit scales are also the same incrementally. 0°R and −459.76°F are the same temperature point, as is the freezing point of water at 491.67°R and 32°F.

Heat pumps produce heat by reversing the cooling cycle (*Figure 5*). For this reason, heat pumps are sometimes called reverse-cycle air conditioners. The basic operating principle of a heat pump is that there is some heat in the air, even though the air may be very cold. In fact, the temperature would have to be –460°F for a total absence of heat to exist. In the heating mode, a special valve, known as a reversing valve, switches the compressor input and output so that the condenser operates as the evaporator and the evaporator becomes the condenser. Because of this role reversal, the coils in a heat pump are referred to as the outdoor coil and indoor coil instead of the condenser and evaporator.

The relationship between temperature and pressure is critical to mechanical refrigeration. As you study the process, you will learn that the same refrigerant can be very cold at one point in the system (the evaporator input) and very hot at another (the condenser input). These two points are often only inches apart. This is possible because of pressure changes caused by the compressor and expansion device. In addition to the circulation of refrigerant, air must also circulate. Fans at the condenser and evaporator move air across the condenser and evaporator coils.

This is a simple explanation of the refrigeration cycle. It is meant to provide a basic idea of how an air conditioner works. Later in the training program, this subject is explored in greater detail. The relationship between temperature and pressure will also be studied in depth. It is the key to understanding and troubleshooting mechanical refrigeration systems.

The refrigeration cycle is the same in all refrigeration equipment, from the small air conditioner in a car to the system that cools the largest office building. The difference is in the size and construction of the components, piping, and the amount and type of refrigerant.

1.4.0 Refrigeration

The mechanical refrigeration cycle used for comfort cooling is also used in product refrigeration equipment such as the coolers and freezers found in supermarkets and convenience stores (*Figure 6*). The most significant difference between comfort cooling and refrigeration is the operating temperatures (*Figure 7*). Because of the low temperatures at which refrigeration equipment operates, it uses different refrigerants than comfort cooling systems. Moreover, there is a distinction between operating temperatures and refrigerants for coolers and freezers.

Refrigeration equipment based on the mechanical refrigeration cycle is widely used in commercial and industrial applications. Warehouses and distribution centers are often equipped with large coolers and freezers. Food processing plants, such as meat packing plants, dairies, and frozen food processors use refrigeration equipment on a large scale. Refrigeration equipment is also needed to bring perishable food to market in ships and trucks.

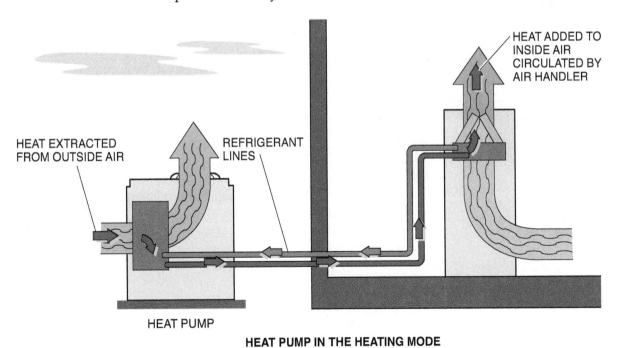

HEAT EXTRACTED FROM OUTSIDE AIR

REFRIGERANT LINES

HEAT ADDED TO INSIDE AIR CIRCULATED BY AIR HANDLER

HEAT PUMP

HEAT PUMP IN THE HEATING MODE

03101-13_F05.EPS

Figure 5 Heat pump heating cycle.

Hermetic and Semi-Hermetic (Serviceable) Compressors

Hermetically sealed compressors are typically used in residential and light commercial air conditioners and heat pumps. Semi-hermetic compressors, also known as serviceable compressors, are used in large-capacity refrigeration or air conditioning chiller units. Semi-hermetic compressors can be partially disassembled for repair in the field. Hermetic compressors are sealed and cannot be repaired in the field.

03101-13_SA02.EPS

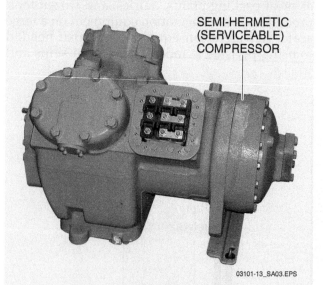

SEMI-HERMETIC (SERVICEABLE) COMPRESSOR

03101-13_SA03.EPS

03101-13_F06.EPS

Figure 6 Retail refrigeration equipment.

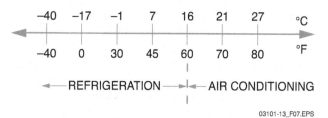

TEMPERATURE RANGE

| −40 | −17 | −1 | 7 | 16 | 21 | 27 | °C |
| −40 | 0 | 30 | 45 | 60 | 70 | 80 | °F |

← REFRIGERATION → ← AIR CONDITIONING

03101-13_F07.EPS

Figure 7 Comfort cooling and refrigeration temperature ranges.

Another important difference between comfort cooling and refrigeration is the need to defrost the evaporator coil in order to eliminate frost buildup. The buildup occurs because the below-freezing temperatures at which the coil operates cause condensate water to freeze on the coil. This eventually blocks airflow and prevents normal operation. Defrost can be accomplished in a number of ways, including using electric heaters and pumping hot refrigerant gas through the coil. In refrigerators that operate at above-freezing temperatures, defrost can be accomplished by letting the ambient air temperature melt the frost buildup during the normal compressor off-cycle. Freezers and ice makers are more likely to use the electric heater or hot gas method.

Dual-Purpose Systems

It is very common to see a furnace with a cooling coil mounted on it. This approach allows the furnace blower and ductwork to be used for both heating and cooling. The cooling coil on the furnace serves as the evaporator. A condensing unit containing the compressor and condensing coil is installed outdoors to provide the necessary heat transfer. A two-speed blower fan is used in the furnace because cooling requires a higher volume of airflow than heating.

03101-13_SA05.EPS

COOLING COIL

FURNACE

03101-13_SA04.EPS

Additional Resources

ABC's of Air Conditioning. Syracuse, NY: Carrier Corporation.

Refrigeration and Air Conditioning: An Introduction to HVAC/R, Larry Jeffus and David Fearnow. New York, NY: Pearson.

Fundamentals of HVAC/R, Carter Stanfield and David Skaves. New York, NY: Pearson.

1.0.0 Section Review

1. In a hydronic heating system, the heat transfer medium is _____.

   a. air
   b. water
   c. natural gas
   d. ammonia

2. Ultraviolet light is one method used to _____.

   a. heat residences
   b. control the growth of bacteria
   c. create air movement
   d. produce a cooling effect

3. In the mechanical refrigeration cycle, the purpose of the evaporator is to _____.

   a. convert refrigerant from high pressure to low pressure
   b. transfer heat from the refrigerant to the outdoor air
   c. transfer heat from the indoor air to the refrigerant
   d. provide the pressure difference to circulate refrigerant

4. A defrost cycle is required in _____.

   a. refrigeration systems
   b. gas heating systems
   c. comfort cooling systems
   d. ventilation equipment

2.0.0 GUIDING PRINCIPLES FOR HVACR SERVICE TECHNICIANS

Objective

Describe the principles that guide HVACR installation and service techniques.

   a. Identify common safety principles and organizations.
   b. Describe the importance of LEED construction and energy management.
   c. Describe trade licensing and certification requirements.
   d. Identify important codes and permits.

Trade Terms

Bond: Short for surety bond. It is a funded guarantee that a contractor will perform as agreed.

Carbon monoxide (CO): A common byproduct of combustion processes, CO is a colorless, tasteless, and odorless gas that is lighter than air and quite toxic. CO reacts with blood hemoglobin to form a substance that significantly reduces the flow of oxygen to all parts of the body. In some countries, CO is responsible for the majority of fatal air poisoning events.

Chlorofluorocarbon (CFC) refrigerant: A class of refrigerants that contains chlorine, fluorine, and carbon. CFC refrigerants have a very adverse effect on the environment.

Easement: A portion of a property that is set aside for public utilities or municipalities.

Hydrochlorofluorocarbon (HCFC) refrigerant: A class of refrigerants that contains hydrogen, chlorine, fluorine, and carbon.

International Building Code (IBC): A series of model construction codes. These codes set standards that apply across the country. This is an ongoing process led by the International Code Council (ICC).

Reclaimed: Used refrigerant that has been remanufactured to bring it up to the standards required of new refrigerant.

Recovery: The removal and temporary storage of refrigerant in containers approved for that purpose.

Recycling: Circulating recovered refrigerant through filtering devices that remove moisture, acid, and other contaminants.

Sustainable construction: Construction that involves minimum impact on land, natural resources, raw materials, and energy over the building's life cycle. It uses material in a way that preserves natural resources and minimizes pollution.

There is more to becoming a successful HVACR service technician than just learning the technical aspects of the systems. The successful technician always follows good safety practices; is conscious of the need to conserve energy and protect the environment; and follows applicable codes and standards.

2.1.0 Safety

The subject of on-the-job safety as defined in OSHA regulations, was covered extensively in the NCCER Module 00101, *Basic Safety*. In addition to the general safety practices covered in that module, there are safety concerns specific to HVACR work. These concerns include:

- *Working at heights* – HVACR installation and service work often require the technician to work from a ladder, install or service equipment on a roof (including sloped roofs), or climb on a cooling tower. In all situations of that type, fall protection and ladder safety regulations must be followed.
- *Working with refrigerants and oils* – There are a number of issues associated with refrigerants and oils:
  - In a confined or enclosed space, refrigerant accidentally released from a system can displace oxygen and potentially cause suffocation.
  - Since many common refrigerants boil at temperatures well below 0°F when at atmospheric pressure, serious injuries can result from contact with refrigerants when they are released from their normally pressurized containers or refrigerant circuits.
  - Some rarely used refrigerants contain propane or butane, which are combustible.
  - Exposure to substantial amounts of ammonia refrigerant can cause eye, skin, and mucous membrane irritation.
  - When exposed to an open flame, some refrigerants produce phosgene gas, which can sicken anyone who breathes the fumes. Refrigerant removed from a system after a severe compressor burnout can produce the same result.

– When oil comes into contact with oxygen, an electric arc could cause an explosion under certain conditions. Oxygen must never be used to purge a refrigeration system.

• *Working around gas furnaces* – Natural gas and propane are flammable and explosive. Always check gas supply lines for leaks. In addition, the burning of these gases in the furnace produces deadly carbon monoxide (CO). A cracked heat exchanger or improperly installed vent pipe can cause these gases to be released into the conditioned space.

• *Working around oil furnaces* – Oil is flammable and its fumes are explosive. Each attempt to restart an oil burner will inject a small quantity of oil into the fire pot. A homeowner may try to restart a furnace several times before calling for help. To avoid a possible explosion, always check the fire pot and clean out the accumulated oil before attempting to start the burner.

2.2.0 Energy Conservation and LEED Construction Principles

HVACR systems play an important role in efforts to reduce toxic emissions that come from the burning of fossil fuels and to reduce dependence on foreign oil. High-efficiency furnaces, heat pumps, and comfort air conditioning systems are available for both commercial and residential installations.

A number of manufacturers have voluntarily partnered with the Environmental Protection Agency (EPA) in an ENERGY STAR® program to market units that exceed minimum efficiency standards. Building designs may also qualify for this program through the use of high-efficiency HVACR systems combined with thermal storage systems.

A high-efficiency system is only effective if it is operating at maximum efficiency. This means performing scheduled maintenance, which includes testing and adjusting for peak performance, as well as replacing defective or underperforming parts.

In an effort to improve energy performance, manufacturers have a number of add-on devices designed to improve the energy efficiency of air conditioning and heating equipment. Some of these devices are designed to recover energy that would otherwise be wasted. The heat recovery ventilator (HRV) shown in *Figure 8* is a good example. In this particular configuration, air drawn from outdoors and exhaust air from the furnace pass through the heat exchanger where heat from the exhaust air is transferred to the incoming cool air before it passes over the furnace heat exchangers. Since the air has been pre-warmed, less energy is needed to raise the air to the temperature needed for heating. Some HRVs work directly through a building ventilation system without going through the heating equipment.

Another example is the ice storage system. Many electric utilities charge a higher rate for electricity during peak-use daytime hours. To reduce the energy costs of air conditioning a building, ice is generated by refrigeration systems at night using low-cost off-peak power. The ice is stored in insulated tanks and is used to aid the building's

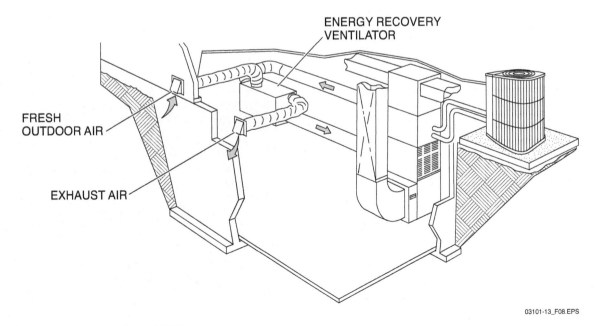

Figure 8 Heat recovery ventilator (HRV).

air-conditioning system during the daytime. This reduces energy consumption during peak usage hours and makes power generation by the utility company more consistent.

2.2.1 LEED Program

Today, the construction and maintenance of buildings is often done in accordance with Leadership in Energy and Environmental Design (LEED) principles. LEED is a rating system established by the US Green Building Council (USGBC). Its purpose is to encourage adoption of sustainable construction standards. It does this through a rating system that considers six specific categories:

- Sustainable sites
- Water efficiency
- Energy and atmosphere
- Material and resources
- Indoor environmental quality
- Innovation and design

The basic concept of LEED is to design and build structures that have minimal impact on the environment. Builders accomplish this goal by using recyclable materials, minimizing waste, creating an energy-efficient building envelope that minimizes heating and cooling waste, ensuring good indoor air quality, and using energy-efficient equipment and appliances. LEED certification for a building is available at silver, gold, and platinum levels, with each level being more demanding.

On the HVACR side, meeting LEED certification standards means selecting energy-efficient equipment and sizing it correctly for the structure and the loads that it must satisfy. This requires a thorough analysis of heating and cooling loads in order to correctly size the system. A system with too much capacity wastes energy and operates inefficiently.

Most LEED-certified commercial buildings also have some form of building management system (BMS) to monitor and control the operation of the HVAC system along with other vital building systems. A properly programmed and implemented BMS can significantly reduce the energy consumption of the HVAC systems, which often consume more energy in a facility than any other system or component.

One important thing to note is that LEED certification goes beyond the design and construction of the building. Once completed, the building and its equipment must also be maintained in accordance with the level of LEED certification awarded.

Proper installation and sealing of ductwork plays an important role in the efficient operation of heating and cooling equipment. If the supply ductwork leaks, conditioned air can escape into unconditioned spaces, resulting in wasted energy. Leaks in ductwork on the return side can result in energy losses too, as well as dust and pollutants being drawn into the heating ducts. A special type of mastic is used to seal duct seams against leaks (*Figure 9*). Mastic is a flexible sealant that can be painted on or applied as a mastic-backed tape to seal duct joints.

Another area where the HVACR technician can have a meaningful impact on energy savings and air quality is in proper maintenance of the equipment. One area in particular, that has an effect on indoor air quality, is the maintenance of air filters. The return ducts on HVAC systems are equipped with air filters that must be periodically cleaned or replaced. The quality of these filters ranges from inexpensive fiberglass filters that capture large dust particles to electronic air cleaners that capture all but microscopic pollutants. In between are a variety of washable and replaceable filters that must be periodically serviced. The HVACR technician can show building occupants how to properly service these filters, or provide these services personally. Failure to perform this maintenance can result in damage to the equipment, and have a negative effect on indoor air quality.

2.3.0 Licensing and Certification

HVACR installers and service technicians work on equipment that affects the health and safety of the people who occupy the buildings in which the equipment is installed. For example, gas furnaces are supplied by natural gas or propane, both of

Figure 9 Duct joints sealed with mastic.

which are explosive. The byproducts of combustion can kill people if the byproducts escape from a system. These byproducts contain carbon monoxide (CO), so venting and maintenance of the furnaces must be performed by knowledgeable technicians. In order to ensure that the people who install and service this equipment are qualified to do so, states and municipalities require them to be licensed.

Of the 50 states in the United States, 37 of them, along with the District of Columbia, require some type of license in order to install and service HVACR equipment. In most of the states that do require a license, applicants must pass a state exam and pay an annual fee. In many cases, the state requires that the applicant show evidence of workers compensation and liability insurance. In several states, contractors must post a bond.

Most states that require licenses focus on mechanical and general contractors as opposed to individual technicians who work for those contractors. However, several states require journeymen to be licensed. At least one state even requires apprentices to be licensed.

A mechanical contractor is a company that specializes in the installation, construction, maintenance, and repair of HVACR equipment. Projects on which mechanical contractors work are often large-scale commercial or industrial installations. However, many mechanical contracting firms focus solely on the residential market. General contractors are construction contractors who may work on various construction projects that include the installation of HVACR equipment.

In some states, contractors who perform HVACR work must obtain a general contractor or contractor license. In other states, there is a distinction between licenses for general contractors and mechanical contractors. Licensing requirements are often based on the type of work the licensee will be doing. For example, some states have separate licenses for residential and commercial work, while others issue licenses for work on systems of different sizes or projects in excess of a certain value.

Some states issue master licenses in addition to journey licenses. A master licensee may be required by law to supervise the work of journeymen. This means that a contracting firm must have at least one person holding a master license.

The important thing to take away from this discussion is that the state or locality in which you work probably will require you to obtain some type of license, pay a fee for it, and take a test in order to obtain it.

2.3.1 EPA Certification

Scientific studies have shown that the chlorine in chlorofluorocarbon (CFC) and hydrochlorofluorocarbon (HCFC) refrigerants can damage the ozone layer that protects Earth and its inhabitants from the sun's ultraviolet rays. One of the suspected effects is an increased rate of skin cancer. In addition to the toxic effect of chlorine, there is evidence that the release of refrigerant to the atmosphere contributes to global warming, a condition known as the greenhouse effect. The heat trapped in the atmosphere leads to a gradual increase in Earth's temperature. Refrigerants, especially CFCs, are viewed as major contributors to global warming. Many refrigerants traditionally used in air conditioning equipment were CFCs and HCFCs. HCFCs are still in use, while CFC refrigerants have now been virtually eliminated in the United States. Common examples of these refrigerants are:

- R-12, a CFC, was used for many years in automobile air conditioners and in domestic/commercial refrigerators. This refrigerant is no longer manufactured, but supplies of reclaimed refrigerant are still available. It has one of the highest levels of chlorine content, making it more likely to damage the ozone layer.
- R-22 is an HCFC refrigerant. It was used for many years in residential and light commercial air conditioners and some refrigeration applications. As of January 2010, manufacturers can no longer manufacture HVAC equipment that contains R-22. New R-22 refrigerant cannot be manufactured after January 2020. R-22 contains a very small percentage of chlorine compared to R-12.

Table 1 shows the ozone depletion potential (ODP) and the global warming potential (GWP) of various refrigerants. This subject is covered in much greater depth in future training modules. Note that the refrigerants with no atmospheric life recorded are blends of two or more existing refrigerants. This is due to the fact that the various components of the blend separate once they enter the atmosphere. Carbon dioxide, the final entry in the table, has a GWP of 1, and is the basis for the GWP values for all of the others. For example, HCFC-22, with a GWP of 1,700, has 1,700 times the potential to create a greenhouse effect than carbon dioxide. CFC-12 is the basis for the ODP of all other refrigerants in the table.

Refrigerant typically comes in cylinders packaged in boxes like the ones shown in *Figure 10*. They are color-coded to represent the type of refrigerant. For example, the container for R-404A is bright orange.

In the 1980s, many nations of the world made a commitment to phase out these chemicals. Those countries agreed to take steps in the interim to prevent the discharge of refrigerants into the atmosphere. In 1990, the US Congress passed the Clean Air Act, which calls for the phase out of the most toxic refrigerants, the eventual elimination of all CFCs and HCFCs, and the strict control and labeling of refrigerants. As a result of the requirement to phase out CFCs and HCFCs, a number of new, environmentally friendly refrigerants have been developed. The US Environmental Protection Agency (EPA) is responsible for implementing and enforcing this law, which has a significant impact on the HVACR trade. For example, it has imposed the following restrictions on refrigerants, regardless of the chlorine content:

- Anyone releasing these refrigerants to the atmosphere is subject to a stiff fine, and possibly a prison term.
- Anyone handling these refrigerants must have EPA-sanctioned certification. Without the EPA-sanctioned card that shows you have completed the training, you cannot even buy refrigerants. The training needed to obtain EPA certification is included in this curriculum. There are four

Table 1 Common Refrigerant Environmental Impact Data

Refrigerant	Ozone Depletion Potential (ODP)	Global Warming Potential (GWP)	Atmospheric Life, in Years
CFC-11 (Trichlorofluoromethane)	1.0	4,000	45
CFC-12 (Dichlorodifluoromethane)	1.0	8,500	100
CFC-13 (Chlorotrifluoromethane)	1.0	11,700	640
HCFC-22 (Chlorodifluoromethane)	0.05	1,700	12
HFC-32 (Difluoromethane)	0	543	4.9
CFC-113 (Trichlorotrifluoroethane)	0.8	5,000	85
CFC-114 (Dichlorotetrafluoroethane)	1.0	9,300	300
HCFC-123 (Dichlorotrifluoroethane)	0.02	93	1.3
HCFC-124 (Chlorotetrafluoroethane)	0.02	480	5.8
HFC-125 (Pentafluoroethane)	0	3,400	29
HFC-134a (Tetrafluoroethane)	0	1,300	14
HFC-143a (Trifluoroethane)	0	4,300	52
HFC-152a (Difluoroethane)	0	120	1.4
HCFC-401A	0.37	1,100	NA
HCFC-402A	0.02	2,600	NA
HFC-404A	0	3,300	NA
HFC-407A	0	2,000	NA
HFC-407C	0	1,600	NA
HFC-410A	0	1,725	NA
R-717 Ammonia	0	0	NA
R-718 Water	0	0	NA
R-729 Air	0	0	NA
R-744 Carbon Dioxide	0	1	50–200

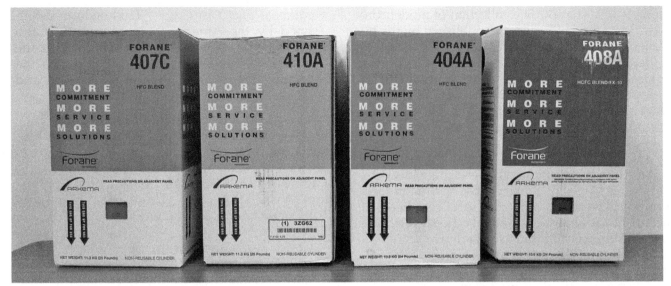

Figure 10 Refrigerant cylinders.

levels of EPA certification, based on the type of work. The levels range from a Type I certification that allows a person to work on systems containing less than 5 pounds of high-pressure refrigerant, to a Universal certification which allows work on any type of system containing refrigerant.

- Records must be kept on all transactions involving these refrigerants. These records must include purchase, use, reprocessing, and disposal.

Before a sealed refrigeration system can be opened for repair, the refrigerant it contains must be identified. With few exceptions, all refrigerants must be recovered using approved recovery equipment, and stored in approved containers. *Figure 11* shows a refrigerant recovery system in use. When the repair is complete and the system is resealed, the same refrigerant may be returned to the system. It may also be used in another system belonging to the same owner. Recovered refrigerant should, however, be recycled before reuse. Recycling removes most moisture and impurities that could damage the system. Some refrigerant recovery units have a built-in recycling capability.

If the refrigerant is badly contaminated or no longer needed, it can be reclaimed. This is done at remanufacturing centers where the refrigerant is returned to the standards of purity that govern new refrigerants. Reclaimed refrigerant can be resold on the open market, but cannot be classified as new refrigerant.

In addition to federal regulations, there may be state regulations that apply to refrigerants. State regulations may be stricter. It is critical that everyone in the HVACR industry understands and follows the EPA regulations regarding the handling, storage, and labeling of refrigerants. Failure to do so could be very costly to both the worker and the employer.

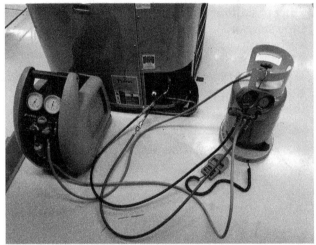

Figure 11 Refrigerant recovery equipment.

2.4.0 Codes and Permits

Construction work is governed by building codes, which are enforced by local municipalities. The purpose of a building code is to regulate the health and safety aspects of building construction in a community. Building codes regulate new construction, for example, by establishing limits on the height and floor area of buildings, by specifying the separation between buildings, and by requiring specific setbacks for buildings and equipment from property boundaries and easements. Local codes are based on the International Building

Code (IBC). The IBC is a collection of model construction codes managed and published by the International Code Council (ICC).

Before any construction work can begin, the contractor must apply to the local government for a permit. In order to obtain the permit, the contractor must submit a description of the project, including drawings, if applicable. In some locations, an HVACR contractor must obtain a permit in order to install new equipment. Once the permit is pulled, inspectors will perform inspections at various stages of construction. A final inspection must be passed in order for a certificate of occupancy (CO) to be issued. The HVACR system for a building is an important factor in the inspection process. Here are examples of the HVACR-related inspection items that might be on the list of a municipal building inspector:

- Furnace venting
- Gas piping for furnaces
- Duct installations
- Pipe supports for refrigerant lines and hydronic system piping
- Condensate drains
- Equipment placement
- Locations of piping for furnace combustion air intake and exhaust.

> **NOTE**
>
> In locations where hurricanes or tornadoes are a threat, proper anchorage of outdoor equipment is a special concern. Areas with earthquake potential also have special requirements.

In addition to building codes, there are other national codes that have been incorporated into state and municipal building codes by adoption. These codes include the following:

- *National Electrical Code (NEC®)* – This code is published by the National Fire Protection Association (NFPA) as *NFPA 70®*. The *NEC®* establishes the minimum standard for electrical wiring. This code is rigidly enforced. For that reason, electrical wiring installations for HVACR systems are nearly always performed by a licensed electrician.

- *National Fuel Gas Code* – This code is developed by the American Gas Association and published as *NFPA 54* by the NFPA. The code covers the installation of gas piping and the venting of appliances such as gas furnaces and water heaters.

In addition to national codes, localities may adopt standards developed by national and international organizations such as the American Society of Refrigeration and Air Conditioning Engineers (ASHRAE), Air Conditioning Contractors' Association (ACCA), and Sheet Metal and Air Conditioning Contractors' National Association (SMACNA).

Codes and standards are not the same thing. The main difference is that codes can be enforced by law because they are incorporated into ordinances. Standards are produced by associations of engineers, contractors, and manufacturers. Their purpose is to establish criteria for the design, manufacture, and installation of equipment. Standards are often included by reference in contractual documents such as drawings and specifications. If they are so incorporated, it is up to the contractors to make sure their employees are familiar with, and comply with, the referenced standards.

A standard can take on the force of law if it is adopted by state or municipal governments as part of a building code structure. An example is *ASHRAE Standard 62.1*, which deals with indoor air quality. Some states have adopted this standard as part of their building code, which means that the municipal inspector's checklists will include such items as duct sealing, air filter efficiency, use of outdoor air for ventilation, and moisture and humidity control.

The takeaway message is that it is important for HVACR installers to become familiar with code requirements for the areas in which they work. Inspectors will not hesitate to order work to be re-done if it does not meet code requirements.

There's an App for Everything

HVACR technicians can benefit from applications that are available for smart phones and tablets. Here are a few examples of such apps: duct-sizing calculator; refrigerant temperature-pressure charts; refrigerant charging charts; pipe-sizing charts; and load-calculating aids. Apps like these are available through smart application packages such as HVACR Buddy® and HVACR Toolkit.

03101-13_SA06.EPS

Additional Resources

Your Role in the Green Environment, NCCER. Current edition. New York, NY: Pearson.

2.0.0 Section Review

1. Phosgene gas can be produced when some refrigerants are exposed to _____.

   a. air
   b. oil
   c. water
   d. a flame

2. Mastic is a material used to _____.

   a. test for gas leaks
   b. seal ductwork
   c. insulate refrigerant lines
   d. lubricate bearings

3. A mechanical contractor is a company that _____.

   a. services automobile mechanical systems
   b. specializes in the installation and repair of air conditioning systems
   c. manufactures air conditioning equipment
   d. constructs entire buildings

4. Building permits are usually issued by the _____.

   a. local government
   b. state
   c. Federal Government
   d. building architect

SECTION THREE

3.0.0 CAREERS IN HVACR

Objective

Identify career paths available in the HVACR trade.

   a. Identify the responsibilities and characteristics needed to be a successful HVACR technician.

   b. Identify residential, commercial, and industrial career opportunities.

   c. Describe opportunities provided by equipment manufacturers.

Trade Terms

Chiller: A high-volume hydronic cooling unit.

On the job learning (OJL): Learning obtained while working under the supervision of a journeyman.

Since ancient times, people have sought ways to make the buildings in which they live, work, and play more comfortable. The members of the HVACR craft are skilled workers who install, maintain, and repair the equipment that makes this possible. Working in the HVACR industry is challenging and rewarding because environmental technology is constantly changing. Technical advances in HVACR systems are made every day in advanced computerized controls, greater operating efficiency, and improved packaging.

The HVACR industry offers many opportunities for advancement. The training you are receiving can qualify you to become an installer, troubleshooter, sales technician, system design specialist, and eventually even the owner of your own HVACR service business.

3.1.0 Employee Responsibilities and Characteristics

In order to be successful, the professional must be able to use current trade materials, tools, and equipment to finish the task quickly and efficiently. A technician must be able to adjust methods to meet each situation. The successful technician must keep abreast of technical advancements and continually gain the skills to use them. A professional never takes chances with regard to their personal safety or the safety of others. The sections that follow describe some characteristics that distinguish a professional employee.

3.1.1 Professionalism

The word *professionalism* is a broad term that describes the desired overall behavior and attitude expected in the workplace. Professionalism involves a number of qualities, including honesty; productivity; safety; civility; cooperation; teamwork; clear and concise communication; being on time and prepared for work; and regard for the impact of behavior on co-workers. It can be demonstrated in a variety of ways every minute you are in the workplace.

3.1.2 Honesty

Honesty and personal integrity are important traits of the successful professional. Professionals pride themselves in performing a job well, and in being punctual and dependable. Each job is completed in a professional way, and never by cutting corners or reducing materials. A valued professional maintains work attitudes and ethics that protect property such as tools and materials belonging to employers, customers, and other trades from damage or theft at the shop or job site.

Honesty and success go hand-in-hand. It is not simply a choice between good and bad, but a choice between success and failure. Dishonesty always catches up with you. Whether you are stealing materials, tools, or equipment from the job site or simply lying about your work, it will not take long for your employer to find out. Of course, you can always go and find another employer, but this option will ultimately run out.

Honesty also means giving a fair day's work for a fair day's pay and doing what you say you will do. Employers place a high value on an employee who is strictly honest. Customers also place a high value on honesty, and tend to return to those they can trust when they need help.

3.1.3 Loyalty

Employees expect employers to look out for their interests, to provide them with steady employment, and to promote them to better jobs as openings occur. Employers feel that they, too, have a right to expect their employees to be loyal to them—to keep the company's interests in mind, to speak well of the company to others, to keep any minor troubles strictly within the plant or office, and to keep all matters that pertain to the business absolutely confidential. Both employers and employees should keep in mind that loyalty is not something to be demanded; rather, it is something to be earned. Loyalty between an employer and an employee requires both parties to work in each other's best interest.

3.1.4 Willingness to Learn

Every business has its own way of doing things. Employers expect their workers to be willing to learn their system. Sometimes, a change in safety regulations or the purchase of new equipment makes it necessary for even experienced employees to learn new methods. Employees often resent having to accept improvements because of the retraining that is involved. However, methods must be kept up to date in order to meet competition and show a profit. It is this profit that enables the owner to continue in business and to provide jobs for the employees.

3.1.5 Willingness to Take Responsibility

Every employee has the responsibility for working safely and must take that responsibility seriously. In addition, employers expect their employees to see what needs to be done, and then do it. It is frustrating for a supervisor to have to ask again and again that a certain job be done. Once the responsibility has been delegated, the employee should continue to perform the duties without further direction.

Regardless of your skill level, everyone makes mistakes from time to time. In fact, the expensive errors are often made by workers with the most experience, since those workers are in a position of higher responsibility. All workers must take responsibility for their mistakes and accept their part in correcting the error. A highly experienced craftworker said that, "one major difference in an amateur and a professional is that the professional knows how to correct his mistakes."

3.1.6 Willingness to Cooperate

To cooperate means to work effectively with others. In the modern business world, cooperation is the key to getting things done. Employees must be able to work as a member of a team with the employer, supervisor, fellow workers, and even customers in a common effort to get the work done efficiently, safely, and on time. People can work well together only if there is some understanding about what work is to be done, when and how it will be done, and who will do it. Rules and regulations are a necessity in the working world and employees must embrace them to function effectively as a member of the team.

3.1.7 Tardiness and Absenteeism

Tardiness means being late for work, and absenteeism means being off the job for one reason or another. Consistent tardiness and frequent absences are an indication of poor work habits, unprofessional conduct, and a lack of commitment.

Failure to get to work on time results in lost time and resentment on the part of those who do come on time. In addition, it may lead to penalties, including dismissal. Although it may be true that a few minutes out of a day are not very important, you must remember that a principle is involved. It is your obligation to be at work at the time indicated. In fact, arriving a little early indicates your interest and enthusiasm for your work,

Ethical Principles for Members of the Construction Trades

- *Honesty* – Be honest and truthful in all dealings. Conduct business according to the highest professional standards. Faithfully fulfill all contracts and commitments. Do not deliberately mislead or deceive others.
- *Integrity* – Demonstrate personal integrity and the courage of your convictions by doing what is right even where there is pressure to do otherwise. Do not sacrifice your principles because it seems easier.
- *Loyalty* – Be worthy of trust. Demonstrate fidelity and loyalty to companies, employers and sponsors, co-workers, and trade institutions and organizations.
- *Fairness* – Be fair and just in all dealings. Do not take undue advantage of another's mistakes or difficulties. Fair people are open-minded and committed to justice, equal treatment of individuals, and tolerance for and acceptance of diversity.
- *Respect for others* – Be courteous and treat all people with equal respect and dignity.
- *Obedience* – Abide by laws, rules, and regulations relating to all personal and business activities.
- *Commitment to excellence* – Pursue excellence in performing your duties, be well informed and prepared, and constantly try to increase your proficiency by gaining new skills and knowledge.
- *Leadership* – By your own conduct, seek to be a positive role model for others.

which is appreciated by employers. The habit of being late is another thing that stands in the way of promotion.

It is sometimes necessary to take time off from work. No one should be expected to work when sick or when there is serious trouble at home. However, workers must not get into the habit of letting unimportant and unnecessary matters keep them from the job. This results in lost production and hardship on those who try to carry on the work with less help. If it is necessary to miss work, at least phone the office early in the morning so they have time to make alternate arrangements. If you do not call, it leaves those at work uncertain about what to expect. They have no way of knowing whether you have merely been held up and will be in later, or whether immediate steps should be taken to assign your work to someone else.

Employers do not tolerate habitual lateness or absenteeism. In order to control it, they resort to docking pay, demotion, and even dismissal. In fairness to the company and to those workers who do show up for work, an employer is sometimes forced to discipline those who do not follow the rules.

3.2.0 Career Opportunities

Career opportunities in the HVACR trade are many and varied. There is a large existing base of HVACR systems that need service, repair, and replacement. In addition, every time a new residential, commercial, or industrial building is constructed, it contains one or more HVACR system elements.

To get an idea of how vast the HVACR trade is, picture the town in which you live. Then think about the fact that almost every building in town contains some form of equipment to provide heating and cooling, as well as air circulation and purification. Expand that view to include the entire country and you realize that there are millions of heating, air conditioning, and air management systems. New ones are being added every day; old ones are wearing out and being repaired or replaced. From that perspective, the opportunities in the craft appear limitless.

Figure 12 provides an overview of career opportunities in the HVACR trade. For the purposes of this discussion, it is convenient to view the HVACR industry as having the following three segments:

- *Residential/light commercial* – Residential/light commercial companies sell, install, and service residential and light commercial equipment and systems such as furnaces and packaged air conditioners.

- *Commercial/industrial* – Commercial/industrial companies are generally mechanical contractors that install and maintain systems for large office buildings, factories, apartment complexes, shopping malls, and so forth.
- *Manufacturing* – Manufacturing companies build and market HVACR systems and equipment.

3.2.1 Residential/Light Commercial

At this level, you might find anything from a one-person installation and service business to a firm with 100 or more employees, including heating and air conditioning specialists, installers, sheet metal workers, and sales engineers. In such businesses, HVACR specialists may work alone or with a partner. They typically respond to service calls from homes or small businesses. They may also install furnaces and air conditioning equipment sold by their firm's sales engineer. In other firms, one group may do installations while another group handles troubleshooting and maintenance. At this level, a technician is expected to work with a wide variety of products from many different manufacturers. Systems can consist of anything from a window air conditioner to complete, centralized heating/air-conditioning systems with as much as 25 tons of cooling capacity.

Local or regional distributors provide equipment, parts, special tools, and other services for the firms that sell, install, and service HVACR equipment. The distributor needs salespeople who know HVACR equipment. The distributor may also provide engineering support and service training for its dealers. Distributorships are often affiliated with a single manufacturer.

> **NOTE**
>
> Many community-based firms are subsidiaries of nationwide firms known as consolidators. A consolidator is an umbrella organization that provides centralized management, purchasing, training, and other functions that give small, local companies the power of a large, national company.

3.2.2 Commercial/Industrial

Large commercial and industrial systems have many components and may require thousands of feet of ductwork and piping. Such systems are designed by engineers and architects. Many HVACR craftworkers are required for these projects and an individual is more likely to specialize. For example, where a single wall thermostat often controls residential systems, a large com-

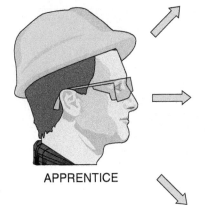

RESIDENTIAL/LIGHT COMMERCIAL

- Installer
- Service Technician
- Distributor Sales

- Service Supervisor
- Sales Engineer
- Owner
- Inspector
- Code Enforcement Officer
- Estimator

COMMERCIAL/INDUSTRIAL

- Installer
- Air System Technician
- AC Service Technician
- Refrigeration Service Technician

- Foreman
- Technical Specialist
- Training Specialist
- Operating Engineer
- Project Manager
- Estimator

APPRENTICE

MANUFACTURING

- Test Technician
- Engineering Assistant
- Assembler

- Supervisor
- Technical Specialist
- Trainer
- Field Service Engineer

03101-13_F12.EPS

Figure 12 Career opportunities in the HVACR trade.

mercial system often has central computer controls that are installed and serviced by a system control specialist. These systems, commonly known as building management systems (BMS), require specialized skills to program and maintain. In many cases, the technician must be intimately familiar with the specific software package being used. Others may specialize in working with large steam boilers; still others may choose to become experts in installing and servicing large capacity hydronic cooling units known as chillers.

Companies that install such equipment are often large construction firms or mechanical contractors that may work anywhere in the world. They are likely to do only the installation and let the building owner contract with another firm for maintenance. Many companies that have large facilities employ their own HVACR maintenance people.

3.2.3 HVACR Training

The Department of Labor's Office of Apprenticeship sets the minimum standards for training programs across the country. These programs rely on mandatory classroom instruction and on-the-job learning (OJL). They require at least 144 hours of classroom instruction per year and 2,000 hours of OJL per year. In a typical four-year HVACR apprenticeship program, trainees spend 576 hours in classroom instruction and 8,000 hours in OJL before receiving certificates issued by registered apprenticeship programs.

Modern Air Conditioning

Dr. Willis Carrier, founder of Carrier Corporation, is credited with the invention of modern air conditioning. In 1902, he developed a system that could control both humidity and temperature using a non-toxic, non-flammable refrigerant. Air conditioning started out as a means of solving a problem in a printing facility where heat and humidity were causing paper shrinkage. Later systems served similar purposes in textile plants. The concept wasn't applied to comfort air conditioning until about 20 years later when Carrier's centrifugal chillers began to be installed in department stores and movie theaters.

Large Commercial Chiller Unit

As the name implies, chillers use chilled water as a cooling medium. The chiller acts as the evaporator. Chillers are often combined with a cooling tower that acts as the condensing unit. Water flowing over the cooling tower absorbs the heat extracted from the indoor air. This photo shows a large commercial chiller unit.

03101-13_SA07.EPS

To address the training needs of the professional communities, NCCER developed a four-year HVACR training program. NCCER uses the minimum Department of Labor standards as a foundation for a comprehensive curriculum that provides in-depth classroom and OJL experience.

This NCCER curriculum provides trainees with industry-driven training and education. It adopts a purely competency-based teaching approach. This means that trainees must show the instructor that they possess the knowledge and skills needed to safely perform the hands-on tasks that are covered in each module.

When the instructor is satisfied that a trainee has the required knowledge and skills for a given module, that information is sent to NCCER and recorded in the National Registry. The National Registry can then confirm training and skills for workers as they move from state to state, company to company, or even within a company.

Whether you enroll in an NCCER program or another apprenticeship program, make sure you work for an employer or sponsor who supports a nationally standardized training program that includes credentials to confirm your skill development.

Apprentice training goes back thousands of years. Its basic principles have not changed over time. First, it is a means for a person entering the craft to learn from those who have mastered the craft. Second, it focuses on learning by doing; real skills versus theory. Some theory is presented in the classroom. However, it is always presented in a way that helps the trainee understand the purpose behind the skill that is to be learned.

Apprenticeship Standards – All apprenticeship standards prescribe certain work-related or on-

the-job learning. This OJL is broken down into specific tasks in which the apprentice receives hands-on training. In addition, a specified number of hours are required in each task. The total amount of OJL for an HVACR apprenticeship program is traditionally 8,000 hours, which amounts to four years of training. In a competency-based program, it may be possible to shorten this time by testing out of specific tasks through a series of performance exams.

In a traditional program, the required OJL may be acquired in increments of 2,000 hours per year. Layoffs or illness may affect the duration. The apprentice must log all work time and turn it in to the apprenticeship committee so that accurate time control can be maintained.

The classroom instruction and work-related training does not always run concurrently due to such reasons as layoffs, type of work needed to be done in the field, etc. Furthermore, apprentices with special job experience or coursework may obtain credit toward their classroom requirements. This reduces the total time required in the classroom while maintaining the total 8,000-hour OJL requirement. These special cases depend on the type of program and the regulations and standards under which it operates.

Informal OJL provided by employers is usually less thorough than that provided through a formal apprenticeship program. The degree of training and supervision in this type of program often depends on the size of the employing firm. A small contractor may provide training in only one area, while a large company may be able to provide training in several areas.

For those entering an apprenticeship program, a high school or technical school education is desirable. Courses in shop, mechanical drawing, and general mathematics are helpful. Manual dexterity, good physical condition, and quick reflexes are important. The ability to solve problems quickly and accurately and to work closely with others is essential. You must also have a high concern for safety.

The prospective apprentice must submit certain information to the apprenticeship committee. This may include the following:

- Aptitude test (General Aptitude Test Battery or GATB Form Test) results (usually administered by the local Employment Security Commission)
- Proof of educational background (candidate should have school transcripts sent to the committee)
- Letters of reference from past employers and friends
- Proof of age
- If the candidate is a veteran, a copy of Form DD214
- A record of technical training received that relates to the construction industry and/or a record of any pre-apprenticeship training
- High school diploma or General Equivalency Diploma (GED)

The apprentice must do the following:

- Wear proper safety equipment on the job
- Purchase and maintain tools of the trade as needed and required by the contractor
- Submit a monthly on-the-job learning report to the committee
- Report to the committee if a change in employment status occurs
- Attend classroom instruction and adhere to all classroom regulations such as attendance requirements

Youth Apprenticeship Programs – A Youth Apprenticeship Program is also available that allows students to begin their apprentice training while still in high school. A student entering the program in eleventh grade may complete as much as two years of an NCCER program by high school graduation. In addition, the program, in cooperation with local craft employers, allows students to work in the trade and earn money while still in school. Upon graduation, the student can enter the industry at a higher level and with more pay than someone just starting the apprenticeship program.

This training program is similar to the one used by NCCER, learning centers, contractors, and colleges across the country. Students are recognized through official transcripts and can enter the next year of the program wherever it is offered. They may also have the option of applying the credits at a two-year or four-year college that offers degree or certification programs in the construction trades.

Apprenticeships in the United States

In 2008, more than 500,000 apprentices received registered apprenticeship training in the United States.

An Ancient Apprenticeship Contract

How did these old apprenticeships work? Here's one example. Historical records include a contract between a young Greek named Heracles and a weaver. In exchange for food, a tunic (a long shirt), and 20 holidays a year, Heracles worked for the weaver for five years. After two and a half years, he was paid 12 drachmas. In his fifth year, his pay was scheduled to double.

In addition to learning to become a journeyman weaver, Heracles most likely worked seven long days each week and had to clean the shop, build fires, pick up supplies, make deliveries, and do whatever other jobs the master weaver needed to have done. It's also likely that any damage he caused came out of his promised pay.

So how much pay is 12 drachmas? There is no way of knowing for certain, but the word *drachma* means handful—not much pay for almost three years of hard work. Yet Heracles was willing to put in the time to become a skilled master in his chosen trade. As a master weaver, Heracles could look forward to a better income and a better life.

3.3.0 Opportunities in Manufacturing

There are thousands of HVACR manufacturers. Some of the larger ones cover the entire HVACR spectrum, while others focus on a particular product or market. For example, one may make window air conditioners, another gas furnaces, and yet another might make only heavy commercial equipment such as chillers. Some companies work in niche markets, manufacturing specialized equipment such as that used for flash freezing of food products. Regardless of their market, manufacturers employ a variety of HVACR specialists such as:

- Test technicians
- Engineering assistants
- Training specialists
- Instruction book writers
- Field service technicians

A major advantage of working in such a company is that the jobs generally pay good salaries, have benefits, and offer opportunities for advancement. Some manufacturers have multiple locations, so the opportunities for advancement are sometimes broad-based. Many manufacturing companies, especially the larger ones, offer a variety of in-house training courses, as well as tuition reimbursement for college courses.

Testing technicians often work in quality control labs. They may test manufactured equipment to make sure it performs to specifications. They may also test equipment returned under warranty to determine what caused it to fail. An engineering assistant works with equipment designers and may help build and test prototypes of new designs. Large manufacturers, and those that manufacture specialized equipment offer training to their distributors and clients. A training specialist develops and delivers the training courses offered by the manufacturer.

New equipment is always accompanied by installation instructions. Many companies provide operation and maintenance handbooks as well. This material is prepared by technical writers who work with engineers to produce the instruction books.

Manufacturers that design and build specialized equipment or large engineered systems generally have a cadre of field service technicians who perform periodic maintenance and are called upon to troubleshoot and repair systems in the event of a failure. Needless to say, if specialized HVACR equipment fails, there is usually an urgent need to get it back on line. The technicians who perform this work are highly trained, able to work under stress, and available to travel on short notice.

Child Labor Laws

Federal law establishes the minimum standards for workers under the age of 18. Some municipal jurisdictions may enforce stricter regulations. Employers are required to abide by the laws that apply to them.

The Child Labor Provisions of the Fair Labor Standards Act forbid employers from using illegal child labor, and also forbid companies from doing business with any other business that does. DOL investigates alleged abuses of the law. In such cases, employers have to provide proof of age for their employees.

In addition to the Child Labor Provisions, employers in the construction trades are required to follow DOL's *Child Labor Bulletin No. 101, Child Labor Requirements in Nonagricultural Occupations Under the Fair Labor Standards Act. Bulletin No. 101* does the following:

- Explains the coverage of the Child Labor Provisions
- Identifies minimum age standards
- Lists the exemptions from the Child Labor Provisions
- Sets out employment standards for 14- and 15-year-old workers
- Defines the work that can be performed in hazardous occupations
- Provides penalties for violations of the Child Labor Provisions
- Recommends the use of age certificates for employees

More Than Just Candles

Wax has actually been used for several thousand years to create precise molds and patterns in a process known as investment casting. This process is used to form parts such as titanium turbine blades used in jet engines. Precise copies of turbine blades are first created in wax, and then dipped into a ceramic slurry to apply a coating about ¼" thick. Heat then hardens the ceramic material as the melted wax flows out. The now-hollow ceramic mold is filled with molten titanium.

A new wax mold must be used for each and every blade. The molds must be stored in an environment maintained at very precise temperature and humidity levels because even the smallest change in the environmental conditions can cause dimensional changes in the wax. Highly specialized HVACR systems, such as those manufactured by Kathabar Dehumidification Systems, Inc., are capable of maintaining these critical environments. These units heat, cool, humidify, and dehumidify as necessary to maintain precise set points. Without this unique equipment to maintain the highest level of mold accuracy, the extreme levels of reliability found in jet engines today would not be possible.

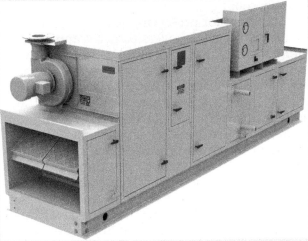

03101-13_SA08.EPS

3.0.0 Section Review

1. Keeping your employer's interests in mind is an example of _____.
   a. honesty
   b. loyalty
   c. cooperation
   d. taking responsibility

2. The minimum number of annual classroom hours required for an apprentice program is _____.
   a. 50
   b. 75
   c. 112
   d. 144

3. Engineering assistants are most likely to be employed by _____.
   a. residential service companies
   b. mechanical contractors
   c. industrial refrigeration contractors
   d. equipment manufacturers

Summary

The HVACR craft involves equipment used for heating, cooling, and purifying indoor air, as well as equipment for preserving food and other perishable products. It covers furnaces and air conditioners used in homes and businesses, as well as the heating and cooling systems used in large office buildings and industrial complexes. It includes equipment used in retail, commercial, and industrial enterprises for the preservation and packaging of food and other perishables.

One of the most serious issues affecting the craft is the damage that can be done to Earth's ozone layer by the improper release of refrigerants into the atmosphere. There are severe penalties for improper refrigerant disposal. HVACR technicians form an important line of defense in the fight to improve energy efficiency and reduce emission of harmful chemicals into Earth's atmosphere. Federal regulations govern the handling and disposal of refrigerants used in HVACR systems. Technicians are required to receive training and obtain an EPA certificate for this purpose.

Because of the widespread use of HVACR equipment, there are many career opportunities in the craft. Jobs are available with small local firms, large industrial and commercial contractors, and the manufacturing firms that build and market HVACR equipment. The apprentice program provides an opportunity to learn the trade through a combination of hands-on training and related classroom learning. The Youth Apprentice Program allows students to begin their training while still in high school.

1. In a common household forced-air furnace, heat is transferred from the _____.

    a. heat exchangers to the air
    b. natural gas or oil to the conditioned space
    c. air to the heat exchangers
    d. refrigerant to the outdoor air

2. Water is often used in heating systems as a _____.

    a. fuel
    b. refrigerant
    c. heat exchange medium
    d. heat exchanger

3. Ventilation is concerned with _____.

    a. preventing outdoor air from entering a building
    b. maintaining a constant air temperature
    c. removing moisture from indoor air
    d. the introduction of fresh air to control indoor air quality

4. The growth of bacteria and other harmful organisms can be reduced or eliminated by _____.

    a. humidification
    b. cooling air
    c. ultraviolet light
    d. common air filters

5. The term *mechanical refrigeration cycle* refers to _____.

    a. the process by which circulating refrigerant absorbs heat in one location and moves it to another location
    b. the process by which refrigerant moves through a compressor
    c. mobile refrigeration units
    d. the process by which refrigerant is recycled to remove impurities

6. A device that is able to extract heat from the air through the use of the refrigeration cycle is a(n) _____.

    a. furnace
    b. evaporator
    c. heat pump
    d. expansion device

7. The expansion device in the mechanical refrigeration cycle _____.

    a. raises the refrigerant pressure entering the evaporator
    b. lowers the refrigerant pressure entering the evaporator
    c. raises the refrigerant pressure at the condenser inlet
    d. lowers the refrigerant pressure at the condenser outlet

8. In a refrigeration system, heat is transferred from the indoor air to the refrigerant at the _____.

    a. compressor
    b. furnace
    c. evaporator
    d. condenser

9. A leak in the vent pipe of a gas furnace is a concern because it can _____.

    a. release carbon monoxide
    b. release flammable gas
    c. release carbon dioxide
    d. cause heat loss

10. Which of these companies is likely to install HVACR equipment during the construction of a large office building?

    a. Mechanical contractor
    b. Distributor
    c. Manufacturer
    d. General contractor

11. In order to be resold on the open market, a refrigerant must have been _____.

    a. reclaimed
    b. recycled
    c. recovered
    d. refurbished

12. Releasing CFC and HCFC refrigerants or certain substitutes to the atmosphere is _____.

    a. okay, provided that the refrigerant has been recycled
    b. okay, provided that the refrigerant has been reclaimed
    c. prohibited by federal law
    d. prohibited in some states

13. The primary purpose of a building code is to regulate the quality of all mechanical installations on a commercial site.

   a. True
   b. False

14. Which of the following is a function performed by a distributor?

   a. Repair of residential heating systems
   b. Sale of equipment and parts
   c. Manufacture of equipment
   d. Installation of mechanical systems

15. Development and delivery of training programs is typically a function of a _____.

   a. mechanical contractor
   b. distributor
   c. manufacturer
   d. residential service company

Trade Terms Quiz

Fill in the blank with the correct trade term that you learned from your study of this module.

1. A substance that is _____ is considered poisonous.

2. Used refrigerant that has been reprocessed to bring it up to the standards required of new refrigerant is called _____.

3. A class of refrigerants that contains hydrogen, chlorine, fluorine, and carbon is called _____.

4. A(n) _____ is a heat exchanger that transfers heat from the air flowing over it to the cooler refrigerant flowing through it.

5. A class of refrigerants that contains chlorine, fluorine, and carbon is called _____.

6. The removal and temporary storage of refrigerant in containers approved for that purpose is called _____.

7. A liquid metering device is also known as a(n) _____.

8. A(n) _____ is a heat exchanger that transfers heat from the refrigerant flowing inside it to the air or water flowing over it.

9. The process by which a circulating refrigerant absorbs heat from one location and transfers it to another location is called a(n) _____.

10. The movement of heat energy from a warmer substance to a cooler substance is known as _____.

11. The process of circulating recovered refrigerant through filtering devices that remove moisture, acid, and other contaminants is known as _____.

12. The use of machinery to provide cooling is called _____.

13. A substance harmful to your health is said to be _____.

14. A high-volume hydronic cooling unit is called a(n) _____.

15. The _____ is a series of model construction codes that set standards that apply across the country.

16. The mechanical device that converts low-pressure, low-temperature refrigerant gas into high-temperature, high-pressure refrigerant gas in a refrigeration system is a(n) _____.

17. The funded performance guarantee that a contractor puts up is called a _____.

18. A portion of property set aside for use by a utility is called a(n) _____.

19. A device that can produce heat by reversing the mechanical refrigeration cycle is a(n) _____.

20. Systems that use water as a heat transfer medium are known as _____ systems.

21. Documented learning that is obtained while working is called _____.

22. The construction process that focuses on minimizing environmental impact is known as _____.

23. A potentially-deadly, colorless, tasteless, and odorless gas that is a common byproduct of combustion processes is _____.

Trade Terms

Bond	Easement	Hydronic	On the job learning (OJL)
Carbon monoxide (CO)	Evaporator	International Building	Reclaimed
Chiller	Expansion device	Code (IBC)	Recovery
Chlorofluorocarbon (CFC)	Heat pump	Mechanical refrigeration	Recycling
refrigerant	Heat transfer	Mechanical refrigeration	Sustainable construction
Compressor	Hydrochlorofluorocarbon	cycle	Toxic
Condenser	(HCFC) refrigerant	Noxious	

 03101 **Introduction to HVAC**

Joseph Pietrzak

Senior Program Manager
Duke Energy

How did you get started in the construction industry (i.e., took classes in school, a summer job, etc.)?
Getting started in the construction industry came naturally to me as I was always drawn to mechanical systems. When I was young, I would often take things apart to see how they worked and then try to put them back together again (hoping they still worked).

After moving from New York to Florida, I had an opportunity to attend school full-time for six months to learn Air Conditioning and Refrigeration. Upon completing this course, I started a job the very next day. I worked and learned the trade for about five years, and then I decided to take the state exam for my Class A license.

Shortly after this, I was hired by Florida Power's Demand-Side Management Department. The goal was to reduce demand and kilowatt-hour usage by their 1.7 million-customer base. In 2000, I started working in technology development-type programs, which included the development of alternative energy platforms. We built the first two hydrogen-fueling stations in the state to service 18 vehicles. In addition, I did a lot of work with solar power generation and systems. I am currently the Senior Solar Program Manager for the following five SunSense programs: Residential PV Incentives, Commercial PV Incentives, Solar Water Heating, Solar for Schools, and Solar for Low Income Families. The key to my success has always been the determination to never stop learning.

Who or what inspired you to enter the industry (i.e., a family member, school counselor, etc.)? Why?
My initial inspiration to enter the HVACR industry came through a desire to provide for my family. After moving to Florida, I noted a great demand for HVACR technicians and I thought it would be a great career to pursue. I loved working on mechanical systems, there was always a demand in the job market, and I was confident that I would be able to advance my career as long as I was honest, dependable, and worked hard.

What do you enjoy most about your career?

The thing that I enjoy the most is the diversity and variety that it offers. My career has evolved over the years and I currently work for the largest electric utility in the nation managing all the solar programs in Florida. I love the interaction with the customers and contractors and I thoroughly enjoy the research portion where we are asked to validate information and data.

Why do you think training and education are important in construction?
I think training and education are important because workers must develop and measure their competency so that they can perform at a safe and effective level, for both themselves and the people around them. Technology advances rapidly and constantly, so it is important to continue with education and training in order to provide the best possible service to all consumers.

Why do you think credentials are important in construction?
Credentials validate an individual's level of experience and verify that the skills have been observed by a competent instructor.

How has training/construction impacted your life and career (i.e., advancement opportunities, better wages, etc.)?
I take advantage of every training opportunity available through my employer. Training enhances my skills and level of experience. Certifications are eventually reviewed by potential bosses who may find that they document just the skill set needed, and an opportunity for advancement is created. Training has personally impacted my life by allowing me to advance in my career, earn higher wages, and participate in a variety of new, innovative programs that help others and the environment.

Would you recommend construction as a career to others? Why?
Yes, I would recommend construction as a career to others because you can always earn a living working with your hands. The trades offer a wide variety of unique opportunities.

What does craftsmanship mean to you?
Craftsmanship, to me, means doing every job to the best of my ability and demonstrating honesty, integrity, and hard work in the process.

Trade Terms Introduced in This Module

Bond: Short for *surety bond*. It is a funded guarantee that a contractor will perform as agreed.

Carbon monoxide (CO): A common byproduct of combustion processes, CO is a colorless, tasteless, and odorless gas that is lighter than air and quite toxic. CO reacts with blood hemoglobin to form a substance that significantly reduces the flow of oxygen to all parts of the body. In some countries, CO is responsible for the majority of fatal air poisoning events.

Chiller: A high-volume hydronic cooling unit.

Chlorofluorocarbon (CFC) refrigerant: A class of refrigerants that contains chlorine, fluorine, and carbon. CFC refrigerants have a very adverse effect on the environment.

Compressor: In a refrigeration system, the mechanical device that converts low-pressure, low-temperature refrigerant gas into high-temperature, high-pressure refrigerant gas.

Condenser: A heat exchanger that transfers heat from the refrigerant flowing inside it to the air or water flowing over it.

Easement: A portion of a property that is set aside for public utilities or municipalities.

Evaporator: A heat exchanger that transfers heat from the air flowing over it to the cooler refrigerant flowing through it.

Expansion device: Also known as the liquid metering device or metering device. Provides a pressure drop that converts the high-temperature, high-pressure liquid refrigerant from the condenser into the low-temperature, low-pressure liquid refrigerant entering the evaporator.

Heat pump: A comfort air conditioner that is able to produce heat by reversing the mechanical refrigeration cycle.

Heat transfer: The transfer of heat from a warmer substance to a cooler substance.

Hydrochlorofluorocarbon (HCFC) refrigerant: A class of refrigerants that contains hydrogen, chlorine, fluorine, and carbon.

Hydronic: A heating or air conditioning system that uses water as a heat transfer medium.

International Building Code (IBC): A series of model construction codes. These codes set standards that apply across the country. This is an ongoing process led by the International Code Council (ICC).

Mechanical refrigeration: The use of machinery to provide cooling.

Mechanical refrigeration cycle: The process by which a circulating refrigerant absorbs heat from one location and transfers it to another location.

Noxious: Harmful to health.

On the job learning (OJL): Learning obtained while working under the supervision of a journeyman.

Reclaimed: Used refrigerant that has been remanufactured to bring it up to the standards required of new refrigerant.

Recovery: The removal and temporary storage of refrigerant in containers approved for that purpose.

Recycling: Circulating recovered refrigerant through filtering devices that remove moisture, acid, and other contaminants.

Sustainable construction: Construction that involves minimum impact on land, natural resources, raw materials, and energy over the building's life cycle. It uses material in a way that preserves natural resources and minimizes pollution.

Toxic: Poisonous.

Additional Resources

This module presents thorough resources for task training. The following resource material is suggested for further study.

ABC's of Air Conditioning. Syracuse, NY: Carrier Corporation.

Fundamentals of HVAC/R, Carter Stanfield and David Skaves. New York, NY: Pearson.

Refrigeration and Air Conditioning: An Introduction to HVAC/R, Larry Jeffus and David Fearnow. New York, NY: Pearson.

Your Role in the Green Environment, NCCER. Current edition. New York, NY: Pearson.

Figure Credits

Courtesy of Kathabar Dehumidification Systems, SA08

Section Review Answer Key

Answer	Section Reference	Objective
Section One		
1. b	1.1.0	1a
2. b	1.2.0	1b
3. c	1.3.0	1c
4. a	1.4.0	1d
Section Two		
1. d	2.1.0	2a
2. b	2.2.1	2b
3. b	2.3.0	2c
4. a	2.4.0	2d
Section Three		
1. b	3.1.3	3a
2. d	3.2.3	3b
3. d	3.3.0	3c

NCCER CURRICULA — USER UPDATE

NCCER makes every effort to keep its textbooks up-to-date and free of technical errors. We appreciate your help in this process. If you find an error, a typographical mistake, or an inaccuracy in NCCER's curricula, please fill out this form (or a photocopy), or complete the online form at **www.nccer.org/olf**. Be sure to include the exact module ID number, page number, a detailed description, and your recommended correction. Your input will be brought to the attention of the Authoring Team. Thank you for your assistance.

Instructors – If you have an idea for improving this textbook, or have found that additional materials were necessary to teach this module effectively, please let us know so that we may present your suggestions to the Authoring Team.

NCCER Product Development and Revision

13614 Progress Blvd., Alachua, FL 32615

Email: curriculum@nccer.org
Online: www.nccer.org/olf

❏ Trainee Guide ❏ Lesson Plans ❏ Exam ❏ PowerPoints Other _____

Craft / Level: _____ Copyright Date: _____

Module ID Number / Title: _____

Section Number(s): _____

Description: _____

_____

_____

_____

Recommended Correction: _____

_____

_____

_____

_____

Your Name: _____

Address: _____

_____

Email: _____ Phone: _____

Trade Mathematics

OVERVIEW

Math is an essential skill required to advance in the HVACR profession. Math is used when cutting and fitting pipe, sizing and installing ductwork, and when calculating electrical values such current flow.

Module 03102

Trainees with successful module completions may be eligible for credentialing through the NCCER Registry. To learn more, go to **www.nccer.org** or contact us at **1.888.622.3720.** Our website has information on the latest product releases and training, as well as online versions of our *Cornerstone* magazine and Pearson's product catalog.

Your feedback is welcome. You may email your comments to **curriculum@nccer.org,** send general comments and inquiries to **info@nccer.org,** or fill in the User Update form at the back of this module.

This information is general in nature and intended for training purposes only. Actual performance of activities described in this manual requires compliance with all applicable operating, service, maintenance, and safety procedures under the direction of qualified personnel. References in this manual to patented or proprietary devices do not constitute a recommendation of their use.

03102 V5

Objectives

When you have completed this module, you will be able to do the following:

1. Convert units of measurement from the inch-pound system to the metric system, and vice-versa.
   a. Identify units of measure in the inch-pound and metric systems.
   b. Convert length, area, and volume values.
   c. Convert weight values.
   d. Convert pressure and temperature values.
2. Solve basic algebra equations.
   a. Define algebraic terms.
   b. Demonstrate an understanding of the sequence of operations.
   c. Solve basic algebraic equations.
3. Identify and describe geometric figures.
   a. Describe the characteristics of a circle.
   b. Identify and describe types of angles.
   c. Identify and describe types of polygons.
   d. Calculate various values associated with triangles.

Performance Tasks

This is a knowledge-based module; there are no performance tasks.

Trade Terms

Absolute pressure
Area
Atmospheric pressure
Barometric pressure
Coefficient
Constant
Exponent
Force

Mass
Newton (N)
Polygon
Unit
Vacuum
Variable
Volume

Industry Recognized Credentials

If you are training through an NCCER-accredited sponsor, you may be eligible for credentials from NC-CER's Registry. The ID number for this module is 03102. Note that this module may have been used in other NCCER curricula and may apply to other level completions. Contact NCCER's Registry at 888.622.3720 or go to **www.nccer.org** for more information.

Contents

1.0.0 Using the Metric System 1
  1.1.0 Units of Measure 1
  1.2.0 Length, Area, and Volume 2
    1.2.1 Length 3
    1.2.2 Area 3
    1.2.3 Volume 4
    1.2.4 Wet Volume Measurements 6
  1.3.0 Weight Conversions 6
  1.4.0 Pressure and Temperature 6
    1.4.1 Absolute Pressure 10
    1.4.2 Static Head Pressure 10
    1.4.3 Vacuum 11
    1.4.4 Temperature 12
    1.4.5 Temperature Conversions 12
2.0.0 Solving Problems Using Algebra 16
  2.1.0 Definition of Terms 16
    2.1.1 Mathematical Operators 16
    2.1.2 Equations 16
    2.1.3 Variables 16
    2.1.4 Constants 16
    2.1.5 Coefficients 17
    2.1.6 Powers and Roots 17
  2.2.0 Sequence of Operations 18
  2.3.0 Solving Algebraic Equations 18
    2.3.1 Rules of Algebra 19
3.0.0 Working With Geometric Figures 22
  3.1.0 Characteristics of a Circle 22
  3.2.0 Angles 22
  3.3.0 Polygons 23
  3.4.0 Working With Triangles 26
    3.4.1 Right Triangle Functions 26
    3.4.2 Right Triangle Calculations Using the Pythagorean Theorem 27
    3.4.3 Converting Decimal Feet to Feet and Inches and Vice Versa 27
Appendix Conversion Factors 36

Figures and Tables

Figure 1   Common inch-pound and metric system values 2
Figure 2   Comparison of inches to centimeters 4
Figure 3   Duct shapes 4
Figure 4   Converting the area of a square from inch-pound
            to metric units 5
Figure 5   Converting the area of a rectangle from inch-pound to
            metric units 5
Figure 6   Converting the area of a circle from metric to
            inch-pound units 6
Figure 7   Volume conversion for a rectangular prism 7
Figure 8   Calculating the volume of a cylindrical tank or cylinder in cm 8
Figure 9   Converting cylinder volume from metric to inch-pound units 8
Figure 10 Common metric prefixes used with volumes 9
Figure 11 Tire pressure label 9
Figure 12 Closeup of pressure gauge 10
Figure 13 Comparison of temperature scales 14
Figure 14 Sample temperature unit conversions 14
Figure 15 Perpendicular and parallel lines 22
Figure 16 A circle 22
Figure 17 An angle 23
Figure 18 Types of angles 24
Figure 19 Adjacent, complementary, and supplementary angles 24
Figure 20 Common polygons 25
Figure 21 Triangles 26
Figure 22 Common labeling of angles and sides in a right triangle 27

Table 1   Inch-Pound and Metric System Units of Measure 2
Table 2   Common Units in the Inch-Pound System 3
Table 3   Metric System Prefixes 3
Table 4   Length Conversion Multipliers 4
Table 5   Volume Relationships 8
Table 6   Weight Equivalents 9
Table 7   Pressure Conversions 11

1.0.0 USING THE METRIC SYSTEM

Objective

Convert units of measurement from the inch-pound system to the metric system, and vice versa.

    a. Identify units of measure in the inch-pound and metric systems.
    b. Convert length, area, and volume values.
    c. Convert weight values.
    d. Convert pressure and temperature values.

Trade Terms

Absolute pressure: The total pressure that exists in a system. Absolute pressure is expressed in pounds per square inch absolute (psia). Absolute pressure = gauge pressure + atmospheric pressure.

Area: The amount of surface in a given plane or two-dimensional shape.

Atmospheric pressure: The standard pressure exerted on Earth's surface. Atmospheric pressure is normally expressed as 14.7 pounds per square inch (psi) or 29.92 inches of mercury.

Barometric pressure: The actual atmospheric pressure at a given place and time.

Coefficient: A multiplier (e.g., the numeral 2 as in the expression 2b).

Force: A push or pull on a surface. In this module, force is considered to be the weight of an object or fluid. This is a common approximation.

Mass: The quantity of matter present.

Newton (N): The amount of force required to accelerate one kilogram at a rate of one meter per second.

Unit: A definite standard of measure.

Vacuum: Any pressure that is less than the prevailing atmospheric pressure.

Volume: The amount of space contained in a given three-dimensional shape.

More than 95 percent of the world uses the metric system of measurement. Many HVACR products used in the United States are now manufactured in other countries. The dimensions, weights, temperatures, and pressures in the installation, operating, and maintenance instructions provided with some of these products are stated using metric system values. Even if the literature has been converted to show inch-pound values, the mounting hardware is likely to remain in metric terms.

The use of the metric system in the United States is becoming more common. Many manufacturers publish their technical manuals and instrument data sheets displaying all values, as in the past, in inch-pound system units, but also putting the metric equivalents in parentheses behind the inch-pound units. In some cases, only the metric units may be listed, so it is a good time to become familiar with both and understand how to convert from one to the other. *Figure 1* shows some metric system quantities that have become widely recognized.

1.1.0 Units of Measure

Most work in science and engineering is based on the exact measurement of physical quantities. A measurement is simply a comparison of a quantity to some definite standard measure of dimension called a unit. Whenever a physical quantity is described, the units of the standard to which the quantity was compared, such as a foot, a liter, or a pound, must be specified. A number alone is not enough to describe a physical quantity.

The importance of specifying the units of measurement for a number used to describe a physical quantity should be apparent because the same physical quantity may be measured using a variety of different units. For example, length may be measured in inches, feet, yards, miles, millimeters, centimeters, meters, and kilometers.

Today, both the inch-pound and metric systems are widely used in engineering, construction, and technical work. Therefore, it is necessary to have some degree of understanding of both systems of units. HVACR work is concerned with the following values, which can be expressed in both the inch-pound and metric systems:

- Dimensions and distances
- Weight
- Volume
- Pressure
- Temperature

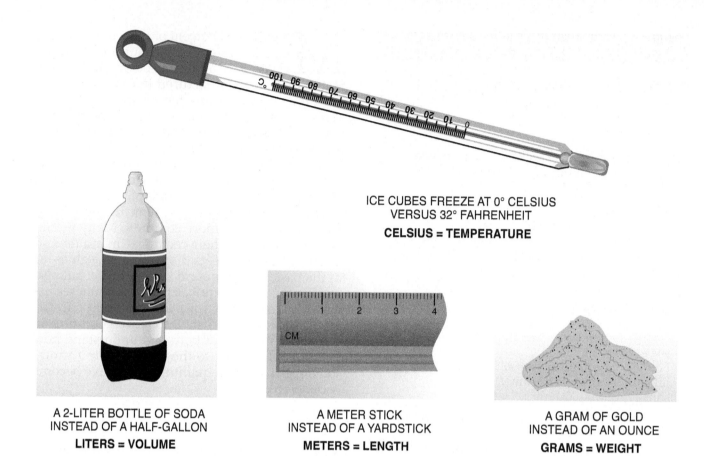

ICE CUBES FREEZE AT 0° CELSIUS
VERSUS 32° FAHRENHEIT
CELSIUS = TEMPERATURE

A 2-LITER BOTTLE OF SODA
INSTEAD OF A HALF-GALLON
LITERS = VOLUME

A METER STICK
INSTEAD OF A YARDSTICK
METERS = LENGTH

A GRAM OF GOLD
INSTEAD OF AN OUNCE
GRAMS = WEIGHT

Figure 1 Common inch-pound and metric system values.

Table 1 shows how these values are stated in both the inch-pound and metric systems.

Once it is understood, the metric system is actually much simpler to use than the inch-pound system because it is a decimal system in which prefixes are used to denote powers of ten. The older inch-pound system, on the other hand, requires the use of conversion factors that must be memorized. For example, one mile is 5,280 feet, and 1 inch is 1/12 of a foot. In contrast, a centimeter is one-one hundredth of a meter and a kilometer is 1,000 meters. An additional advantage of the metric system is that it is not necessary to add and subtract fractions. *Table 2* lists some of the more common units in the inch-pound system.

The metric system prefixes are listed in *Table 3*. From this table, it can be seen that the use of the metric system is logically arranged, and that the name of the unit also represents an order of magnitude (via the prefix) that foot and pound cannot.

The most common metric system prefixes are mega- (M), kilo- (k), centi- (c), milli- (m), and micro- (µ). Even though these prefixes may seem difficult to understand at first, most people are probably using them regularly. For example, you have probably seen the terms *mega*watts, *kilo*meters, *centi*meters, *milli*volts, and *micro*amps.

1.2.0 Length, Area, and Volume

In the field, numbers usually represent physical quantities. In order to give meaning to these quantities, measurement units are assigned to the numbers to represent quantities such as length, area, and volume.

Table 1 Inch-Pound and Metric System Units of Measure

Value	Inch-Pound System	Metric System
Dimensions	Inches and feet	Centimeters and meters
Weight	Ounces and pounds	Grams and kilograms
Volume–dry	Cubic inches and cubic feet	Cubic centimeters and cubic meters
Volume–liquid	Quarts and gallons	Liters
Pressure	Pounds per square inch	Pascal and bar
Temperature	Degrees Fahrenheit (F)	Degrees Celsius (C) (Centigrade)

Table 2 Common Units in the Inch-Pound System

Unit	Equivalent
12 inches (in)	1 foot
1 yard (yd)	3 ft
1 mile (mi)	5,280 ft
16 ounces (oz)	1 pound (lb)
1 ton	2,000 lb
1 minute	60 seconds (sec)
1 hour (hr)	3,600 sec
1 US gallon (gal)	0.1337 cubic foot (cu ft)

1.2.1 Length

Length typically refers to the long side of an object or surface. With liquids, the measurement of the level of fluid in a tank is basically a measurement of length from the surface of the fluid to the bottom of the tank. Length can be expressed in either inch-pound or metric units; that is, inches or centimeters.

Table 4 shows the relationships of the most common units of length. *Figure 2* compares inches to centimeters.

The multipliers in *Table 4* are used to make the conversions from one system to the other.
Example:

An installation plan for an HVACR system manufactured in Europe requires a thermostat to be mounted 122 centimeters (cm) above the floor, but your measuring tape is calibrated in inches. How many inches above the floor should the thermostat be placed?

$$1 \text{ cm} = 0.3937''$$

$$122 \text{ cm} = 122 \times 0.3937''$$

$$122 \text{ cm} = 48.0314''$$

$$= 48'' \text{ (rounded off)}$$

or (changing to feet)

$$122 \text{ cm} = 122 \times 0.03281'$$

$$122 \text{ cm} = 4.00282'$$

$$= 4' \text{ (rounded off)}$$

> **NOTE**
>
> The *Appendix* contains listings of common conversion factors and common formulas used in construction. Also, some scientific calculators have the capability of converting English units to metric units and vice versa.

1.2.2 Area

Area is the measurement of the surface of a two-dimensional object. In HVACR work, it is necessary to know the area of a duct in order to determine the amount of airflow it supports. Duct can be rectangular, square, or round (*Figure 3*).

The area of any rectangle is equal to the length multiplied by the width. In the inch-pound system, the unit of area is the square foot (ft^2) or square inch (in^2). *Figure 4* shows the application of this concept to the square. When converting areas from one measurement system to the other, every dimension must be converted. So, to convert the dimensions of the square shown in *Figure 4*, both the length and the width must be converted, as shown. The area can then be calculated using the metric values.

Length

If the manufacturer's product data sheet states that a fan coil unit is 1.22 meters high and 0.46 meter wide, what size opening (in feet) is needed to get this unit inside a building?

Table 3 Metric System Prefixes

Prefix		Unit		
micro- (µ)	1/1,000,000	0.000001	10^{-6}	One-millionth
milli- (m)	1/1,000	0.001	10^{-3}	One-thousandth
centi- (c)	1/100	0.01	10^{-2}	One-hundredth
deci- (d)	1/10	0.1	10^{-1}	One-tenth
deka- (da)	10	10.0	10^{1}	Tens
hecto- (h)	100	100.0	10^{2}	Hundreds
kilo- (k)	1,000	1,000.0	10^{3}	Thousands
mega- (M)	1,000,000	1,000,000.0	10^{6}	Millions
giga- (G)	1,000,000,000	1,000,000,000.0	10^{9}	Billions

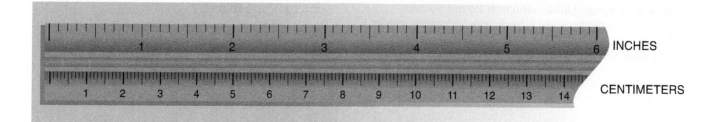

Figure 2 Comparison of inches to centimeters.

Table 4 Length Conversion Multipliers

Unit	Centimeter	Inch	Foot	Meter	Kilometer
1 millimeter	0.1	0.03937	0.003281	0.001	0.000001
1 centimeter	1	0.3937	0.03281	0.01	0.00001
1 inch	2.54	1	0.08333	0.0254	0.0000254
1 foot	30.48	12	1	0.3048	0.0003048
1 meter	100	39.37	3.281	1	0.001
1 kilometer	100,000	39,370	3,281	1,000	1

The same principle applies to a rectangle, as shown in *Figure 5*.

The area of a circular duct is found using the following formula:

$$Area = \pi r^2$$

Where:

A = the area of the circle

π = a constant of 3.14159 (usually abbreviated to 3.14)

r = the radius (distance from the center to the edge of the circle)

When converting the dimensions of a circle, only the radius has a measured dimension that must be converted. In the example shown in *Figure 6*, centimeters must be converted into inches.

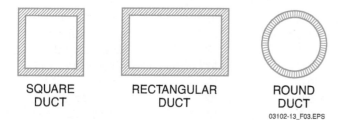

SQUARE DUCT   RECTANGULAR DUCT   ROUND DUCT

03102-13_F03.EPS

Figure 3 Duct shapes.

1.2.3 Volume

Volume is the amount of space occupied by a three-dimensional object. The volume of a cube or rectangular prism, such as a room in a building, is the product of three dimensions: length, width, and height. In the familiar inch-pound system, the most common units of volume are cubic feet (ft^3) or cubic inches (in^3). There are a number of situations in which an HVACR technician would need to calculate volume.

The Meter

Originally, the meter was defined as $\frac{1}{10,000,000}$ of Earth's meridional quadrant (the distance from the North Pole to the equator). This was how the meter got its name. This distance was etched onto a metal bar that is kept in France. In 1866, the United States legalized the use of the metric system and placed an exact copy of the metal bar into the US Bureau of Standards. In October 1983, the world's scientists redefined the meter to avoid the minor potential of error associated with the metal bar in France. Now a meter is defined as being equal to the distance that light travels in $\frac{1}{299,792,458}$ of a second, which is equivalent to 39.37 inches.

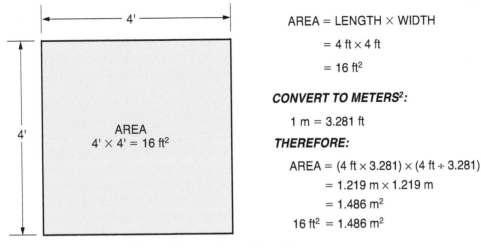

AREA = LENGTH × WIDTH

= 4 ft × 4 ft

= 16 ft²

CONVERT TO METERS²:

1 m = 3.281 ft

THEREFORE:

AREA = (4 ft × 3.281) × (4 ft ÷ 3.281)

= 1.219 m × 1.219 m

= 1.486 m²

16 ft² = 1.486 m²

Figure 4 Converting the area of a square from inch-pound to metric units.

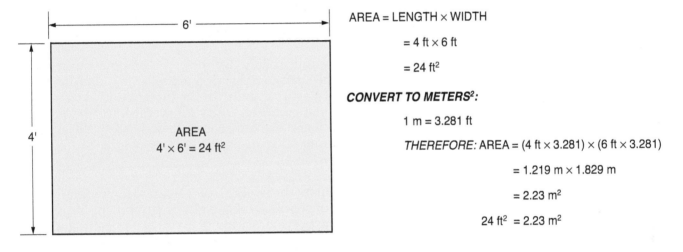

AREA = LENGTH × WIDTH

= 4 ft × 6 ft

= 24 ft²

CONVERT TO METERS²:

1 m = 3.281 ft

THEREFORE: AREA = (4 ft × 3.281) × (6 ft × 3.281)

= 1.219 m × 1.829 m

= 2.23 m²

24 ft² = 2.23 m²

Figure 5 Converting the area of a rectangle from inch-pound to metric units.

Here are some examples:

- In order to calculate the cooling or heating load a given space represents, it is necessary to determine the volume of the space. When calculating the load for a house, the estimator would need to know the volume of each room. Room volume is also a consideration in determining the number of air changes that occur in a given time.
- Assume that the oil needs to be drained from a semi-hermetic compressor. You know the compressor holds 2.5 quarts of oil, but you don't know if the capacity of your container is more than 2.5 quarts. The same principle applies if the oil has been drained into an un-calibrated container and you need to determine how much oil was removed.

- In a hydronic cooling or heating system, it is sometimes necessary to mix glycol (antifreeze) with the water in a specific proportion. To do so, it is necessary to first calculate the amount of water the pipes and equipment components can hold.

Figure 7 shows the three-dimensional measurement of a rectangular prism (space) such as a room. Its volume is calculated by multiplying length × width × height. When converting these two volumes from one measurement system to the other, remember that every dimension must be independently converted. So, to convert the dimensions of the box, the length, width, and height must be converted as shown in *Figure 7.*

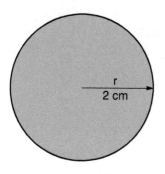

AREA = πr^2

= (3.14) (2 cm × 2 cm)

= 3.14 × 4 cm²

= 12.56 cm²

CONVERT TO INCHES²:

1 in = 2.54 cm

THEREFORE: RADIUS IN INCHES

= 2 ÷ 2.54

= 0.7874 in

AREA = (3.14) (0.7874 in × 0.7874 in)

= 3.14 × 0.62 in²

= 1.95 in²

12.56 cm² = 1.95 in²

Figure 6 Converting the area of a circle from metric to inch-pound units.

This method works for finding the volume of box-shaped containers, but will not work for tanks and other cylindrical containers, as shown in *Figure 8*. The volume of a cylindrical tank, such as a refrigerant cylinder, can be simplified by treating it as the area of the circle times its depth (or height).

> **NOTE**
>
> The radius of a cylinder or circle equals ½ the diameter. In order to determine the radius of a round container, it is first necessary to determine the diameter, then divide by 2.

In the example of the cylindrical tank, both the radius and the height must be converted to inch-pound values as shown in *Figure 9* in order to calculate the volume.

1.2.4 Wet Volume Measurements

When the capacity of a space is calculated in cubic feet or cubic meters, it is considered dry volume. When liquid volume is involved, such as determining the amount of a fluid that would fill a container, it is classified as wet volume because liquid measures are used. Common wet measures in the inch-pound system include the pint, quart, and gallon.

The metric system also uses wet measuring units. The liter is the most common measure. A liter is about 5 percent greater than a quart. Knowing the wet measures for a substance allows easy handling and measuring of fluids, since the fluid conforms to the shape of the container.

Table 5 shows the volume relationships between the liter and the dry volume of the cubic meter in the metric system, and the pint and gallon in the inch-pound system.

Figure 10 shows that the same metric prefixes that apply to the meter can be used with the liter.

1.3.0 Weight Conversions

Weight is actually the force an object exerts on the surface of Earth due to its mass and the pull of Earth's gravity. In the inch-pound system, the most common units for weight are the pound (lb) and the ounce (oz). *Table 6* shows the weight units used in both the inch-pound and metric systems.

In HVACR work, metric conversions can involve the weight of refrigerant for charging purposes or the weight of equipment for lifting purposes. Imagine that you are installing a new rooftop packaged unit and the equipment specs give the weight of the unit as 1,200 kg. How do you select the equipment needed to raise the unit onto the roof when the rigging equipment is rated in pounds?

Table 6 shows that 1 kg = 2.205 pounds, so multiplying 1,200 by 2.205 yields 2,646 pounds. Therefore, any slings or lifting equipment used to raise the unit must have a rated capacity greater than 2,646 pounds.

1.4.0 Pressure and Temperature

Pressure and temperature are covered together because they are interdependent. If either the pressure or the temperature changes in a closed system, such as an HVACR system or an automobile tire, the other is affected in proportion. The temperature-pressure relationship is the key to the refrigeration cycle on which most air conditioning and refrigeration systems are based. When testing or troubleshooting a system, technicians measure temperature and pressure values at key points in a system to determine if the system is operating correctly.

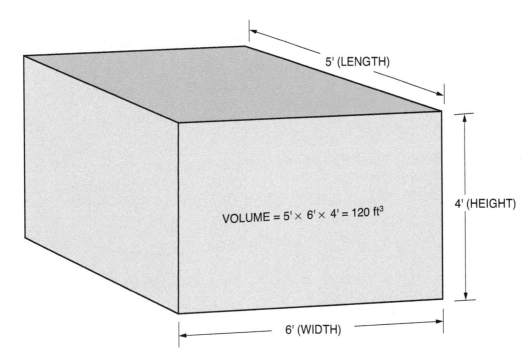

VOLUME = LENGTH × WIDTH × HEIGHT

= 5 ft × 6 ft × 4 ft

= 120 ft³

CONVERT TO METERS³:

1 m = 3.281 ft

$$\frac{1\ m}{3.281\ ft} = 0.305$$

5 ft ÷ 3.281 = 1.524 m

6 ft ÷ 3.281 = 1.829 m

4 ft ÷ 3.281 = 1.219 m

THEREFORE: VOLUME

= 1.524 m × 1.829 m × 1.219 m

= 3.4 m³

120 ft³ = 3.4 m³

Figure 7 Volume conversion for a rectangular prism.

For clarification, if there are 32 pounds of air in the tires of your car, it really means that the air in the tires exerts 32 more pounds of force per square inch on the inside of the tires than the air pressure of the atmosphere exerts on the outside of the tires.

Both liquids and gases are capable of exerting pressure. Gases are compressed under pressure and expand when the pressure is lowered. Liquids are generally considered to be incompressible. The refrigeration cycle refrigerant exists as both a gas and a liquid in different parts of the system.

Tanks of refrigerant provide a good example of liquids and gases exerting pressure. The pressure inside a refrigerant tank depends on the type of refrigerant in the drum and the ambient temperature surrounding the drum. For every refrigerant, there is a specific temperature/pressure relationship. For example, a drum of R-410A refrigerant sitting in the back of an open truck with the sun shining on it may have a drum temperature of 105°F. A temperature/pressure chart for R-410A, would show that this corresponds to a pressure of about 339.9 psig. This means that the drum has a pressure of about 340 pounds pushing outward against its walls for each square inch of drum surface area. At 32°F, the pressure is only 101.6 psig.

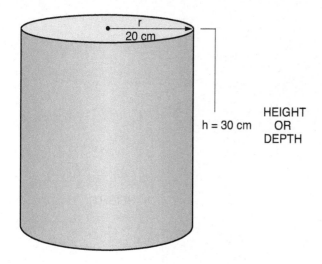

VOLUME = $(\pi r^2) \times h$
   or
   = AREA OF THE CIRCLE × HEIGHT

$$\underbrace{\qquad\qquad}_{\text{AREA}} \quad \underbrace{\qquad}_{\text{HEIGHT}}$$

= (3.14) (20 cm) (20 cm) × (30 cm)

= (3.14) (400 cm²) × (30 cm)

= (1,256 cm²) × (30 cm)

= 37,680 cm³

Figure 8 Calculating the volume of a cylindrical tank or cylinder in cm.

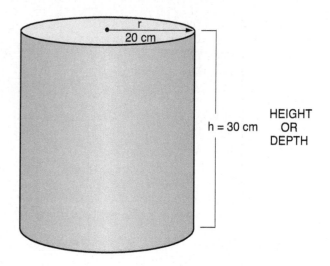

VOLUME = $(\pi r^2) \times h$

= (3.14 × 20²) × h

= (3.14 × 400 cm²) × 30 cm

= 1,256 cm² × 30 cm

= 37,680 cm³

CONVERT TO INCHES³:

 1 cm = 0.3937 in

 20 cm = 7.874 in

 30 cm = 11.811 in

THEREFORE: VOLUME

   = [(3.14) (7.874²)] × 11.811 in

   = (3.14 × 62) × 11.811 in

   = 194.68 in² × 11.811 in

   = 2,299.36 in³

37,680 cm³ = 2,299.36 in³

Figure 9 Converting cylinder volume from metric to inch-pound units.

> NOTE
>
> Even though the atmosphere is not visible and cannot be felt, it still has mass because it consists of moisture and gases (such as oxygen). Scientists have determined that the atmosphere exerts a pressure of 14.7 pounds per square inch (psi) at sea level. At altitudes above sea level, the pressure decreases because there is less atmosphere. In Denver, Colorado, which is roughly 5,000 feet above sea level, the pressure is 12.1 psi.

Table 5 Volume Relationships

Unit	Cubic Meter	Gallon	Liter	Pint
Cubic meter (m³)	1	264.172	1,000	2,113.38
US gallon (gal)	0.0037	1	3.785	8
Liter (L)	0.001	0.264	1	2.113
US pint (pt)	0.0005	0.125	0.4731	1

Pounds per square inch (psi), the most common unit of pressure in the inch-pound system, is equivalent to newtons per square meter (N/m²) in the metric system. One N/m² is also referred to as one pascal, after Blaise Pascal, a seventeenth-century French philosopher who contributed greatly to mathematics and engineering. The pascal, or Pa, is equal to a force of one newton exerted on an area of one square meter. It would take 200,000 pascals (200 kPa) to inflate an ordinary automobile tire to about 28 psi. Automobile tire pressure is typically stated in kilopascals and psi (*Figure 11*).

In HVACR work, where much higher pressures are involved, the bar (b) is a more convenient measure of pressure when metric values must be used. One bar is equal to 14.5 psi and 100 kilopascals. The bar is the metric value often used on HVACR pressure gauges (*Figure 12*) along with psig units. Bars are also used by weather forecasters to describe changes in atmospheric pressure. You are probably familiar with barometric pressure readings from watching the weather information on the evening news. You will also see the term *millibar*, which is 0.001 bar (b).

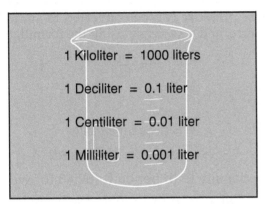

03102-13_F10.EPS

Figure 10 Common metric prefixes used with volumes.

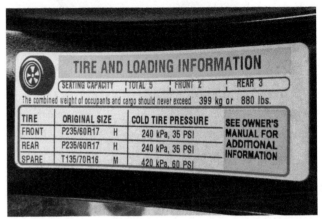

03102-13_F11.EPS

Figure 11 Tire pressure label.

Table 6 Weight Equivalents

Unit	Kilogram	Pound	Ounce	Gram
1 kilogram (kg)	1	2.205	35.27	1,000
1 pound (lb)	0.4536	1	16	453.6
1 ounce (oz)	0.02835	0.0625	1	28.35
1 gram (g)	0.001	0.002205	0.03527	1

Volume

Air conditioning equipment located outdoors is mounted on a pad in order to provide stability and protection. While there are prefabricated pads for smaller units, large units often require a poured concrete pad. The concrete pad on which this unit is to be installed measures 13' × 6' × 6". To order or mix the concrete, the installers had to figure out how much concrete was required. What was the volume of concrete needed to construct this pad expressed in cubic feet, cubic yards, and cubic meters?

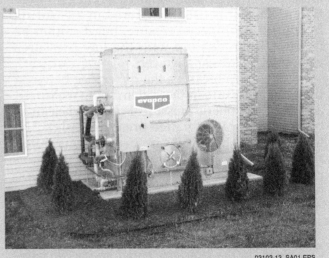

03102-13_SA01.EPS

Figure 12 Close-up of pressure gauge.

The next section relates these metric units to the more familiar pressure units of the inch-pound system and introduces the concept of pressure measurement at the same time.

1.4.1 Absolute Pressure

The standard atmospheric pressure exerted on the surface of Earth is 14.696 psi taken at sea level with the air at 70°F. For most practical applications, this number is usually rounded to 14.7 psi. Atmospheric pressure is also expressed as 29.9213 inches of mercury, which is equal to 14.696 psi. The rounded-off value for inches of mercury (in. Hg) is 29.92.

Weather conditions usually cause a slight variation in the atmospheric pressure. The actual atmospheric pressure is known as the barometric pressure. It is usually ignored in measuring hydraulic machinery pressure, but it cannot be ignored when dealing with the low pressures generated by fans and blowers.

Most gauges measure the difference between the actual pressure in the system being measured and the atmospheric pressure. Pressure measured by gauges is called gauge pressure (psig). The total pressure that exists in the system is called the absolute pressure (psia). Absolute pressure is equal to gauge pressure plus the atmospheric pressure. In other words:

psia = gauge pressure (psig) + 14.7

Example:
A steam boiler pressure gauge reads 295 psig. Find the absolute pressure.

psia = psig + 14.7
psia = 295 + 14.7
psia = 309.7

Boiler pressure is sometimes specified in terms of atmospheres, where one atmosphere is equal to 14.696 (14.7) psi (see *Table 7*).

1.4.2 Static Head Pressure

Municipal water systems usually define pressure in terms of the static head or height in feet from the point of use to elevated water reservoirs. Gauge pressure can be converted to static head pressure, and vice versa, using the formula:

$$P = \frac{hd}{144}$$

Where:

P = pressure (psig)

h = height in feet (head)

d = density in lb per cu ft (62.43 for water)

144 = used to convert square feet into square inches

For example, if the water level in a reservoir is maintained at 150 feet above the point of draw, which is the level from which water is drawn. What is the water pressure at the point of draw resulting from this head?

$$P = \frac{hd}{144} = \frac{150 \times 62.43}{144} = 65 \text{ psig}$$

Other terms are necessary to measure extremely low pressures, such as those developed by blowers and fans. These pressures are often measured in inches of water (in. H_2O). Sometimes it is necessary to convert from one measure to another. Consult *Table 7* to solve the following problems.
Examples:

1. A steam generator operates at a pressure of 320 atmospheres. What is this pressure in terms of psig?

psig = 320 × 14.7 = 4,704 psig

2. Blower and fan pressures are usually measured in inches of water (in. H_2O) because small differences in pressure can be readily detected. Due to the low pressures involved, the exact atmospheric pressure is usually measured to determine the actual discharge pressure. Calculate the discharge pressure in

psia if the measured discharge pressure of a blower is 56.55 in. H_2O and the barometric pressure is 28.49 in. Hg.

Step 1 Find the blower discharge pressure in psig:

$$psig = in. \ H^2O \times 0.0361$$

$$= 56.55 \times 0.0361$$

$$= 2.041455$$

Step 2 Find the actual atmospheric pressure in psi.

$$psi = in. \ Hg \times 0.491$$

$$= 28.49 \times 0.491$$

$$= 13.98859$$

Step 3 Find the absolute pressure of the blower discharge.

Absolute pressure = gauge pressure + actual atmospheric pressure

$$= 2.041455 + 13.98859$$

$$= 16.030045 \text{ or approx. } 16.03 \text{ psia}$$

1.4.3 Vacuum

In HVACR work, a vacuum is any pressure that is less than the prevailing atmospheric pressure. It is usually measured in terms of inches of mercury (in. Hg). Actual barometric pressure must always be determined in measuring a vacuum. Absolute pressure in a vacuum is calculated using the following formula:

Absolute vacuum pressure = barometric pressure – vacuum gauge reading

Example:
A vacuum gauge attached to a line reads 17.2 in. Hg. The barometric pressure reads 29.85 in. Hg. What is the absolute pressure in the line?

Absolute vacuum pressure = barometric pressure – vacuum gauge reading

$$= 29.85 - 17.2$$

$$= 12.65 \text{ in. Hg}$$

Tire Pressure

If you are an auto racing fan, you probably know that slight variations in tire pressure can affect the handling performance of a race car. A crew chief might change tire pressure by just half a psi during a pit stop, for example. Compressed air contains moisture, which will heat up as the tire gets hotter and cause the tire pressure to increase. For that reason, race cars use nitrogen instead of compressed air in their tires because nitrogen is a dry gas and does not react to temperature.

03102-13_SA02.EPS

Table 7 Pressure Conversions

Multiply	By	To Obtain
Atmospheres	14.7	Pounds per square inch
Pounds per square inch	0.0680	Atmospheres
Pounds per square inch	0.069	Bar
Bar	14.5	Pounds per square inch
Pounds per square inch	27.68	Inches of water (H_2O)
Pounds per square inch	2.31	Feet of water (H_2O)
Pounds per square inch	2.04	Inches of mercury (Hg)
Inches of water	0.0361	Pounds per square inch
Feet of water	0.433	Pounds per square inch
Inches of mercury	0.491	Pounds per square inch

1.4.4 Temperature

Temperature is the intensity level of heat and is usually measured in degrees Fahrenheit or degrees Celsius. Temperature is measured in degrees on a temperature scale. In order to establish the scale, a substance is needed that can be placed in reproducible conditions. The substance used is water. The point at which water freezes at atmospheric pressure is one reproducible condition and the point at which water boils at atmospheric pressure is another. The four temperature scales commonly used today are the Fahrenheit scale, Celsius scale, Rankine scale, and Kelvin scale (see *Figure 13*). On the Fahrenheit scale, the freezing temperature of water is 32°F and the boiling temperature is 212°F. On the Celsius scale, the freezing temperature of water is 0°C and the boiling temperature is 100°C. The temperatures at which these fixed points occur were established by the inventors of the scales.

The Rankine scale and the Kelvin scale are based on the theory that at some extremely low temperature, no molecular activity occurs. The temperature at which this condition occurs is called absolute zero, the lowest temperature possible. Both the Rankine and Kelvin scales have their zero degree points at absolute zero. On the Rankine scale, the freezing point of water is 491.7°R and the boiling point is 671.7°R. The increments on the Rankine scale correspond in size to the increments on the Fahrenheit scale; for this reason, the Rankine scale is sometimes called the absolute Fahrenheit scale. Both the Rankine and Fahrenheit scales are part of the English system of measurement.

On the Kelvin scale, the freezing point of water is 273°K and the boiling point is 373°K. The increments on the Kelvin scale correspond to the increments on the Celsius scale; for this reason, the Kelvin scale is sometimes called the absolute Celsius scale. Both the Kelvin and Celsius scales are part of the metric system of measurement.

The scales of primary importance in HVACR work are the Fahrenheit scale and the Celsius scale. The Rankine scale and the Kelvin scale are used primarily in scientific applications.

1.4.5 Temperature Conversions

Because both the Fahrenheit and Celsius scales are used in the HVACR industry, it is sometimes necessary to convert between the two. An HVACR

Gauge Manifold Set

The gauge manifold set shown here is one of the most common items of service equipment. It is used to monitor system pressures in air conditioning and refrigeration systems. The gauge on the right is called a compound gauge because it reads below atmospheric pressure in inches of mercury (in. Hg) and above atmospheric pressure (psig). The left-hand gauge reads high pressures up to several hundred psig. Remember, in order to convert readings shown on the gauges (psig) to absolute pressure readings (psia), 14.7 must be added to the reading.

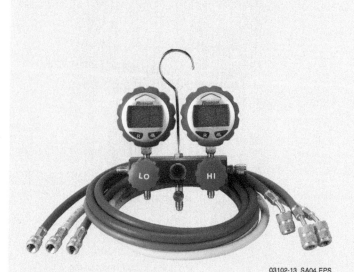

03102-13_SA04.EPS

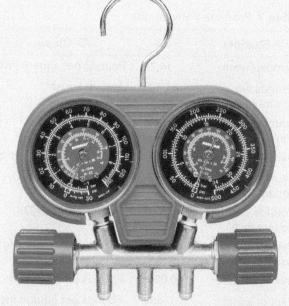

03102-13_SA03.EPS

technician must be familiar with these conversions.

On the Fahrenheit scale, there are 180 degrees between the freezing temperature and boiling temperature of water. On the Celsius scale, there are 100 degrees between the freezing and boiling temperatures of water. The relationship between the two scales can be expressed as follows:

Fahrenheit range (freezing to boiling)
Celsius range (freezing to boiling)

$$= \frac{180°}{100°} = \frac{9}{5}$$

Therefore, one degree Fahrenheit is $\frac{5}{9}$ of one degree Celsius and conversely, one degree Celsius is $\frac{9}{5}$ of one degree Fahrenheit. Thus, to convert a Fahrenheit temperature to a Celsius temperature, it is necessary to subtract 32° (since 32° corresponds to 0°), then multiply by $\frac{5}{9}$. To convert a Celsius temperature to a Fahrenheit temperature, it is necessary to multiply by $\frac{9}{5}$, and then add 32°C.

$$°C = \frac{5}{9}(°F - 32°)$$

$$°F = \left(\frac{9}{5} \times °C\right) + 32°$$

Practice these calculations to become more comfortable with using them. *Figure 14* shows two examples.

NOTE

The *Appendix* contains temperature conversion formulas.

Creating and Measuring a Vacuum

Any air or moisture (noncondensables) trapped in an air conditioning or refrigeration system must be removed before the system can be charged with refrigerant. This requires that a vacuum be drawn on the system using a vacuum pump, such as the one shown here. The vacuum pump creates a pressure differential between the system and the pump. This causes air and moisture vapor trapped in the system at a higher pressure to move into a lower-pressure (vacuum) area created in the vacuum pump. When the vacuum pump lowers the pressure (vacuum) in the system enough, as determined by the ambient temperature of the system, liquid moisture trapped in the system boils and changes into a vapor. Like free air, this water vapor is then pulled out of the system, processed through the vacuum pump, and exhausted to the atmosphere. The level of vacuum present in the system can be measured using a vacuum gauge, such as the one shown.

03102-13_SA05.EPS

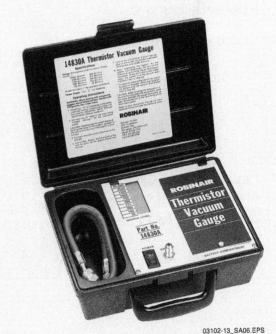

03102-13_SA06.EPS

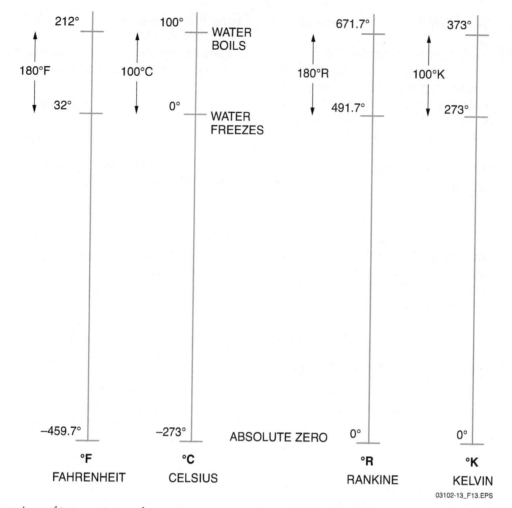

Figure 13 Comparison of temperature scales.

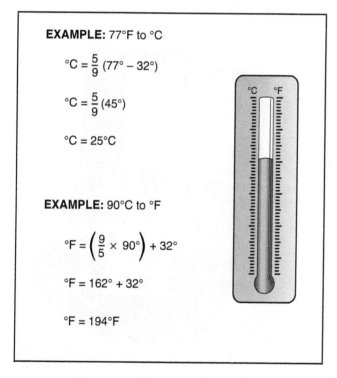

EXAMPLE: 77°F to °C

$$°C = \frac{5}{9}(77° - 32°)$$

$$°C = \frac{5}{9}(45°)$$

$$°C = 25°C$$

EXAMPLE: 90°C to °F

$$°F = \left(\frac{9}{5} \times 90°\right) + 32°$$

$$°F = 162° + 32°$$

$$°F = 194°F$$

Figure 14 Sample temperature unit conversions.

Degrees Centigrade

Because there are 100 degrees on a standard portion of the Celsius scale, Celsius temperature measurements are sometimes referred to as degrees Centigrade.

Temperature Conversions

Depending on the application, refrigeration systems operate to cool a refrigerated area to temperatures between –40°F and +60°F. Comfort cooling systems operate to maintain temperatures in the conditioned space from +60°F to +80°F. What are the corresponding temperature ranges for refrigeration and air conditioning systems when expressed as Celsius temperatures?

Digital Thermometers

Thermometer with digital readouts can provide temperature readings in both Celsius and Fahrenheit. The round-face thermometer is a temperature probe with a digital readout. The one on the right is an infrared thermometer that measures temperature using a laser beam. Many analog thermometers are also calibrated for both scales.

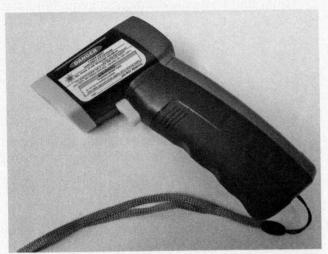

03102-13_SA08.EPS

03102-13_SA07.EPS

1.0.0 Section Review

1. In the metric system, liquid volume is stated in terms of _____.

   a. cubic feet
   b. in^3
   c. liters
   d. pascals

2. What is the area in square feet of a circle with a diameter of 3 meters?

   a. 7.06 ft^2
   b. 9.42 ft^2
   c. 28.26 ft^2
   d. 76.1 ft^2

3. A condensing unit that weighs 240 kg would equal _____.

   a. 240 pounds
   b. 109 pounds
   c. 768 pounds
   d. 529 pounds

4. If the refrigerant pressure at 10°C equals 9.88 bar, what are the corresponding pressure and temperature values in the inch-pound system?

   a. 50°F and 143.3 psi
   b. 40°F and 143.3 psi
   c. 23°F and 24.4 psi
   d. 50°F and 24.4 psi

SECTION TWO

2.0.0 SOLVING PROBLEMS USING ALGEBRA

Objectives

Solve basic algebraic equations.
  a. Define algebraic terms.
  b. Demonstrate an understanding of the sequence of operations.
  c. Solve basic algebraic equations.

Trade Terms

Coefficient: A multiplier (e.g., the numeral 2 as in the expression 2b).

Constant: An element in an equation with a fixed value.

Exponent: A small figure or symbol placed above and to the right of another figure or symbol to show how many times the latter is to be multiplied by itself (e.g., $b^3 = b \times b \times b$).

Variable: An element of an equation that may change in value.

Algebra is the mathematics of defining and manipulating equations containing symbols instead of numbers. The symbols may be either constants or variables. They are connected to each other with mathematical operators such as +, −, ×, and ÷. Knowing how to do calculations in algebra is an important skill for HVACR technicians. Basic algebra is necessary to get through the training program and is essential for doing airflow, electrical current, and voltage calculations. Algebraic calculations are used in many types of troubleshooting.

As in all fields of mathematics, an understanding of algebra requires the knowledge of some basic rules and definitions.

2.1.0 Definition of Terms

This section defines basic algebraic terms, including operators, equations, variables, constants, and coefficients.

2.1.1 Mathematical Operators

Mathematical operators define the required action using a symbol. Common operators include the following:

+ Addition
− Subtraction
× or · Multiplication
÷ or / Division

2.1.2 Equations

An equation is a collection of numbers, symbols, and mathematical operators connected by an equal sign (=). Some examples of equations are:

$$2 + 3 = 5$$
$$P = EI$$
$$\text{Volume} = L \times W \times H$$

2.1.3 Variables

A variable is an element of an equation that may change in value. For example, here is the simple equation for the area of a rectangle:

$$\text{Area} = L \times W$$

If the area is 12 and the length (L) is 6, the equation would read as follows:

$$12 = 6 \times W$$

In this case, it is easy enough to see that the width (W) is equal to 2 ($12 = 6 \times 2$). What is the width if the length is equal to 3?

$$12 = 3 \times W$$

In this case, the width is equal to 4 ($12 = 3 \times 4$). Therefore, in these two equations ($12 = 6 \times W$ and $12 = 3 \times W$), W must be considered a variable because it may change, depending on the value of L.

2.1.4 Constants

A constant is an element of an equation that does not change in value. For example, consider the following equation:

$$2 + 5 = 7$$

In this equation, 2, 5, and 7 are constants. The number 2 will always be 2, 5 will always be 5, and 7 will always be 7, no matter what the equation. Constants also refer to accepted values that represent one element of an equation and do not change from situation to situation. One of the most common constants is pi (π). It has an approximate value of 3.14 and represents the ratio of the circumference to the diameter in a circle. It is used in most calculations involving circles.

2.1.5 Coefficients

A coefficient is a multiplier. Consider the following equation:

$$\text{Area} = \text{L} \times \text{W}$$

In this equation, L is the coefficient of W. It can also be written as LW, without the multiplication sign. No multiplication symbol is required when the intended relationship between symbols and letters is clear. For example:

- 2L means two times L (2 is the coefficient of L)
- IR means I times R (I is the coefficient of R)

2.1.6 Powers and Roots

Many mathematical formulas used in electronics and construction require that the power or root of a number be found. A power (or exponent) is a number written above and to the right of another number, which is called the base. For example, the expression y^x means to take the value of y and multiply it by itself x number of times. The xth root of a number y is another number that when multiplied by itself x times returns a value of y. Expressed mathematically:

$$(x\sqrt{y})^x = y$$

Squares and Square Roots – The need to find squares and square roots is common in HVAC mathematics. A square is the product of a number or quantity multiplied by itself. For example, the square of 6 means 6 × 6. To denote a number as squared, simply place the exponent 2 above and to the right of the base number. For example:

$$6^2 = 6 \times 6 = 36$$

The square root of a number is the divisor which, when multiplied by itself (squared), gives the number as a product. Extracting the square root refers to a process of finding the equal factors which, when multiplied together, return the original number. The process is identified by the radical symbol [√]. This symbol is a shorthand way of stating that the equal factors of the number under the radical sign are to be determined. Finding the square roots is necessary in many calculations, including calculating loads and determining the airflow in a duct.

For example, $\sqrt{16}$ is read as the square root of 16. The number consists of the two equal factors 4 and 4. Thus, when 4 is raised to the second power or squared, it is equal to 16. Squaring a number simply means multiplying the number by itself.

The number 16 is a perfect square. Numbers that are perfect squares have whole numbers as the square roots. For example, the square roots of perfect squares 4, 25, 36, 121, and 324 are the whole numbers 2, 5, 6, 11, and 18, respectively.

Squares and square roots can be calculated by hand, but the process is very time consuming and subject to error. Most people find squares and square roots of numbers using a calculator. To find the square of a number, the calculator's square key $[x^2]$ is used. When pressed, it takes the number shown in the display and multiplies it by itself. For example, to square the number 4.235, enter 4.235, press the $[x^2]$ key, and then read 17.935225 on the display.

Similarly, to find the square root of a number, the calculator's square root key $[\sqrt{}]$ or $[\sqrt{x}]$ is used. When pressed, it calculates the square root of the number shown in the display. For example, to find the square root of the number 17.935225, enter 17.935225, press the $[\sqrt{}]$ or $[\sqrt{x}]$ key, then read 4.235 on the display. Note that on some calculators, the $[\sqrt{}]$ or $[\sqrt{x}]$ key must be pressed before entering the number.

Other Powers and Roots – It is sometimes necessary to find powers and roots other than squares and square roots. This can easily be done with

Powers and Roots

Powers and roots can be easily calculated using a scientific calculator, such as the one shown here.

03102-13_SA09.EPS

a calculator. The powers key [y^x] raises the displayed value to the xth power. The order of entry must be y, [y^x], then x. For example, to find the power 2.86^3, enter 2.86, press the [y^x] key, enter 3, press the [=] key, then read 23.393656 on the display.

The root key [$x\sqrt{y}$] is used to find the xth root of the displayed value y. The order of entry is y [INV] [$x\sqrt{y}$]. The [INV] or [2nd F] key, when pressed before any other key that has a dual function, causes the second function of the key to be operated. For example, to find the cube root of 1,500 or ($\sqrt[3]{1,500}$), enter 1,500, press the [INV] [$x\sqrt{y}$] keys, enter 3, press the [=] key, then read 11.44714243 on the display. Note that on some calculators, the [$x\sqrt{y}$] key must be pressed before entering the number.

The answer can easily be checked by using the [y^x] function to raise the answer on the display to a power of 3. For example, $11.44714243^3 = 1,500$.

2.2.0 Sequence of Operations

Complicated equations must be solved by performing the indicated operations in a prescribed sequence. This sequence is: multiply, divide, add, and subtract (MDAS). For example, the following equation can result in a number of answers if the MDAS sequence is not followed:

$$3 + 3 \times 2 - 6 \div 3 = x$$

To come up with the correct result, this equation must be solved in the following order:

Step 1 Multiply: $3 + 3 \times 2 - 6 \div 3$

Step 2 Divide: $3 + 6 - 6 \div 3$

Step 3 Add: $3 + 6 - 2$

Step 4 Subtract: $9 - 2$

Result: 7 = x

2.3.0 Solving Algebraic Equations

Some equations may include several variables. Solving these equations means simplifying them as much as possible and, if necessary, separating the desired variable so that it is on one side by itself, with everything else on the other side. Problems such as these are known as algebraic expressions. When an algebraic expression appears in an equation, the MDAS sequence also applies. For example:

$$P = R - [5(3A + 4B) + 40L]$$

The parentheses represent multiplication, so they are worked on first. When working with multiple sets of parentheses or brackets, always begin by eliminating the innermost symbols first, then working your way to the outermost symbols. Thus, in the above equation, the expressions within parentheses (3A + 4B) with the coefficient of 5 are multiplied first, giving:

$$P = R - [15A + 20B + 40L]$$

The brackets also represent multiplication, so they are worked on next. The minus sign is the same as a coefficient of –1, so each term within the brackets is multiplied by –1, giving:

$$P = R - 15A - 20B - 40L$$

At this point, the equation has been simplified as much as possible, but what happens when the same equation is applied to a real-life situation? Suppose you have just installed ductwork in five identical apartments in a complex, and you want to determine the profit on the job. If this equation were written out in longhand, it would look like this:

Profit (P) is equal to the payment received (R) minus five apartments times three pieces of one type of ductwork in each apartment at a certain cost per piece (A) plus four pieces of a second type of ductwork in each apartment at a certain cost per piece (B) plus forty hours of labor times an hourly rate (L).

It makes a lot more sense to simply write it in algebraic form:

$$P = R - [5(3A + 4B) + 40L]$$

or

$$P = R - 15A - 20B - 40L$$

Now, plug in numbers for the known values. Say that R = $1,500, A = $10, B = $15, and L = $15. This results in:

$$P = 1,500 - (15 \times 10) - (20 \times 15) - (40 \times 15)$$

Multiplying yields:

$$P = 1,500 - 150 - 300 - 600$$

Result:

$$P = \$450$$

2.3.1 Rules of Algebra

There are a few simple rules that, once memorized, help simplify and solve almost any equation you encounter as an HVACR technician.

Rule 1:

If the same value is added to or subtracted from both sides of an equation, the resulting equation is valid. For example, consider the following equation:

$$5 = 5$$

If 3 is added to each side of the equation, the resulting equation remains valid (both sides are still equal to each other).

$$5 + 3 = 5 + 3$$
$$8 = 8$$

In the same way, if the same number is subtracted from both sides of an equation, the resulting equation is valid. For example, consider the following equation:

$$5 = 5$$

If 4 is subtracted from both sides of the equation, the resulting equation remains valid.

$$5 - 4 = 5 - 4$$
$$1 = 1$$

Moving variables from one side of an equation to another is done in the same way as when moving constants. Recall that when an equation is solved for one particular variable, it means that the variable should be on one side of the equation by itself. For example, consider the following pressure equation used frequently in the trade when working with system pressures in air conditioning and steam boiler systems:

Absolute pressure = gauge pressure + 14.7

To solve this equation for gauge pressure, 14.7 must be moved to the other side of the equation with absolute pressure. To do so, subtract 14.7 from both sides of the equation:

Absolute pressure – 14.7
= gauge pressure + 14.7 – 14.7

It should be clear that the + 14.7 and – 14.7 on the right cancel each other, leaving:

Absolute pressure – 14.7 = gauge pressure

or

Gauge pressure = absolute pressure – 14.7

The equation has been solved for gauge pressure. To take this new equation and solve it for absolute pressure again, simply add 14.7 to each side:

Gauge pressure + 14.7
= absolute pressure – 14.7 + 14.7

Again, the + 14.7 and – 14.7 on the right cancel each other out, leaving:

Gauge pressure + 14.7 = absolute pressure

or

Absolute pressure = gauge pressure + 14.7

Rule 2:

If both sides of an equation are multiplied or divided by the same value, the resulting equation is valid. This example uses Ohm's law, which is used by HVACR technician to calculate voltage, current, and resistance values for electrical circuits. This equation is as follows:

$$E = IR$$

Where:

$$E = \text{voltage}$$
$$I = \text{current}$$
$$R = \text{resistance}$$

If you know the voltage (E) and the current (I), but need to find the resistance (R), how do you rearrange the equation? To solve this equation for R, I must be moved to the other side of the equation with E. To do so, divide both sides by I:

$$E \div I = IR \div I$$

The two on the right cancel each other out, leaving:

$$E \div I = R$$

or

$$R = E \div I$$

The equation has been solved for resistance. To take this new equation and solve it for E again, each side is multiplied by I:

$$I \times R = E \div I \times I$$

The two on the right cancel each other out, leaving:

$$IR = E$$

or

$$E = IR$$

Rule 3:

Like terms may be added and subtracted in a manner similar to constant numbers. For example, given the equation:

$$2A + 3A = 15$$

The like terms (2A and 3A) may be added directly:

$$5A = 15$$

Dividing both sides of the equation by 5, leaves:

$$5A \div 5 = 15 \div 5$$

$$A = 3$$

These rules may be used repeatedly until an equation is in the desired form. For example, take a look at the following pressure equation:

$$p = hd \div 144$$

Where:

$$p = pressure$$

$$h = height$$

$$d = density$$

Solve the equation for density (d).

Step 1 Remove the fraction by multiplying both sides by 144.

$$P \times 144 = hd \div 144 \times 144$$

The two 144s on the right side cancel each other, leaving:

$$144P = hd$$

Step 2 Divide both sides by h.

$$144P \div h = d \times h \div h$$

or

$$d = 144P \div h$$

The equation has been solved for density (d).

Fan Airflow Versus Speed

The performance of all fans and blowers is governed by three rules called the Fan rules. One of these rules states that the amount of air delivered by a fan in cubic feet per minute (cfm) varies directly with the speed of the fan as measured in revolutions per minute (rpm). This is expressed mathematically as:

$$\text{New cfm} = \frac{\text{new rpm} \times \text{existing cfm}}{\text{existing rpm}}$$

Given the equation above for calculating a new cfm, how would you solve the equation to determine the new rpm, assuming you know the new cfm, the existing cfm, and the existing rpm?

Additional Resources

Mathematics for Carpentry and the Construction Trades, Alfred P. Webster, Upper Saddle River NJ: Prentice Hall.

Math for the Technician, Leo A. Meyer, Haywood CA: Lama Books.

2.0.0 Section Review

1. In an equation, a value that is subject to change is a(n) _____.
   a. coefficient
   b. variable
   c. exponent
   d. constant

2. When solving an equation, the first step is to _____.
   a. multiply
   b. divide
   c. add
   d. subtract

3. Given the equation E = IR, rewrite it to solve for I.
   a. I = E × R
   b. I = E − R
   c. I = E ÷ R
   d. I = E + R

3.0.0 WORKING WITH GEOMETRIC FIGURES

Objectives

Identify and describe geometric figures.
  a. Describe the characteristics of a circle.
  b. Identify and describe types of angles.
  c. Identify and describe types of polygons.
  d. Calculate various values associated with triangles.

Trade Terms

Polygon: A shape formed when three or more straight lines are joined in a regular pattern.

Geometry is the study of various figures. It consists of two main fields: plane geometry and solid geometry. Plane geometry is the study of two-dimensional figures such as squares, rectangles, triangles, circles, and polygons. Solid geometry is the study of figures that occupy space, such as cubes, spheres, and other three-dimensional objects. The focus of this section is on the elements of plane geometry.

Geometric objects such as polygons and triangles are formed by lines. A line that forms a right angle (90 degrees) with one or more lines is said to be perpendicular to those lines (*Figure 15*). The distance from a point to a line is the measure of the perpendicular line drawn from that point to the line. Two or more straight lines that are the same distance apart at all points are said to be parallel. Parallel lines do not intersect.

3.1.0 Characteristics of a Circle

As shown in *Figure 16*, a circle is a curved line that connects with itself and has these other properties:

- All points on a circle are the same distance (equidistant) from the point at the center.
- The distance from the center to any point on the curved line, called the radius (r), is always the same.
- The shortest distance from any point on the curve through the center to a point directly opposite is called the diameter (d). The diameter is equal to twice the radius (d = 2r).

- The distance around the outside of the circle is called the circumference. It can be determined by using the equation: circumference = πd, where π is a constant equal to approximately 3.14 and d is the diameter (2r).
- A circle is divided into 360 parts with each part called a degree; therefore, one degree = $\frac{1}{360}$ of a circle. The degree is also the unit of measurement commonly used to measure the size of an angle.

3.2.0 Angles

Two straight lines meeting at a point known as the vertex form an angle (*Figure 17*). The two lines are the sides, or rays, of the angle. The angle is the amount of opening that exists between the rays

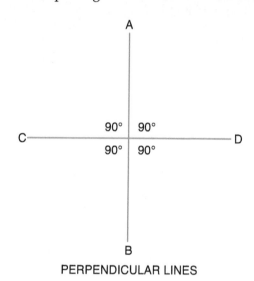

PERPENDICULAR LINES

EQUAL DISTANCE AT ALL PERPENDICULARS

PARALLEL LINES

03102-13_F15.EPS

Figure 15 Perpendicular and parallel lines.

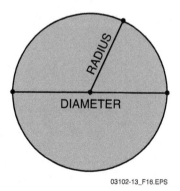

03102-13_F16.EPS

Figure 16 A circle.

and is measured in degrees. There are two ways commonly used to identify angles. One is to assign a letter to the angle, such as angle D shown in *Figure 17*. This is written ∠D. The other way is to name the two end points of the rays and put the vertex letter between them (e.g., ∠ABC). When the angle measure is shown in degrees, it should be written inside the angle, if possible. If the angle is too small to show the measurement, it may be placed outside of the angle, with an arrow to the inside.

The common types of angles are shown in *Figures 18* and *19*.

- *Right angle* – A right angle has rays that are perpendicular to each other. The measure of this angle is always 90 degrees.
- *Straight angle* – A straight angle does not look like an angle at all. The rays of a straight angle lie in a straight line, and the angle measures 180 degrees.
- *Acute angle* – An acute anble is less than 90 degrees.
- *Obtuse angle* – An obtuse angle is greater than 90 degrees, but less than 180 degrees.
- *Adjacent angles* – When three or more rays meet at the same vertex, the angles formed are said to be adjacent (next to) one another. In *Figure 19*,

the angles ∠ABC and ∠CBD are adjacent angles. The ray BC is said to be common to both angles.
- *Complementary angles* – Two adjacent angles that have a combined total measure of 90 degrees. In *Figure 19*, ∠DEF is complementary to ∠FEG.
- *Supplementary angles* – Two adjacent angles that have a combined total measure of 180 degrees. In *Figure 19*, ∠HIJ is supplementary to ∠JIK.

3.3.0 Polygons

A polygon is formed when three or more straight lines are joined in a regular pattern. Some of the most familiar polygons are shown in *Figure 20*. As shown, they have common names that generally refer to their number of sides. When all sides of a polygon have equal length and all internal angles are equal, it is called a regular polygon.

Each of the boundary lines forming the polygon is called a side of the polygon. The point at which any two sides of a polygon meet is called a vertex of the polygon. The perimeter of any polygon is equal to the sum of the lengths of each of the sides. The sum of the interior angles of any

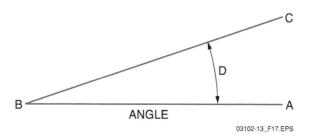

Figure 17 An angle.

Geometry

In the HVACR field, geometry has its greatest applications in the field of sheet metal layout. It is used to determine angles for transition fittings, radius elbows, and many other sheet metal layout challenges. Geometry also comes into play in determining the offset angles for piping that must be directed around obstacles.

Inside and Outside Pipe Diameters

The different sizes of pipe and tubing used in HVACR applications are expressed in terms of their inside diameter (ID) or their outside diameter (OD). The inside diameter is the distance between the inner walls of a pipe. Typically, the ID is the standard measure for the copper tubing used in heating and plumbing applications. The outside diameter is the distance between the outer walls of a pipe. The OD is typically the standard measure for the copper tubing used in air conditioning and refrigeration (ACR) applications. For example, a roll of soft copper tubing like the one shown here is commonly available in sizes from ⅛" OD to ⅞" OD.

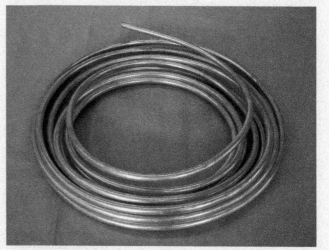

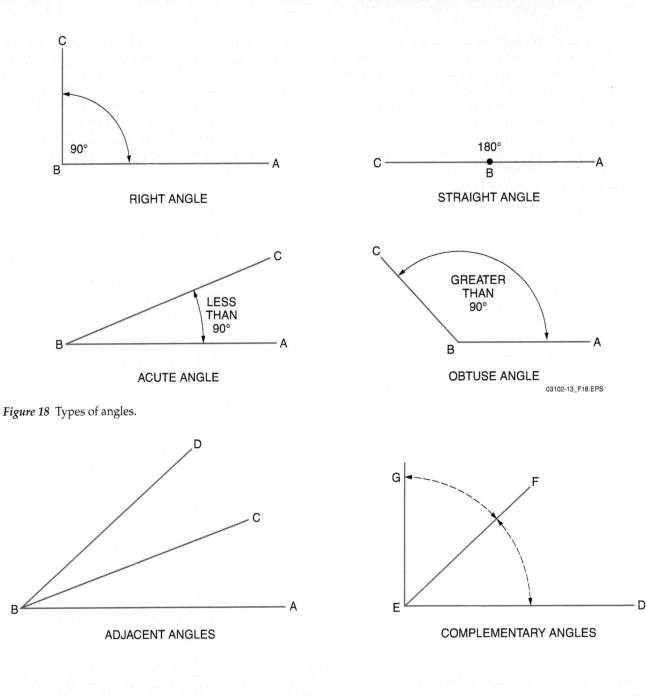

Figure 18 Types of angles.

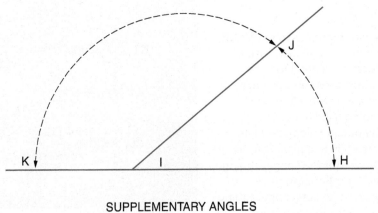

Figure 19 Adjacent, complementary, and supplementary angles.

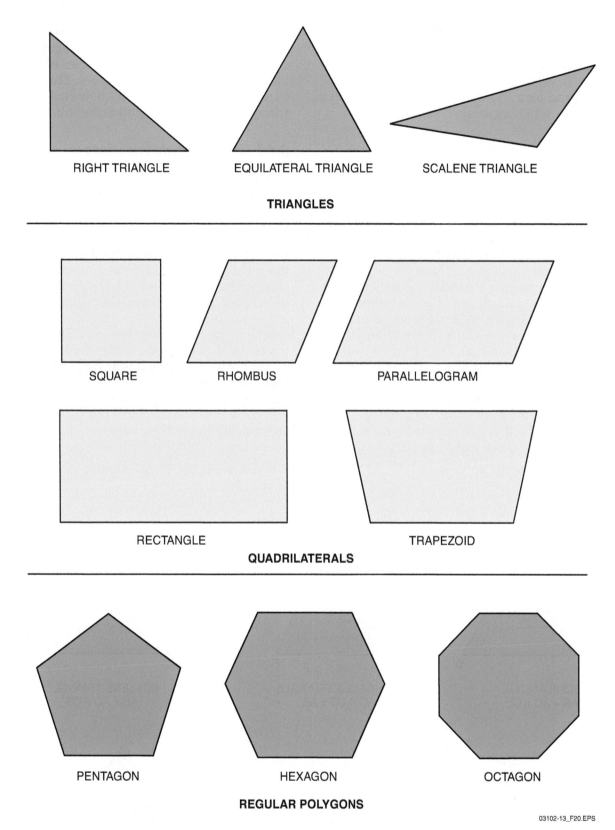

RIGHT TRIANGLE   EQUILATERAL TRIANGLE   SCALENE TRIANGLE

TRIANGLES

SQUARE   RHOMBUS   PARALLELOGRAM

RECTANGLE   TRAPEZOID

QUADRILATERALS

PENTAGON   HEXAGON   OCTAGON

REGULAR POLYGONS

03102-13_F20.EPS

Figure 20 Common polygons.

polygon is equal to (n – 2) × 180 degrees, where n is the number of sides.

For example, the sum of the interior angles for a square is 360 degrees [(4 – 2) × 180 degrees = 360 degrees] and for a triangle is 180 degrees [(3 – 2) × 180 degrees = 180 degrees].

3.4.0 Working With Triangles

As mentioned previously, triangles are three-sided polygons. *Figure 21* shows three different types of triangles. A regular polygon with three equal sides is called an equilateral triangle. Two types of irregular triangles are the isosceles (having two sides of equal length) and the scalene (having all sides of unequal length). An important fact to remember about triangles is that the sum of the three angles of any triangle equals 180 degrees. As shown, all three sides of an equilateral triangle are equal. In such a triangle, the three angles are also equal. The isosceles triangle has two equal sides with the angles opposite the equal sides also being equal.

Triangles are also classified according to their interior angles. If one of the three interior angles is 90 degrees, the triangle is called a right triangle. If one of the three interior angles is greater than 90 degrees, the triangle is called an obtuse triangle. If each of the interior angles is less than 90 degrees, the triangle is called an acute triangle. The sum of the three interior angles of any triangle is always equal to 180 degrees. This is helpful to remember whenever you know two angles of a triangle and need to calculate the third.

3.4.1 Right Triangle Functions

One of the most common triangles is the right triangle. Since it has one 90-degree angle (right angle), the other two angles are acute angles. They are also complementary angles because their sum equals 90 degrees. The right triangle has two sides perpendicular to each other, thus forming the right angle. To aid in writing equations, the sides and angles of a right triangle are labeled as shown in *Figure 22*. Normally, capital (uppercase) letters are used to label the angles, and lowercase letters are used to label the sides. The third side, which is always opposite the right angle (C), is called the hypotenuse. It is always longer than either of the other two sides. The other sides can be remembered as a for altitude and b for base. Note that the letters used to label the sides and angles are opposite each other. For example, side a is opposite angle A, and so forth.

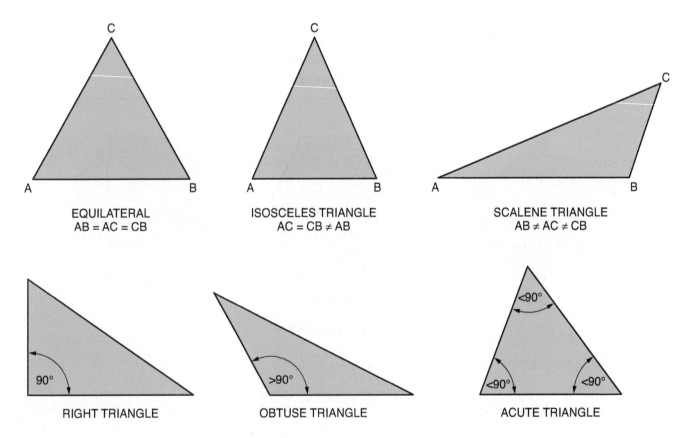

Figure 21 Triangles.

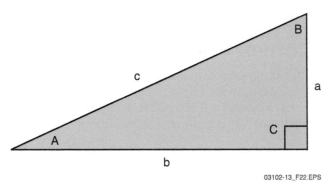

Figure 22 Common labeling of angles and sides in a right triangle.

3.4.2 Right Triangle Calculations Using the Pythagorean Theorem

If the lengths of any two sides of a right triangle are known, the length of the third side can be determined using a rule called the Pythagorean theorem. It states that the square of the hypotenuse (c) is equal to the sum of the squares of the remaining two sides (a and b). Expressed mathematically:

$$c^2 = a^2 + b^2$$

The formula may be rearranged to solve for the unknown side as follows:

$$a = \sqrt{c^2 - b^2}$$
$$b = \sqrt{c^2 - a^2}$$
$$c = \sqrt{a^2 + b^2}$$

For example, a right triangle has an altitude (side a) equal to 8' and a base (side b) equal to 12'. To find the length of the hypotenuse (side c), proceed as follows:

$$c = \sqrt{a^2 + b^2}$$
$$c = \sqrt{8^2 + 12^2}$$
$$c = \sqrt{64 + 144}$$
$$c = \sqrt{208}$$
$$c = 14.422$$

To determine the actual length of the hypotenuse using the formula above, it is necessary to calculate the square root of the sum of the sides squared. Fortunately, this is easy to do using a scientific calculator. On many calculators, you simply key in the number and press the square root [√] key. On some calculators, the square root does not have a separate key. Instead, the square root function is the inverse of the [x^2] key, so you have to press [INV] or [2nd F], depending on the calculator, followed by [x^2], to obtain the square root.

3.4.3 Converting Decimal Feet to Feet and Inches and Vice Versa

When using trigonometric functions to calculate numerical values for lengths, distances, and angles, the answers are normally expressed as a decimal. Construction drawings and specifications often express such measurements in feet and inches. For this reason, it is often necessary to convert between these two measurement systems. Conversion tables are available in many reference books that can be used for this purpose. However, in case conversion tables are not readily available, you should become familiar with the methods for making such conversions mathematically.

To convert values given in decimal feet into equivalent feet and inches, use the following procedure. For this example, 45.3646' will be converted to feet and inches.

Step 1 Subtract 45' from 45.3646' = 0.3646'.

Step 2 Convert 0.3646' to inches by multiplying 0.3646' by 12 = 4.3752".

Step 3 Subtract 4" from 4.3752" = 0.3752".

Step 4 Convert 0.3752" into eighths of an inch by multiplying 0.3752" by 8 = 3.0016 eighths or, when rounded off, ⅜. Therefore, 45.3646' = 45'-4⅜".

To convert values given in feet and inches (and inch-fractions) into equivalent decimal feet val-

Trigonometry

Fifty years before Pythagoras, Thales (a Greek mathematician) figured out how to measure the height of the pyramids using a technique that later became known as trigonometry. Trigonometry recognizes that there is a relationship between the size of an angle and the lengths of the sides in a right triangle.

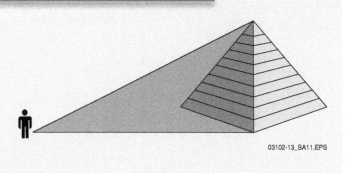

ues, use the following procedure. For example, convert 45'-4⅜" to decimal feet.

Step 1 Convert the inch-fraction to a decimal. This is done by dividing the numerator of the fraction (top number) by the denominator of the fraction (bottom number). For example, ⅜" = 0.375.

Step 2 Add the 0.375 to 4" to obtain 4.375".

Step 3 Divide 4.375" by 12 to obtain 0.3646'.

Step 4 Add 0.3646' to 45' to obtain 45.3646'. Therefore, 45'-4⅜" = 45.3646'.

Converting Dimensions

When using a calculator to determine the length of the hypotenuse for a right angle, the answer shown on the calculator is 6.354'. What is the length of the hypotenuse when expressed in feet and inches?

3-4-5 Rule

The 3-4-5 rule is based on the Pythagorean theorem and has been used in building construction for centuries. It is a simple method for laying out or checking 90-degree angles (right angles) and requires only the use of a tape measure. The numbers 3-4-5 represent dimensions in feet that describe the sides of a right triangle. Right triangles that are multiples of the 3-4-5 triangle are commonly used (e.g., 9-12-15, 12-16-20, 15-20-25, etc.). The specific multiple used is determined by the relative distances involved in the job being laid out or checked.

An example of the 3-4-5 rule using the multiples of 15-20-25 is shown here. In order to square or check a corner as shown in the example, first measure and mark 15' down the line in one direction, then measure and mark 20' down the line in the other direction. The distance measured between the 15' and 20' points must be exactly 25' to ensure that the angle is a perfect right angle.

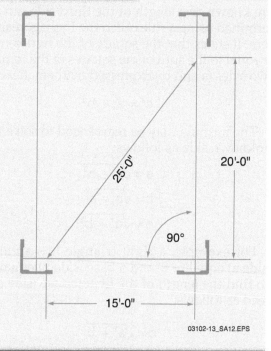

03102-13_SA12.EPS

Additional Resources

Mathematics for Carpentry and the Construction Trades. Alfred P. Webster. Upper Saddle River NJ: Prentice Hall.

Math for the Technician. Leo A. Meyer. Haywood CA: Lama Books.

3.0.0 Section Review

1. The formula for calculating the circumference of a circle is _____.

   a. πr^2
   b. 2π
   c. πd
   d. $I \times R$

2. An angle that is greater 90 degrees but less than 180 degrees is a(n) _____.

   a. right angle
   b. acute angle
   c. supplementary angle
   d. obtuse angle

3. Which of the following is *not* considered a polygon?

   a. triangle
   b. circle
   c. rectangle
   d. trapezoid

4. If a right triangle has a hypotenuse of 15" and a height of 12", what is the length of the base?

   a. 9 inches
   b. 19.2 inches
   c. 81 inches
   d. 9 feet

SUMMARY

This module built on the knowledge gained in NCCER Module 00102, *Introduction to Construction Math.* It covered mathematics important to the HVACR technician, with an emphasis on the metric system. It also introduced basic concepts concerning powers and roots, algebra, geometry, and trigonometry. HVACR technicians use math functions in their day-to-day work activities. For example, it may be necessary to perform any one of the following calculations during an installation or service call:

- Calculate the area of rectangular or round ductwork to determine airflow volume.
- Calculate the current load that an electrical device places on an electrical service.
- Determine the required offset angle for piping or ductwork.
- Convert temperature readings from Fahrenheit to Celsius.

Increasing emphasis is being placed on the use of the metric system as world markets become more integrated. The vast majority of the world's countries use the metric system. HVACR products manufactured in those countries are increasingly finding their way into US markets, so it is important for HVACR technicians to be able to work with metric system values.

1. One meter is equal to _____.
   a. 0.3937 feet
   b. 39.37 feet
   c. 3.281 feet
   d. 3,281 feet

2. A room has a length of 3.568 meters and a width of 2.438 meters. The area of this room in square feet is _____.
   a. 20 ft^2
   b. 71.36 ft^2
   c. 93.64 ft^2
   d. 384 ft^2

3. A duct has inside dimensions of 16" × 12". The area, in ft^2, is equal to _____.
   a. 1.33 ft^2
   b. 1.92 ft^2
   c. 13.3 ft^2
   d. 28 ft^2

4. The value of π, to 2 decimal places, is _____.
   a. 1.00
   b. 3.14
   c. 3.33
   d. 3.45

5. How many liters of liquid can a 55-gallon drum hold?
   a. 55
   b. 110
   c. 175
   d. 208

6. A crate containing a boiler is marked as weighing 175 pounds. What is its weight in kilograms?
   a. 68.5 kilograms
   b. 79.38 kilograms
   c. 96.33 kilograms
   d. 125 kilograms

7. Find the absolute pressure if the gauge pressure is 164 psig.
   a. 14.7 psia
   b. 149.3 psia
   c. 164 psia
   d. 178.7 psia

8. A temperature of 0°C is equal to _____.
   a. −17.8°F
   b. 32°F
   c. 100°F
   d. 144°F

9. A temperature of 68°F is equal to _____.
   a. 20°C
   b. 36°C
   c. 41°C
   d. 100°C

10. The square root of 12,500, which is expressed mathematically as $\sqrt{12,500}$, is equal to _____.
    a. 1.11803
    b. 11.1803
    c. 111.803
    d. 1118.03

11. 6^2 is the same as _____.
    a. 4
    b. 8
    c. 12
    d. 36

12. Given the equation $P = E^2/R$, solve for R.
    a. $R = E^2/P$
    b. $R = E^2 \times P$
    c. $R = P/E^2$
    d. $R = \sqrt{PE}$

13. The distance from the center of a circle to any point on the curved line is called the _____.
    a. circumference
    b. radius
    c. diameter
    d. cord

14. A triangle in which all sides are of unequal length is called a(n) _____.

    a. equilateral triangle
    b. isosceles triangle
    c. right triangle
    d. scalene triangle

15. Which of the following is *not* a form of the Pythagorean theorem?

    a. $c^2 = a^2 + b^2$
    b. $a = \sqrt{c^2 - b^2}$
    c. $b = \sqrt{c^2 - a^2}$
    d. $c = \sqrt{a^2 - b^2}$

Trade Terms Quiz

Fill in the blank with the correct trade term that you learned from your study of this module.

1. A(n)_____ is a definite standard of measure or quantity.

2. A(n) _____ is a small figure or symbol placed above and to the right of another figure or symbol to show how many times the latter is to be multiplied by itself.

3. The actual atmospheric pressure at a given place and time is called the _____.

4. _____ is the amount of space contained in a given three-dimensional shape.

5. A push or pull on a surface is called _____.

6. A(n) _____ is an element in an equation with a fixed value.

7. The standard pressure exerted on Earth's surface is called _____.

8. _____ is the total pressure that exists in a system.

9. An element of an equation that may change in value is called a(n) _____.

10. is the quantity of matter present.

11. A multiplier is also called a(n) _____.

12. Any pressure that is less than the prevailing atmospheric pressure is referred to as a(n) _____.

13. The unit of force required to accelerate one kilogram at a rate of one meter per second is a(n) _____.

14. The amount of surface contained in a given size plane or two-dimensional shape is referred to as .

15. A shape formed when three or more straight lines are joined in a regular pattern is called a(n) _____.

Trade Terms

Absolute pressure	Constant	Polygon
Area	Exponent	Unit
Atmospheric pressure	Force	Vacuum
Barometric pressure	Mass	Variable
Coefficient	Newton (N)	Volume

Corky Dotson
Project Manager
Lee Company

How did you get started in the construction industry (i.e., took classes in school, a summer job, etc.)?

When I was in high school, they offered a new General Metals class. Given the description of the class, I thought it would be interesting so I signed up for it. It was mostly about sheet metal layout, welding, and constructing various objects from metal. After attending the class for a while, I found it fascinating to be able to take a drawing and create a functional object. I loved the way you could take math, simple geometry, and radial line development to bring a flat drawing to life.

Who or what inspired you to enter the industry (i.e., a family member, school counselor, etc.)? Why?

After taking the General Metals class for two years, I thought the HVAC trade would be an interesting occupation, and I could get paid to do something I liked. The timing just happened to be right. I graduated high school in the summer of 1978 and a friend told me about an open helper position at a local sheet metal shop. The company was Lee Company and I went to apply for the job. I started at the very bottom but learned as I went. I also went back for additional training to become a sheet metal layout man for the company. I am still with Lee Company today.

What do you enjoy most about your career?

One thing that I find extremely satisfying came to my attention in the very beginning. I truly enjoy the feeling of satisfaction I get at the end of the day when I see what has been produced. I can step back and admire the quality of the work and know that I overcame any obstacles that the task presented. No matter what position I have been in to this point in my career, I could always step back and see what has been accomplished and take pride in the part I played in the finished product.

Why do you think training and education are important in construction?

Education is valuable not only to make you better at a chosen profession, but it can also lead you down a path toward an occupational choice. Ongoing training keeps you abreast of the latest techniques in your field. The educational process never really ends. Whether you are learning in the classroom or learning from the execution of your trade, new knowledge builds a foundation for the next thing in line. You can never learn it all.

Why do you think credentials are important in construction?

Credentials are important not only to track your own accomplishments, but also for any potential employer to be able to gauge your knowledge of the trade. Most employers perceive a trade worker with credentials as someone striving to master his or her craft. When you take your own credentials seriously, you will be taken seriously as a professional.

How has training/construction impacted your life and career (i.e., advancement opportunities, better wages, etc.)?

Construction has not impacted my life. It is my life. I have been in this trade since I graduated high school, and I am still doing it with the same company today. As far as advancement opportunities are concerned, it has been very good to me. As I have advanced in skill, the company trusted me enough to offer me more responsibilities and higher wages. You have to pay your dues in any industry or career field. It shouldn't be about chasing a dollar. It is about establishing your skills and demonstrating loyalty to your company. At that point, the money will come to you. So many people in the trades tend to jump from job to job and never reap the sweetest fruits of their effort.

Would you recommend construction as a career to others? Why?
I would most definitely recommend construction as a trade, I find it fulfilling to be able to supply a service that brings comfort and convenience to a building for others. Although your work is rarely visible after the installation, its presence in the background affects everyday life in the workplace and home environments. It is a trade others may not appreciate, but it creates internal pride. You know you have made a difference that others may take for granted.

What does craftsmanship mean to you?
Craftsmanship has a simple definition. When the job is done, a true craftsman would be proud to sign his or her name to it.

Absolute pressure: The total pressure that exists in a system. Absolute pressure is expressed in pounds per square inch absolute (psia). Absolute pressure = gauge pressure + atmospheric pressure.

Area: The amount of surface in a given plane or two-dimensional shape.

Atmospheric pressure: The standard pressure exerted on Earth's surface. Atmospheric pressure is normally expressed as 14.7 pounds per square inch (psi) or 29.92 inches of mercury.

Barometric pressure: The actual atmospheric pressure at a given place and time.

Coefficient: A multiplier (e.g., the numeral 2 as in the expression 2b).

Constant: An element in an equation with a fixed value.

Exponent: A small figure or symbol placed above and to the right of another figure or symbol to show how many times the latter is to be multiplied by itself (e.g., $b^3 = b \times b \times b$).

Force: A push or pull on a surface. In this module, force is considered to be the weight of an object or fluid. This is a common approximation.

Mass: The quantity of matter present.

Newton (N): The amount of force required to accelerate one kilogram at a rate of one meter per second.

Polygon: A shape formed when three or more straight lines are joined in a regular pattern.

Unit: A definite standard of measure of dimension or quantity.

Vacuum: Any pressure that is less than the prevailing atmospheric pressure.

Variable: An element of an equation that may change in value.

Volume: The amount of space contained in a given three-dimensional shape.

CONVERSION FACTORS AND COMMON FORMULAS

COMMON MEASURES
WEIGHT UNITS
1 ton = 2,000 pounds
1 pound = 16 dry ounces
LENGTH UNITS
1 yard = 3 feet
1 foot = 12 inches
VOLUMES
1 cubic yard = 27 cubic feet
1 cubic foot = 1,728 cubic inches
1 gallon = 4 quarts
1 quart = 2 pints
1 pint = 2 cups
1 cup = 8 fluid ounces
AREA UNIT
1 square yard = 9 square feet
1 square foot = 144 square inches

PREFIX	SYMBOL	NUMBER	MULTIPLICATION FACTOR
giga	G	billion	$1,000,000,000 = 10^9$
mega	M	million	$1,000,000 = 10^6$
kilo	k	thousand	$1,000 = 10^3$
hecto	h	hundred	$100 = 10^2$
deka	da	ten	$10 = 10^1$
			BASE UNITS $1 = 10^0$
deci	d	tenth	$0.1 = 10^{-1}$
centi	c	hundredth	$0.01 = 10^{-2}$
milli	m	thousandth	$0.001 = 10^{-3}$
micro	μ	millionth	$0.000001 = 10^{-6}$
nano	n	billionth	$0.000000001 = 10^{-9}$

WEIGHT UNITS

1 kilogram	=	1,000 grams
1 hectogram	=	100 grams
1 dekagram	=	10 grams
1 gram	=	1 gram
1 decigram	=	0.1 gram
1 centigram	=	0.01 gram
1 milligram	=	0.001 gram

LENGTH UNITS

1 kilometer	=	1,000 meters
1 hectometer	=	100 meters
1 dekameter	=	10 meters
1 meter	=	1 meter
1 decimeter	=	0.1 meter
1 centimeter	=	0.01 meter
1 millimeter	=	0.001 meter

VOLUME UNITS

1 kiloliter	=	1,000 liters
1 hectoliter	=	100 liters
1 dekaliter	=	10 liters
1 liter	=	1 liter
1 deciliter	=	0.1 liter
1 centiliter	=	0.01 liter
1 milliliter	=	0.001 liter

US TO METRIC CONVERSIONS

WEIGHTS

1 ounce	=	28.35 grams
1 pound	=	435.6 grams or 0.4536 kilograms
1 (short) ton	=	907.2 kilograms

LENGTHS

1 inch	=	2.540 centimeters
1 foot	=	30.48 centimeters
1 yard	=	91.44 centimeters or 0.9144 meters
1 mile	=	1.609 kilometers

AREAS

1 square inch	=	6.452 square centimeters
1 square foot	=	929.0 square centimeters or 0.0929 square meters
1 square yard	=	0.8361 square meters

VOLUMES

1 cubic inch	=	16.39 cubic centimeters
1 cubic foot	=	0.02832 cubic meter
1 cubic yard	=	0.7646 cubic meter

LIQUID MEASUREMENTS

1 (fluid) ounce	=	0.095 liter or 28.35 grams
1 pint	=	473.2 cubic centimeters
1 quart	=	0.9263 liter
1 (US) gallon	=	3,785 cubic centimeters or 3.785 liters

TEMPERATURE MEASUREMENTS

To convert degrees Fahrenheit to degrees Celsius, use the following formula: $C = 5/9 \times (F - 32)$.

METRIC TO US CONVERSIONS

WEIGHTS

1 gram (G)	=	0.03527 ounces
1 kilogram (kg)	=	2.205 pounds
1 metric ton	=	2,205 pounds

LENGTHS

1 millimeter (mm)	=	0.03937 inches
1 centimeter (cm)	=	0.3937 inches
1 meter (m)	=	3.281 feet or 1.0937 yards
1 kilometer (km)	=	0.6214 miles

AREAS

1 square millimeter	=	0.00155 square inches
1 square centimeter	=	0.155 square inches
1 square meter	=	10.76 square feet or 1.196 square yards

VOLUMES

1 cubic centimeter	=	0.06102 cubic inches
1 cubic meter	=	35.31 cubic feet or 1.308 cubic yards

LIQUID MEASUREMENTS

1 cubic centimeter (cm^3)	=	0.06102 cubic inches
1 liter (1,000 cm^3)	=	1.057 quarts, 2.113 pints, or 61.02 cubic inches

TEMPERATURE MEASUREMENTS

To convert degrees Celsius to degrees Fahrenheit, use the following formula: $F = (9/5 \times C) + 32$.

PRESSURE:
Absolute pressure = gauge pressure + atmospheric pressure. The atmospheric pressure at sea level is typically accepted as 14.7 psi.

Gauge pressure to static pressure: P = hd/144, where P = the pressure in psi; h = the height of the column in feet; d = the density of the liquid in pounds per cubic foot.

CIRCLE:
Area = πr^2, where r is the radius
Circumference = πd, where d is the diameter

SQUARES/RECTANGLES:
Area = length $\times$ width
Volume = length $\times$ width $\times$ height

TEMPERATURE CONVERSION:
°C = 5/9 (°F – 32°)
°F = (9/5 $\times$ °C) + 32°

SEQUENCE OF OPERATIONS:
PEMDAS = parentheses, exponents (powers and/or roots), multiplication, division, addition, and subtraction

AIR FLOW VOLUME CHANGE:
New cfm = new rpm $\times$ existing cfm/existing rpm

TRIANGLES:
Area = (ab)/2, where a is the base length and b is the height

Pythagorean theorem for right triangles:
$c^2 = a^2 + b^2$, or $c = \sqrt{a^2 + b^2}$, where c is the hypotenuse of the triangle. The hypotenuse is the side opposite the right angle. Sides a and b are adjacent to the angle.

PYRAMID:
Volume = (Ah)/3, where A is the area of the base and h is the height

CYLINDER:
Volume = $\pi r^2 h$, where r is the radius of the base and h is the height

CONE:
Volume = $(\pi r^2 h)/3$, where r is the radius of the base and h is the height

SPHERE:
Volume = $(4\pi r^3)/3$, where r is the radius

Additional Resources

This module is presented as a thorough resource for task training. The following resource material is suggested for further study.

Mathematics for Carpentry and the Construction Trades. Alfred P. Webster. Upper Saddle River NJ: Prentice Hall.

Math for the Technician. Leo A. Meyer. Haywood CA: Lama Books.

Metric-conversions.org: Metric Conversion Charts and Calculators.

Figure Credits

Courtesy of Ritchie Engineering Company, Inc., YELLOW JACKET Products Division, Figure 12

Courtesy of SPX Service Solutions, SA03–SA06

Courtesy of Extech Instruments Corp., a FLIR Company, SA07

Answer	Section Reference	Objective Reference
Section One		
1. c	1.1.0	1a
2. d*	1.2.3; Figure 6	1b
3. d*	1.3.0	1c
4. a*	1.4.0	1d
Section Two		
1. b	2.1.3	2a
2. a	2.2.0	2b
3. c*	2.3.1	2c
Section Three		
1. c	3.1.0	3a
2. d	3.2.0	3b
3. b	3.3.0	3c
4. a*	3.4.2	3d

*Calculations for these answers are provided below.

Section Review Calculations

Section 1.0.0

2. 3 meter diameter = 1.5 meter radius

   1.5 meters × 3.281' per meter = 4.922' radius

   Area = πr^2

   Area = 3.14×4.922^2

   Area = 3.14 × 24.226

   Area = 76.1 ft^2

3. 1 kg = 2.205 lb

   240 × 2.205 = 529.2 lb

4. °F = $\frac{9}{5}$ × C + 32

   = 1.8 × 10 + 32

   = 50°F;

   1 bar = 14.5 psi

   9.88 bar = 143.3 psi

Section 2.0.0

3. Divide both sides by R, leaving

   E/R = I × R ÷ R

   The two Rs on the right of the equation cancel each other, leaving

   E/R = I

   I = E ÷ R

Section 3.0.0

4. $b = \sqrt{c^2 - a^2}$

   $b = \sqrt{225 - 144}$

   b = 9"

Fan Airflow Versus Speed

The performance of all fans and blowers is governed by three rules called the Fan rules. One of these rules states that the amount of air delivered by a fan in cubic feet per minute (cfm) varies directly with the speed of the fan as measured in revolutions per minute (rpm). This is expressed mathematically as:

$$\text{New cfm} = \frac{\text{new rpm} \times \text{existing cfm}}{\text{existing rpm}}$$

Given the equation above for calculating a new cfm, how would you solve the equation to determine the new rpm, assuming you know the new cfm, the existing cfm, and the existing rpm?

NCCER CURRICULA — USER UPDATE

NCCER makes every effort to keep its textbooks up-to-date and free of technical errors. We appreciate your help in this process. If you find an error, a typographical mistake, or an inaccuracy in NCCER's curricula, please fill out this form (or a photocopy), or complete the online form at **www.nccer.org/olf**. Be sure to include the exact module ID number, page number, a detailed description, and your recommended correction. Your input will be brought to the attention of the Authoring Team. Thank you for your assistance.

Instructors – If you have an idea for improving this textbook, or have found that additional materials were necessary to teach this module effectively, please let us know so that we may present your suggestions to the Authoring Team.

NCCER Product Development and Revision

13614 Progress Blvd., Alachua, FL 32615

Email: curriculum@nccer.org
Online: www.nccer.org/olf

❏ Trainee Guide   ❏ Lesson Plans   ❏ Exam   ❏ PowerPoints   Other _____

Craft / Level: _____   Copyright Date: _____

Module ID Number / Title: _____

Section Number(s): _____

Description: _____

_____

_____

_____

_____

Recommended Correction: _____

_____

_____

_____

_____

Your Name: _____

Address: _____

_____

Email: _____   Phone: _____

Basic Electricity

OVERVIEW

Like most other appliances, an HVACR system needs electricity in order to operate. Most of the problems an HVACR technician encounters when faced with a service call involve the electrical system, so an understanding of electrical theory, components, and circuits is essential. Technicians must have a firm foundation to get started, allowing them to understand the circuits in complex systems and continue to grow throughout their careers.

Module 03106

Trainees with successful module completions may be eligible for credentialing through the NCCER Registry. To learn more, go to **www.nccer.org** or contact us at 1.888.622.3720. Our website has information on the latest product releases and training, as well as online versions of our *Cornerstone* magazine and Pearson's product catalog.

Your feedback is welcome. You may email your comments to **curriculum@nccer.org**, send general comments and inquiries to **info@nccer.org**, or fill in the User Update form at the back of this module.

This information is general in nature and intended for training purposes only. Actual performance of activities described in this manual requires compliance with all applicable operating, service, maintenance, and safety procedures under the direction of qualified personnel. References in this manual to patented or proprietary devices do not constitute a recommendation of their use.

03106 V5

Objectives

When you have completed this module, you will be able to do the following:

1. Describe the fundamentals of electricity.
   a. State how electrical power is created and distributed.
   b. Describe the difference between alternating current and direct current.
   c. Identify general electrical safety practices.
   d. Describe the OSHA requirements and procedures related to electrical lockout/tagout.
2. Explain basic electrical theory.
   a. Define voltage, current, resistance, and power and describe how they are related.
   b. Use Ohm's law to calculate the current, voltage, and resistance in a circuit.
   c. Use the power formula to calculate how much power is consumed by a circuit.
   d. Describe the differences between series and parallel circuits and calculate circuit loads for each type.
3. Identify the electrical measuring instruments used in HVACR work and describe their uses.
   a. Describe how voltage is measured.
   b. Describe how current is measured.
   c. Describe how resistance is measured.
4. Identify electrical components used in HVACR systems and describe their functions.
   a. Identify and describe various load devices and explain how they are represented on circuit diagrams.
   b. Identify and describe various control devices and explain how they are represented on circuit diagrams.
   c. Identify and describe the types of electrical diagrams used in HVACR work.

Performance Tasks

Under the supervision of the instructor, you should be able to do the following:

1. Use the proper instrument to measure voltage in an energized circuit.
2. Use the proper instrument to measure current in an energized circuit.
3. Use the proper instrument to measure resistance.
4. Use a multimeter to check circuit continuity.
5. Assemble and test low- and high-voltage series and parallel circuits using a transformer and selected control and load devices.

Trade Terms

Alternating current (AC)	Digital meter	Loads	Time-delay fuses
Ammeter	Direct current (DC)	Motor starters	Solenoid coil
Ampere (A)	Electromagnet	Multimeter	Thermistor
Analog meter	Ground fault	Ohms (Ω)	Transformer
Arc	Inductive	Pilot duty	Volts (V)
Clamp-on ammeter	In-line ammeter	Power	Voltage
Conductor	Inrush current	Rectifier	Watts (W)
Contactors	Insulator	Relay	
Continuity	Ladder diagram	Resistance	
Current	Line duty	Short circuit	

Industry Recognized Credentials

If you are training through an NCCER-accredited sponsor, you may be eligible for credentials from NCCER's Registry. The ID number for this module is 03106. Note that this module may have been used in other NCCER curricula and may apply to other level completions. Contact NCCER's Registry at 888.622.3720 or go to **www.nccer.org** for more information.

Contents

1.0.0 Fundamentals of Electricity 1
  1.1.0 Power Generation and Distribution 1
  1.2.0 AC and DC Voltage 3
  1.3.0 Electrical Safety 7
    1.3.1 General Safety Practices 8
  1.4.0 OSHA Lockout/Tagout Rule 9
    1.4.1 Lockout/Tagout Procedure 9
    1.4.2 Restoring Power 10
2.0.0 Basic Electrical Theory 11
  2.1.0 Current, Voltage, Resistance, and Power 11
  2.2.0 Ohm's Law 12
  2.3.0 Electrical Power 13
  2.4.0 Electrical Circuits 14
    2.4.1 Series Circuits 15
    2.4.2 Parallel Circuits 15
    2.4.3 Series-Parallel Circuits 16
3.0.0 Electrical Measuring Instruments 19
  3.1.0 Measuring Voltage 20
    3.1.1 Voltage Testers 20
  3.2.0 Measuring Current 21
  3.3.0 Measuring Resistance 22
    3.3.1 Megohmmeters 23
4.0.0 Electrical Components in HVACR Systems 26
  4.1.0 Loads 26
    4.1.1 Motors 26
    4.1.2 Electric Heaters 27
  4.2.0 Control Devices 28
    4.2.1 Switches 28
    4.2.2 Fuses and Circuit Breakers 30
    4.2.3 Solenoid Coils 31
    4.2.4 Relays, Contactors, and Motor Starters 31
    4.2.5 Transformers 33
    4.2.6 Overload Protection Devices 34
    4.2.7 Thermistors 34
    4.2.8 Electronic Controls 35
  4.3.0 Electrical Diagrams 37

Figures and Tables

Figure 1   Coal-fired power plant. 2
Figure 2   Nuclear power plant. 2
Figure 3   Dam providing hydroelectric power. 2
Figure 4   Electrical power distribution. 3
Figure 5   Residential power distribution. 4
Figure 6   Magnetism. 5
Figure 7   A turbine-generator under construction. 6
Figure 8   One complete cycle of AC voltage. 6
Figure 9   Ground fault circuit interrupter (GFCI). 8
Figure 10   Condensing unit with wall-mounted disconnect switch. ... 9
Figure 11   Current flow. 12
Figure 12   Ohm's law triangle. 13
Figure 13   Ohm's law and power formula visual aid. 14
Figure 14   A basic electrical circuit. 14
Figure 15   Series and parallel circuits. 15
Figure 16   Parallel circuit with two loads. 16
Figure 17   Parallel circuit with three loads. 16
Figure 18   Series-parallel circuit. 16
Figure 19   Digital and analog multimeters. 19
Figure 20   Multimeter with a range dial. 20
Figure 21   Checking voltage across a load device. 20
Figure 22   Voltage tester. 20
Figure 23   Clamp-on ammeter/multimeter. 21
Figure 24   Clamp-on ammeter in use. 21
Figure 25   Measuring low currents. 21
Figure 26   In-line ammeter test setup. 22
Figure 27   Ohmmeter connection for continuity testing. 22
Figure 28   Continuity tester. 23
Figure 29   Handheld, battery-operated megohmmeter. 24
Figure 30   Examples of electric motors. 27
Figure 31   Motor wiring-diagram symbol examples. 27
Figure 32   Electric resistance heater assembly. 27
Figure 33   Switch contact arrangements. 28
Figure 34   Fused power disconnect. 28
Figure 35   Electronic wall thermostat. 29
Figure 36   Pressure switches. 29
Figure 37   Fuses. 30
Figure 38   Circuit breakers. 31
Figure 39   Solenoid coil and valve. 31
Figure 40   Relays. 32
Figure 41   Contactor and relay in a typical HVACR circuit. 33
Figure 42   Control transformer. 34
Figure 43   Transformer winding ratio. 34
Figure 44   Overload devices. 35
Figure 45   Thermistor application. 35

Figures and Tables (continued)

Figure 46 Programmable, Wi-Fi- enabled electronic thermostat. 36

Figure 47 Examples of electronic equipment controls.. 36

Figure 48 Microprocessor-based programmable controller............................. 37

Figure 49 Examples of electrical diagrams... 38

Figure 50 Example of equipment diagram with notes and legend. 39

Table 1 Effects of Current on the Human Body 8

1.0.0 FUNDAMENTALS OF ELECTRICITY

Objective

Describe the fundamentals of electricity.

a. State how electrical power is created and distributed.
b. Describe the difference between alternating current and direct current.
c. Identify general electrical safety practices.
d. Describe the OSHA requirements and procedures related to electrical lockout/tagout.

Trade Terms

Alternating current (AC): An electrical current that changes direction on a cyclical basis.

Ampere (A): The basic unit of measurement for electrical current, represented by the letter A.

Conductor: A material through which it is relatively easy to maintain an electrical current.

Current: The rate or volume at which electrons flow in a circuit. Current (I) is measured in amperes.

Direct current (DC): An electric current that flows in one direction. A battery is a common source of DC voltage.

Electromagnet: A coil of wire wrapped around a soft iron core. When current flows through the coil, magnetism is created.

Loads: Devices that convert electrical energy into another form of energy (heat, mechanical motion, light, etc.). Motors are the most common significant loads in HVACR systems.

Power: The rate of doing work, or the rate at which energy is dissipated. Electrical power is measured in watts.

Rectifier: A device that converts AC voltage to DC voltage.

Resistance: An electrical property that opposes the flow of current through a circuit. Resistance (R) is measured in ohms.

Transformer: Two or more coils of wire wrapped around a common core. Used to raise and lower voltages.

Volts (V): The unit of measurement for voltage, represented by the letter V. One volt is equivalent to the force required to produce a current of one ampere through a resistance of one ohm.

Voltage: The driving force that makes current flow in a circuit. Voltage, often represented by the letter E, is measured in volts. Also known as *electromotive force (emf)*, *difference of potential*, or *electrical pressure*.

Air conditioning, heating, and refrigeration systems use electricity as a source of power for compressors, fan motors, and other loads. Even a simple gas- or oil-fired furnace needs electricity to operate. Electrical switching devices, sensors, and indicators are used in the control circuits of all systems, in addition to major loads such as fan motors. Some control systems are quite complex. Many systems also use microprocessor controls, and some systems used in commercial buildings are managed from a central computer.

Many of the problems an HVACR technician encounters on a service call involve the electrical system at some level. Therefore, it is important to understand electrical circuits and be able to interpret complex circuit diagrams. Without these skills, it is impossible to accurately and reliably diagnose system problems.

1.1.0 Power Generation and Distribution

There are many sources of electrical power. A great deal of the electricity is generated at coal-fired power plants like the one shown in *Figure 1*. Nuclear power plants (*Figure 2*) are also a major source of electricity. Steam produced in power plants is used to drive huge turbine-generators, which produce the electricity. Other sources of electricity include turbine-generators powered by natural gas, as well as hydroelectric power plants (*Figure 3*) where water flowing over dams is used to drive the turbines.

The electrical power produced at the source travels through long-distance transmission lines (*Figure 4*). The voltage may be as high as 765,000 volts (V). High voltages are used because they result in lower line losses over the long distances that the electricity must travel. A device known as a transformer is used to lower the voltage to more usable levels as it reaches electrical substations and eventually homes, offices, and factories.

The voltage in a typical residential home is usually about 240V. At the wall outlet where small household appliances are plugged in (such as televisions and toasters), the voltage is about 120V (*Figure 5*). Larger electrical loads such as electric stoves, clothes dryers, and air conditioning systems usually require the full 240V. Commercial

Figure 1 Coal-fired power plant.

Figure 2 Nuclear power plant.

Figure 3 Dam providing hydroelectric power.

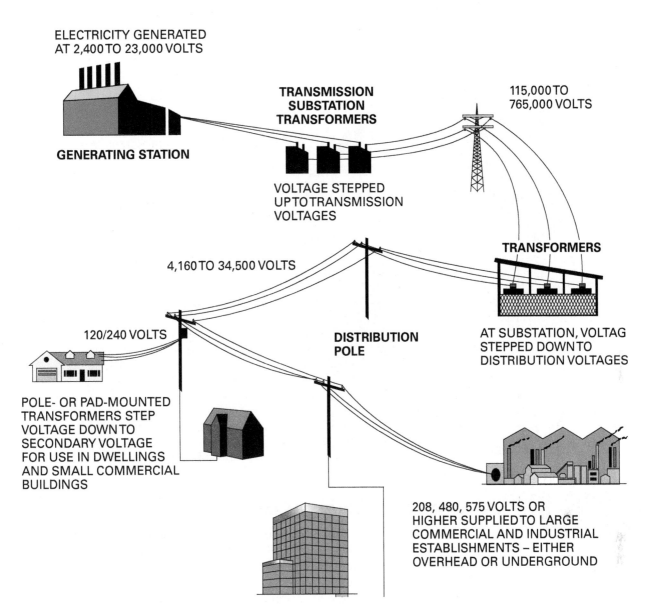

ELECTRICITY GENERATED
AT 2,400 TO 23,000 VOLTS

GENERATING STATION

TRANSMISSION
SUBSTATION
TRANSFORMERS

VOLTAGE STEPPED
UP TO TRANSMISSION
VOLTAGES

115,000 TO
765,000 VOLTS

TRANSFORMERS

AT SUBSTATION, VOLTAG
STEPPED DOWN TO
DISTRIBUTION VOLTAGES

4,160 TO 34,500 VOLTS

120/240 VOLTS

DISTRIBUTION
POLE

POLE- OR PAD-MOUNTED
TRANSFORMERS STEP
VOLTAGE DOWN TO
SECONDARY VOLTAGE
FOR USE IN DWELLINGS
AND SMALL COMMERCIAL
BUILDINGS

208, 480, 575 VOLTS OR
HIGHER SUPPLIED TO LARGE
COMMERCIAL AND INDUSTRIAL
ESTABLISHMENTS – EITHER
OVERHEAD OR UNDERGROUND

Figure 4 Electrical power distribution.

buildings and factories may receive anywhere from 208V to 575V. This depends on the amount of power their machines consume and their voltage requirements.

1.2.0 AC and DC Voltage

The kind of electric current produced by a battery is known as direct current (DC). Automobiles, cell phones, laptops, calculators, and flashlights are good examples of devices that use DC voltage. All of these examples are battery-powered and must be able to move from place to place. The electricity supplied by the local utility is alternating current (AC). Almost all HVACR devices use AC power. Many modern units use DC motors or electronic circuit boards that do require DC voltage. However, instead of using batteries that need

to be replaced or recharged, these units contain special electronic circuits called rectifiers that convert AC power to DC. The rectifier is usually built into the component itself. The charger that allows a cell phone or other portable device to be plugged into a wall socket for charging is also a rectifier, coupled with a transformer to also reduce the voltage.

One of the basic principles of AC components is that the flow of an electric current through a conductor produces a magnetic field around the conductor. If the conductor is coiled around an iron bar, the strength of the magnetic field is multiplied. The result is an electromagnet (*Figure* 6) that attracts and repels other magnetic objects, just like an iron magnet. This is the basis on which electric motors, transformers, and similar components operate. The opposite is also true;

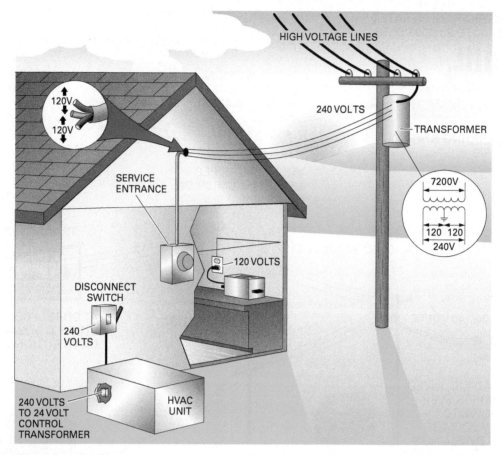

Figure 5 Residential power distribution.

Alternative Energy Sources

To reduce dependence on fossil fuels such as oil and coal, initiatives are being taken all over the world to develop alternatives. From these initiatives, many secondary or alternative methods of producing electrical power have emerged. Among the most common alternative sources are wind and solar energy. The use of power-generating facilities that burn organic material known as biomass is also becoming more common. Biomass describes any material that will decompose, such as wood, plants, leaves, or garbage, and can be burned. Methane gas produced by decomposition in landfills is also being captured and used as an energy source.

Figure Credit: Courtesy of DOE/NREL

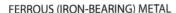
FERROUS (IRON-BEARING) METAL

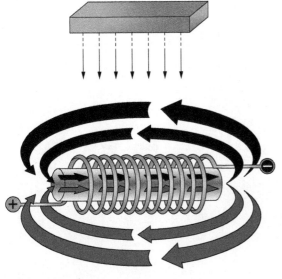

Figure 6 Magnetism.

that is, if a conductor is rotated in a magnetic field, a voltage is induced into the conductor. This is the principle on which a generator works.

The electricity produced by a power generating station and distributed across the country to consumers is AC. In a typical power plant, coal or nuclear materials produce steam, which in turn causes the shaft of a giant turbine (*Figure 7*) to rotate. The turbine shaft is connected to the rotor of a generator, which also rotates. As the rotor turns, a current is induced into the armature. As the armature changes position in relation to the magnetic field, the amount of current increases and decreases (alternates) in a pattern shown in *Figure 8*. So, while DC is a straight-line voltage, AC is constantly changing. The rate at which it changes is so rapid, that it appears to the user as a constant (effective) voltage, which is 70.7 percent of the maximum (peak) voltage induced.

Electromagnets

Electromagnets can be extremely powerful. For example, they are powerful enough to lift a vehicle at the scrap yard. One advantage of an electromagnet is that it can be turned on and off. Electromagnets are often used at scrap yards and processing facilities that recycle steel and other metals. When handling scrap, aluminum and other non-ferrous materials are not attracted to the magnet, providing a basic level of sorting. The same principle of creating a magnetic field around an electrically energized winding applies to many components that are used in HVACR electrical circuits.

Figure Credit: © AngelPet/Shutterstock.com

Figure 7 A turbine-generator under construction.

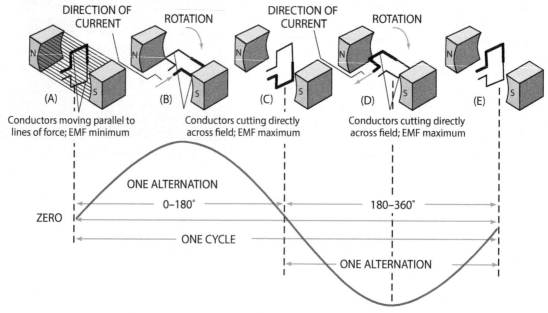

Figure 8 One complete cycle of AC voltage.

Transformers

The voltage produced by the motor-generator set in a power station may be in the 13,000V range. Large transformers like the one shown in image A, are used to step up the voltage to tzhe level carried by the high-voltage transmission lines. The voltage is carried over the high-voltage transmission lines to substations where other transformers step down the voltage to the level required for local distribution. Pole transformers like the one shown in image B, as well as pad-mounted models for utility and industrial purposes, step the voltage down further to the level needed for homes and businesses.

(A) Large Transformer

(B) Pole Transformer

1.3.0 Electrical Safety

Working with electrical devices always involves some degree of danger. Technicians must be constantly alert to the potential for electric shock and take the necessary precautions. The amount of electrical current that passes through the human body determines the outcome of an electrical shock. The higher the voltage, the greater the current, and the greater the chance of receiving a fatal shock. Electrical safety is thoroughly covered in the NCCER Module 00101, "Basic Safety". Some key points are presented here to reinforce the importance of safety when working on electrical equipment.

Electrical current flows along the path of least resistance to return to its source. If a person comes in contact with a live conductor and a grounded object, that person becomes a load in the electrical circuit. The potential for shock increases dramatically if the skin is damp. Broken skin (a cut, for example) will also reduce the resistance of the human skin. Even currents of less than one-third of an ampere (A) can severely injure or kill a person. *Table 1* describes the effect of different levels of current flowing through the human body.

WARNING!

Voltages of 600V or more are almost ten times as likely to kill as lower voltages. On the job, you spend most of your time working on or near lower voltages. However, lower voltages can also kill. For example, portable, electrically operated hand tools powered by 120V can cause severe injuries or death if the frame or case of the tool becomes energized. Many electrical accidents can be prevented by following proper safety practices, including the use of ground fault circuit interrupters and proper grounding.

Did You Know?

Electrical Safety

Every year in the United States, there are many deaths resulting from electrocution. Electrocution is one of the top four leading causes of death in the construction workplace, making it one of OSHA's Fatal Four hazards. According to OSHA, there were 81 deaths in 2015 resulting from electrocution in the construction industry alone—that's three every two weeks. It doesn't have to be that way, and you can be part of the change.

Table 1 Effects of Current on the Human Body

Current Value	Typical Effects
1mA	Perception level. Slight tingling sensation.
5mA	Slight shock. Involuntary reactions can result in other injuries.
6 to 30mA	Painful shock, loss of muscular control.
50 to 150mA	Extreme pain, respiratory arrest, severe muscular contractions. Death possible.
1,000mA to 4,300mA	Ventricular fibrillation, severe muscular contractions, nerve damage. Typically results in death.

This information has been included in this training module to help you gain respect for the environment where you work and to stress how important safe working habits really are. Since electricity is not something that can be seen, technicians must give constant attention to the existence of energized electrical circuits and recognize the potential for serious injury.

> **NOTE**
> OSHA regulations state that employees have a duty to follow the safety rules put in place by their employers. The amount an employee can lawfully collect from worker's compensation or disability insurance due to an injury may be restricted if the employee was in violation of safety rules when injured. A company also may terminate employment or take other disciplinary action if safety rules are violated.

1.3.1 General Safety Practices

HVACR technicians work with potentially deadly levels of electricity every day. They can do so because good safety practices have become second nature to them. The following are some general safety practices to follow whenever you are working with electricity:

- Unless it is essential to work with the power on, always de-energize the system at the source. Lockout and tag out the disconnect switch in accordance with company or site procedures and OSHA requirements. The OSHA requirements are presented in the next section.

- Use a voltmeter to verify that the power to the unit is disabled. Remember that even though the power may be switched off, there is still electrical potential at the input side of the disconnect switch or circuit breaker.
- Use personal protective equipment (PPE) such as rubber gloves when necessary.
- Use insulated hand tools and double-insulated power tools.
- Use a ground fault circuit interrupter (GFCI) when using power tools (*Figure 9*).
- Short wiring to ground before touching de-energized wires.
- If possible, do not kneel directly on the ground when making voltage measurements.
- Remove all metal jewelry such as rings, watches, and bracelets.
- When it is necessary to test an energized circuit, keep one hand outside the unit if possible. This reduces the risk of completing a circuit through your upper body that might cause current to pass through your heart.
- When working with electric power tools, inspect the tool and its power cord to make sure they are safe to use.

The *National Electrical Code*® (*NEC*®), when used together with the electrical code for your local area, defines the minimum requirements for the installation of electrical systems. The *NEC*® specifies the minimum provisions necessary for protecting people and property from electrical hazards. An electrician generally installs the electrical circuit from the building's electrical service to the disconnecting device for the HVACR equipment. When installing an HVACR system, the HVACR technician may be required to connect the power from the disconnecting device to the HVACR unit. When doing so, be sure to follow all applicable code requirements.

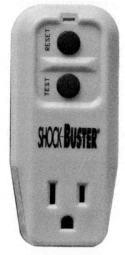

Figure 9 Ground fault circuit interrupter (GFCI).

1.4.0 OSHA Lockout/Tagout Rule

In addition to many other safety rules, OSHA enforces 29 *CFR* 1926.417, *Lockout and Tagging of Circuits*. This rule covers the specific procedure to be followed for the "servicing and maintenance of machines and equipment in which the unexpected energization or startup of the machines or equipment, or releases of stored energy, could cause injury to employees. This standard establishes minimum performance requirements for the control of such hazardous energy."

The purpose of the OSHA procedure is to make sure that machinery is isolated from all potentially hazardous energy sources, and locked out and tagged before workers perform any service or maintenance activities where the unexpected energization, startup, or release of stored energy could cause injury.

> **WARNING!**
> The OSHA procedure provides only the minimum requirements for the lockout/tagout procedure. Employees should also consult the required lockout/tagout procedures established by the employer. Remember, lives depend on this procedure. It is critical that the correct procedures be used.

1.4.1 Lockout/Tagout Procedure

To prepare for a lockout/tagout, make a survey to locate and identify all isolating devices (*Figure 10*). Be certain which switch(es), valve(s), or other energy-isolating devices apply to the equipment to be locked and tagged.

The following procedure outlines the general steps for lockout/tagout (note that this procedure is provided only as an example). Each employer

Figure 10 Condensing unit with wall-mounted disconnect switch.

will designate who is qualified to conduct the procedure. Not every employee has the proper training to do so.

> **WARNING!**
> It is essential to remember there is often more than one type of energy applied to a system or its equipment. Even a single type of energy may have more than one source, such as a packaged HVAC unit with one electrical power source for the unit and a separate power source for an installed electric-heating package. Failure to recognize and render all energy sources safe can result in serious injury or death.

Step 1 Notify all affected employees of the lockout/tagout and why it is necessary. The authorized employee will know the type of energy that the machine or equipment uses and the associated hazards.

Step 2 If the machine or equipment is running, shut it down using the normal procedure. For example, press the stop button or open the toggle switch.

Step 3 Operate the switch, valve, or other energy-isolating device(s) so that the equipment is isolated from its energy source(s). Stored energy must be dissipated or restrained by repositioning, blocking, or bleeding down. Examples of stored energy are springs, elevated machine members, rotating flywheels, hydraulic systems, and air, gas, steam, or water pressure.

Step 4 Lockout and tag the energy isolating devices with assigned individual lock and tag.

Step 5 Use a voltmeter that has been tested on a known source to confirm that electrical power has been de-energized. After making sure that no personnel are exposed, operate the start button or other normal operating controls to make certain the equipment will not operate. This will confirm that the energy sources have been disconnected. The equipment is now locked and tagged.

> **CAUTION**
> Return operating controls to their neutral or Off position after confirming that all energy sources have been disabled.

All equipment must be locked and tagged to protect against accidental operation when such operation could cause injury to personnel. Never try to operate any switch, valve, or energy-isolating device when it is locked and tagged.

1.4.2 Restoring Power

After service and/or maintenance is complete, power can be restored to the system. Use the following basic procedure to restore power to the equipment:

Step 1 Check the area around the machines or equipment to make sure that all personnel are at a safe distance.

Step 2 Remove all tools from the machine or equipment and reinstall any guards.

Step 3 Confirm that all equipment operating controls are in the neutral or Off position.

Step 4 Double-check that all personnel are in the clear, and remove all lockout/tagout devices.

Step 5 Turn on the disconnect switch or other energy-isolating device to restore energy to the equipment.

Additional Resources

Principles of Electric Circuits, Thomas L. Floyd. 9th Edition. New York, NY: Pearson.

NCCER Module 26102, *Electrical Safety*.

NCCER Module 75121, *Electrical Safety*.

NCCER Module 26501, *Managing Electrical Hazards*.

1.0.0 Section Review

1. The voltage supplied to a commercial building from a power transformer may be as high as _____.

    a. 765,000V
    b. 2,400V
    c. 575V
    d. 120V

2. The kind of voltage supplied by the local utility is _____.

    a. AC
    b. DC
    c. resistive
    d. inductive

3. Which of the following is a recommended safety practice when performing measurements on live electrical circuits?

    a. Kneel on the ground when using a voltmeter.
    b. Keep both hands inside the unit while working.
    c. Avoid wearing gloves of any kind around energized circuits.
    d. Use insulated hand tools and double-insulated power tools.

4. Employees should always consult the required lockout/tagout procedures established by their employer because _____.

    a. an employer's requirements are more important than OSHA's requirements
    b. OSHA's requirements represent only the minimum requirements
    c. employers are not required to follow OSHA
    d. OSHA requirements do not apply to employers

2.0.0 BASIC ELECTRICAL THEORY

Objective

Explain basic electrical theory.
  a. Define voltage, current, resistance, and power and describe how they are related.
  b. Use Ohm's law to calculate the current, voltage, and resistance in a circuit.
  c. Use the power formula to calculate how much power is consumed by a circuit.
  d. Describe the differences between series and parallel circuits and calculate circuit values for each type.

Performance Tasks

5. Assemble and test low- and high-voltage series and parallel circuits using a transformer and selected control and load devices.

Trade Terms

Arc: A visible flash of light and the release of heat that occurs when an electrical current crosses an air gap.

Continuity: A continuous current path. The absence of continuity indicates an open circuit.

Inductive: Of a load, able to become electrically charged from being near another electrically charged body, or to become magnetized by being within an existing magnetic field. The process itself is called *induction*.

Insulator: A device or substance that inhibits the flow of current; the opposite of conductor.

Ohms (Ω): The basic unit of measurement for electrical resistance, represented by the symbol Ω.

Relay: A magnetically operated device consisting of a coil and one or more sets of contacts.

Thermistor: A semiconductor device that changes resistance with a change in temperature.

Watts (W): The unit of measure for power consumed by a load.

An electrical circuit consists of four main elements: a voltage source, a load device that uses the electricity produced by the source, a control device that can be used to turn the circuit on and off, and conductors to carry the current. When the circuit is completed, an electrical current will flow in the circuit, causing the load device to perform work. The source produces a voltage that causes current to flow, while the load offers resistance to current flow and consumes power. Voltage, current, resistance, and power are the building blocks of electricity.

2.1.0 Current, Voltage, Resistance, and Power

An electrical current is caused by the movement of electrons from one point to another. Electrons are negatively charged particles that exist in all matter. When a difference in the number of electrons exists between two points, electrons flow away from the point with more electrons to the point with fewer electrons. This difference in electrical potential between the two points is called *voltage*. Voltage must be present for electrons to flow.

Note that when observing electron flow, current flows from negative to positive. However, electricians and electrical engineers conventionally view current as flowing in the opposite direction, from positive to negative. This concept is an artifact from the time when Benjamin Franklin first began experimenting with electricity. Even so, it works particularly well with electronic and microprocessor theory, and is now the generally accepted view for current flow.

Lightning is a good example of natural electron flow. A storm cloud has a negative charge with respect to Earth. Lightning occurs when the difference in potential between the cloud and Earth becomes so great that the air between them acts as a conductor.

In the common 12VDC car battery, a chemical reaction causes one of the poles to be negative with respect to the other. If a conductor is connected between the negative (–) and positive (+) poles of the battery, electrons flow from the negative pole to the positive pole (*Figure 11*).

To make use of the potential energy of the battery, resistive devices such as a light bulbs are connected between the two battery poles. Resistance is the property that causes a material to resist or impede current flow. A resistive device in a circuit is an electrical load. When the switch closes, the moving electrons flow through the bulb, which converts the electrical energy into light. The bulb consumes the electrical energy in the process. The rate of energy consumption is known as *power*, and is expressed in watts (W). For example, a 60W light bulb is designed to consume 60 watts of power.

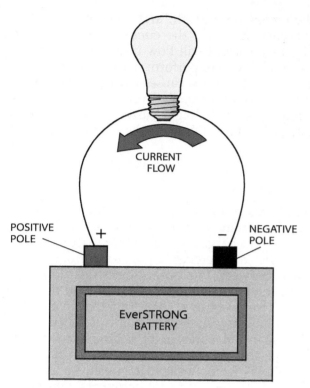

Figure 11 Current flow.

Current is expressed in amperes (abbreviated amps or A), voltage is expressed in volts (V), and resistance is expressed in ohms (Ω). The AC voltages HVACR technicians work with usually range from 24VAC to 600VAC. The occasional DC voltages HVACR technicians encounter will generally be low voltages in the 5VDC to 15VDC range. Higher DC voltages are found in variable frequency drive (VFD) controls that can vary the speed of AC-powered motors. Occasionally, technicians may encounter a very high-voltage DC circuit, such as the ignition circuit for a gas or oil furnace, which can produce a 10,000V spark. This high voltage is needed to allow the voltage to jump across a gap, creating an arc.

Currents in HVACR circuits may be as low as a few microamps (millionths of an amp). Currents in the milliamp, or mA (thousandths of an amp), range are more common. Some circuits, especially motor circuits, may consume very high currents in the range of 100A or more. From a safety point of view, the important thing to remember is that even currents in the milliamp range can be hazardous.

Resistances may be large values expressed in thousands of ohms (kilohms or kΩ) or millions of ohms (megohms or MΩ). In some cases, a resistance value may be only a few ohms. For example, a thermistor (temperature-sensitive resistor) used as a temperature-sensing devices falls into this category.

Voltage

Eighteenth-century physicist Alessandro Volta theorized that when certain objects and chemicals come into contact with each other, they produce an electric current. Believing that electricity came from contact between metals only, Volta coined the term *metallic electricity*. To demonstrate his theory, Volta placed two discs, one of silver and the other of zinc, into a weak acidic solution. When he linked the discs together with wire, electricity flowed through the wire. Thus, Volta introduced the world to the battery, known at the time as the *Voltaic pile*. Volta needed a unit to measure the strength of the electric push. The volt, named after Volta, is that measure. Voltage has other names as well, including *electromotive force* (*emf*) and *potential difference*. Voltage can be viewed as the potential to do work.

Voltage, current, resistance, and power are mathematically related. If any two of the values are known, the other two can be calculated. For example, if it is known how much voltage is available and how much power the load consumes, the amount of current the circuit will draw can be calculated. This can be done using the formulas discussed in the following sections.

2.2.0 Ohm's Law

Ohm's law is a formula for calculating voltage, current, and resistance. The formula was derived by Georg Ohm, a German physicist, in 1827. The values in the formula are expressed as E, I, and R, respectively. The letter E is used to represent voltage based on Ohm's original concept of voltage as electromotive force (E). Similarly, the letter I for current is based on the intensity of electron flow.

Figure 12 shows the three variations of Ohm's law. The pyramid provides a way to remember the formula visually. If you know two of the values, cover the unknown value with your finger to see how to solve the problem—by multiplication or division of the two knowns. For example, if the voltage and resistance are known, covering the I shows that E is divided by R to determine current (I).

For example, in a 120V circuit containing a 60Ω load, the current flow is 2A. This can be determined by using Ohm's law, as follows:

$$I = \frac{E}{R} = \frac{120V}{60\Omega} = 2A$$

The pyramid also shows that current (I) is multiplied by resistance (R) to determine voltage (E).

Lightning

A lightning strike can carry a current of 20,000 amps and reach temperatures of 54,000°F. It is no wonder that being in the path or vicinity of a lightning strike is often deadly.

Figure Credit: NASA

Note that Ohm's law applies only to resistive circuits. Because of the nature of alternating current, circuits containing motors, relay coils, and other inductive loads do not act the same as purely resistive loads.

	LETTER SYMBOL	UNIT OF MEASUREMENT
CURRENT	I	AMPERES (A)
RESISTANCE	R	OHMS (Ω)
VOLTAGE	E	VOLTS (V)

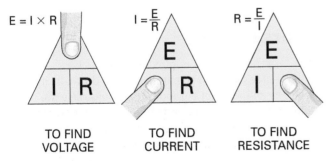

$E = I \times R$ $I = \dfrac{E}{R}$ $R = \dfrac{E}{I}$

TO FIND VOLTAGE   TO FIND CURRENT   TO FIND RESISTANCE

Figure 12 Ohm's law triangle.

2.3.0 Electrical Power

The power consumed by most load devices determines their rating, rather than the amount of resistance they offer. For example, an electric heating element may have a rating of 5,000W, or 5kW. The equations based on Ohm's law will not help to determine how much current the heater will draw, because two variables in the Ohm's law formula—resistance and current—are missing. In this case, the power formula must be used. *Figure 13* shows the variations of the power formula, as well as Ohm's law variants since they are closely related. In each quadrant of the power formula, the three variations of the equations that determine the value in white are shown. The version of the power formula needed looks like this:

$$I = \frac{P}{E}$$

In a 240V circuit, the 5,000W heater would draw 20.8A (5,000W ÷ 240V). This is a common electric heating element, and a common voltage at

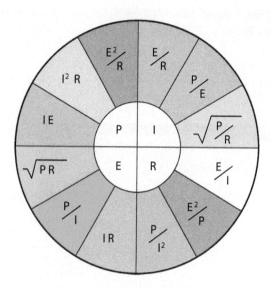

Figure 13 Ohm's law and power formula visual aid.

which they operate. For a load like this, the wire must be sized to accommodate this much current and more, following the wire sizing rules of the *NEC*®. If the heater were to operate at 120V, the power formula indicates that it would draw 41.7A (5,000W ÷ 120V). That means the wire would have to be sized to accommodate this load, resulting in significantly larger and more costly wiring. Overloaded electrical conductors are a frequent cause of fires, especially when they are overloaded consistently. Properly applied overcurrent protection devices, such as circuit breakers and fuses, prevent this from happening.

Electric motors are often rated in equivalent horsepower instead of watts. A swimming pool pump, for example, might be powered by a 2 horsepower (hp) electric motor. There is a simple conversion from horsepower to watts: 1 hp = 746W. Therefore, a 2 hp motor would consume 1,492W. In a 120V circuit, it would draw more than 12A as shown here:

$$I = \frac{P}{E} = \frac{1,492W}{120V} = 12.43A$$

The ability to perform this type of calculation is important because adding new loads to existing circuits is often something an HVACR technician or estimator must consider.

2.4.0 Electrical Circuits

A basic electrical circuit is shown in *Figure 14*. A simple electrical circuit is a closed loop that contains a voltage source, a load, and conductors to carry current. It often contains a switching device to open and close the circuit as desired.

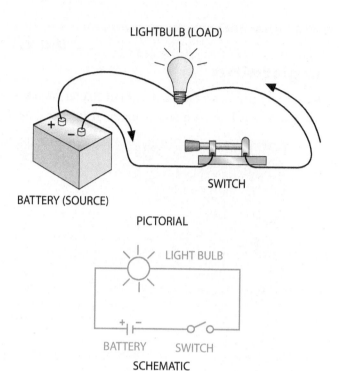

Figure 14 A basic electrical circuit.

An electrical conductor is a material that readily carries an electric current. Most metals are good electrical conductors as well as thermal conductors. Copper and aluminum are the most common electrical conductors used to wire AC circuits of all types. Gold and silver are better conductors, but are too expensive to use, except in micro-miniature electronic circuits. The minerals found in water also make it an excellent conductor. That is why it is dangerous to work with electricity or use electrical appliances in or around a wet environment.

An insulator is the opposite of a conductor. It inhibits the flow of electricity. Rubber, fiberglass, and porcelain are good insulators. Tools used in electrical trades are often insulated with rubber to help protect against electrical shock. Special tools used by power line workers, called *hot sticks*, are usually constructed of fiberglass.

Figure 15 shows simple examples of series and parallel circuits. The most common by far is the parallel circuit, in which all the loads are connected across the source.

Horsepower

The term *horsepower* originates from the time when horses were used extensively to do work. It was determined that a good horse could raise the equivalent of 33,000 pounds to a height of one foot in one minute using a pulley system. This can be stated as 33,000 foot-pounds per minute of work.

2.4.1 Series Circuits

A series circuit provides only one path for current flow. For that reason, the current is the same at any point in the circuit. The total resistance of the circuit is equal to the sum of the individual resistances. The 12V series circuit in *Figure 15 (A)* has two 30Ω loads. The total resistance (R_T) is therefore 60Ω. The total amount of current (I_T) flowing in the circuit is 0.2A, which is calculated as follows:

$$I_T = \frac{E_T}{R_T} = \frac{12V}{60Ω} = 0.2A$$

If there were five 30Ω loads, the total resistance would be 150Ω. The current flow is the same through all the loads. However, the voltage measured across any load (voltage drop) varies depending on the resistance of that load. The sum of the voltage drops equals the total voltage applied to the circuit. In the case of the five 30Ω loads, each would drop one-fifth of the applied voltage (since all five of the loads are equal).

Circuits containing only loads in series are uncommon in HVACR work. An important trait of a series circuit is that if the circuit is open at any point, no current will flow. For example, if five light bulbs are connected in series and one of them blows, all five lights will go off because the continuity of the circuit has been broken.

2.4.2 Parallel Circuits

In a parallel circuit, each load is connected directly to the voltage source. Therefore, the voltage drop is the same through all loads because each load sees an equal applied voltage. The source sees the circuit as two or more individual circuits containing one load each.

In the parallel circuit in *Figure 15*, the source sees three circuits, each containing a 30Ω load. The current flow through any load varies depending on the resistance of that load. Thus, the total current drawn by the circuit is the sum of the individual currents.

The total resistance of a parallel circuit is calculated differently from that of a series circuit. In a parallel circuit, the total resistance is less than the smallest of the individual resistances. For example, each of the three 30Ω loads draws 0.4A at 12V; therefore, the total current is 1.2A. This is calculated as follows:

$$I = \frac{E}{R} = \frac{12V}{30Ω} = 0.4A$$

0.4A per load × 3 loads = 1.2A

Now Ohm's law can be used to calculate the total resistance (R_T):

$$R_T = \frac{E_T}{I_T} = \frac{12V}{1.2A} = 10Ω$$

This example is simple because all the resistances have the same value. When the resistances are different, the current must be calculated for each load individually. The sum of the individual currents is equal to the total current.

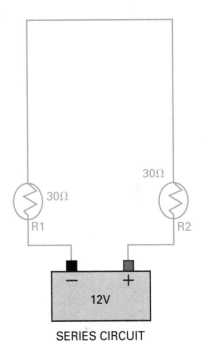

Figure 15   Series and parallel circuits.

Parallel circuits have an advantage over series circuits because they offer multiple paths. Therefore, if the continuity of a path is broken, current can still flow in the other paths. In other words, parallel circuits continue working even if one load branch of the circuit opens. Household circuits are wired in parallel. This is why series circuits are very rarely found in HVACR systems and equipment. Most such circuits are found in packaged electronic devices.

The following formulas can be used to convert parallel resistances to a single resistance value:

- For two resistances connected in parallel:

$$R_T = \frac{R1 \times R2}{R1 + R2}$$

- For three or more resistances connected in parallel:

$$R_T = \frac{1}{\dfrac{1}{R1} + \dfrac{1}{R2} + \dfrac{1}{R3}}$$

For examples of these two formulas in use, see *Figure 16* and *Figure 17*.

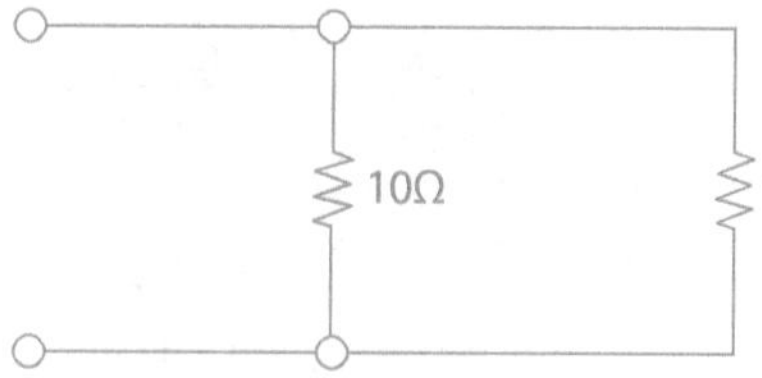

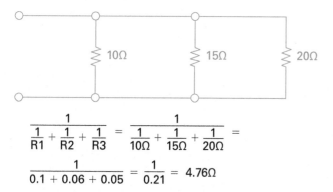

$$\frac{R1 \times R2}{R1 + R2} = \frac{10 \times 15}{10 + 15} = \frac{150}{25} = 6$$

Figure 16 Parallel circuit with two loads.

$$\frac{1}{\dfrac{1}{R1} + \dfrac{1}{R2} + \dfrac{1}{R3}} = \frac{1}{\dfrac{1}{10\Omega} + \dfrac{1}{15\Omega} + \dfrac{1}{20\Omega}} =$$

$$\frac{1}{0.1 + 0.06 + 0.05} = \frac{1}{0.21} = 4.76\Omega$$

Figure 17 Parallel circuit with three loads.

2.4.3 Series-Parallel Circuits

Some electronic circuits, known as *series-parallel circuits* or *combination circuits*, contain a hybrid arrangement (*Figure 18*). However, you will rarely find loads connected in this arrangement.

To determine the total resistance of a series-parallel circuit, the parallel loads must be converted to their equivalent series resistance. The load resistances are then added to determine total circuit resistance.

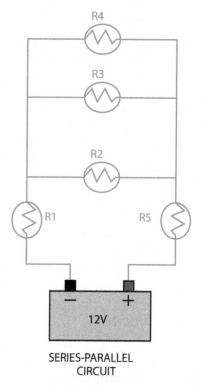

SERIES-PARALLEL
CIRCUIT

Figure 18 Series-parallel circuit.

> **NOTE**
>
> Because series-parallel circuits are uncommon, you will not have to determine their characteristics very often. However, these calculations are presented in more detail in Module 26104, "Electrical Theory," from NCCER's *Electrical Level One.*

Is It a Series Circuit?

When the term *series circuit* is used, it refers to the way the loads are connected. The same is true for parallel and series-parallel circuits. You will rarely, if ever, find loads in an HVACR system connected in series, or in a series-parallel arrangement. The simple circuit shown here illustrates this point. At first glance, you might think it is a series-parallel circuit. On closer examination, you can see that there are only two loads, and they are connected in parallel. Therefore, it is a parallel circuit. The control devices are wired in series with the loads, but only the loads are considered in determining the type of circuit.

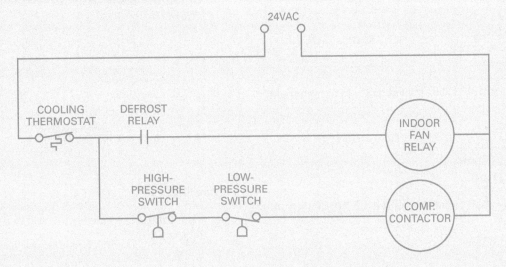

Additional Resources

Principles of Electric Circuits, Thomas L. Floyd. 9th Edition. New York, NY: Pearson.

NCCER Module 26104, *Electrical Theory*.

2.0.0   Section Review

1. Which of the following is measured in watts?

    a. Current
    b. Voltage
    c. Resistance
    d. Power

2. The formula I × R is used to solve for what electrical value?

    a. Voltage
    b. Current
    c. Resistance
    d. Power

3. How much current would a 1.5 hp, 240V motor draw?

    a. 0.0063A
    b. 0.21A
    c. 4.66A
    d. 160A

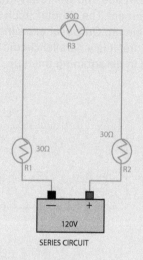

Figure SR01

4. The total resistance of the circuit shown in *Figure SR01* is _____.

    a. 3Ω
    b. 10Ω
    c. 60Ω
    d. 90Ω

3.0.0 ELECTRICAL MEASURING INSTRUMENTS

Objective

Identify the electrical measuring instruments used in HVACR work and describe their uses.

a. Describe how voltage is measured.
b. Describe how current is measured.
c. Describe how resistance is measured.

Performance Tasks

1. Use the proper instrument to measure voltage in an energized circuit.
2. Use the proper instrument to measure current in an energized circuit.
3. Use the proper instrument to measure resistance.
4. Use a multimeter to check circuit continuity.

Trade Terms

Ammeter: A test instrument for measuring electrical current.

Analog meter: A meter that uses a needle to indicate a value on a scale.

Clamp-on ammeter: An ammeter with operable jaws that are placed around a conductor to sense the magnitude of the current flow.

Digital meter: A meter that provides a direct numerical reading of the value measured.

In-line ammeter: A current-reading meter that is connected in series with the circuit under test.

Multimeter: A test instrument capable of reading voltage, current, and resistance. Also known as a *volt-ohm-milliammeter (VOM)*.

Test meters are used to measure voltage, current, and resistance. The most common test meter is the volt-ohm-milliammeter (VOM), commonly known as a multimeter. The multimeter combines a voltmeter, ohmmeter, and ammeter into a single instrument. *Figure 19* shows both digital and analog multimeters. An analog meter is so-called because a pointer moves in proportion to the value being measured, and the value is determined by reading a scale displayed behind the needle. A digital meter displays the result numerically on the screen. Digital models have become the most popular.

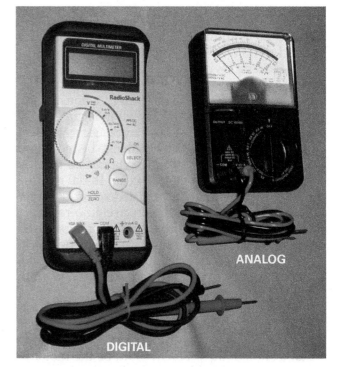

Figure 19 Digital and analog multimeters.

In the field, a multimeter or clamp-on ammeter (which can be either self-contained or an accessory for a multimeter) is the commonly used test instrument for voltage, resistance, and current measurements. Many multimeters are also capable of reading low current values in the milliamp range, without an accessory. Accessory ammeters that connect to a multimeter expand its capabilities to measure much higher current values. This capability is essential in the HVACR trade. Self-contained ammeters are also readily available, and many of those are also capable of reading voltage and resistance. This makes them very versatile instruments. It is not unusual for a technician to have a standard multimeter as well as a clamp-on ammeter that can serve as a multimeter. The range of a digital meter can usually be selected for the desired level of precision, but many also determine the best range when each reading is taken.

Analog Meters

Analog meters, which require the reader to interpret a scale on the meter face have been largely replaced by direct-reading digital meters. Analog meters are useful when it is necessary to see the magnitude of fluctuations in the value being measured. The user can see how much the needle moves from side to side. One reason that digital meters are often preferred is because the analog needle movement is relatively sensitive to rough handling.

3.1.0 Measuring Voltage

Voltage measurements are very commonly performed by HVACR technicians. All multimeters have a selector switch that is used to select the test function (voltage, current, or resistance), as well as the range of values within a given function. A multimeter with a range dial is shown in *Figure 20*. The various ranges provide better more accurate measurements. If the range of the value to be measured is unknown, the highest range should be used to avoid damaging the meter. Once an initial reading is taken, the range can be adjusted accordingly for greater precision. Some meters are auto-ranging; that is, they automatically adjust the range based on the voltage being measured. When using a voltmeter, it must be connected in parallel with (across) the component or circuit to be tested, as shown in *Figure 21*.

If a circuit is not operating as it should, the voltmeter is used to determine if the correct voltage is available. For example, if a motor is not running when it should, the problem could be the motor itself, or it could lack the proper voltage due to a control device or a blown fuse. Motors are controlled by switching devices. If the switching device fails, the motor will not receive power and will not function. A motor should never be replaced without first checking and confirming the appropriate voltage is being applied to its wiring leads. This is essentially true for all electrically powered devices. The supply voltage must always be checked carefully, since the power source must be energized and the controls set for the device to operate.

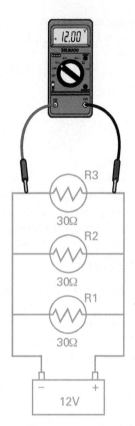

Figure 21 Checking voltage across a load device.

3.1.1 Voltage Testers

A voltage tester, shown in *Figure 22*, is a very simple tool used to check for the presence of voltage. It can be used as a troubleshooting tool or as a safety device to make sure the voltage is turned off before touching any terminals or conductors. When the probes are touched to the circuit, the light on the instrument will turn on if a voltage is present. Instruments like these are available in several voltage ranges, so it is important to know something about the circuit you are checking and the voltage you expect from it.

Note that this type of compact, simple instrument is not generally used for work requiring accuracy. Technicians use these to detect the simple presence or absence of voltage. For most troubleshooting tasks, a multimeter is far more effective and informative.

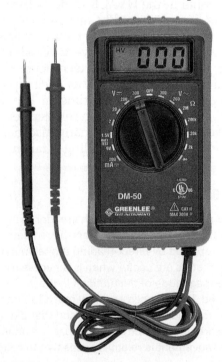

Figure 20 Multimeter with a range dial.

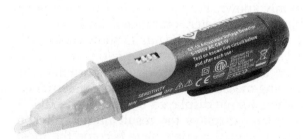

Figure 22 Voltage tester.

3.2.0 Measuring Current

Clamp-on ammeters like the one shown in *Figure 23* are used by HVACR technicians to measure AC current. The jaws of the ammeter are placed around a conductor. Current flowing through the wire creates a magnetic field, which induces a proportional current in the ammeter jaws. This current is read on the meter movement and appears as a digital readout or, on an analog meter, as a deflection of the meter needle. The meter shown in *Figure 23* is also a multimeter.

Not all ammeters have the same capacity to measure current. When the measurement of high current levels is expected, make sure the range of the ammeter is suitable for the task. Note that the jaws of the clamp-on ammeter must only be placed around one conductor at a time, as shown in *Figure 24*. Placing the jaws around two separate conductors results in an inaccurate reading.

When very low currents are being measured with a clamp-on ammeter, it may be necessary to wrap a single conductor multiple times around the meter jaws, as illustrated in *Figure 25*. The reading must then be divided by the number of times the conductor passes through the jaws to determine the current flow. For example, if the wire passes through the meter jaws 5 times (as shown) and the reading is 5A, the actual current is 1A (5A ÷ 5 wraps of wire). This is done when the reading is expected to be near or below the accurate range of the meter to make the reading more accurate.

In-line current measurements, using probes at each end of the measured circuit, are much less common. This practice is used to measure very small current values, which is not required frequently. If a standard multimeter is used instead of a clamp-on ammeter, its in-line ammeter must be connected in series with the circuit. To do so, the circuit must be de-energized while the ammeter is connected. The current then passes through the meter. This is shown in *Figure 26*. If the ammeter was placed across a closed circuit, some of the current would flow through the existing circuit instead of the meter, and the current reading would be incorrect.

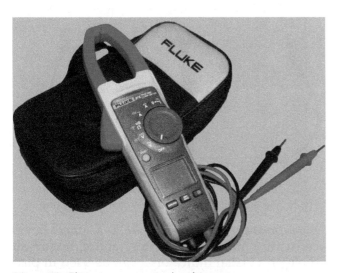

Figure 23 Clamp-on ammeter/multimeter.

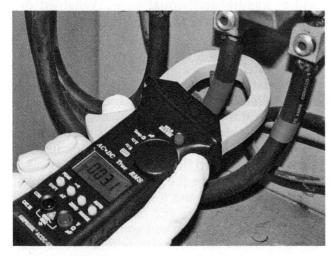

Figure 24 Clamp-on ammeter in use.

CONDUCTOR

$$\text{CURRENT FLOW} = \frac{\text{METER READING (AMPS)}}{\text{NUMBER OF WRAPS AROUND AMMETER JAWS}}$$

Figure 25 Measuring low currents.

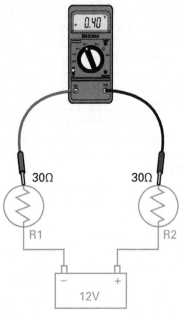

SERIES CIRCUIT

Figure 26 In-line ammeter test setup.

In addition to good safety practices, remember the following guidelines when measuring current:

- If the ammeter jaws are dirty or misaligned, a meter will not read correctly. The jaws must meet and interlock completely.
- When using an analog ammeter, always start at a higher range than necessary to avoid damaging the meter. Then adjust to a range that provides the needed level of detail or accuracy.
- Do not clamp the ammeter jaws around two different conductors at the same time; an inaccurate reading will result.

3.3.0 Measuring Resistance

An ohmmeter contains an internal battery that acts as a voltage source. Therefore, resistance measurements are always made with the system power shut off. While ohmmeters are sometimes used to measure the resistance of a load (such as when testing motor windings), they are more often used to check continuity in a circuit. A wire or closed switch offers virtually no resistance to current flow. With the ohmmeter connected as in *Figure 27*, and the three switches closed, the current produced by the ohmmeter battery will flow unopposed and the meter will show zero resistance. This means the circuit has continuity (it is continuous). If a switch is open, however, there is no path for current and the meter will see infinite resistance (it does not have continuity).

> **CAUTION**
>
> Never measure resistance in a live (energized) circuit. The voltage present in a live circuit may permanently damage the meter. Higher-quality meters are equipped with a fuse or circuit breaker to protect the meter if this occurs. A circuit in parallel with the circuit you are testing can also provide a current path that will make it appear as though the circuit you are testing is good, even when it is not. This problem can be avoided by isolating any single device being tested from the rest of the circuit by physically disconnecting it.

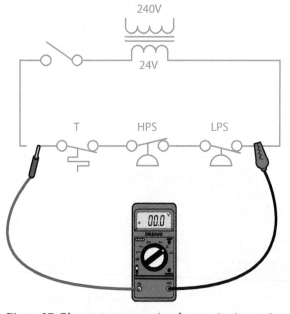

Figure 27 Ohmmeter connection for continuity testing.

A continuity tester (*Figure 28*) is a simple device consisting mainly of a battery and either an audible or a visual indicator. It can be used in place of an ohmmeter to test the continuity of a wire and to identify individual wires contained in a conduit or other raceway. Like the simple voltage tester, the continuity tester does not provide any level of accuracy. It can only report that a circuit is open or closed.

The tester can be used to test wiring that has been pulled through conduit and terminated in some way at the opposite end. To test the continuity of a wire, strip the insulation off the end of the wire to be tested at one end of the conduit run, then connect (short) the wire to the metal conduit. At the other end of the conduit run, clip the alligator clip lead of the tester to the conduit and touch the probe to the end of the wire under test. If the tester's audible alarm sounds or the indicator light comes on, there is continuity. Note again

---

Figure 28 Continuity tester.

that this only indicates that there is continuity between the two points being tested (not the actual value of any resistance). If there is no audible or visual indication, the wire is open.

To identify individual wires in a conduit run, leave the wire connected to the conduit at one end of the run, as outlined above. Touch the tester probe to the wires in the conduit one at a time, with the other test lead connected to the conduit, until the tester audible alarm sounds or the indicator lights. Then put matching identification tags on both ends of the wire. Continue this procedure until all the wires have been identified. This procedure can be very helpful to identify individual wires that are all the same size and color in a conduit run. Of course, the conduit must be metallic (conductive) for this test to work.

3.3.1 Megohmmeters

Megohmmeters are special ohmmeters that are designed to read very-high resistance values. The name comes from the fact that the reading is generally in megohms (MΩ), or millions of ohms. A common ohmmeter is not capable of reading such high resistance values.

In the HVACR field, megohmmeters are most often used to test the condition of motor windings, including the motor in sealed compressors. Megohmmeters develop high DC voltages and apply that voltage to a motor winding. A common ohmmeter connected between a motor winding lead and the motor shell will typically show a completely open circuit. A megohmmeter, however, can report the many millions of ohms of resistance there should be between the windings

Around the World

Test Instruments—Old and New

Early electricians used individual meters to test circuit parameters. Today, those instruments are primitive compared to cutting-edge combination multimeters and clamp-on ammeters with digital readouts and memory functions.

and ground. This reading helps to indicate the potential for a grounded wiring to occur, providing a means of predicting the longevity of the motor. Motor windings that are well insulated will show an extreme amount of resistance to ground to the point where current cannot follow the path.

When oil and refrigerant in a circuit become contaminated, the compressor motor's resistance to ground can decrease. This is due to the breakdown of the winding insulation through chemical attack. Lower than normal readings, therefore, often indicate that the motor winding insulation has broken down to some degree, and/or that the oil has contaminants.

To test a compressor or motor winding, it is best to operate the compressor until it reaches its normal operating temperature. As soon as the compressor is shut down, power is disconnected from the system and the power leads to the compressor are physically disconnected. With the compressor electrically isolated from everything else, one probe of the megohmmeter is connected to a winding connection, while the other is connected to the motor or compressor body. Both connections must be very secure and provide good electrical contact; some paint may need to be removed from the point of connection on the compressor body to ensure a good electrical path. The megohmmeter is then energized, applying a high DC voltage to the circuit.

Handheld, battery-operated megohmmeters (*Figure 29*) have become more popular in recent years, with many capable of generating as much as 1,000VDC (1kVDC) from the battery alone. Other megohmmeters can generate up to 15kVDC. Some of these megohmmeters have a hand crank to manually develop the voltage. For general HVACR compressor and motor testing, the handheld, battery-operated models provide sufficient test voltages. Their outward appearance is very much the same as any other multimeter.

Megohmmeters are valuable tools that can aid both troubleshooting and preventive maintenance programs. For valuable and critical compressors, readings can be taken when the compressor is new and then compared to future readings on a timed basis. This process will reveal any trend toward an electrical failure of a compressor. Generally, any reading over 100MΩ indicates sound winding insulation. Values between 50MΩ and 100MΩ are cause for concern. As values fall below 50MΩ, significant steps may be required to clean up the oil and refrigerant to remove contaminants. A reading of 20MΩ and below indicates serious contamination or damaged motor windings and the possibility of a failure in the near future. For motors not used in compressors, there is no refrigerant or oil involved. The motor will either fail eventually, or is replaced proactively before it does. A motor related to a critical application or life safety is usually tested regularly and proactively replaced when the megohmmeter readings fall.

> **WARNING!**
>
> In order to safely use a megohmmeter, it is essential to carefully follow manufacturer's instructions for both the megohmmeter and the compressor motor.

> **WARNING!**
>
> When testing a live circuit, use an insulated alligator clip on the common meter lead. That way, only one hand (the one holding the other probe) is in the unit at the same time when power is applied. This helps prevent an electric shock that can travel from hand-to-hand, passing through the heart.

Figure 29 Handheld, battery-operated megohmmeter.

Principles of Electric Circuits, Thomas L. Floyd. 9th Edition. New York, NY: Pearson.

Electronics Fundamentals: Circuits, Devices, and Applications, Thomas L. Floyd. New York, NY: Pearson.

NCCER Module 26112, *Electrical Test Equipment*.

3.0.0 Section Review

1. When testing a parallel circuit with a voltmeter, _____.

   a. the circuit must be broken to avoid an erroneous reading
   b. the power switch for the test unit must be turned off
   c. a clamp-on voltmeter must always be used
   d. the meter is placed in parallel with the component or circuit

2. Which type of test meter must be connected in series with the circuit under test?

   a. Voltmeter
   b. Ohmmeter
   c. Ammeter
   d. Electromechanical

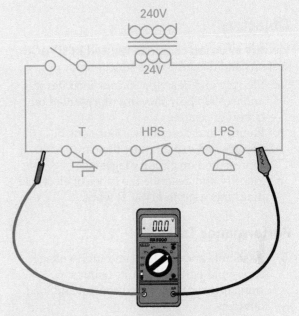

Figure SR02

3. The ohmmeter in *Figure SR02* should read _____.

   a. the resistance of the 24V transformer winding
   b. zero resistance (closed circuit)
   c. infinite resistance (open circuit)
   d. the resistance of the three switches

SECTION FOUR

4.0.0 ELECTRICAL COMPONENTS IN HVACR SYSTEMS

Objective

Identify electrical components used in HVACR systems and describe their functions.

   a. Identify and describe various load devices and explain how they are represented on circuit diagrams.

   b. Identify and describe various control devices and explain how they are represented on circuit diagrams.

   c. Identify and describe the types of electrical diagrams used in HVACR work.

Performance Tasks

  5. Assemble and test low- and high-voltage series and parallel circuits using a transformer and selected control and load devices.

Trade Terms

Contactors: Control devices consisting of a coil and one or more sets of contacts used as switching devices in high-voltage circuits.

Ground fault: An unintentional, electrically conducting connection between an energized conductor of an electrical circuit and another conductor, such as metal objects, earth, or an equipment frame. The fault current passes through the grounding system, as well as a person or other conductive surface in the path.

Inrush current: A significant rise in electrical current associated with energizing inductive loads such as motors.

Ladder diagram: A simplified schematic diagram in which the load lines are arranged like the rungs of a ladder between vertical lines representing the voltage source.

Line duty: A protective device connected in series with the supply voltage.

Motor starters: Magnetic switching devices used to control heavy-duty motors.

Pilot duty: A protective device that opens the motor control circuit, which then shuts off the motor.

Short circuit: The bypassing of a load by a conductor, causing a very high current flow.

Time-delay fuses: Fuses with a built-in time delay to accommodate the inrush current of inductive loads.

Solenoid coil: An electromagnetic coil used to control a mechanical device such as a valve or relay contacts.

Circuit diagrams, often known as *wiring diagrams* or *wiring schematics*, use symbols to represent the electrical components and connections. In this section, you will learn about electrical components and how they are shown on wiring schematics. Electrical components generally fall into two categories: load devices and control devices. Some devices, such as relays, contain both an energy-consuming element (load) and a control element (switch contacts). Relays are treated as control devices, since that is the function they perform. The symbols commonly used on wiring schematics are shown in the *Appendix*.

4.1.0 Loads

Any device that consumes electrical energy and does work in the process is a load. Loads convert electrical energy into some other form of energy, such as light, heat, or mechanical energy. Loads have resistance and consume power. An electric motor is an example of a load, as well as the burner on an electric stove. As electrical current flows through a burner element, it is converted into heat. Even the coil of a relay, although it consumes little power, is a load.

4.1.1 Motors

Electric motors convert electrical energy into mechanical energy. Motors are the most common loads found in HVACR systems and are used primarily to drive compressors and fans. Cooling systems use motors to turn the condenser and evaporator fans and to run the compressor. Furnaces use them to turn their induced-draft fans and blower fans.

Figure 30 shows examples of three common motors. *Figure 31* shows how the same motors are depicted on schematic wiring diagrams. Because it is likely that several motors will appear on a typical HVACR circuit diagram, letter codes such as IFM (indoor fan motor) and OFM (outdoor fan motor) are used to identify the role of a motor. Other components that appear on circuit diagrams are also identified by letter codes. The

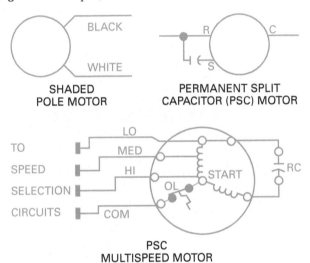

MULTISPEED PSC MOTOR

SHADED POLE MOTOR

PERMANENT SPLIT CAPACITOR (PSC) MOTOR

Figure 30 Examples of electric motors.

BLACK

WHITE

SHADED POLE MOTOR

R   C

S

PERMANENT SPLIT CAPACITOR (PSC) MOTOR

TO
SPEED
SELECTION
CIRCUITS

LO
MED
HI
COM

OL

START

RC

PSC MULTISPEED MOTOR

Figure 31 Motor wiring-diagram symbol examples.

Electric Heat in Heat Pump Systems

Heat pump systems often contain electric resistance heaters. Local codes may even require that a heat pump have enough supplemental heat capacity to support up to 100 percent of the heating demand, in the event the heat pump fails. A heat pump system with 15,000W (15kW) of supplementary electric heaters would draw about 65A in a 230V electrical system. However, most heat pump systems operate the electric heat in stages, rather than operate all 15kW at once. However, there is always the potential that all of it will be needed, and the electrical service must be able to accommodate the full load, along with the rest of the structure's electrical needs.

letter codes are generally defined in a legend found on the wiring diagram.

In most residential and commercial systems, the compressor motor is the largest single electrical load. In a residential system, the indoor blower motor and outdoor condenser fan motor may draw only a few amps of current, while the compressor may consume 5 to 10 times as much.

4.1.2 Electric Heaters

Electric heaters are also called *resistance heaters* because they are made of high-resistance material that consumes a large amount of power, which is converted into heat. The burners of an electric stove are also electric resistance heaters. The common diagram symbol for an electric heater is the same symbol used for a resistor. An electric resistance heater assembly and diagram are shown in *Figure 32*.

On occasion, when replacing a fossil-fuel furnace with a heat pump, it is necessary to increase the capacity of the electrical service. This can be a significant expense for the building owner. The salesperson who specifies the system must take such factors into account when pricing the job.

A heat pump system can also use a fossil-fuel furnace as its source of supplemental/emergency heat. This helps prevent the additional cost of electrical upgrades. When a furnace is used to supplement the heat pump, only one of the two units can operate in the heating mode at any one time. However, supplemental electric heat can operate as needed right along with the heat pump.

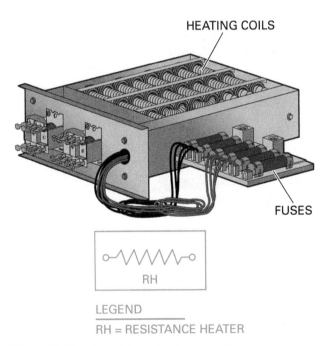

HEATING COILS

FUSES

RH

LEGEND
RH = RESISTANCE HEATER

Figure 32 Electric resistance heater assembly.

 03106 **Basic Electricity**

Module Three 27

4.2.0 Control Devices

There are many types of control devices. Their primary function is to control the operation of loads. Many control devices also control the action of other controls. Anyone working on HVACR systems will encounter these devices on a daily basis. Although some control devices operate at the same voltage as the loads, most residential and commercial control systems operate at the low voltage of 24VAC. Some of today's systems use low-voltage DC to both control and transfer information based on the voltage returned to a solid-state control.

4.2.1 Switches

Switches stop and start the flow of power to other control devices or loads. Electrical switches are classified according to the force used to operate them. Examples include manual, magnetic, temperature, light, moisture, and pressure switches.

The simplest type of switch is one that makes (closes) or breaks (opens) a single electrical circuit. More complicated switches control several circuits. The switching action is described by the number of poles (electrical circuits to the switch) and the number of throws (circuits fed by the switch). *Figure 33* shows some of the common switch arrangements, pictured in the following order:

1. Single-pole, single-throw (SPST)
2. Double-pole, single-throw (DPST)
3. Double-pole, double-throw (DPDT)

One good example of a manual switch is a power disconnect switch. It is usually mounted near the HVACR unit in an accessible location.

It allows the technician to de-energize the system for service. As shown in the photo and the schematic diagram in *Figure 34*, some disconnects are fused so that power is removed at the unit in case of an electrical overload. The *National Electrical Code®* (*NEC®*) and many local building codes require that a power disconnect device be located within sight of an air conditioning unit. Some, like the one shown in *Figure 34*, also have a handle on the outside of the enclosure to turn the power on and off. Others have a plug assembly that is pulled out to turn the power off.

A thermostatic switch opens and closes in response to changes in temperature. While a cooling thermostat closes its contacts on a temperature rise, a heating thermostat closes its contacts on a temperature fall. Thermostats are primary control devices in heating, cooling, and refrigeration systems. The thermostat shown in *Figure 34* is a heating/cooling model, which means that separate switches open and close simultaneously to control the equipment.

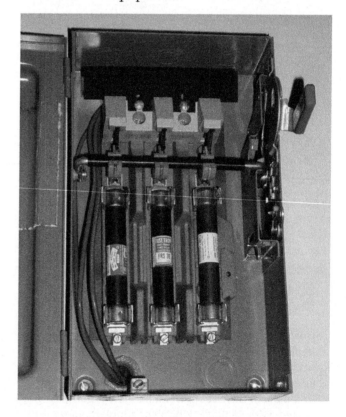

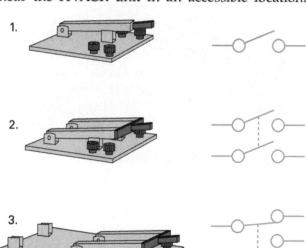

Figure 33 Switch contact arrangements.

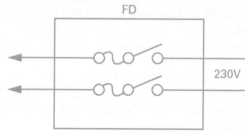

Figure 34 Fused power disconnect.

The thermostat shown in *Figure 35* is an electronic model. This thermostat senses temperature using special temperature-sensitive resistors called *thermistors*. The electronic circuitry translates the resistance to a temperature, and then determines if heating or cooling is needed. Thermistors are covered in more detail later in this module.

Other temperature-controlled switches may provide protection from heat for motors, or act as safety switches in heating equipment.

Pressure switches are used in a variety of ways. One common use is to shut off a system compressor if the refrigerant pressure becomes too high or low. Pressure switches can also control fan motors or pumps, as well as many other devices. *Figure 36* shows some examples of pressure switches. There are many possible forms and uses.

Light-sensitive switches are sometimes used to sense flame. If the flame in an oil furnace goes out, for example, the light-sensing switch shuts off the flow of oil to the combustion chamber. Light-sensitive switches are also used in commercial or retail stores to sense that the building is occupied and start the HVAC system. When the first employee of the day turns on the lights, the switch enables the controls for normal operation. Once the lights are turned off for a short time, the system returns to an unoccupied status and maintains an energy-saving set of temperatures. The light-sensing portion of these switches generates a small amount of electricity from the light, indicating the presence or absence of light.

Moisture-sensitive switches are used in humidistats. A strand of hair or special man-made material that expands and contracts with changes in moisture, can be used to activate the equipment. Solid-state humidity-sensing devices have replaced most other materials in humidistats today.

Case History

Faulty Disconnect Switch

A service technician was in a crawl space performing pre-winter maintenance on an oil furnace. He had de-energized the furnace using the disconnect switch, but did not verify the absence of power using a meter. Unfortunately, the disconnect was wired improperly, allowing current to flow to the furnace even when the switch was in the Off position. As a result, the technician was electrocuted.

The Bottom Line: Don't rely solely on disconnect switches. Always verify the absence of power by using a reliable voltmeter that is constantly tested on known voltage sources.

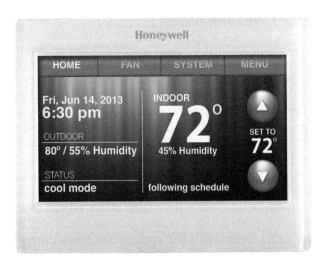

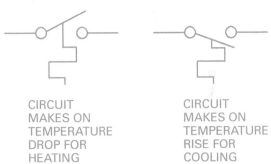

CIRCUIT MAKES ON TEMPERATURE DROP FOR HEATING

CIRCUIT MAKES ON TEMPERATURE RISE FOR COOLING

Figure 35 Electronic wall thermostat.

HPS   OPENS ON PRESSURE INCREASE

LPS   OPENS ON PRESSURE DROP

LEGEND

HPS = HIGH-PRESSURE SWITCH
LPS = LOW-PRESSURE SWITCH

Figure 36 Pressure switches.

4.2.2 Fuses and Circuit Breakers

Fuses and circuit breakers protect components and wiring against damage from current surges or short circuits. A short circuit occurs when current flow bypasses a load. An example would be when conductors touch due to damaged insulation. Because there is no load, current flow is uninhibited and can quickly create enough heat to start a fire.

As shown in *Figure 37*, a fuse contains a metal link that melts when the current exceeds the rated capacity of the fuse (e.g., 20A), which is its design capacity to withstand current. Because the link is in series with the circuit, the circuit opens when the strip melts. The fuse will blow more quickly in the presence of a short circuit than it will in the presence of a current overload. Once a fuse has blown, it must be replaced.

Motors draw far more current during startup than they do when they are running normally.

This is due to the energy required to make the rotor and any connected machinery turn from a dead stop. It is a characteristic associated with most inductive loads. Even relay coils exhibit this elevated current draw when they are first energized. Once a motor has reached its normal operating speed, its current drops sharply. The initial current surge, known as inrush current, can easily be six times that of its normal running current. It is important to note that inrush current occurs and recedes very quickly, often in the blink of an eye.

To accommodate the inrush current when a motor starts and protect the motor, different fuses are used. Otherwise, the fuse would blow each time a motor tries to start, since the inrush current usually exceeds the fuse's current limit. Special delayed-opening fuses, called time-delay fuses or *slow-blow fuses*, are used for HVACR equipment and similar inductive loads.

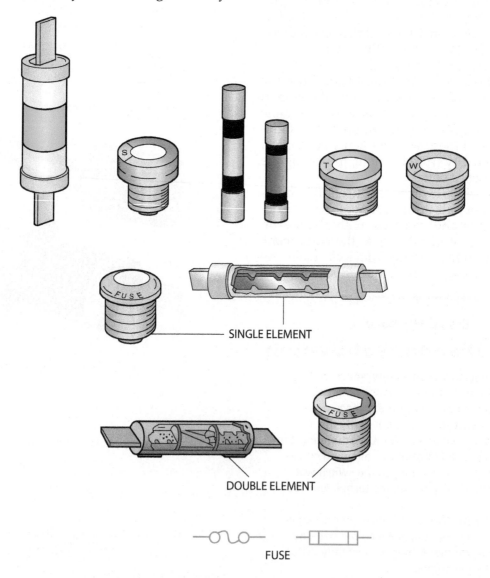

SINGLE ELEMENT

DOUBLE ELEMENT

FUSE

Figure 37 Fuses.

These fuses will not blow unless the current surge is sustained. Note that resistive loads, such as electric heaters, do not experience an inrush of current when they are first energized and do not require this type of fuse.

Circuit breakers (*Figure 38*) serve the same basic purpose as fuses. The big advantage of circuit breakers is that they operate like switches and can be reset when they trip. Fuses must be replaced once they open. If a circuit breaker trips more than once, however, it is a sign of a recurring problem. Special HACR (heating, air conditioning, and refrigeration) circuit breakers are used for air conditioning equipment for the same reason that time-delay fuses are used. They provide the necessary time delay for inductive loads to start and current flow to stabilize.

A thermal-trip circuit breaker contains a spring-loaded metal element that opens when high current flow develops excessive heat, causing it to trip. Magnetic-trip breakers contain a small coil of wire that develops a strong enough magnetic field when the current is excessive to open the contacts magnetically.

A problem known as a ground fault occurs when a live conductor touches another conducting substance, such as the metal frame of a unit. GFCI devices are installed to detect and quickly isolate such problems for safety's sake.

4.2.3 Solenoid Coils

A solenoid coil (*Figure 39*) works like an electromagnet. When current flows through the coil, magnetism is produced and used to attract or repel a magnetic shaft that passes through its core. Solenoid coils are used to open and close valves and position switching devices.

4.2.4 Relays, Contactors, and Motor Starters

Relays, contactors, and motor starters are magnetically controlled switches consisting of a solenoid-style coil and one or more sets of movable contacts that open and close in response to current flow in the coil. In other words, they are electrically operated switches.

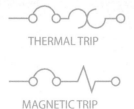

THERMAL TRIP

MAGNETIC TRIP

Figure 38 Circuit breakers.

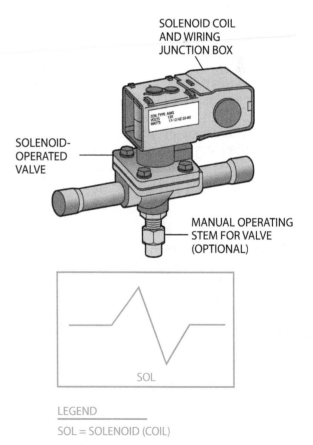

SOLENOID COIL
AND WIRING
JUNCTION BOX

SOLENOID-
OPERATED
VALVE

MANUAL OPERATING
STEM FOR VALVE
(OPTIONAL)

SOL

LEGEND

SOL = SOLENOID (COIL)

Figure 39 Solenoid coil and valve.

Figure 40 shows different types of relays in a schematic form. When current flows through the relay coil, a magnetic field is created, just like a solenoid coil mounted on a valve. The magnetic field acting on a magnetic armature draws the contacts together. Some contacts close when the relay is energized, providing a path for current. They are referred to as *normally open* (NO) contacts because they default to being open when the relay is de-energized. Other contacts open when the relay is energized. These are called *normally closed* (NC) contacts, since they default to being closed when the relay is de-energized. NC contacts are often shown on a drawing with a slash through them.

The coil and contacts of a relay are most often shown at different locations on a wiring diagram. This is because the relay coil is usually a part of the low-voltage control circuit, powered by the transformer, while the contacts are part of a line-voltage load circuit. On the drawing, the coil and contacts have the same name or identifier. The identifier *CR* is used in *Figure 40*, which stands for *control relay*. When there are a number of control relays, they are numbered CR1, CR2, and so on. The acronym DFR is often used to identify a defrost relay in a heat pump, while FR usually identifies a fan relay. The same pole and throw descriptions used for switches also apply to relays and contactors.

A time-delay relay uses a thermal element or digital control to delay its response to an event. Delays can range from a few seconds to hours. A typical use in HVACR is to keep a blower running for 1–3 minutes after the heating burners or air conditioning unit has shut down. The blower extracts residual heat or cooling effect, slightly improving efficiency. Similarly, a time-delay relay

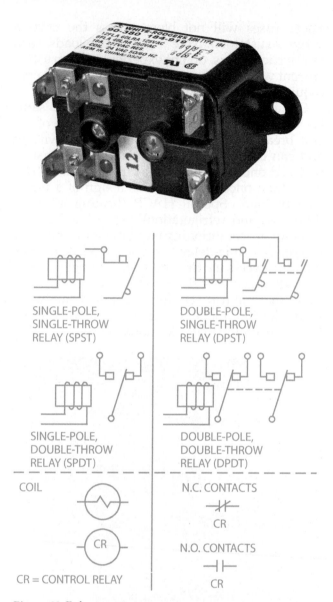

Figure 40 Relays.

Plug-In Relays

Relays are the most common switching devices found in air conditioning and heating systems. This photo shows several types of relays designed to plug into matching bases. The wiring is connected to the base assembly, and the relay simply plugs into it. No wiring connections or tools are needed to replace the relay. Since many relays of this type have a clear housing, allowing the technician to see the contacts inside, they are sometimes referred to as *ice cube relays*.

might delay the start of a blower until the furnace heat exchangers have warmed up. This keeps cool air from being blown into the conditioned space.

A contactor is basically a heavy-duty relay designed to carry larger currents through its contacts. They are commonly used to control the flow of power to compressors and similar electrical loads.

Motor starters are used to start larger motors. Some motors require external protection, depending on their capacity in horsepower. The *NEC®* determines what motors and motor applications require a motor starter. A motor starter combines a contactor with overload protection devices. When excessive current is encountered, a built-in overload relay opens the circuit to the contactor coil and stops the motor. This characteristic alone makes a motor starter different from a standard contactor.

Figure 41 shows an example of how a relay and contactor might be used in a control circuit. The coils of both the contactor and the relay are shown on the 24V side of the diagram, while the contacts of both are shown on the line-voltage side of the diagram. The coil of the relay, labeled as IFR (indoor fan relay) can be energized two ways. When the FAN switch is in the AUTO position, the relay is energized when the thermostat is switched to ON and the cooling switch closes on a rise in temperature. Changing the FAN switch from AUTO to ON will energize the relay coil immediately. The contactor coil is also energized when the thermostat is switched to ON and the thermostat closes. The NO contacts of the relay and contactor close when they are energized. Once the IFR contacts close, the indoor fan starts. When the contactor contacts close, both the compressor and the outdoor fan motor (OFM) start. Remember that diagrams like this one show the position of the switches and contacts when de-energized unless otherwise stated.

4.2.5 Transformers

A transformer (*Figure 42*) is used to raise (step up) or lower (step down) a voltage. Step-down transformers are very common in HVACR work. Large step-up transformers are typically used in the power industry, boosting the voltage of the power-generating facility for transmission to distant consumers. Small step-up transformers are used in the HVACR field to generate an igniting spark for oil and gas burners. Beyond this however, step-up transformers are rarely needed in HVACR applications. The vast majority of transformers encountered are step-down transformers. A common control transformer for a residential system steps the 120VAC or 240VAC line voltage down to 24VAC. Low-voltage circuits are less hazardous to work with and are easily insulated. Low-voltage components are also less expensive to build.

A transformer usually consists of two or more coils of wire wound around a common iron core. When current passes through one coil, called the *primary winding*, the magnetic field developed cuts through the other coil, called the *secondary winding*. The first magnetic field creates, or induces, current in the secondary winding. This process is known as *induction*. Depending on the number of turns of wire in the secondary winding, the voltage induced in the secondary is stepped down or stepped up from that in the primary.

The number of turns in the primary and secondary windings determine exactly what happens to the voltage. Refer to *Figure 43*. Note the number of turns drawn on each side of the two vertical bars, which represent the iron core. There are 15 turns on the line-voltage side, and 3 on the low-voltage side. This is a ratio of 15:3, or 5:1. Therefore, the secondary voltage will be one-fifth of the voltage applied to the primary winding. When 120V is applied, a 24V output is generated in the secondary winding.

Control transformers are sized by a rating known as the *volt-amp rating (VA rating)*. The VA rating is the product of the output voltage of the transformer multiplied by the current it is capable

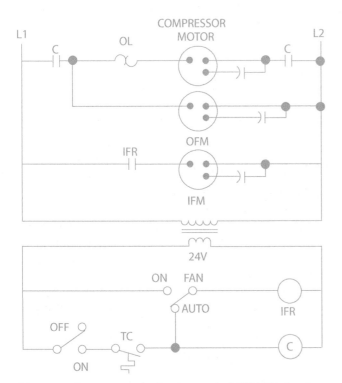

Figure 41 Contactor and relay in a typical HVACR circuit.

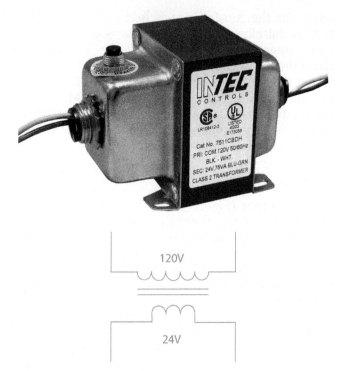

Figure 42 Control transformer.

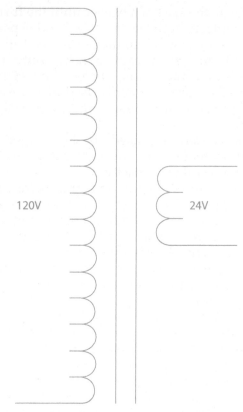

Figure 43 Transformer winding ratio.

of sustaining. For example, the 75VA transformer in *Figure 42* can sustain a current load of roughly 3.1A (75VA ÷ 24V = 3.125A). To find the current any transformer can sustain, divide the VA rating by the design input voltage.

To protect the transformer, many control circuits contain a fuse or circuit breaker in the secondary circuit wiring. The transformer in *Figure 42* has an integral circuit breaker; the black button is used to reset the tiny breaker. If the transformer is fused by the manufacturer, the fuse is usually buried beneath the outer insulation and can only be accessed by destroying the material. These fuses are generally soldered into the winding and are challenging to replace. The fuse is primarily added for fire protection and not for convenient field repair, although it can be done.

4.2.6 Overload Protection Devices

Overload devices stop the flow of current when safe current or temperature limits are exceeded. *Figure 44* shows examples of external overload devices that are often used on small compressors. Devices that are actuated (activated) by excessive temperatures are referred to as *thermal overloads*. Current overloads also develop heat when the current is excessive, but the heat is developed within the device itself. A thermal overload device reacts to excessive temperatures adjacent to it. Overload devices are often embedded in the windings of larger motors, including compressor motors. Overload devices are also component parts of motor starters.

Protective devices are placed into two categories. Line duty devices are directly in line with the voltage source powering the load. In pilot duty devices, a set of contacts in the control circuit open when an overload condition occurs.

Some overload devices reset automatically when the cause of the overload has been removed. Others must be manually reset, usually by pressing a button on the device. If the overload device is embedded in the motor windings, it must reset automatically.

4.2.7 Thermistors

A thermistor is a special type of resistor in which the resistance changes in response to the surrounding temperature. A thermistor serves as an input to an electronic control device such as a thermostat. When the resistance of the thermistor changes beyond a setpoint, the control circuit reacts by opening or closing a set of contacts. *Figure 45* shows a thermistor used to measure temperature on the inlet side of an air damper. The thermistor is an input to a controller that opens and closes the damper depending on the temperature.

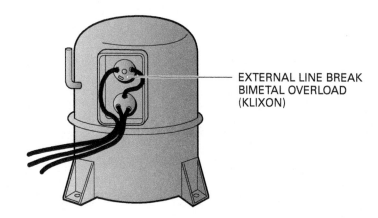

EXTERNAL LINE BREAK
BIMETAL OVERLOAD
(KLIXON)

CURRENT AND TEMPERATURE (THERMAL OVERLOAD DEVICE)

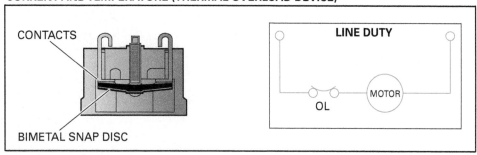

CONTACTS

LINE DUTY

BIMETAL SNAP DISC

OL

MOTOR

CURRENT

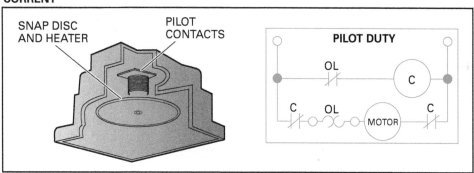

SNAP DISC
AND HEATER

PILOT
CONTACTS

PILOT DUTY

OL

C

C   OL

MOTOR

C

Figure 44 Overload devices.

THERMISTOR

Figure 45 Thermistor application.

4.2.8 Electronic Controls

Modern comfort-system thermostats are a perfect example of electronic controls that now dominate many aspects of HVACR system control. The thermostat shown in *Figure 46* is fully programmable. In general, this means that operating schedules for time and temperature can be made for each day of the week. In addition, an installer menu allows it to be configured for a variety of applications and operating preferences. It is also Wi-Fi enabled, allowing the user to access basic thermostat settings and functions from any Internet-connected device such as a smartphone, tablet, or computer.

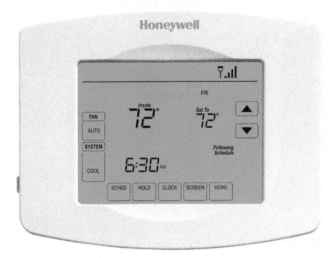

Figure 46 Programmable, Wi-Fi- enabled electronic thermostat.

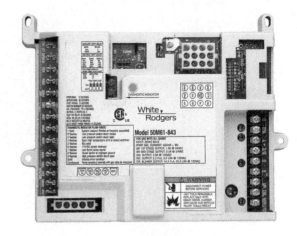

Figure 47 Examples of electronic equipment controls.

In many of today's HVACR products, electromechanical control devices like switches and relays have been replaced by electronic controls. Electronic circuits use solid-state timing, switching, and sensing devices to control loads and protective circuits. Electronic circuits operate on DC voltage rectified from an AC power input to the board. Electronic circuits are reliable because they have no moving parts. Most electronic circuits consist of micro-miniature components mounted on sealed modules or printed circuit (PC) boards as shown in *Figure 47*. The copper ribbon circuits found on the board serve the same purpose as wiring, providing the conductive path between components. However, note that traditional electromechanical relays are often found on the boards. The green PC board in *Figure 47* has several black relays mounted on the top of the board. These relays have DC coils, but are otherwise like any other AC-operated relay used to complete a control circuit or power a light electrical load. The white PC board shown is partially sealed for protection. Many such controls can be field-configured to provide specific functions using microswitches or tiny jumpers across terminals.

Programmable controllers (*Figure 48*) provide a wider range of programmability than the average PC board or electronic control. They are equipped with microprocessors like those of a standard computer. A programmable controller can receive a large amount of data and then make decisions based on its programming. They offer versatility and precision in handling multiple control actions and processes.

Microprocessor-controlled systems may be self-diagnosing. When there is a problem, the microprocessor evaluates information from sensors distributed around the system to determine the source. It may also test itself in various ways. A digital readout, flashing light code, or other means of communication is then used to tell the technician what function is inoperative and/or what component is not responding as expected.

PC boards are usually treated as black boxes. In other words, if the technician finds that a function of the board or controller has failed, the entire device is replaced rather than field-repaired. Unlike conventional circuits, it is not necessary to analyze the circuit to figure out which component has failed.

Some electronic controls and boards can be turned in for partial credit against a new part, and the manufacturer will refurbish the failed device. Electronic controls are often condemned without proper testing, sometimes because their operation is misunderstood. It is essential to have the right information on hand when testing or troubleshooting any electronic device.

Figure 48 Microprocessor-based programmable controller.

4.3.0 Electrical Diagrams

HVACR technicians encounter a variety of different types of electrical diagrams. The diagrams provided in this section will allow you to make a connection between various components and their appearances on electrical diagrams.

An electrical diagram is like a road map. If it can be read and interpreted, a technician can determine how the components are supposed to act when the system is running properly. It also enables a technician to quickly identify in what are to begin troubleshooting. Diagrams supplied by manufacturers come in a variety of formats. *Figure 49* provides two examples, both of which might be provided inside a piece of equipment. It shows a connection diagram and a ladder diagram.

A connection diagram shows how the wiring is physically connected. It also may show the color of each wire, which is not generally shown on ladder diagrams. The wiring diagram is helpful for troubleshooting or tracing wiring.

In a ladder diagram, the wire colors and physical connection information are removed and the pictorial views of some components are replaced with standard electrical diagram symbols. This diagram makes it easier to envision the flow of power. The ladder diagram is arranged to make it easy to trace circuits. The power source is presented as the upright legs of the ladder, and each load line and its related control devices is represented as a rung of the ladder. Electrical diagrams of all types generally have legends that identify each of the components by their abbreviations.

The diagrams provided by the manufacturer may include a variety of features. *Figure 50* is an example of wiring information from a manufacturer. They often include ladder diagrams, connection diagrams, and/or component-location diagrams. The legend and notes are provided at the bottom. Similar diagrams are often pasted on the inside of a unit access panel.

Hopscotch Troubleshooting

Each switch or set of relay contacts in a circuit represents a condition that must be met in order to energize the compressor contactor (CC). This must occur before the compressor can operate. If the thermostat is calling for cooling and the compressor is not running, there is a good chance that one of the conditions is not being met. When you are troubleshooting the circuit with a voltmeter, start by placing your meter probes across the entire circuit, as shown in the illustration here. This verifies that voltage is applied to the circuit. Then, move the hot (red) probe to the next component in the series chain (position 2) and read the voltage. Keep doing this until the meter registers no voltage. The last component you jumped (or its related wiring) will be the defective component. This technique is sometimes referred to as hopscotch troubleshooting.

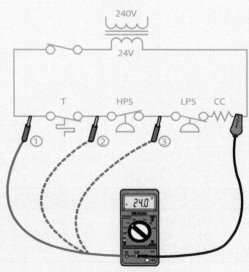

HOPSCOTCH TROUBLESHOOTING

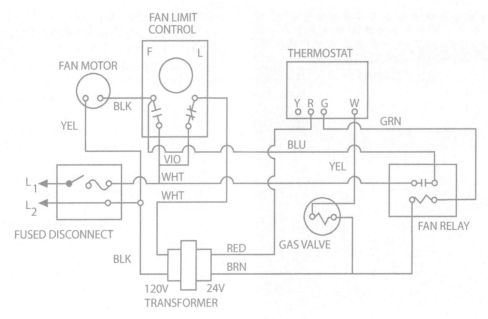

CONNECTION DIAGRAM

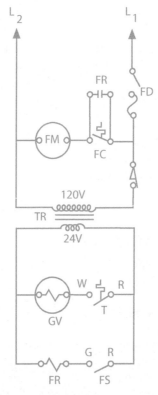

LADDER DIAGRAM

LEGEND							
FC	FAN LIMIT CONTROL		FR	FAN RELAY		GV	GAS VALVE
FD	FUSED DISCONNECT		FS	THERMOSTAT FAN SWITCH		TR	TRANSFORMER
FM	FAN MOTOR						

Figure 49 Examples of electrical diagrams.

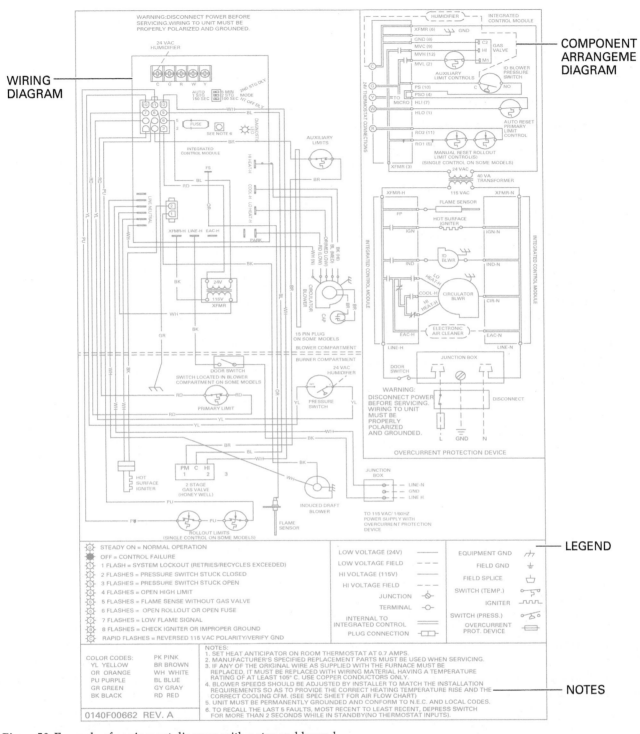

Figure 50 Example of equipment diagram with notes and legend.

Additional Resources

Principles of Electric Circuits, Thomas L. Floyd. 9th Edition. New York, NY: Pearson.

Electronics Fundamentals: Circuits, Devices, and Applications, Thomas L. Floyd. New York, NY: Pearson.

NCCER Module 26210, *Circuit Breakers and Fuses*.

NCCER Module 26211, *Control Systems and Fundamental Concepts*.

NCCER Module 12406, *Programmable Logic Controllers*.

4.0.0 Section Review

1. Which of the following devices is a load that converts electrical energy into mechanical energy?

   a. Transformer
   b. Electric heater
   c. Motor
   d. Light bulb

2. Which of the following is a switching device commonly used to control the flow of power to a compressor?

   a. Contactor
   b. Relay
   c. Transformer
   d. Solenoid coil

3. The type of circuit diagram that has the power source shown as upright legs with connections between them is a _____.

   a. component location diagram
   b. wiring diagram
   c. ladder diagram
   d. simplified schematic

SUMMARY

Electricity is produced by power generating stations and carried over high-voltage transmission lines to local substations, where it is stepped down to usable voltage levels. It is then transferred to customers on local distribution lines. HVACR technicians must frequently test and troubleshoot electrical circuits. To do this safely and effectively, they must understand the relationships among the primary electrical values of voltage, current, resistance, and power. The mathematical principles that makes this possible are demonstrated through Ohm's law and the power equation.

All HVACR systems contain electrical circuits. To install and service HVACR systems, the technician must know how the electrical components work, how to read circuit diagrams, and how to use electrical test equipment. When working with electricity, following proper safety practices is extremely important. Failure to develop good safety habits and follow safety rules can result in injury or death to yourself or your co-workers.

The electrical components used in HVACR systems consist of load devices and control devices. Load devices include motors and electric resistance heaters. There are many types of controls, including switches, relays, and thermostats, just to name a few. Some switching devices are circuit-protection components, like pressure safety switches and overload relays designed to protect the compressor from damage. An HVACR technician must be able to recognize each of these devices and identify them on electrical diagrams.

1. Central air conditioning systems in homes generally require a power source of _____.

   a. 1,500 volts
   b. 750 volts
   c. 240 volts
   d. 120 volts

2. What characteristic do devices that use DC instead of AC power often have in common?

   a. They often operate at a much higher voltage than AC devices.
   b. They usually require more current when they operate.
   c. They are battery-powered and mobile.
   d. They are usually used to produce supplemental heat.

3. The worst effect a current of 150mA is likely to have on the human body is _____.

   a. a slight tingling sensation
   b. a slight shock and involuntary action
   c. painful shock and loss of muscular control
   d. extreme pain and possible death

4. One milliamp is equal to _____.

   a. $\frac{1}{10}$ of an amp
   b. $\frac{1}{100}$ of an amp
   c. $\frac{1}{1,000}$ of an amp
   d. $\frac{1}{1,000,000}$ of an amp

5. Current is measured in _____.

   a. watts
   b. volts
   c. amperes
   d. ohms

6. How much current is drawn by a 120V circuit containing a 1,500W load?

   a. 125mA
   b. 0.08A
   c. 12.5A
   d. 180,000A

7. How much power is consumed by a 120V parallel circuit that contains a 100Ω load and an 80Ω load and draws 2.7A?

   a. 44W
   b. 80W
   c. 120W
   d. 324W

8. If a load is supplied by 120V and draws 5A of current, how much power does it consume?

   a. 24W
   b. 115W
   c. 125W
   d. 600W

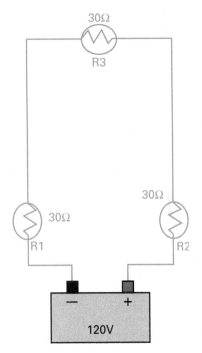

Figure RQ01

9. The circuit shown in *Figure RQ01* is a(n) _____.

   a. series circuit
   b. parallel circuit
   c. open circuit
   d. short circuit

10. If a circuit containing five 50Ω resistors has a total resistance of 10Ω, the circuit is a(n) _____.

   a. series circuit
   b. parallel circuit
   c. series-parallel circuit
   d. combination circuit

11. Which of the following statements about voltage testing is true?

   a. When accuracy is required, use a voltage tester.
   b. Voltage meters are essentially the same as multimeters.
   c. A multimeter is better than a voltage tester for troubleshooting.
   d. Voltage testers generally offer recording functions.

12. Placing the jaws of a clamp-on ammeter around two separate conductors will result in _____.

   a. an inaccurate reading
   b. a more accurate reading
   c. doubling the reading
   d. damaging the meter

13. A continuity tester operates like a(n) _____.

   a. ammeter
   b. voltmeter
   c. ohmmeter
   d. speedometer

14. Which of these components is considered a load?

   a. A switch
   b. A thermistor
   c. A thermostat
   d. An electric heater

15. A DPDT switch can control _____.

   a. 1 circuit
   b. 2 circuits
   c. 3 circuits
   d. 4 circuits

16. Fuses are rated based on _____.

   a. their physical size
   b. how much current they can withstand
   c. their full load amps (FLA) rating
   d. their locked rotor amps (LRA) rating

Figure RQ02

17. The symbol shown in *Figure RQ02* represents a _____.

   a. transformer
   b. solenoid coil
   c. thermal trip circuit breaker
   d. magnetic trip circuit breaker

18. A device that acts as an electromagnet to open or close a valve is a _____.

   a. pressure switch
   b. thermostat
   c. solenoid coil
   d. thermistor

19. A transformer is used to _____.

   a. transform electrical energy into mechanical energy
   b. raise or lower a voltage
   c. absorb excess current in a circuit
   d. increase the resistance of a circuit

20. The purpose of a legend on a circuit diagram is to _____.

   a. show the history of the drawing
   b. identify the abbreviated diagram components
   c. show the physical locations of components
   d. identify the wire colors used on the diagram

Trade Terms Quiz

Fill in the blank with the correct trade term that you learned from your study of this module.

1. An unbroken circuit path has _____.

2. The amount of electrical energy used by an electrical load device is called _____.

3. The type of electrical diagram in which the loads are arranged between two vertical lines is known as a(n)_____.

4. A material that readily carries electricity is a(n) _____.

5. Magnetic control devices used for switching higher-current circuits are called _____.

6. A control device that opens a control circuit for a motor or similar load, is known as a(n) _____ device.

7. Resistance in a DC circuit is measured in _____.

8. A meter that uses a needle to indicate a value is a(n) _____.

9. An electromagnetic coil used to open or otherwise move a mechanical device is called a(n) _____.

10. Electrical power consumption is measured in _____.

11. If the electricity changes direction on a cyclical basis, it is known as _____.

12. The type of electrical devices that draw significantly more current on startup or when first energized are those that are _____.

13. The type of ammeter that is placed in series with the load is the _____.

14. A device that converts AC voltage into DC voltage is a(n) _____.

15. Magnetic switching devices that usually include overload protection and are used to control motors are called _____.

16. A device with two coils of wire wrapped around a common core is a(n) _____.

17. Electromotive force (EMF) is another name for _____.

18. When the current flow bypasses a load, it is called a(n) _____.

19. The rate at which electrons flow in a circuit is known as _____.

20. When a protective device is in series with the load, it is known as a(n) _____ device.

21. The _____ is the unit of measure for current flow in an electrical circuit.

22. A(n) _____ is a test meter able to read voltage, current, and resistance.

23. The device used to sense current using jaws that are snapped around a conductor is the _____.

24. An electrical measuring device that provides a direct numerical reading is a(n) _____.

25. Electrical devices that converts electrical energy from one form to another, such as heat or motion, are referred to as _____.

26. Fuses designed to allow time for inductive loads like motors to start, accommodating the inrush current, are called _____.

27. A battery is a common source of _____.

28. The opposition to the flow of electricity is called _____.

29. When current flows through a single coil of wire wrapped around an iron core, a(n) _____ can be created.

30. A magnetically-operated switching device used in low-current circuits is the _____.

31. A(n) _____ is a material that impedes the flow of current. It is the opposite of a conductor.

32. A(n) _____ is used to measure current.

33. Voltage is measured in _____.

34. A semiconductor device that changes resistance as the result of a change in temperature is called a(n) _____.

35. When electric current jumps across an air gap, it creates a(n) _____.

36. When an inductive load like a motor is energized and begins to rotate, the _____ is much higher than the normal current once it reaches full speed.

37. When an energized conductor touches the frame of a motor, a(n) _____ results.

Trade Terms

Alternating current (AC)	Digital meter	Loads	Solenoid coil
Ammeter	Direct current (DC)	Motor starters	Thermistor
Ampere (A)	Electromagnet	Multimeter	Time-delay fuses
Analog meter	Ground fault	Ohms (Ω)	Transformer
Arc	Inductive	Pilot duty	Volts (V)
Clamp-on ammeter	In-line ammeter	Power	Voltage
Conductor	Inrush current	Rectifier	Watts (W)
Contactors	Insulator	Relay	
Continuity	Ladder diagram	Resistance	
Current	Line duty	Short circuit	

John Stronkowski

Director of Education
Industrial Management & Training Institute

How did you get started in the construction industry (i.e., took classes in school, a summer job, etc.)?

I worked in the trades to pay my way through college. At that time, it was a means to an end.

Who or what inspired you to enter the industry (i.e., a family member, school counselor, etc.)? Why?

My family members worked in various trades. My grandfather was a Master Carpenter and that served as an inspiration to me.

What do you enjoy most about your career?

I enjoy the teaching aspect the most. It allows me to share my knowledge and experience, and hopefully impact the lives of my students.

Why do you think training and education are important in construction?

Properly trained people in the trades are crucial to the safety and welfare of the general public. Mistakes and shortcuts have significant costs beyond money.

Why do you think credentials are important in construction?

Documented credentials represent a properly trained tradesperson who is up-to-date with today's technology.

How has training/construction impacted your life and career (i.e., advancement opportunities, better wages, etc.)?

In my dual role as a multiple-licensed tradesperson and a Roman Catholic priest, the construction field has allowed me to be in charge of all major construction projects for the Diocese of Bridgeport, CT. With my vast knowledge of different yet related systems, I am able to provide added benefits to the church without extraordinary costs to our parishioners.

Would you recommend construction as a career to others? Why?

Yes, definitely. The trades offer highly rewarding careers. When you start a construction job and see it through to the end, it gives you a sense of accomplishment.

What does craftsmanship mean to you?

It is professionalism at its highest level and being proud of the work accomplished.

SCHEMATIC SYMBOLS

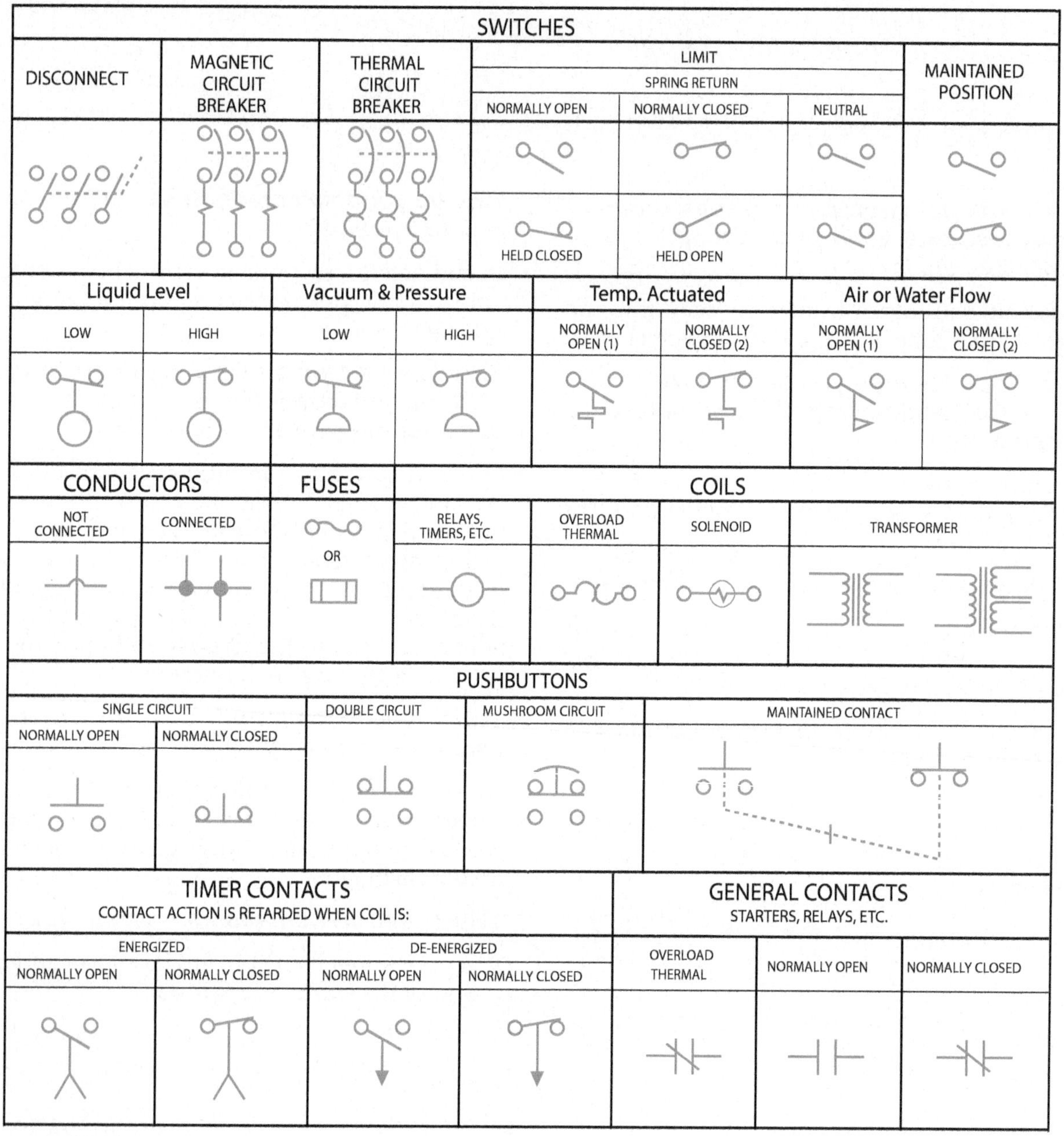

SWITCHES						
DISCONNECT	MAGNETIC CIRCUIT BREAKER	THERMAL CIRCUIT BREAKER	LIMIT			MAINTAINED POSITION
			SPRING RETURN			
			NORMALLY OPEN	NORMALLY CLOSED	NEUTRAL	
			HELD CLOSED	HELD OPEN		

Liquid Level		Vacuum & Pressure		Temp. Actuated		Air or Water Flow	
LOW	HIGH	LOW	HIGH	NORMALLY OPEN (1)	NORMALLY CLOSED (2)	NORMALLY OPEN (1)	NORMALLY CLOSED (2)

CONDUCTORS		FUSES	COILS			
NOT CONNECTED	CONNECTED	OR	RELAYS, TIMERS, ETC.	OVERLOAD THERMAL	SOLENOID	TRANSFORMER

PUSHBUTTONS				
SINGLE CIRCUIT		DOUBLE CIRCUIT	MUSHROOM CIRCUIT	MAINTAINED CONTACT
NORMALLY OPEN	NORMALLY CLOSED			

TIMER CONTACTS CONTACT ACTION IS RETARDED WHEN COIL IS:				GENERAL CONTACTS STARTERS, RELAYS, ETC.		
ENERGIZED		DE-ENERGIZED		OVERLOAD THERMAL	NORMALLY OPEN	NORMALLY CLOSED
NORMALLY OPEN	NORMALLY CLOSED	NORMALLY OPEN	NORMALLY CLOSED			

(1) Make on rise
(2) Make on fall

COILS

AUTOMATIC TRANSFORMER	REACTORS		ADJUSTABLE
	IRON CORE	AIR CORE	
			SHOWN WITH IRON CORE

DIODE RECTIFIERS		MOTORS & COMPRESSORS		
HALF-WAVE	FULL-WAVE	THREE-PHASE	SINGLE-PHASE	
			TWO LEADS	CAPACITOR START/RUN

RESISTORS

FIXED	VARIABLE
	OR

THERMOCOUPLE	LAMPS	BATTERY	GROUND		CAPACITOR	
			ELECTRICAL	MECHANICAL	FIXED	ADJUSTABLE
						X-SIDE NEAR GROUND

Trade Terms Introduced in This Module

Alternating current (AC): An electrical current that changes direction on a cyclical basis.

Ammeter: A test instrument used to measure current flow.

Ampere (A): The basic unit of measurement for electrical current, represented by the letter A.

Analog meter: A meter that uses a needle to indicate a value on a scale.

Arc: A visible flash of light and the release of heat that occurs when an electrical current crosses an air gap.

Clamp-on ammeter: An ammeter with operable jaws that are placed around a conductor to sense the magnitude of the current flow.

Conductor: A material through which it is relatively easy to maintain an electrical current.

Contactors: Control devices consisting of a coil and one or more sets of contacts used as switching devices in high-voltage circuits.

Continuity: A continuous current path. The absence of continuity indicates an open circuit.

Current: The rate or volume at which electrons flow in a circuit. Current (I) is measured in amperes.

Digital meter: A meter that provides a direct numerical reading of the value measured.

Direct current (DC): An electric current that flows in one direction. A battery is a common source of DC voltage.

Electromagnet: A coil of wire wrapped around a soft iron core. When current flows through the coil, magnetism is created.

Ground fault: An unintentional, electrically conducting connection between an energized conductor of an electrical circuit and another conductor, such as metal objects, earth, or an equipment frame. The fault current passes through the grounding system, as well as a person or other conductive surface in the path.

Inductive: Of a load, able to become electrically charged from being near another electrically charged body, or to become magnetized by being within an existing magnetic field. The process itself is called *induction*.

In-line ammeter: A current-reading meter that is connected in series with the circuit under test.

Inrush current: A significant rise in electrical current associated with energizing inductive loads such as motors.

Insulator: A device or substance that inhibits the flow of current; the opposite of conductor.

Ladder diagram: A simplified schematic diagram in which the load lines are arranged like the rungs of a ladder between vertical lines representing the voltage source.

Line duty: A protective device connected in series with the supply voltage.

Loads: Devices that convert electrical energy into another form of energy (heat, mechanical motion, light, etc.). Motors are the most common significant loads in HVACR systems.

Motor starters: Magnetic switching devices used to control heavy-duty motors.

Multimeter: A test instrument capable of reading voltage, current, and resistance. Also known as a *volt-ohm-milliammeter (VOM)*.

Ohms (Ω): The basic unit of measurement for electrical resistance, represented by the symbol Ω.

Pilot duty: A protective device that opens the motor control circuit, which then shuts off the motor.

Power: The rate of doing work, or the rate at which energy is dissipated. Electrical power is measured in watts.

Rectifier: A device that converts AC voltage to DC voltage.

Relay: A magnetically operated device consisting of a coil and one or more sets of contacts.

Resistance: An electrical property that opposes the flow of current through a circuit. Resistance (R) is measured in ohms.

Short circuit: The bypassing of a load by a conductor, causing a very high current flow.

Solenoid coil: An electromagnetic coil used to control a mechanical device such as a valve or relay contacts.

Thermistor: A semiconductor device that changes resistance with a change in temperature.

Time-delay fuses: Fuses with a built-in time delay to accommodate the inrush current of inductive loads.

Transformer: Two or more coils of wire wrapped around a common core. Used to raise and lower voltages.

Volts (V): The unit of measurement for voltage, represented by the letter V. One volt is equivalent to the force required to produce a current of one ampere through a resistance of one ohm.

Voltage: The driving force that makes current flow in a circuit. Voltage, often represented by the letter E, is measured in volts. Also known as *electromotive force (emf)*, *difference of potential*, or *electrical pressure*.

Watts (W): The unit of measure for power consumed by a load.

Additional Resources

This module presents thorough resources for task training. The following resource material is suggested for further study.

Principles of Electric Circuits, Thomas L. Floyd. 9th Edition. New York, NY: Pearson.
Electronics Fundamentals: Circuits, Devices, and Applications, Thomas L. Floyd. New York, NY: Pearson.
NCCER Module 26102, *Electrical Safety*.
NCCER Module 75121, *Electrical Safety*.
NCCER Module 26501, *Managing Electrical Hazards*.
NCCER Module 26104, *Electrical Theory*.
NCCER Module 26112, *Electrical Test Equipment*.
NCCER Module 26210, *Circuit Breakers and Fuses*.
NCCER Module 26211, *Control Systems and Fundamental Concepts*.
NCCER Module 12406, *Programmable Logic Controllers*.
Wi-Fi® is a registered trademark of Wi-Fi Alliance, **www.wi-fi.org**

Figure Credits

Section Review Answer Key

Answer	Section Reference	Objective
Section One		
1. c	1.1.0	1a
2. a	1.2.0	1b
3. d	1.3.1	1c
4. b	1.4.0	1d
Section Two		
1. d	2.1.0	2a
2. a	2.2.0; Figure 12	2b
3. c	2.3.0	2c
4. d	2.4.1	2d
Section Three		
1. d	3.1.0	3a
2. c	3.2.0	3b
3. c	3.3.0	3c
Section Four		
1. c	4.1.1	4a
2. a	4.2.4	4b
3. c	4.3.0	4c

Section Review Calculations

2.0.0 SECTION REVIEW

Question 3

First, convert the motor's horsepower to watts:

Power in watts = 1.5 hp × 746 W/hp
Power in watts = 1,119W

Use the power in watts and the applied (total) voltage with the power formula to determine the current:

$$I = \frac{P}{E} = \frac{1,119W}{240V} = 4.66A$$

The current is **4.66A**.

Question 4

Since this is a series circuit, add all of the resistances to determine the total resistance (R_T):

$$R_T = R1 + R2 + R3$$
$$R_T = 30\Omega + 30\Omega + 30\Omega$$
$$R_T = 90\Omega$$

The total resistance is **90Ω**. Craftsmanship

NCCER CURRICULA — USER UPDATE

NCCER makes every effort to keep its textbooks up-to-date and free of technical errors. We appreciate your help in this process. If you find an error, a typographical mistake, or an inaccuracy in NCCER's curricula, please fill out this form (or a photocopy), or complete the online form at **www.nccer.org/olf**. Be sure to include the exact module ID number, page number, a detailed description, and your recommended correction. Your input will be brought to the attention of the Authoring Team. Thank you for your assistance.

Instructors – If you have an idea for improving this textbook, or have found that additional materials were necessary to teach this module effectively, please let us know so that we may present your suggestions to the Authoring Team.

NCCER Product Development and Revision
13614 Progress Blvd., Alachua, FL 32615

Email: curriculum@nccer.org
Online: www.nccer.org/olf

❏ Trainee Guide   ❏ Lesson Plans   ❏ Exam   ❏ PowerPoints   Other _____

Craft / Level: _____ Copyright Date: _____

Module ID Number / Title: _____

Section Number(s): _____

Description: _____

_____

_____

_____

Recommended Correction: _____

_____

_____

_____

Your Name: _____

Address: _____

Email: _____ Phone: _____

Introduction to Heating

OVERVIEW

Most homes and businesses have some type of heating system. Installing and servicing furnaces is a big responsibility. Because flame and combustible fuels are involved, there is always the potential for fire or explosion. If furnaces are properly installed according to the manufacturer's instructions, and periodically inspected and serviced by qualified technicians, they will operate satisfactorily for many years.

Module 03108

Trainees with successful module completions may be eligible for credentialing through the NCCER Registry. To learn more, go to www.nccer.org or contact us at 1.888.622.3720. Our website has information on the latest product releases and training, as well as online versions of our *Cornerstone* newsletter and Pearson's product catalog.

Your feedback is welcome. You may email your comments to **curriculum@nccer.org**, send general comments and inquiries to **info@nccer.org**, or fill in the User Update form at the back of this module.

This information is general in nature and intended for training purposes only. Actual performance of activities described in this manual requires compliance with all applicable operating, service, maintenance, and safety procedures under the direction of qualified personnel. References in this manual to patented or proprietary devices do not constitute a recommendation of their use.

03108 V5

From *HVAC Level One, Trainee Guide*, NCCER.

INTRODUCTION TO HEATING

Objectives

When you have completed this module, you will be able to do the following:

1. Explain the fundamental concepts of heating and combustion.
   a. Describe the heat-transfer process.
   b. Identify gas fuels and their combustion characteristics.
2. Describe the role of forced-air gas furnaces in residential heating.
   a. Describe the types of gas furnaces and how they operate.
   b. Identify and describe the equipment and controls used in gas furnaces.
   c. Describe the basic installation and maintenance requirements for gas furnaces.
3. Describe hydronic and electric heating systems.
   a. Describe the operation of hydronic heating systems.
   b. Describe the operation of electric heating equipment.

Performance Tasks

Under the supervision of the instructor, you should be able to do the following:

1. Identify the components of induced-draft and condensing furnaces and describe their functions.
2. Perform common maintenance tasks on a gas furnace, including air filter replacement and temperature measurements.

Trade Terms

Annual Fuel Utilization Efficiency (AFUE)
Atomized
Balance point
Combustion
Condensing furnace
Conduction
Convection
Dual-fuel system
Electronically commutated motor (ECM)
Flame rectification
Fusible link
Geothermal heat pump
Heat exchanger
Hot surface igniter (HSI)
Induced-draft furnace
Infiltration
Manometer
Multipoise furnace
Natural-draft furnace
Orifices
Primary air
Radiation
Redundant gas valve
Secondary air
Sectional boiler
Standing pilot
Spud
Thermocouple

Industry Recognized Credentials

If you are training through an NCCER-accredited sponsor, you may be eligible for credentials from NCCER's Registry. The ID number for this module is 03108. Note that this module may have been used in other NCCER curricula and may apply to other level completions. Contact NCCER's Registry at 888.622.3720 or go to **www.nccer.org** for more information.

Contents

1.0.0   Heating Fundamentals 1
  1.1.0   Heat Transfer 1
    1.1.1   Conduction 1
    1.1.2   Convection 2
    1.1.3   Radiation 2
    1.1.4   Humidity 3
    1.1.5   Temperature 4
    1.1.6   Heat Measurement 4
  1.2.0   Combustion 4
    1.2.1   Complete Combustion 4
    1.2.2   Incomplete Combustion 4
    1.2.3   Combustion Efficiency 5
    1.2.4   Flames 6
    1.2.5   Fuels 7
2.0.0   Gas Furnaces 10
  2.1.0   Types of Gas Furnaces 10
  2.2.0   Gas Furnace Components and Controls 13
    2.2.1   Heat Exchangers 16
    2.2.2   Fans and Motors 17
    2.2.3   Air Filters 18
    2.2.4   Gas Valve Assemblies 20
    2.2.5   Manifold and Orifices 22
    2.2.6   Gas Burners 22
    2.2.7   Ignition Devices 23
    2.2.8   Gas Furnace Safety Controls 24
  2.3.0   Gas Furnace Installation and Maintenance 25
    2.3.1   Furnace Venting 27
    2.3.2   Combustion Air 27
    2.3.3   Gas Furnace Maintenance 27
3.0.0   Hydronic and Electric Heating Systems 31
  3.1.0   Hydronic Heating Systems 31
    3.1.1   Major Boiler Components 31
    3.1.2   Boiler Components and Controls 34
    3.1.3   Other Components of a Hydronic Heating System 34
  3.2.0   Electric Heating 37
    3.2.1   Electric Furnaces 37
    3.2.2   Heat Pumps 39

Figures and Tables

Figure 1   Methods of heat transfer. 2
Figure 2   Tubular gas-fired radiant heater. 3
Figure 3   Combustion analysis equipment. 5
Figure 4   Upflow furnace. 11

Figure 5   Horizontal furnace. 12

Figure 6   Low-boy furnace. 13

Figure 7   Counterflow furnace. 13

Figure 8   A multipoise furnace in three possible positions. 14

Figure 9   Condensing furnace components. 15

Figure 10 Heat exchangers. 16

Figure 11 Furnace air flow. 17

Figure 12 Gas-fired unit heater with a propeller fan. 18

Figure 13 Air filters. 19

Figure 14 Redundant gas valve. 20

Figure 15 Solenoid-operated gas valve. 21

Figure 16 Diaphragm-operated gas valve. 21

Figure 17 Gas manifold. 22

Figure 18 Combustion air supply. 22

Figure 19 Gas burners. 22

Figure 20 Gas-furnace ignition devices. 23

Figure 21 Energized HSI. 24

Figure 22 Example of a high-temperature limit switch. 24

Figure 23 Flame rollout switch. 24

Figure 24 Combustion airflow pressure-sensing switch. 25

Figure 25 Natural- and induced-draft furnace venting. 26

Figure 26 Combustion air and vent terminations for a
condensing furnace. 27

Figure 27 Adjusting manifold pressure. 28

Figure 28 Temperature rise measurements. 29

Figure 29 Residential hydronic-heating system with
baseboard radiation. 32

Figure 30 A standard gas-fired hot water boiler. 32

Figure 31 Cutaway of a condensing boiler. 32

Figure 32 Wall-mounted condensing boiler. 32

Figure 33 Aquastats. 33

Figure 34 Pressure-relief valve. 34

Figure 35 Hydronic circulators. 34

Figure 36 Simplified zoned hydronic-heating system. 35

Figure 37 Automatic air vents. 36

Figure 38 Water-pressure regulator and pressure-
relief valve combination. 37

Figure 39 Electric furnace. 37

Figure 40 Electric resistance-heating elements. 38

Figure 41 Schematic diagram of a heat pump system. 39

Figure 42 Water-source heat pump closed-loop examples. 41

Figure 43 Balance point calculations. 41

Table 1   Heating Values of Common Fuels 7

SECTION ONE

1.0.0 HEATING FUNDAMENTALS

Objective

Explain the fundamental concepts of heating and combustion.

a. Describe the heat-transfer process.
b. Identify gas fuels and their combustion characteristics.

Trade Terms

Annual Fuel Utilization Efficiency (AFUE): HVAC industry standard for defining furnace efficiency, expressed as a percentage of the total heat available from a fuel. The AFUE goes beyond the specific thermal efficiency of a unit in that it accounts for both the peak measurement of fuel-to-heat conversion efficiency and the losses of efficiency that occur during startup, shutdown, etc.

Atomized: Broken into tiny pieces or fragments, such as a liquid being broken into tiny droplets to create a fine spray.

Combustion: The process by which a fuel is ignited in the presence of oxygen.

Condensing furnace: A furnace that contains a secondary heat exchanger that extracts latent heat by condensing exhaust (flue) gases.

Conduction: In the context of heat transfer, the process through which heat is directly transmitted from one substance to another when there is a difference of temperature between regions in contact.

Convection: The movement caused within a fluid (air or water, for example) by the tendency of hotter fluid to rise and colder, denser material to fall, resulting in heat being transferred.

Heat exchanger: A device that is used to transfer heat from a warm surface or substance to a cooler surface or substance.

Hot surface igniter (HSI): A ceramic device that glows when electric current flows through it; used to ignite the fuel/air mixture in a gas furnace.

Induced-draft furnace: A furnace in which a motor-driven fan draws air from the surrounding area or from outdoors to support combustion and create a draft in the heat exchanger.

Natural-draft furnace: A furnace in which the natural flow of air from around the furnace provides the air to support combustion and venting of combustion byproducts.

Orifices: Precisely drilled holes that control the flow of gas to the burners.

Radiation: In the context of heat transfer, the direct transmission of heat through electromagnetic waves through space or another medium. No contact between the two substances is required.

People rely on heating systems in homes, schools, and places of business to keep the temperature in a comfort range. There are many types of heating systems. Most of them burn natural gas or fuel oil; many others use electricity as an energy source. In this module, you will learn the basic principles of heating. You will also learn about various types of heating systems. Before beginning to study heating systems, it is important to first understand basic heating terms and processes. The basic processes involved in heating are heat transfer and combustion.

1.1.0 Heat Transfer

For heat transfer to occur, there must be a difference in temperature between two objects. A larger difference in temperatures results in more heat being transferred, and at a more rapid pace. Heat always moves from a warmer region to a cooler region. Any object, including the human body, gives off heat if the air around it is cooler than the object. The reason humans feel cold in the winter is because the body is losing heat to the cooler air around it. To keep the heat from escaping, there are two choices: wear layers of insulation (clothing), or warm the surrounding air so that the body stops giving up heat. In hot weather, your body feels hot because it cannot transfer heat to air that is a higher temperature. At that point, the body must rely on removing heat through the evaporation of moisture from the skin.

The three basic methods of heat transfer are conduction, convection, and radiation. Heating systems use one or more of these methods to warm the air in the conditioned space, as shown in *Figure 1*. Refer to this figure for examples of heat transfer as they are described in detail in the sections that follow.

1.1.1 Conduction

Conduction is the flow of heat from one part of a material to another part or to a substance in direct contact with it. The rate at which a material transmits heat is known as its thermal conductivity. The amount of heat transmitted by conduction between two surfaces is determined by the size of the area of contact, thickness, and conductivity of the materials, as well as the temperature difference between the two. One good example of conduction is the transfer of heat from hot coffee to a cold cup. The cup becomes warmer while the coffee becomes colder.

1.1.2 Convection

Convection is natural air motion due to warmer air rising and denser, cooler air falling. For convection to occur, there must be a difference in temperature between the source of heat and the surrounding air. The greater the difference in temperature, the greater the movement of air by convection will be. The greater the movement of air, the greater the heat transfer will be.

Hot water heating terminals, such as the one shown in the convection example of *Figure 1*, rely on convection. The room air in contact with the baseboard heater is warmed, causing it to rise. As it rises, cool air falls in to replace it and consistent convection is now established. The air leaving the terminal gives up its heat to the cooler room air, furnishings, and walls. As it cools, it falls back toward the floor and the process continues.

1.1.3 Radiation

Radiation is the transfer of heat through space by wave motion. Heat passes from one object to another without warming the space between them. The amount of heat transferred by radiation depends upon the area of the radiating body, the temperature difference between the two objects, and the distance between the heat source and the object being heated.

The heat from the sun or a fireplace is a good example of radiant heat. The heat from radiation

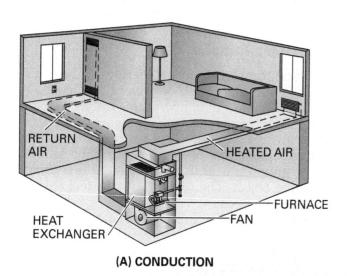

(A) CONDUCTION

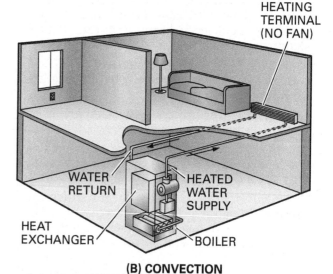

(B) CONVECTION

(C) RADIATION

Figure 1 Methods of heat transfer.

heat sources tends to be intense nearest the source and it heats only solid objects that are directly in its path. When you are sitting at a campfire, for example, the front of your body can be hot while your back is cold.

One of the most common and effective applications of radiant heat is the tubular gas-fired radiant heating system (*Figure 2*). This system is highly effective in large, open areas such as automotive garages and warehouses. Heating the air is not a priority in these areas. Humans and other solid objects, however, can be kept comfortably warm. This significantly limits the heat loss when large doors are opened and a great deal of air is exchanged.

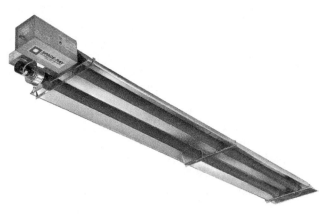

Figure 2 Tubular gas-fired radiant heater.

Furnace Controls

Most modern gas furnaces have solid-state control modules that monitor and manage furnace operation. These modules use information from sensing devices to diagnose problems and direct the technician to a faulty component. The behavior of indicator lights provides the operating status or information related to a failure.

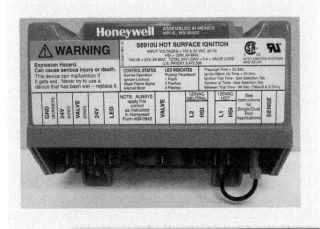

Heat Transfer Experiments

Here are some simple experiments you can do to demonstrate heat transfer theories:

- *Conduction* – Take a piece of copper pipe and attach a thermometer to one end. When the temperature reading has stabilized, apply heat to the other end of the pipe and watch the temperature rise. Then try it with a piece of steel pipe and see if the thermometer seems to rise at the same rate.
- *Convection* – Obtain a light plastic bag. While holding the bag upside down, fill it with hot air from a hair dryer. Let the bag go and watch it rise. As the air cools, the bag will begin to descend.
- *Radiation* – Obtain two thermometers. Place one in the shade and one in direct sunlight. Note how much higher the temperature reading is on the thermometer in direct sunlight.

1.1.4 Humidity

Heat transfer is affected by the moisture content of the air, which is referred to as humidity. You often hear the term *relative humidity*. This term describes the volume of moisture in the air relative to its water-holding capacity. A relative humidity of 50 percent means that the air contains half the moisture it is capable of holding at its present temperature. Warm air can hold more water than cool air. As a result, if the temperature of a volume of air is increased, without any change in its actual moisture content, the relative humidity falls. For example, 80°F (27°C) with a relative humidity of 50 percent contains 76 grains of moisture per pound of dry air. Raising the temperature of the air to 85°F (29°F), without adding or removing any moisture, results in a relative humidity of 42 percent. This is because the 85°F air can hold more moisture than air at 80°F, so it contains a lower percentage of the total it can contain.

The body feels colder when the humidity is low, since the evaporation of water from the skin occurs at an increased rate, causing a cooling effect. Therefore, homes in many climates are equipped with humidifiers to add moisture during winter months when the air is dry, increasing comfort.

1.1.5 Temperature

Temperature is defined as the presence or absence of heat measured on a numerical scale. It can be accurately measured using a thermometer. There are two thermometer scales in frequent use. One scale is the English system where temperature is measured in Fahrenheit. The other scale is called the Celsius (centigrade) scale. It is important to understand both scales, because both are used across the world.

The Celsius thermometer is the easier of the two to understand. The freezing point of water on the Celsius scale is 0°C and the boiling point of water is 100°C. The freezing point of water on the Fahrenheit scale is 32°F and the boiling point of water is 212°F. To reduce confusion about which scale is involved, a capital C is placed after a Celsius degree indication, and a capital F is placed after the number of Fahrenheit degrees.

There are times when it may be necessary to change from one scale to the other. These conversions can be done using the following formulas:

$$\text{Degrees Celsius} = \tfrac{5}{9} \times (F - 32°)$$
$$\text{Degrees Fahrenheit} = (\tfrac{9}{5} \times C) + 32°$$

Where:
F = Temperature in Fahrenheit
C = Temperature in Celsius

For example, we can convert 32°F to Celsius as shown here:

$$C = \tfrac{5}{9} \times (F - 32°)$$
$$C = \tfrac{5}{9} \times (32° - 32°)$$
$$C = \tfrac{5}{9} \times 0°$$
$$C = 0°C$$

1.1.6 Heat Measurement

The unit of measurement of heat in the inch-pound system is the British thermal unit (Btu). One Btu is the amount of heat required to raise the temperature of one pound of water by 1°F.
It is also important to be aware of the rate of heat change; that is, how fast or how slowly a piece of heating equipment produces heat. The term for this is *Btus per hour (Btuh)*. For example, the capacity of a furnace may be stated as 110,000 Btuh.

Furnaces are selected based on their Btu rating. Actually, there are two ratings. The input rating is the actual amount of heat in Btus produced by the furnace, while the output rating is the number of usable Btus. The difference is a function of the efficiency rating of the furnace. For example, if a furnace with an input rating of 150,000 has an efficiency rating of 90 percent, its output rating is 135,000 Btus, or 90 percent of

150,000. The remaining 10 percent is mostly lost as the hot byproducts of combustion are vented to the outdoors.

1.2.0 Combustion

Combustion is the burning of fuel to create heat. During combustion, oxygen is combined with a fuel to release the stored energy in the form of heat. There are three components necessary for combustion to take place:

- *Fuel* – The fuel can be a gas, such as natural gas; a liquid, such as fuel oil; or a solid, such as coal. Two elements that all fuels have in common are carbon and hydrogen.
- *Heat (a source of ignition)* – A pilot burner, electric spark, or hot surface igniter (HSI) can be used to ignite a gas burner, while a continuous electric arc is typically used to ignite fuel oil.
- *Oxygen* – Oxygen must be present for the combustion of fuel to take place.

Combustion is a chemical reaction between fuel, heat, and oxygen; therefore, all three components must be present for combustion to occur. Natural gas will burn in its natural state as long as a source of ignition and oxygen are both present. Fuel oil is atomized (converted to a fine spray) before it is burned. This ensures an effective mixture of oxygen and fuel.

1.2.1 Complete Combustion

Complete combustion takes place when carbon combines with oxygen to form carbon dioxide (CO_2). CO_2 is nontoxic and can be exhausted to the atmosphere. Hydrogen combines with oxygen to form water vapor (H_2O), which is also harmless and can be exhausted to the atmosphere.

1.2.2 Incomplete Combustion

Incomplete combustion results from a lack of oxygen and causes undesirable products to form. These undesirable products include carbon monoxide (CO), pure carbon (soot), and aldehyde. Both CO and aldehyde are toxic substances. Soot causes coating of the heating surfaces of the furnace and reduces heat transfer.

While CO is odorless, aldehydes have a distinct sharp odor that irritates the membranes of the nose and throat. If aldehydes are present, it is virtually certain that CO is also present, although the absence of aldehydes does not mean that CO is absent. The odor of aldehydes helps alert both consumers and service technicians to a severe problem. Since CO is odorless, tasteless, colorless,

Carbon Monoxide

The symptoms of CO inhalation include confusion, lightheadedness, vertigo, and headaches. Flu-like symptoms are often reported, leading people to believe their health issue is likely viral or bacterial. Serious cases can easily lead to death. CO can also have severe effects on the unborn child of a pregnant woman. Long-term exposure to low levels of the gas can lead to memory loss and depression. CO mainly causes adverse effects in humans by combining with blood hemoglobin in the lungs. CO poisoning can often be prevented with the use of carbon monoxide detectors.

HVAC technicians must take CO very seriously and maintain furnaces in peak operating condition. Consumers must be able to depend on their technician for protection against this invisible hazard. HVAC technicians can further help customers by encouraging them to install CO detectors. In some areas, the law requires CO detectors.

and non-irritating, it can be very difficult to detect without special instruments. As a result, it is a significant cause of fatal poisoning in some countries.

Fuel-burning devices must be adjusted so that complete combustion of the fuel takes place. Sufficient air must be provided for proper combustion to take place and to eliminate the hazards of incomplete combustion. Furnaces are normally adjusted to provide from 5 to 50 percent excess air to guard against the possibility of incomplete combustion.

1.2.3 Combustion Efficiency

When fuel is burned in a furnace, a certain amount of heat is lost in the hot gases, known as flue gases, that are vented through the chimney. This function is necessary for the disposal of the products of combustion, but the loss must be kept to a minimum to allow the furnace to operate at its peak efficiency. Air entering the furnace combustion chamber at room temperature or lower is heated to flue gas temperatures that range from 100°F to 600°F, depending upon the design and adjustments of the furnace.

If the amouwnt of heat lost is 20 percent, the furnace efficiency would be 80 percent. Charts are available for determining the efficiency of a furnace based on the temperature and the CO_2 content of the flue gases. Knowing the efficiency of the furnace makes it possible for a technician to calculate the furnace output. Combustion analysis

equipment (*Figure 3*) can be used by service technicians to analyze the content of the flue gases from both gas- and oil-burning systems to determine the actual combustion efficiency.

The current industry standard for defining furnace efficiency is Annual Fuel Utilization Efficiency (AFUE). AFUE takes into account operating efficiency as well as combustion efficiency. The National Appliance Energy Conservation Act of 1987 required that all gas furnaces built after 1991 have an AFUE of no less than 78 percent. Most commonly, a natural-draft furnace that relies on

Figure 3 Combustion analysis equipment.

convection for combustion air has an AFUE of less than 78 percent. At this point in time, most natural-draft furnaces have reached the end of their life cycle; very few remain in service in the United States. With the addition of special features such as electronic ignition and vent dampers, it was possible to increase the AFUE of a natural-draft furnace to an AFUE of 80 percent. At this writing, the AFUE requirements for furnaces, with a few exceptions, remains at 78 percent. However, changes have been discussed for some years now and legislation could eventually require higher efficiencies.

In the 1990s, most all natural-draft furnaces on the market were replaced by the induced-draft furnace, which uses a fan to draw in and exhaust combustion air. Induced-draft furnaces have AFUEs that range from 78 to 85 percent, depending on the kinds of efficiency options installed. These furnaces are still manufactured in large numbers. The increase in efficiency primarily results from the heat exchanger having more surface area in contact with the heated air. With a fan to help force air through, the air path through the heat exchanger can be longer and more complex, resulting in improved heat transfer.

The condensing furnace has the highest efficiency—90 percent or greater. These furnaces use a condensing coil as a secondary heat exchanger to extract the latent heat from the combustion byproducts before they are vented outdoors. In 2017, some condensing furnaces were reaching a 98.5 percent efficiency level, which means that all but 1.5 percent of the heat produced is delivered to the conditioned space. In addition to secondary heat exchangers, these ultrahigh-efficiency furnaces use advanced heat exchangers, smart gas valves, and advanced electronic controls to achieve an exceptional level of efficiency. Since the vast majority of the heat is removed from the byproducts of combustion by design, the vent piping can be constructed of plastic. This is certainly never possible with natural- or induced-draft models, which expel much air at a significantly higher temperature.

Over the years since the original standards were invoked in the early 1990s, the DOE has made a number of proposed changes and decisions to efficiency standards. Due to opposition and, in some cases, litigation, the DOE was unable to apply many proposed changes. At this writing, the AFUE standards for non-weatherized and weatherized furnaces remain much the same. Certain furnace products, such as those used in mobile homes, do need to meet a newer standard of 80 percent AFUE. Work and negotiation between government and industry continues in this area and the extent of any major changes in the required efficiency of furnaces is still to be determined.

1.2.4 Flames

The type of flame and the intensity with which it burns have a direct relationship with the efficiency of the heating unit. Pressure-type oil burners operate with a yellow flame. Natural-draft gas burners have a blue flame with slight yellow/orange tips, while the flame from induced-draft models is blue with very little or no colored tips at all. The difference is mainly due to the way air is mixed with fuel.

A yellow flame in a gas burner is a sign of incomplete combustion. The burner should be checked for a sufficient supply of air. A proper gas flame is produced when approximately 50 percent of the combustion air (primary air) is mixed with the gas before ignition. The balance of air required, called secondary air, is supplied during combustion to the area above the flame. Improper gas flames can be caused by a lack of primary or

GOING GREEN

The ENERGY STAR® Program

ENERGY STAR® is a joint program developed and administered by the US Environmental Protection Agency (EPA) and the US Department of Energy (DOE). The goal is save energy and protect the environment through the use of energy efficient products and practices. Products identified with the ENERGY STAR® logo meet strict energy efficiency guidelines established by the program. Additional resources available through the ENERGY STAR® program provide energy-efficient guidelines for home construction. The appearance of the ENERGY STAR® logo assures consumers that a product meets the specifications of the program.

Figure Credit: Courtesy of the US Environmental Protection Agency

secondary air, and by contact between the flame and a cool surface. It is also possible to provide too much air to a burner, resulting in a flame that is less than ideal, but does not result in incomplete combustion.

1.2.5 Fuels

Fuels are available in three forms: gases, liquids, and solids. *Table 1* shows the heating values of common fuels. Gases include natural gas, manufactured gas, and liquefied petroleum (LP). Liquids include fuel oils. There are six grades of fuel oil, with No. 2 being the most common. Solids include coal and wood. This module focuses on gas, which is the type most likely to be encountered in the HVAC trade.

The three common types of gaseous fuels are described further here:

- *Natural gas* – Natural gas is the most common of the gaseous fuels. It comes from the earth and often accumulates in the upper part of oil wells. It is colorless and nearly odorless. An odorant, such as a sulfur compound, is added so that the presence of a leak can be easily detected. The content of this gas varies by locality and has a bearing on the Btu content, which varies from 900 to 1,200 Btus/ft^3 depending on the locality, but is usually in the range of 1,000 to 1,050 Btus/ft^3. The chief component of natural gas is methane. It also includes other hydrocarbons.
- *Manufactured gas* – Manufactured gases are combustible gases that are normally produced from solid or liquid fuel and are used mostly in industrial processes. They are produced mainly from coal, oil, and other hydrocarbons, and are comparatively low in Btus per cubic foot (500 to 600). They are not considered to be economical space-heating fuels.
- *Liquefied petroleum* – LP is a byproduct of the oil-refining process. It is stored in its liquid form under pressure, but is vaporized when used. There are two types of LP gas, propane and butane. Propane is more useful as a space-heating fuel because it boils at –40°F (–40°C) and can be readily vaporized for heating in a northern climate. Butane vaporizes at about 32°F (0°C). Propane has a heating value of about 2,500 Btus/ft^3, while butane has a heating value of approximately 3,200 Btus/ft^3. Although it is not commonly used in space-heating systems, butane does have the highest Btu content per cubic foot. LP gas is usually propane with a small amount of butane added. When LP gas is used as a heating fuel, the equipment must be designed or modified slightly for its use.

WARNING!

Propane and butane vapors are considered more dangerous than those of natural gas because they have a higher specific gravity. The vapor is heavier than air and tends to accumulate in low spots and near the floor, increasing the danger of an explosion if an ignition source is present.

LP gas is an alternative in areas where natural gas is not available. Since natural gas is delivered in its vapor form by pipeline, many rural areas do not have access to it. Manufacturers of gas appliances make LP conversion kits to adapt natural gas furnaces for LP use. The kits contain replacement burner orifices and a gas pressure regulator spring kit designed for use with LP. However, the use of LP gas may be heavily regulated in some locales.

NOTE

Fuel oil furnaces are fairly common in some parts of the country, notably the northeastern United States. Oil furnaces are covered later in the program.

Table 1 Heating Values of Common Fuels

Fuel	Heat Released
Coal	
Bituminous	12,000 to 15,000 Btus/lb
Anthracite	13,000 to 14,000 Btus/lb
Oil*	
Grade 1	137,000 Btus/gallon
Grade 2	140,000 Btus/gallon
Grade 4**	141,000 Btus/gallon
Grade 5	148,000 Btus/gallon
Grade 6 (Bunker C)	152,000 Btus/gallon
Gas	
Natural***	900 to 1,200 Btus/ft^3
Manufactured	500 to 600 Btus/ft^3
Liquified Petroleum (LP)	1,500 to 3,200 Btus/ft^3
Wood	6,200 Btus/lb (avg)

*Grades are determined by the American Society for Testing and Materials (ASTM).
**This grade is not commonly used.
***Check with local gas company for specific values.

Home Heating Fuels

The US Energy Information Agency tracks information related to the use of heating fuels used in housing across the United States. The graph shown here shows the dominant fuels used from 1970 through 2009, the latest year of available data. A very small percentage of homes are heated through other means, such as wood or solar thermal systems. As shown, natural gas remains the most popular source of heat through the years. The use of electric heat, predominantly by heat pump systems, has grown significantly at the expense of oil heating systems. Of the fuel oil used for home heating, 84 percent is consumed by a handful of states in the northeast corner of the country.

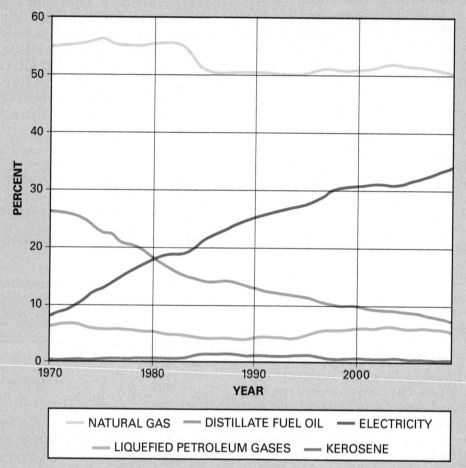

Figure Credit: US Energy Information Administration

High-Efficiency Furnaces

GOING GREEN

The EPA estimates that replacing a 70 percent efficient furnace with a 90 percent furnace can reduce the cost of fuel by more than 25 percent. However, that is not the only benefit. Using a high-efficiency furnace also reduces the amount of CO_2 and other pollutants that result from burning fuel gases.

Additional Resources

Refrigeration and Air Conditioning: An Introduction to HVAC, Larry Jeffus. 4th edition. New York, NY: Pearson.

NCCER Module 03209, *Troubleshooting Gas Heating*.

1.0.0 Section Review

1. The type of heat you would feel from a camp-fire is _____.
   a. convection
   b. conduction
   c. radiant
   d. conductivity

2. The most common gaseous fuel for furnaces is _____.
   a. methane
   b. propane
   c. butane
   d. natural gas

2.0.0 GAS FURNACES

Objective

Describe the role of forced-air gas furnaces in residential heating.

a. Describe the types of gas furnaces and how they operate.
b. Identify and describe the equipment and controls used in gas furnaces.
c. Describe the basic installation and maintenance requirements for gas furnaces.

Performance Tasks

1. Identify the components of induced-draft and condensing furnaces and describe their functions.
2. Perform common maintenance tasks on a gas furnace, including air filter replacement and temperature measurements.

Trade Terms

Electronically commutated motor (ECM): A DC-powered motor able to operate at variable speeds based on the programming of its electronic control module, which is AC-powered and converts AC to three-phase DC. ECMs also operate more efficiently than standard motors.

Flame rectification: A method of proving the existence of a pilot flame by applying an AC current to a flame rod, which is then rectified to a DC current as it flows back to a ground source. Monitoring the DC current flow provides the means of proving a pilot flame has been established.

Infiltration: Air that unintentionally and naturally enters a building through doors, windows, and cracks in the structure.

Manometer: An instrument that measures air or gas pressure by the displacement of a column of liquid.

Multipoise furnace: A furnace that can be configured for upflow, counterflow, or horizontal installation.

Redundant gas valve: A gas control containing two gas valves in series. If one fails, the other is available to shut off the gas when needed.

Spud: A threaded metal device that screws into the gas manifold. It contains the orifice that meters gas to the burners.

Standing pilot: A gas pilot that remains lit continuously.

Thermocouple: A device comprised of two different metals that generates electricity when there is a difference in temperature from one end to the other.

Gas forced-air furnaces are the most common method of heating in the United States. Gas furnaces can be installed in attics, basements, furnace rooms, and crawl spaces.

In a gas forced-air furnace, return air from the space being heated is passed over a heat exchanger where the heat from combustion is transferred to the air. The air is then sent to the space through the supply ductwork. In most furnaces, a fuel such as natural gas or oil is burned to create the heat that warms the heat exchangers.

A forced-air furnace uses a fan to move the air over the heat exchangers and to circulate the air through the distribution system (return and supply ductwork). If cooling is installed with a forced-air furnace, the cooling coil is normally placed in the airstream at the outlet of the furnace. Since the air leaving a cooling coil is generally saturated (at or near 100-percent relative humidity), positioning the coil at the furnace outlet prevents this moist air from flowing through the fan and across the furnace heat exchanger.

2.1.0 Types of Gas Furnaces

There are four basic configurations of forced-air gas furnaces. Each requires a different arrangement of the basic components.

An upflow furnace (*Figure 4*) is used in the basement or in a first-floor equipment room. The blower is located below the heat exchanger. Room air enters at the bottom or lower sides of the unit and exits through the top into the air distribution system.

A horizontal furnace (*Figure 5*) is used in attics or crawlspaces where the height of the furnace must be kept to a minimum. Air enters at one end of the unit through the filter and blower, and is forced horizontally over the heat exchanger, exiting at the opposite end.

The low-boy furnace (*Figure 6*) occupies more floor space and is lower in height than the upflow design. It is best suited for basement installations with limited headroom. The blower is placed alongside the heat exchanger, the return-air plenum is above the blower compartment, and the warm-air outlet or plenum is above the heat exchanger. Low-boy furnaces are almost exclusively oil-fired units.

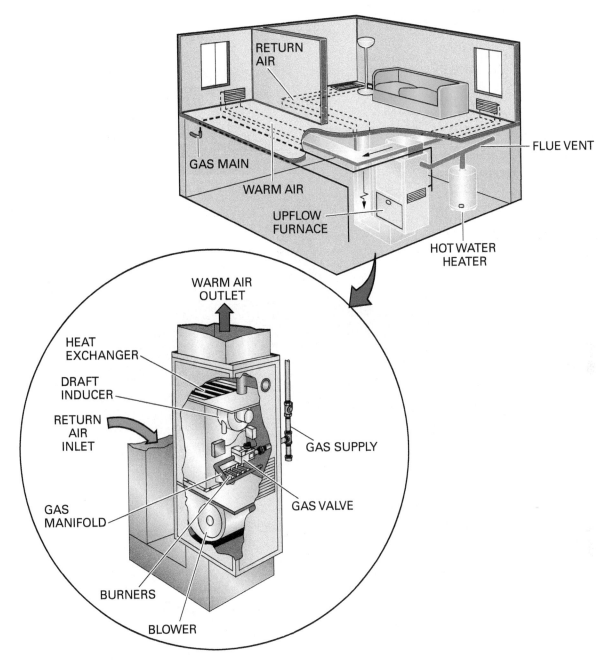

Figure 4 Upflow furnace.

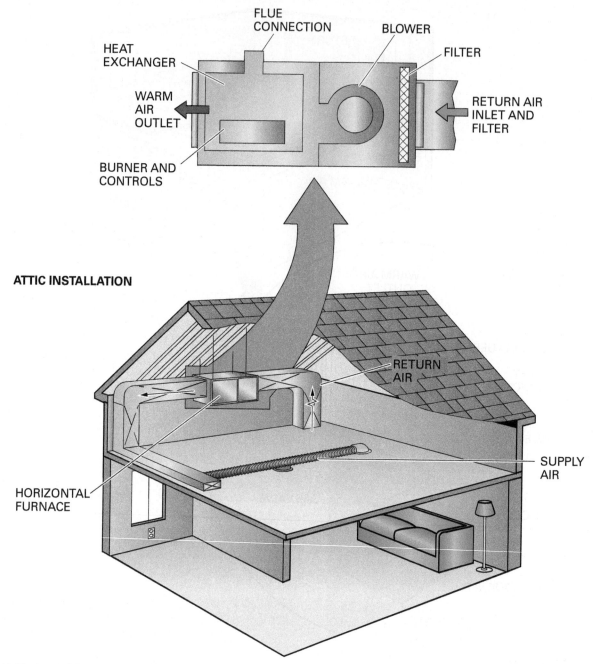

FLUE CONNECTION

HEAT EXCHANGER

BLOWER

FILTER

WARM AIR OUTLET

RETURN AIR INLET AND FILTER

BURNER AND CONTROLS

ATTIC INSTALLATION

RETURN AIR

SUPPLY AIR

HORIZONTAL FURNACE

Figure 5 Horizontal furnace.

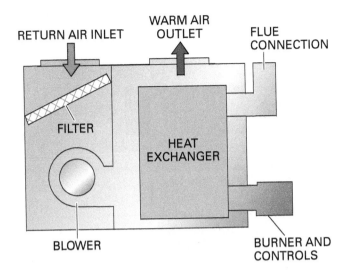

Figure 6 Low-boy furnace.

The counterflow, or downflow, furnace (*Figure* 7) is commonly used in houses with an under-floor air distribution system. The blower is located above the heat exchanger and the return air plenum is connected to the top of the blower area. The warm air supply plenum is connected to the bottom of the enclosure cabinet. The return air duct is generally located in the attic, with a return air grille in the ceiling. The furnace itself works well for garage or first-floor closet installations.

The multipoise furnace was developed by furnace manufacturers to increase the flexibility of a single furnace model. The furnace is designed to allow the furnace to be configured by the installer for upflow, counterflow, or horizontal use. *Figure 8* shows an example of a multipoise furnace applied in three different positions.

Most multipoise furnaces can usually be installed in the upflow position with little or no field modification. They are often shipped from the factory ready for upflow installation because that is the more common configuration. The furnace must be field-converted for counterflow or horizontal airflow applications.

On standard-efficiency furnaces, the conversion is fairly easy. The vent pipe has to be attached to the flue outlet in different ways, depending on the position in which the furnace is installed. Converting a high-efficiency condensing furnace is more complex. Combustion air and vent pipes have different attachment schemes depending on the position in which the furnace is installed. Because high-efficiency condensing furnaces collect condensate from the flue gases (thus the name), the condensate drain and trap must be properly arranged for each application. The furnace manufacturer's conversion procedure must be followed carefully.

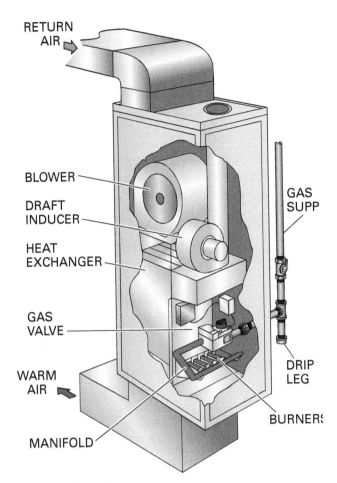

Figure 7 Counterflow furnace.

WARNING!

Failure to precisely follow the manufacturer's installation instructions for multipoise furnaces can be hazardous because flue gas venting and condensate drainage can be affected by improper installation. Improper installation can lead to hazardous conditions for occupants or to furnace failure and expensive rework later.

2.2.0 Gas Furnace Components and Controls

A modern furnace contains many components and control devices, including safety controls that shut off the gas supply if ignition fails to occur. Components and controls in a gas-fired furnace can be conveniently grouped into those involved in the flow of conditioned air; those involved in the flow of fuel gas and combustion air; and those involved in combustion. *Figure 9* shows the interior of a condensing furnace and identifies its major components.

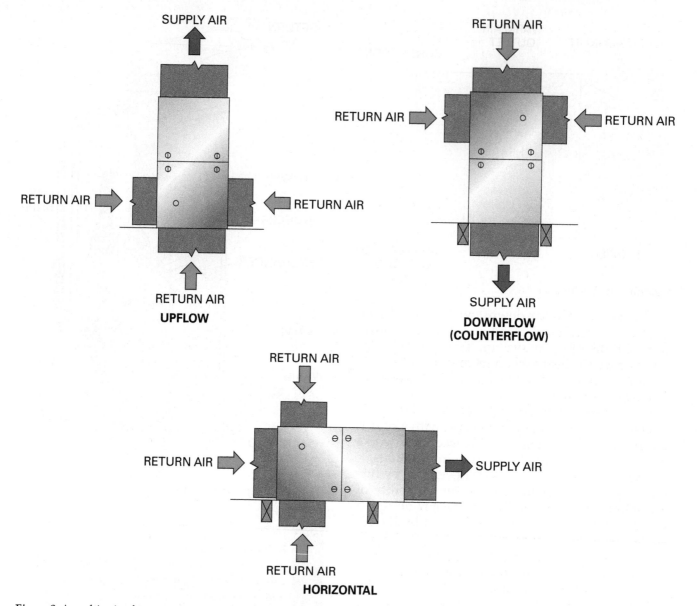

Figure 8 A multipoise furnace in three possible positions.

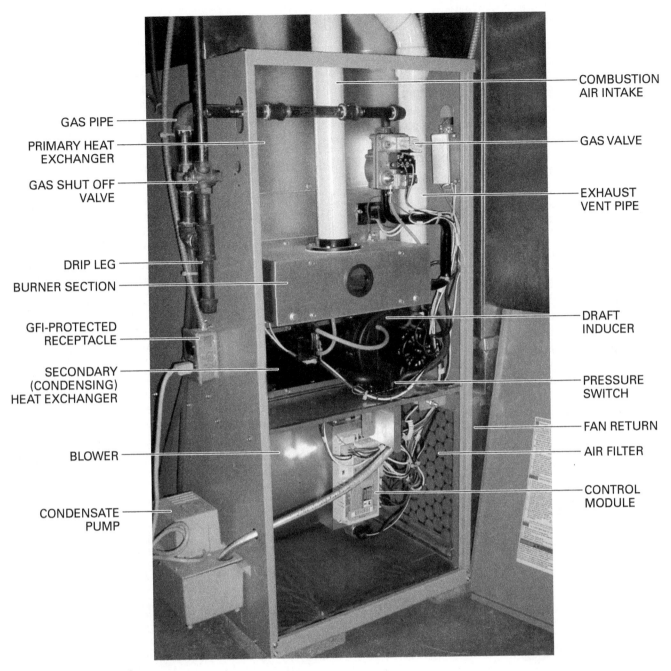

GAS PIPE

PRIMARY HEAT EXCHANGER

GAS SHUT OFF VALVE

DRIP LEG

BURNER SECTION

GFI-PROTECTED RECEPTACLE

SECONDARY (CONDENSING) HEAT EXCHANGER

BLOWER

CONDENSATE PUMP

COMBUSTION AIR INTAKE

GAS VALVE

EXHAUST VENT PIPE

DRAFT INDUCER

PRESSURE SWITCH

FAN RETURN

AIR FILTER

CONTROL MODULE

Figure 9 Condensing furnace components.

2.2.1 Heat Exchangers

The heat exchanger is the part of the furnace where combustion takes place. The heat exchangers are heated by the burning gas. Air flowing over the outside of the heat exchangers in a warm-air furnace picks up the heat as it passes. Of course, the room airflow and the combustion airflow must never be allowed to mix. This can happen when a heat exchanger develops a leak or otherwise deteriorates.

Heat exchangers are usually made of a stamped or rolled aluminized-steel product. There are several types of heat exchangers. The types shown in A, B, and C of *Figure 10* are commonly found in gas-fired furnaces. In these three types, the burners fire directly into the heat exchanger and the hot flue gases flow through the heat exchanger and out the other end into a collector box that directs the gases to the flue piping. Types A and B are often found in induced-draft furnaces, where the induced-draft blower helps push combustion air through the more complex arrangement.

The heat exchangers shown in D and E of *Figure 10* are typical of those found in oil-fired furnaces. The cylindrical heat exchanger (D) is a primary surface. The primary surface is in direct contact with the flame; the secondary surface (E) extracts heat from the hot flue gases. Oil does not burn

as cleanly as gas. Therefore, the long, thin heat exchangers used for gas heating would quickly become clogged if used in oil-fired furnaces.

The number of heat exchanger sections depends on the furnace capacity. In a gas furnace, each heat exchanger section has its own burner. A low-capacity furnace might have two sections fed by two burners; a high-capacity furnace might have five or more sections. The heat exchanger sections terminate in a single collector box that directs the flue gases into the flue pipe.

Condensing furnaces are equipped with a secondary heat exchanger made of stainless steel or plain steel with a corrosion-resistant coating. After the hot air from combustion has passed through the primary heat exchanger, where a great deal of heat has already been extracted, it moves on to the secondary heat exchanger. The secondary heat exchanger often looks like a refrigeration condensing coil. The coil is in the path of the conditioned air flowing through the furnace. This coil is designed to cause the flue gas to achieve maximum contact with the metal in the coil for the longest period of time. Maximum heat is transferred from the flue gas as a result. As the flue gas moves through the coil, the water vapor it contains cools and condenses into liquid. The water comes mainly from the fuel gas itself. The latent heat removed by the condensing process is

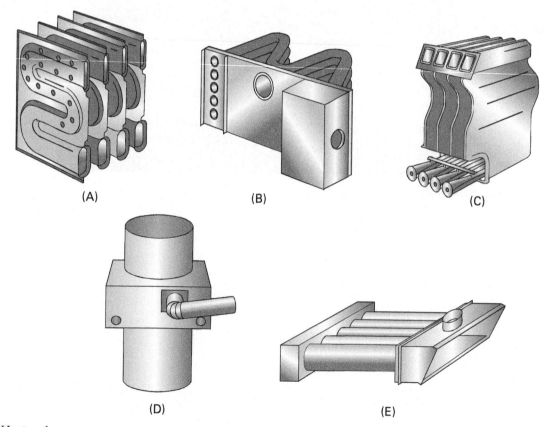

(A)   (B)   (C)

(D)   (E)

Figure 10 Heat exchangers.

also transferred to the conditioned air, along with the heat from the flue gases.

The condensate liquid is often slightly acidic and must be disposed of in accordance with local codes. Condensing furnaces that are designed for installation in multipoise applications may require that the condensate drain be modified. This ensures that the condensate drains from the secondary heat exchanger, regardless of the position in which the furnace is installed.

> **NOTE**
>
> When high-efficiency furnaces are installed in attics, it may be necessary to provide freeze protection to prevent freezing of the condensate and condensate pump.

2.2.2 Fans and Motors

Figure 11 shows the flow of combustion air and conditioned air through a basic furnace. The fan and fan motor, commonly referred to as the blower assembly, are responsible for circulating the air through the ductwork. Most furnaces contain two fans. The blower circulates conditioned air; the induced-draft fan pushes combustion air through the heat exchanger and into the flue piping.

Most induced-draft fans have single-speed motors. However, a variable-speed electronically commutated motor (ECM) may be used to power the draft inducer in deluxe furnaces that have staged or modulating gas valves. The burner input on these furnaces can vary. The ECM enables

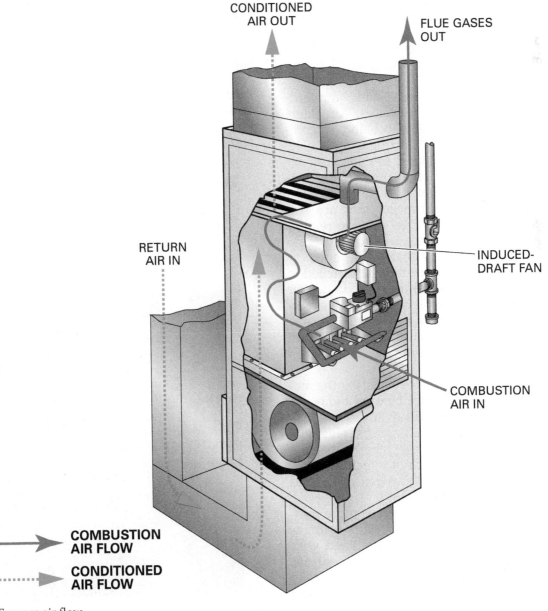

Figure 11 Furnace air flow.

the induced-draft fan to operate at different speeds to more closely match the burner input.

Much of the successful performance of a heating system depends on the proper operation of the blower assembly. It draws return air from the conditioned spaces, then forces it over the heat exchanger and through the supply distribution system to the space to be heated. Outside air is sometimes mixed with the return air to provide adequate indoor air quality. This was rarely necessary in older homes. However, as construction standards improve, introducing some fresh outdoor air into the structure is becoming more common. For commercial installations, it is usually required.

Although the idea is to heat the air passing through to warm a space, the furnace also requires this air to keep from becoming too hot. While an excess amount of air flowing through is of little consequence to furnace operation, insufficient airflow will result in the furnace overheating. This is an important concept to understand that applies to all fossil-fuel and electric-heating equipment.

Centrifugal blower wheels with forward-curved blades are used in furnaces. Air enters through both ends of the wheel and is compressed at the outlet by centrifugal force. The volume of airflow is measured in cubic feet per minute (cfm). One exception to the use of a centrifugal blower wheel is found in some gas-fired unit heaters (*Figure 12*). Propeller fans are still often used in these units. They provide reasonable air-moving performance when there is no ductwork to resists the flow of air.

Blower wheels and motors must be matched to each other and to the system in order to deliver the required amount of air. It is usually necessary to make some adjustment of the air quantities at the time of installation because the resistance of the air distribution system is not the same in each installation.

Blower speeds are changed by altering the motor speed, since most furnace blowers today are direct-drive. In most cases, the motor speed is altered by the use of multispeed motors. This type of motor has a series of electrical taps (connection points) for the windings, each of which provides a different speed. Usually, one speed is selected for heating and another for cooling.

Variable-speed, ECMs have become more common for blowers in furnaces, especially in high-efficiency deluxe models. When combined with electronic room thermostats and furnace controls, they provide more precise control of air volume that translates into quieter and more efficient operation, as well as enhanced indoor comfort.

These motors operate at a variety of different speeds, depending upon the operating mode.

2.2.3 Air Filters

Air filters remove dust, pollen, molds, and other particles from the air circulating through the building. Air filters are commonly located at the return air entrance to the furnace, or at the room return-air inlet. Filters are available in different levels of efficiency (*Figure 13*). Their capacity to capture particles is measured in microns. A micron is $1/1,000$ of a millimeter. There are 25,400 microns in a linear inch. Standard-efficiency filters remove particles of dust, dirt, pollen, and molds as small as 10 microns. That may sound pretty good, but it really means that 50 percent or more of the common particles in the air are getting past the filter. The inexpensive fiberglass filters available in many stores are generally in this class. Most of the permanent washable filters, which are made of metal or polyurethane foam, are in the same efficiency class.

High-efficiency filters capture particles down to about half a micron. That covers mold, pollen, and much of the dust in the air. This class of filters includes self-magnetizing electrostatic filters and bag filters.

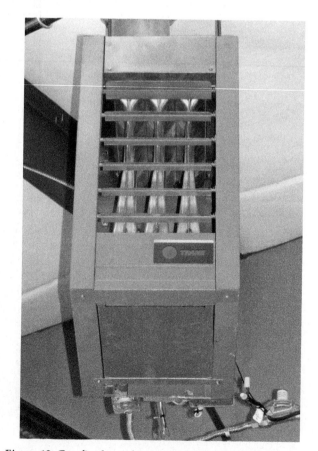

Figure 12 Gas-fired unit heater with a propeller fan.

MERV Ratings for Air Filters

When buying an air filter today, you may see a minimum efficiency rating value (MERV) on the filter label. MERV is part of the American Society of Heating, Refrigerating, and Air-Conditioning Engineers (ASHRAE) *Standard 52.2* that addresses indoor air quality (IAQ). The higher the MERV rating, the more effective the filter is at capturing airborne particles. However, higher efficiency in particle capture can reduce overall system performance. Higher ratings generally mean the filter is more restrictive to airflow, allowing it to capture smaller particles. These more restrictive filters can reduce the amount of airflow through the system. Always follow the furnace manufacturer's MERV rating recommendation when selecting an air filter for a particular furnace, if the information is provided. Filters manufactured by 3M use an MPR (micro-particle performance rating), which relates to the size of particles that the filter is able to capture. For example, a filter with a MERV rating of 6 or 8, would have an MPR rating of 600.

Proper and generous air-distribution system design should allow the use of filters up to a MERV rating of 8–10. A MERV 8 filter typically serves even the most demanding homeowner well. Filters with a MERV rating higher than this are typically found in commercial settings where the added resistance has been factored in to the fan sizing, such as a medical setting. Filters with MERV ratings above 16 are found in surgical suites and clean rooms. They can effectively capture extremely small particles, but offer a substantial amount of resistance to airflow.

(A) CONVENTIONAL FIBERGLASS FILTER (B) ELECTROSTATIC PERMANENT FILTER

(C) PLEATED FILTER

(D) BAG FILTER

(E) HIGH-EFFICIENCY HEPA FILTERS

Figure 13 Air filters.

An electronic air cleaner can capture very fine particles, including those that make up smoke and vapors. Electronic air cleaners can be obtained as standalone units, or may be added to a furnace as an accessory. Electronic air cleaners can produce some ozone, which is classified as a lung irritant when too much is produced. This is a very limited problem, however.

A dirty filter can block airflow, causing a furnace to lose efficiency. Filters need to be replaced or cleaned periodically, depending on the type of filter material. The furnace manufacturer's literature should specify the filter servicing frequency. However, operating conditions truly dictate the need for replacement or cleaning. Note that filters delivered with new furnaces are rarely of good quality.

Turn off the furnace before removing an air filter. The blower should not be allowed to operate with the filter removed. Remember that the filter is not only designed to protect the occupants and help keep the premises cleaner. They are also designed to keep the heat-transfer surfaces clean. Wet evaporator coils can accumulate an astounding amount of debris in a short time.

Some thermostats are equipped with a Check Filter alert. In very rare cases, a sensor that detects the pressure drop across the filter bank activates the message. However, this feature is reserved for more complex commercial systems. In most residential and light commercial cases, a timer simply tracks the elapsed time and activates the warning at the specified interval. Thermostats with this function are almost exclusively programmable, and the period between reminders is an adjustable option.

Disposable filters must be replaced with filters of the same size and type. If the filter is too small for the opening, dirty air can bypass the filter. Air filters must fit snug in their rack to provide proper performance. If a substitute filter offers a much higher resistance to airflow than the original filter, it can cause a troublesome reduction in airflow. Adjustments in blower speed may be required to accommodate them.

2.2.4 Gas Valve Assemblies

The function of a gas burner is to produce a clean flame at the base of the heat exchanger. To do this, the equipment must control and regulate the flow of gas, ensure the proper mixture of gas with air, and ignite the gas under safe conditions. A common gas burner assembly has four major parts or sections: a gas valve, a manifold and orifices,

burners, and an ignition device. Each of these components are discussed separately.

Older systems used a gas supply arrangement in which the gas valve and pressure regulator were separate components. A series of controls in a gas line is sometimes called a gas train. On residential and light commercial systems manufactured since the 1970s, however, these functions, along with a pilot safety control (on pilot-type furnaces) have been combined into a single assembly. Separate gas-pressure regulation is now found mainly on large commercial and industrial gas burners. The modern gas valve is also a redundant gas valve; that is, it contains two independent gas valves in series. If one fails, the other shuts off the gas flow if necessary.

The redundant gas valve assembly shown in *Figure 14* is just one of the many types available. It has a gas outlet for the pilot, as well as a pilot control, which would not be found on the gas control for a direct-ignition furnace.

Gas valves on some high-efficiency furnaces are designed to control two levels, or stages, of heat. These gas valves supply full or partial gas pressure, depending on the heat demand. Others may provide a near-infinite level of gas volume control to precisely match the heat input to the demand of the thermostat. Both automatic and modulating gas valves are described here in more detail.

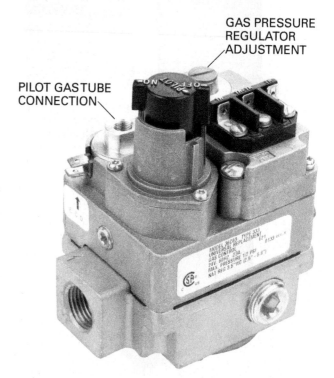

Figure 14 Redundant gas valve.

The principal function of the automatic gas valve is to control gas flow. There are five principal types of automatic gas valves:

- The solenoid-operated valve (*Figure 15*) uses electromagnetic force to operate the valve plunger. When the controls call for heat, the magnetic field of the solenoid coil raises the plunger. Solenoid-operated valves like the one shown here are typically found on large commercial/industrial equipment gas trains.

- The diaphragm-operated valve (*Figure 16*) uses gas pressure above and below the diaphragm to control the device. When the coil is energized (a solenoid coil is still required to operate the valve), the gas supply to the upper section (above the diaphragm) is cut off and the pressure in the upper section is then vented to the atmosphere. The pressure of the gas in the lower section then bends the diaphragm up, allowing gas to flow through the valve. When the coil de-energizes, the upper section is pressurized. With equal pressure above and below the diaphragm, the weight of the diaphragm forces it down to shut off the gas flow.

- Bimetal valve operators have a high-resistance wire wrapped around the diaphragm. When the thermostat calls for heat, current is supplied to the wire, causing it to heat and warp the diaphragm. The warping action opens the valve. The action of this valve is slow and the unit is sometimes referred to as a delayed-action, or slow-opening valve.

- Bulb-type valve operators depend upon the expansion of a liquid-filled bulb to provide the operating force. The sensing bulb is attached to a bellows that expands and contracts, activating the gas flow valve.

The most efficient gas furnaces use two-stage or modulating gas valves to achieve more efficient operation and maximize indoor comfort. Two-stage furnaces fire the burners at a lower input rate on milder days when heating demand is not as great. On very cold days, the burner switches to a higher firing rate, providing more heat to the structure. The higher firing rate also helps the system recover quickly from an energy-saving setback in temperature. The two-stage valve usually has two solenoids controlling two separate valves within one valve body. On a call for low-stage heat, the low-fire solenoid on the valve is energized, providing gas at a lower manifold pressure to the burners. When more heat is required, a high-fire solenoid is energized, supplying gas at a higher manifold pressure to the

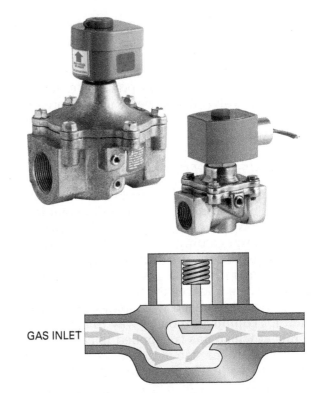

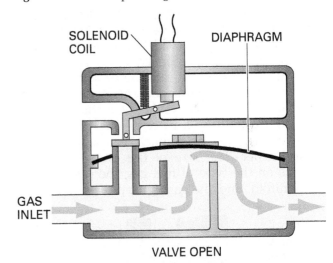

Figure 15 Solenoid-operated gas valve.

Figure 16 Diaphragm-operated gas valve.

burners. The gas valve often contains separate manifold-pressure adjustments for low- and high-fire modes.

Modulating gas valves vary the burner input rate from as low as 40 percent of full input up to 100 percent of input in 1 percent increments. This allows the burner input to more closely match the actual heating load. The modulating valve uses a stepper motor inside the valve to control the position of the gas-valve regulator. Often, a direct-current, pulse-width modulated signal is used to tell the stepper motor to change the position of the regulator. This signal comes from an electronic control that receives heat-load information from the room thermostat. Adjusting manifold pressure on a modulating gas valve is different from adjusting the manifold pressure on a conventional gas valve. It must be done in accordance with the procedure in the manufacturer's installation instructions.

2.2.5 Manifold and Orifices

In a gas furnace, gas fuel is supplied at a low pressure into a burner manifold. The typical manifold pressure for a natural-gas, induced-draft furnace is $3\frac{1}{2}$" of water column (in w.c.). This is equal to only 0.126 psi of pressure. A furnace operating on propane generally has a gas manifold pressure of 10 in w.c. As it leaves the manifold through a small orifice, it is mixed with some of the air required for combustion on its way to the burner. The rate at which gas is supplied to each burner is controlled by the size of the orifice and the applied gas pressure. The pressure is controlled by a pressure regulator, which is usually built into the gas valve assembly.

The burner manifold (*Figure 17*) delivers gas equally to all the burners in a furnace. It is usually made of $\frac{1}{2}$" to 1" black iron pipe, with threaded openings positioned along its length. Spuds are then screwed into the provided openings. (A spud is a threaded metal device that screws into the gas manifold.) The orifice, a precision hole drilled into a piece of brass or similar metal, determines how much gas is delivered to the burner. The size of the orifice selected depends on the type of fuel gas, the pressure in the manifold, and the gas input desired for each

burner. A number stamped on each spud shows the orifice size. If the number is not visible, a numbered drill can be used to determine the size.

Gas burners require some air to be mixed with the gas before combustion. This air is called primary air. Primary air (*Figure 18*) amounts to approximately one-half of the total air required for proper combustion. Too much primary air causes the flame to lift off the burner surface. Too little primary air causes a yellow flame.

Air supplied to the natural-draft burner at the point of combustion is called secondary air. If too little secondary air is present, CO forms. As discussed earlier, most gas burner units are designed to operate on approximately 50 percent excess secondary air supply. For correct flame generation, the proper ratio between primary and secondary air must be maintained. Excessive secondary air is virtually eliminated on induced-draft burners.

2.2.6 Gas Burners

Gas burners can be either single-port (in-shot) or multi-port (*Figure 19*). Single-port burners, which are found in induced-draft furnaces, direct a flame into an opening in the heat exchanger. The flame is actually generated just outside the heat

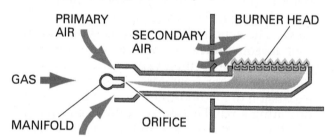

Figure 18 Combustion air supply.

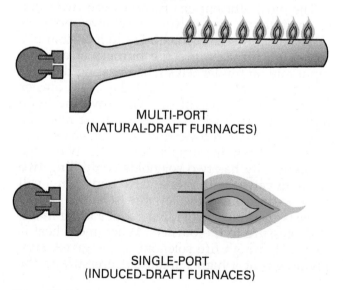

Figure 19 Gas burners.

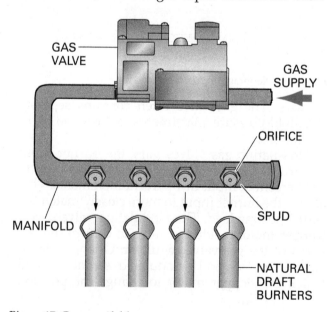

Figure 17 Gas manifold.

Orifice Size

In some situations, it may be necessary to increase or decrease the size of the burner orifice to obtain the correct burner input. The correct way to do this is to replace the orifice with one of the correct size. Never drill out an orifice to increase its size. Drilling in the field can roughen the inside of the orifice, disrupting the smooth flow of gas. The drilling must be performed with great precision, and is best done in a machine shop or manufacturing environment. Also, never peen the orifice with a hammer to decrease its size. This too disrupts the smooth flow of gas.

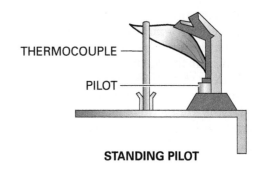

STANDING PILOT

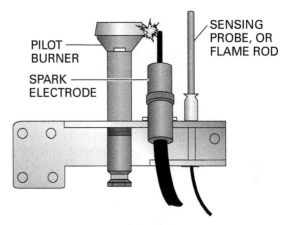

ELECTRIC SPARK IGNITION

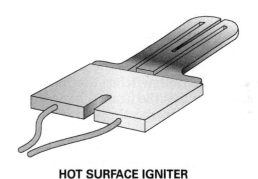

HOT SURFACE IGNITER

Figure 20 Gas-furnace ignition devices.

exchanger opening, in the furnace vestibule. The induced-draft fan is located downstream of the burner, creating a low pressure in the combustion chamber. This reduced pressure draws the flame from the single-port burner into the heat exchanger. Multi-port burners distribute the flame through slots or holes along the length of the burner. Unlike single-port burners, multi-port burners are located inside the heat exchangers.

2.2.7 Ignition Devices

Combustion begins once the gas is ignited. There are several ways that ignition is accomplished. Refer to *Figure 20*.

Older furnaces use a standing pilot, which is a small gas flame that remains on all the time. When gas begins to flow to the main burner, the pilot flame immediately ignites it. The pilot assembly contains a thermocouple that converts the heat from the pilot flame into a small electrical voltage. The thermocouple is connected to the gas-valve control circuit. If the pilot flame stops burning, the voltage from the thermocouple ceases and the gas valve is disabled. This prevents gas from entering the combustion chamber without a source of ignition. A thermopile is used if more voltage is needed. A thermopile consists of thermocouples wired in series. Thermocouples typically produce about 30 millivolts, while thermopiles usually generate about 750 millivolts. Gas furnaces such as wall and floor furnaces that have no blower, and thus no need for a substantial power supply, operate using only a thermopile to operate the unit.

Intermittent ignition saves energy because, unlike the standing pilot, it does not require a continuous gas flow to the pilot. However, a pilot flame is still used to light the main burner. When the thermostat calls for heat, a special transformer in the control circuit produces a high-voltage arc (10,000 volts or more) that ignites the pilot-flame gas. These pilots often use a process known as flame rectification for pilot safety. This process is based on the fact that the pilot flame will produce a tiny current in the microampere range. When this current is sensed, it means that the pilot flame is lit. If the current stops, the control circuit disables the gas valve. On a call for heat, the main gas valve will open only after the pilot flame is proven to exist by the current flow.

Many modern furnaces use a device known as a hot surface igniter (HSI). It is placed next to a burner in place of the pilot. When the thermostat calls for heat, current flows through the igniter, causing it to become extremely hot (*Figure 21*). The intense heat ignites the gas. A flame sensor must be placed near the burners to prove ignition and act as a safety shutoff device. The flame sensor may be separate or, in some cases, integrated into the HSI. HSIs are made of a ceramic material and are therefore fragile. They must be handled carefully.

The early HSIs were very fragile. If they accidentally touched a hard surface while being removed or installed, they were likely to shatter. Manufacturers eventually developed an HSI with a silicon-nitride element. This material is much more durable than the earlier types and is therefore less likely to shatter. Although most technicians do not handle the ceramic portion of the element, as conventional thinking felt the oils from hands would cause damage when heated, manufacturers typically disagree. However, it is always best to minimize touching the element and handle the igniter by its base.

2.2.8 Gas Furnace Safety Controls

Gas furnaces are equipped with several devices that are designed to shut off the furnace if a hazardous condition is sensed. One of these is the high-temperature limit switch. *Figure 22* shows one example of a temperature limit switch. The switch is thermally actuated and is usually mounted above or beside the heat exchangers. Temperature limit switches are located where they can sense that the heat exchanger temperature accurately. Overheating can occur due to a blocked vent or a failed blower motor. Electrically, they are wired in series with the gas valve. If the furnace overheats, the high-temperature limit switch opens, de-energizing the gas valve and

shutting off the gas supply. These switches may reset automatically when the temperature drops below the setpoint. Another high-limit switch is often combined with the fan switch. These switches are called fan-limit switches.

Some counterflow and horizontal furnaces are equipped with manual-reset auxiliary limit switches because of the possibility of reverse airflow occurring if the blower shuts down. Reverse airflow can cause the automatic high-temperature limit switch to reset, even though the conditions that caused it to trip are still present.

Flame rollout switches, such as the one shown in *Figure 23*, are high-limit switches placed in the vestibule the heat exchangers. If there is insufficient combustion air or the vent is restricted, the burner flames can roll out of the combustion chamber and into the furnace vestibule. The

Figure 22 Example of a high-temperature limit switch.

Figure 21 Energized HSI.

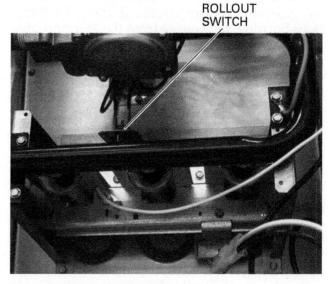
Figure 23 Flame rollout switch.

rollout switch senses this condition and shuts off the power to the gas valve.

A pressure switch, such as the one shown in *Figure 24*, is often used to ensure the presence of combustion airflow through the heat exchangers. Natural-draft furnaces do not require this type of safety device. If the pressure drops below a set point, indicating a failure of the induced-draft fan or an airflow blockage, the switch opens and shuts off the gas valve. The pressure switch must be closed before the gas valve will re-open.

2.3.0 Gas Furnace Installation and Maintenance

Proper installation of a gas furnace is important for the safety of building occupants and reliable operation of the furnace. The following factors are some to consider during the installation of any furnace:

- Location
- Clearances
- Venting
- Combustion air

Applicable national, state, and local codes must be followed when installing a furnace. However, the manufacturers' instructions may be more stringent than the codes. This is done to make sure that codes are not inadvertently violated and to ensure that products are applied in the best way possible. In addition, stringent requirements help protect manufacturers when problems occur with a poor installation.

Proper handling of furnaces is especially important. They are made of sheet metal, which can be damaged if the furnace is dropped or mishandled. Also, some components, such as ceramic igniters, can be easily damaged. Many

Figure 24 Combustion airflow pressure-sensing switch.

parts and pieces can shift position when a furnace is dropped.

A furnace should generally be installed in a location that minimizes the length of duct runs. There should be plenty of space around the furnace to permit easy access for servicing and repair. The manufacturer specifies minimum clearances to combustible materials, and for service access. Flammable materials must not contact the heat exchangers, burners, or any other hot surfaces, such as the flue vent. A suitable location for service and safety is well worth a little extra ductwork.

Furnaces should never be located near a source of aerosol sprays, bleaches, detergents, air fresheners, or cleaning solvents. Even small airborne concentrations of such materials can corrode a furnace. The fumes or vapors are drawn into the combustion chamber along with the combustion air. In such environments, combustion air must be brought in from the outdoors. Manufacturers may refuse to honor warranties if a furnace is exposed to corrosive materials.

To achieve additional heating capacity, as well as a higher volume of airflow, furnaces can be twinned. Two furnaces are placed side-by-side, and their controls are integrated so that they act as a single unit. The heating sections can be operated in stages, although the blower motors should always operate together. It should be noted however, that the merging of the two airstreams typically results in a loss of some total airflow. For example, two twinned furnaces that would each produce 1,200 cfm of airflow alone may produce roughly 2,100 to 2,200 cfm as a twinned pair.

2.3.1 Furnace Venting

The flue gases leaving the furnace normally contain CO_2 and water vapor. CO may be present if combustion is incomplete. CO_2 results if complete combustion occurs. It is not toxic, but it can displace oxygen in an enclosed area if the concentration is large enough. In extreme cases, it can cause asphyxiation. Because of the potential danger from byproducts of combustion, especially CO, flue gases must be vented in accordance with local and national codes.

Natural- and induced-draft furnaces both produce hot flue gases. These gases must be vented through a special metal vent or a masonry chimney (*Figure 25*). It is often necessary to use a double-wall metal vent for the vent connector. If a masonry chimney is used, it must be lined and have the correct dimensions. Even if the chimney has a tile liner, it is sometimes necessary to add a metal liner to meet code requirements. This is

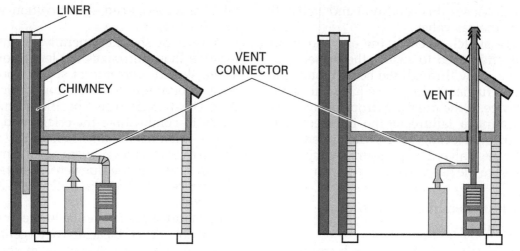

Figure 25 Natural- and induced-draft furnace venting.

especially true if the chimney rises on the outside of the building, because a tile liner takes longer to warm up than a metal liner. The flue gases discharged by the furnace contain water vapor. When this water vapor enters a cold masonry chimney, it condenses and reacts with other combustion products to form compounds that attack the mortar in the chimney, causing it to deteriorate. In addition, the condensate can return to the vent system by draining down the chimney walls, causing corrosion of the vent connector and possibly the heat exchangers.

> **WARNING!**
>
> Condensation is a major concern with induced-draft furnaces. The flue gases from these furnaces are cooler than those of natural-draft furnaces, so condensation is more likely to occur in the vent system. The condensation, in the form of water droplets, can leak back into the furnace and will eventually corrode the heat exchangers. If the heat exchangers are compromised, carbon monoxide can leak into the conditioned spaces, creating a dangerous condition for occupants. These furnaces must be properly and carefully vented in accordance with the manufacturer's instructions and locally accepted codes in order to minimize condensation. For the same reason, a careful inspection of the heat exchangers must be part of any annual service.

When sizing and selecting a vent system, the following factors must be considered:

- The distance from the furnace to the chimney
- The temperature of the flue gases
- Capacity and characteristics of other gas appliances, such as a water heater, that share the same main vent or chimney

Common-Vented Furnace and Water Heater

Common venting of a furnace with a gas water heater is a standard practice. It eliminates the need to have two separate vent connectors or chimneys. An added benefit is that the water heater provides an additional heat source for the vent, keeping it warm and reducing condensation.

The Gas Appliance Manufacturers Association (GAMA) provides tables and instructions that help installers determine how to vent a furnace. Some manufacturers have simplified the tables and instructions for their product lines and provide the information in their installation instructions.

High-efficiency condensing furnaces can be direct-vented through the roof or an outside wall using PVC pipe. Plastic pipe can be used because these furnaces extract so much heat from the combustion process, the flue gas temperatures are well below the melting point of the PVC piping material. During operation, the vent pipe will remain relatively cool to the touch.

For proper condensing-furnace operation, the PVC pipe must be sized and installed according to the furnace installation instructions. Requirements dealing with the diameter of the combustion air and vent pipes, as well as the maximum number of elbows that can be used in the pipe, must be strictly followed. It is also important to note that the combustion air usually must be piped in from the same general location outdoors where the vent line is terminated (*Figure 26*). Vent piping must be pitched for proper condensate drainage and a special fitting may be

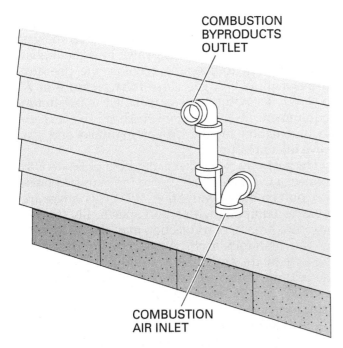

COMBUSTION
BYPRODUCTS
OUTLET

COMBUSTION
AIR INLET

Figure 26 Combustion air and vent terminations for a condensing furnace.

required for vent and combustion air pipe outdoor terminations. Different furnace manufacturers may have different installation requirements.

When installing a condensing furnace to replace a gas furnace that is common-vented with a gas water heater, the venting arrangement for the water heater must then be reconsidered. The existing chimney or vent connector will now only be venting the water heater, while having been sized to vent both units. The existing vent may be oversized for the water heater alone, which may prevent it from venting properly. This situation can be corrected by lining the chimney with a properly sized liner or by installing a new, smaller metal vent.

2.3.2 Combustion Air

Many furnaces obtain the air for combustion from the air surrounding the furnace. This approach is common with natural-draft and induced-draft gas furnaces because the indoor air used for combustion can usually be made up by infiltration. If the furnace is installed in a confined or tightly sealed space such as an equipment closet, the room must be vented to allow for sufficient airflow. However, even if the available space and building construction meet the standards for using indoor air for combustion, there are other factors that must be considered. If corrosive chemicals such as bleach, solvents, construction adhesives, paint stripper, and other household products are used near the furnace, they can contaminate the combustion air, causing corrosion of the heat exchangers. If the furnace cannot be sealed off from the contaminants,

it will be necessary to duct combustion air from outdoors.

High-efficiency condensing furnaces need a lot of combustion air. For that reason, they must have combustion air piped in from the outdoors. The intake is typically located near and below the exhaust vent as shown in *Figure 26*. Care must also be taken when locating the combustion air source for a condensing furnace. For example, it should not be located near a swimming pool because there can be heavy concentrations of chlorine in the air around a pool. Such contaminants in the combustion air can ruin condensing furnace heat exchangers quickly.

> **WARNING!**
> Installation of combustion-air piping requires special attention to local codes. Improper installation can result in incomplete combustion, which is a dangerous condition.

Combustion Air Sources

A fan-assisted (induced-draft) gas furnace installed in an unconfined space, as defined by the *National Fuel Gas Code*, does not require combustion air be ducted from the outdoors. The rule serves as a rule of thumb for making this determination. This rule is derived from the code requiring not less than 50 ft³ of open space for every 1,000 Btuh of the combined input rating for all gas appliances, such as furnace, water heater, or clothes dryer, in the room. This converts to 1 ft³ for every 20 Btuh, or 1/20. The open space is considered to include adjoining rooms that cannot be closed off with doors. Be aware, however, that there is a trend in the industry to use outdoor air regardless of the space available. This trend stems from safety concerns, and is discussed later in the module.

2.3.3 Gas Furnace Maintenance

Gas furnaces require periodic cleaning and inspection of the gas burner assembly, pilot assembly, and automatic gas valve. At the beginning of each heating season, certain maintenance tasks should be performed. For example, air filters should be cleaned or replaced. The blower motor, blower wheel, and heat exchangers should also be cleaned. The heat exchangers can sometimes be cleaned gently with a soft, flexible wire brush attached to a drill. However, this must be done very carefully to avoid damaging the heat exchangers. Cleaning the interior of a heat exchanger is rarely required as long as combustion is complete.

Combustion – The operation of the burners and pilot (if any) should be checked. The burners should produce a steady blue flame with slight yellow/orange tips. A jumpy blue flame that lifts off the burner of a natural-draft furnace may indicate too much draft. A yellow flame indicates a lack of primary air and allows CO to form. A lack of primary air to the burners may also produce other symptoms, including flashback to the spud (also caused by low manifold pressure) and soot deposits on the burners and heat exchangers.

Excessive primary air can cause the flame to lift away from the burner. A flickering or distorted flame could also be a sign of a crack in the heat exchanger. An open crack or hole allows the main blower to affect the flame. If there is any indication of a cracked heat exchanger, shut down and inspect the furnace. A cracked heat exchanger can release CO and other combustion byproducts into the building, posing a serious risk to occupants. A furnace with a known cracked heat exchanger should not be operated. The furnace must be tagged as unsafe and/or inoperable, and the gas supply to it must be disabled.

> **WARNING!**
>
> A furnace that is deemed unsafe to operate due to a cracked heat exchanger or other serious defect that results in an unsafe condition must be locked out and tagged, and the fuel supply physically disconnected from the furnace.

Venting – The byproducts of combustion that are vented to the outdoors include CO_2, nitrogen, and hydrogen. If the furnace heat exchanger or vent pipe is damaged, flue gases may enter the building. While these gases are not generally poisonous themselves, they do replace air in the building. Eventually, occupants may suffocate from a lack of oxygen, or the reduced oxygen supply will result in incomplete combustion at the burner. This leads to the production of deadly CO.

For these reasons, proper venting of the furnace, a thorough inspection of the heat exchangers and flue, and proper adjustment of the flame are essential parts of the service technician's job.

Manifold Pressure – If the burners are not receiving the correct gas pressure, incomplete combustion may occur and the heat exchangers may not warm up enough to prevent condensation. The manufacturer's installation instructions supplied with the furnace specify the correct manifold pressure. For most gas furnaces, the correct manifold pressure is $3\frac{1}{2}$" of water column (w.c.). This is equal to 6.5 mm of mercury (mm Hg). For LP

fuel gas, the manifold pressure is higher, usually around $9\frac{1}{2}$" w.c. (17.7 mm Hg). Of course, the gas pressure supplied to the gas valve must be higher in order for it to properly regulate the pressure to the manifold. A gas valve inlet pressure of 7" to 9" w.c. for natural gas is usually the minimum requirement. It is important to check the furnace documentation for the proper manifold and gas valve inlet pressure.

The best way to check the manifold pressure is to connect a U-tube manometer to the manifold pressure port on the gas valve (*Figure 27*). Gas valves are typically equipped with a port specifically for this purpose. A point of connection may also be located on the manifold itself. A manometer measures pressure by the amount of water that is pushed up the column. Note in *Figure 27* that the manometer reads $1\frac{3}{4}$" w.c. (3.3 mm Hg), up and down. The two values are added, resulting in a pressure reading of $3\frac{1}{2}$" w.c. (6.5 mm Hg). This is a common manifold pressure for natural gas units.

It should be noted that the inlet pressure to most combination gas valves cannot exceed $\frac{1}{2}$ psig, which is equal to 13" to 14" w.c. (24–26 mm Hg). Higher pressures will damage the valve. Also note that there must be sufficient pressure at the inlet of the gas valve to account for any pressure drop through the valve itself and to allow it to operate properly. If only $3\frac{1}{2}$" w.c. is provided at the inlet of the valve, it will not be able to maintain this same pressure in the manifold. This is due to pressure drop as it flows through the valve passages. A gas valve inlet pressure of 7" w.c. for natural gas, and 11" w.c. for propane (13 and 21 mm Hg, respectively), should be sufficient for the gas valve to provide the desired manifold pressure.

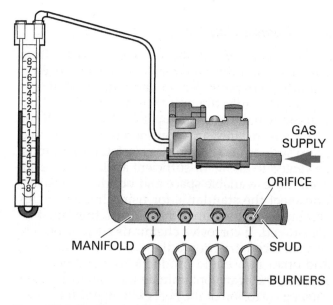

Figure 27 Adjusting manifold pressure.

If the pressure needs to be adjusted, the pressure-regulator adjustment screw on the gas valve is used to make the adjustment. It is usually located on top of the valve, with a cap covering the adjustment screw. The screw changes the amount of pressure applied to a spring, which in turn places pressure on the internal diaphragm.

Most gas valves can be used for both natural and LP gas. The regulator spring in the gas valve can easily be changed, which changes the range of available adjustment. Without first changing the spring, most gas valves cannot be adjusted to a high enough manifold pressure for LP gas operation.

Temperature Rise – The approved temperature rise of a furnace is usually printed on the unit data plate. The temperature rise of a furnace is defined as the temperature difference between the return and supply air streams. In other words, it is the temperature change of the air as it passes over the heat exchanger. For example, the temperature rise stated on a data plate may be shown as 40°F to 70°F (22–30°C). It will always be stated as a range of temperatures.

Measuring the temperature rise of a furnace helps determine that the volume of system airflow is adequate. Note that this refers to the airflow across the heat exchanges and through the air distribution system, not the combustion air supply. If the temperature rise is too low, it usually indicates the airflow is too high. A temperature rise that is higher than the stated range often indicates that the airflow is too low. The problem can also be the result of combustion problems.

To measure the temperature rise, begin by running the furnace for at least 10 minutes after the indoor blower has cycled on. This allows all the furnace and heat exchanger surfaces to reach a normal operating temperature. Then measure and record the return air temperature entering the furnace. Refer to *Figure 28*. The measurement should be taken as close to the furnace inlet as possible. Inserting the probe through a small hole in the furnace return-air plenum is usually a good approach.

The supply air temperature measurement requires some additional thought. If the probe is inserted into the supply duct too close to the furnace, where a line of sight exists between the probe and the heat exchanger, an inaccurate reading will result. The temperature probe will be affected by the radiant heat from the heat exchangers. If the supply duct is routed straight away from the furnace without changes in direction, measure the temperature at least 6' (2 m) away. If a cooling coil is installed on the outlet of the furnace, it will help prevent radiant heat from reaching the probe, and the measurement can be made closer to the unit.

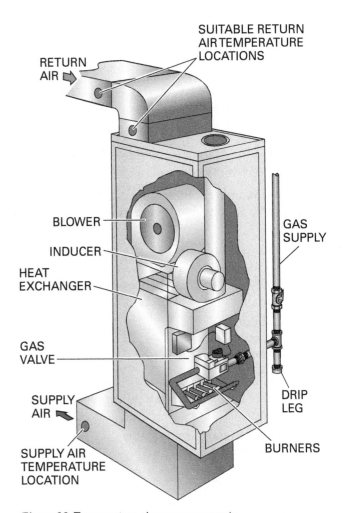

Figure 28 Temperature rise measurements.

However, remember that there is likely some heat loss through the duct walls, so taking the measurement too far away will result in an incorrect reading. If there is an immediate turn in the supply duct, taking the measurement one or two feet away from the turn is usually adequate. The change of direction in the duct will prevent radiant heat from affecting the temperature reading.

The difference in the inlet and outlet temperature represents the temperature rise. If the rise is outside of the stated range, troubleshooting will be necessary to determine the cause.

Always ensure that the blower motor is operating at the desired heating speed when taking temperature-rise measurements. If the fan is switched to On at the thermostat, it typically operates at the speed selected for the cooling mode, which is generally a higher speed than needed for heating. Thus, it would cause inaccurate temperature-rise measurements. Also be sure that the entire air distribution system is attached and in good condition. If the temperature rise is measured with the ductwork detached or incomplete, the airflow values will be higher and the temperature rise will change considerably once the system is complete.

Additional Resources

Refrigeration and Air Conditioning: An Introduction to HVAC, Larry Jeffus. 4th edition. New York, NY. Pearson.

NCCER Module 03202, *Chimneys, Vents, and Flues*.

NCCER Module 03209, *Troubleshooting Gas Heating*.

National Fuel Gas Code, NFPA 54. Boston, MA: National Fire Protection Association. Available at **http://www.nfpa.org**

Proper Venting of Condensing Gas Furnaces. Indianapolis, IN: Bryant Heating and Cooling Systems. Available at **http://dms.hvacpartners.com/docs/1010/Public/05/01-8110-656-25-032008.pdf**

2.0.0 Section Review

1. The type of furnace that has both duct connections on the top is called a(n) _____.

   a. upflow
   b. counterflow
   c. horizontal
   d. low-boy

2. If too little primary air is supplied to a gas burner, the flame will _____.

   a. lift off the burner surface
   b. turn yellow
   c. turn blue
   d. turn orange

3. Gas furnaces should not be vented directly into masonry chimneys because _____.

   a. they are generally too big
   b. mortar causes flue gases to condense
   c. they are usually too far from the furnace
   d. condensation can cause mortar to deteriorate

SECTION THREE

3.0.0 HYDRONIC AND ELECTRIC HEATING SYSTEMS

Objectives

Describe hydronic and electric heating systems.

a. Describe the operation of hydronic heating systems.
b. Describe the operation of electric heating equipment.

Trade Terms

Balance point: An outdoor temperature value that represents an exact match of a structure's heat loss with the capacity of a heat pump to produce heat. At the heating balance point, a heat pump would need to run continuously to maintain the indoor setpoint, without falling below or rising above it.

Dual-fuel system: A heating system typically comprised of a heat pump and a fossil-fuel furnace. Heat pumps supplemented by electric heat are not considered dual-fuel.

Fusible link: An electrical safety device that melts to open a circuit but does not respond to current like a common fuse.

Geothermal heat pump: A heat pump that transfers heat to or from the ground using the earth as a source of heat in the winter and a heat sink in the summer. The system makes use of water as the heat-transfer medium between the earth and the refrigerant circuit.

Sectional boiler: A boiler consisting of two or more similar sections that contain water, with each section usually having an equal internal volume and surface area. Sectional boilers are often shipped in pieces and assembled at the installation site.

Although gas, forced-air furnaces are the most common method of heating in the United States, hydronic heating systems are also popular. In areas of the country with moderate climates, heat pumps, a form of all-electric heat, are commonly used to provide both heating and cooling. Electric heating systems based on resistance heaters are a clean source of heat, but they are not always as cost-effective as gas-fired or hydronic systems. Such systems are typically used in areas where heat is not the highest priority.

3.1.0 Hydronic Heating Systems

So far, this module has focused on forced-air gas-heating systems. Those systems heat air that is circulated throughout the structure. Another type of residential heating system heats water and circulates it throughout the structure through piping. This type of heating system is called a hydronic heating system (*Figure 29*).

The advantages of hydronic heating over forced-air often includes:

- More even temperatures and comfort
- Compact size (pipe is smaller than duct)
- Quieter operation
- Ability to maintain different temperatures in different areas
- Disadvantages of hydronic heating systems include:
  - Higher installed cost when compared to forced-air systems
  - Difficulty in adding air conditioning and other air treatment options

A boiler like the one shown in *Figure 30* is used to heat water in a hydronic heating system. Boilers can be gas- or oil-fired like forced-air furnaces, or they can use electric heating elements immersed in water to heat the water. Gas-fired boilers are also available in high-efficiency condensing versions that extract additional heat from the flue gases. *Figure 31* shows the interior of a high-efficiency condensing boiler.

Newer boiler designs store little or no water, and are designed to provide the needed heating water only on demand. This results in a very compact package that can be wall-mounted (*Figure 32*) as well as free-standing. For larger installations, multiple boilers can be installed and staged to respond based on demand.

Boilers can produce hot water or steam. Steam boilers are rarely installed in new residential applications, due to cost and maintenance requirements. However, if an existing home has a steam boiler that needs to be replaced, it would likely be replaced by a steam boiler because of the differences in piping and other components between steam and hot water systems. This module focuses on residential hydronic systems. Other types of hydronic heating systems are covered in more detail in later modules.

3.1.1 Major Boiler Components

The major components of a boiler include the burner and burner controls, the boiler sections where water is heated, and a pump to circulate the warmed water throughout the structure. The

---

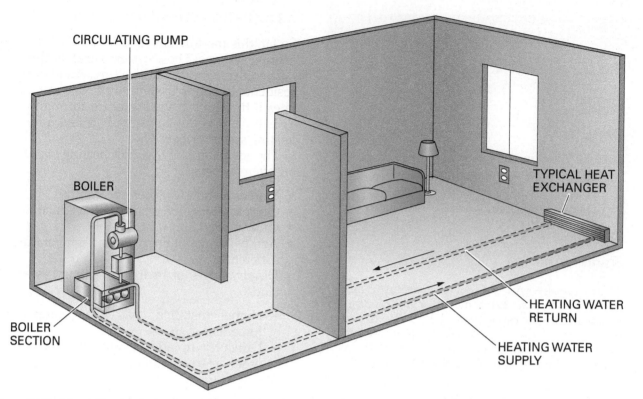

CIRCULATING PUMP

BOILER

TYPICAL HEAT EXCHANGER

BOILER SECTION

HEATING WATER RETURN

HEATING WATER SUPPLY

Figure 29 Residential hydronic-heating system with baseboard radiation.

Figure 30 A standard gas-fired hot water boiler.

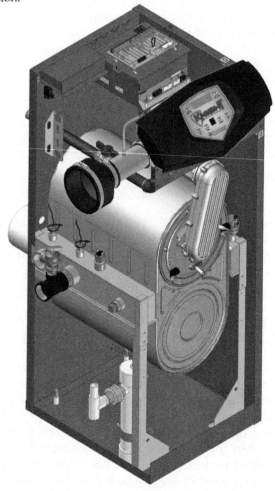

Figure 31 Cutaway of a condensing boiler.

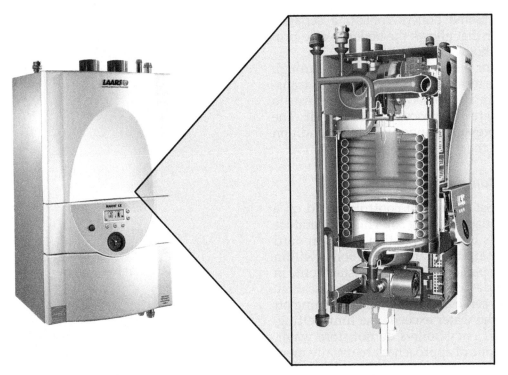

Figure 32 Wall-mounted condensing boiler.

pump may be mounted on the boiler or adjacent to it. The gas-fired burner and the boiler controls are similar in design and function to the burners already discussed under forced-air furnaces.

A standard boiler heat exchanger is usually made of cast iron, but less expensive boilers may have steel heat exchangers. Larger boilers are often created using a number of cast iron sections. This type of boiler is known as a sectional boiler.

Newer, more efficient boilers use finned copper tubing to create a heat exchanger. This construction speeds heat transfer. They hold less water and are able to respond with heat more quickly than cast iron models. Wall-mounted boilers may use heat exchanges of stainless steel. An aluminum-silicon (AlSi) metal alloy is also being used. The burner flame passes over the outside of the boiler sections or heat exchanger, warming the

Around the World

Condensing Boilers

Condensing boilers have long been common in Europe, where systems are designed to operate at lower temperatures (less than 150°F) and fuel efficiency is a higher priority. In a condensing boiler, heat in the flue gases is extracted and used to warm the cool water entering the system. In North America, in the past, the higher water temperatures normally used in heating system design made the use of condensing boilers less desirable. However, improvements in condensing boiler technology have greatly helped overcome this problem, allowing the use of condensing boilers to become more common in North America. They are significantly more efficient than previous designs.

Figure Credit: Courtesy of ECR International Inc.

water within. Some boilers contain a separate heat exchanger that allows the boiler to also provide domestic hot water.

The circulator pump moves the water through the pipes to the terminals. There, the heat is transferred to the air. The cooled water then returns to the boiler, where the process repeats itself. The temperature/pressure gauge allows the condition of the water in the boiler to be monitored.

3.1.2 Boiler Components and Controls

One of the most important controls found on a boiler is the aquastat (*Figure 33*). This temperature switch controls burner operation on a standard boiler to maintain the water in the boiler within a specified temperature range. Aquastats come in a variety of types, depending on the application. Aquastats also serve as limit switches, turning off the burner if the water exceeds the limit setting. Since high-efficiency boilers do not store water, there burner-control approaches are usually more complex.

Water expands when heated, and pressure increases when there is no room for expansion in a closed circuit free of air. If left unchecked, rising pressure in a boiler can cause it to leak or even explode. To prevent this, boilers are equipped with a pressure-relief valve (*Figure 34*). The valve opens and relieves pressure if the pressure becomes too high. These valves are mechanical and have no connection to the electrical control circuit.

A low-water level control is used to ensure that water remains above the minimum needed in the boiler. If the water level is too low, the control prevents the burner from operating by breaking the burner control circuit.

A circulating pump (*Figure 35*) is often factory-installed on the boiler itself. However, some boiler manufacturers ship the pump loose and allow the installer to locate it in a convenient location. Smaller pumps for this purpose are often simply referred to as circulators.

3.1.3 Other Components of a Hydronic Heating System

Other specialized components are needed in a hydronic heating system. *Figure 36* provides an overview of common hydronic system components. In residential applications, the heat is often delivered to the conditioned space by finned baseboard radiators that use convection to transfer heat. Another popular method is the use of tubing embedded in the floor that transfer heat by radiation.

(A) ELECTROMECHANICAL

(B) DIGITAL

Figure 33 Aquastats.

Figure 34 Pressure-relief valve.

(A) STANDARD INLINE CIRCULATOR   (B) ECM-DRIVEN CIRCULATOR

Figure 35 Hydronic circulators.

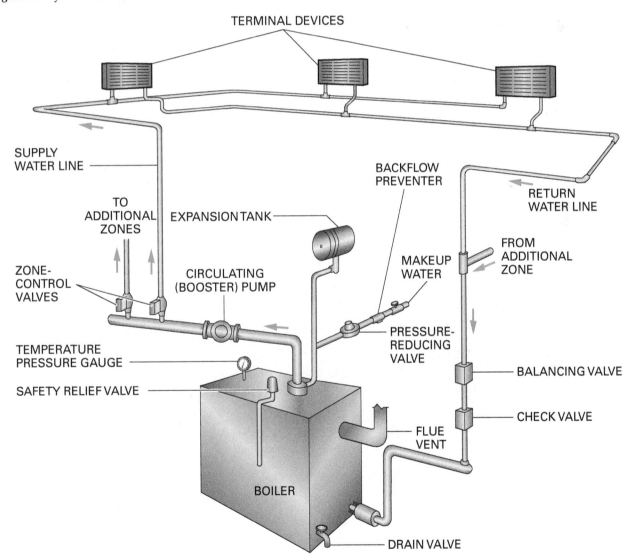

Figure 36 Simplified zoned hydronic-heating system.

One advantage of hydronic systems is that different rooms or zones within the same structure can be kept at different temperatures. This is accomplished by installing a zone control valve in the piping that controls the flow of hot water supplied to each zone. Each zone control valve is opened and closed by its own room thermostat. This allows hot water to flow to the heat-transfer device (baseboard unit, floor piping) in the zone only when needed. Thermostats in other zones open and close zone valves to maintain temperature elsewhere. Here is how a zone valve controls a circulating pump in a typical system:

- A room thermostat for a zone calls for heat, commanding a motor-driven zone valve to open.
- The zone valve is equipped with normally open switch contacts that close when the valve is opened.
- The closure of the switch contacts completes the control circuit to energize the circulating pump.

The pump remains running until the thermostat is satisfied, the zone valve closes, and the control switch in the valve opens the circuit. If other valves have since opened, the pump will remain running until all zones are satisfied.

Air must be eliminated from hydronic systems since it interferes with water circulation in closed systems. When systems are first filled with water, a great deal of air must be purged. However, small amounts of air enter the system any time additional water is added. Air is entrained in water from the municipal water supply. As a result, air elimination is a consistent requirement.

When the system is idle (no flow), air rises to the highest point it can reach in the piping. In large systems, there may be many high points. Air vents are typically positioned at high points in multiple locations. When a manual vent is opened to purge a system, it should be left open until a steady stream of water is obtained. Automatic air vents (*Figure 37*) are installed to eliminate air without constant attention from a user or technician. The outlet of an automatic air vent is often piped to a safe drain.

A make-up water valve connected to the water supply automatically replaces any water lost due to leaks or air venting. It is a pressure-regulating type of valve, and is set to maintain a specified water pressure in the hydronic system. A typical pressure setting is 12 psig. The make-up water valve may also be coupled with a pressure-relief valve (*Figure 38*) to prevent the hydronic system from being over-pressurized by city water pressure. This valve is usually in addition to the pressure-relief valve mounted on the boiler itself. Since many boilers are rated for a maximum water pressure of 30 psig, this is a common setting for both relief valves. Backflow preventers are installed upstream of the make-up water valve to ensure that water from the hydronic system cannot accidently reverse direction and enter the potable water supply.

Since water expands when heated, hydronic systems include some type of expansion tank. The expanding water causes the air inside the tank to be compressed, providing a flexible cushion. This prevents pressure in the system from reaching unsafe levels due to normal expansion from heating.

Figure 37 Automatic air vents.

Figure 38 Water-pressure regulator and pressure-relief valve combination.

All-electric heating can be provided in a number of ways. Heat pumps are the most common all-electric heating systems in use. Self-contained electric heating units are also available.

3.2.1 Electric Furnaces

In an electric furnace, air passes directly over the heated elements without the need for a heat exchanger. There are no byproducts of combustion to vent. From the outside, an electric furnace (*Figure 39*) looks like a simple metal enclosure with access doors.

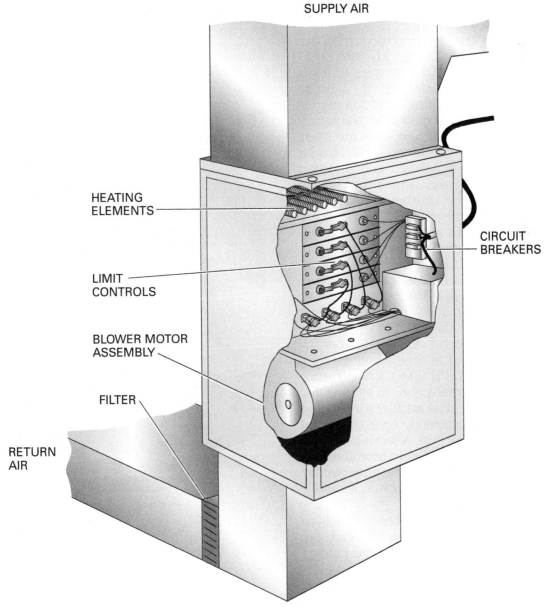

Figure 39 Electric furnace.

Because no fuel is burned in an electric furnace, there is no need for a chimney or vent to carry the products of combustion outdoors. This feature allows for greater operational safety and more installation flexibility. The return air from the conditioned space passes directly over the resistance heaters and into the supply air plenum. The amount of heat supplied by an electric furnace depends upon the number and size of the resistance heaters used in the application. The heating elements are often sequenced on in stages, depending on the demand from the thermostat.

Although heating elements are available in many different ratings, individual elements in central heating systems are typically rated at 4–5 kilowatts (kW). However, remember that the actual heating capacity of a resistance heating element is dependent upon the voltage provided. As the voltage increases, the heating capacity of the element increases along with the wattage.

The major components of an electric furnace are the heating elements, the blower assembly, controls, and the outer cabinet. Like furnaces, a filter is often provided at the return air inlet.

The function of the heating elements (*Figure 40*) is to provide the heat required for the conditioned space. The heating element wires are made of nickel and chromium (Nichrome). The heating element wire is normally spiraled and threaded through a metal holding rack, which has ceramic insulators to prevent the resistance wires from touching the frame of the rack. The elements are similar to those found in small household appliances such as electric clothes dryers and toasters.

The circuit for each heating element may contain a fusible link and a high-temperature safety switch. The switch protects against an overheating condition, and automatically resets when the temperature falls below the setpoint. A common setting for the limit switch is to open at approximately 160°F (71.1°C), and to close when the temperature drops below 125°F (51.6°C). The fusible link is used for added protection. In spite of the name, it is not current-sensitive like a fuse. The fusible link melts to permanently open the circuit when exposed to excessive heat. A technician must replace the link before that element assembly will function. The furnace may also be protected by standard fuses. The fuses protect against excessive current from shorts or grounds.

The blower in an electric furnace is usually a multi- or variable-speed direct-drive blower. It can provide sufficient air for operation of the cooling system as well.

Filters, humidifiers, and cooling coils are added to an electric furnace in much the same manner as with gas and oil furnaces. Therefore, electric

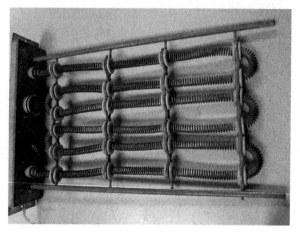

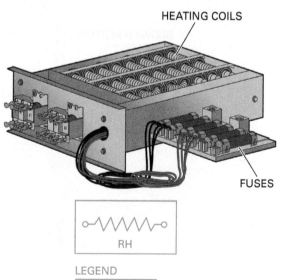

HEATING COILS

FUSES

RH

LEGEND

RH = RESISTANCE HEATER

Figure 40 Electric resistance-heating elements.

furnaces can support year-round comfort like other furnaces.

A residential electric-furnace power supply is usually 208/240V, single-phase, 60 Hertz (Hz) alternating current. This type of power is supplied by three wires: two hot conductors and one grounding conductor. Unlike a 120V power supply, there is no neutral wire. Fused disconnect switches are often provided near the installation, in addition to the circuit breaker found in the electrical panel. All external wiring must conform to the *National Electrical Code*® (*NEC*®).

3.2.2 Heat Pumps

Heat pumps were briefly introduced in Module 03101, "Introduction to HVAC." A heat pump can extract heat from the outdoor air by reversing the mechanical refrigeration cycle. Air-to-air heat pumps are the most common, but water-to-air, water-to-water, and ground-source heat pumps are also popular. Air-to-air heat pumps begin to lose heating capacity as the outdoor temperature falls, so they are used more heavily in areas with moderate climates. Since the temperature of groundwater supplies are far more consistent, water-to-air units provide more consistent capacity throughout the heating season.

A heat pump requires a special valve known as a reversing valve, as shown at the bottom of *Figure 41*. They are also referred to as four-way valves, since they have four piping connections.

In addition to the reversing valve, a heat pump requires a defrost timer to remove frost and ice buildup on the outdoor coil during subfreezing weather. Refrigerant metering devices must be provided for both the indoor and outdoor coils, since each must serve as the evaporator coil in one mode or the other. Note in *Figure 41* that check valves are provided to ensure proper flow through or around the metering device at each coil. In many cases, the metering device and check valve are a single device.

Most heat pumps cannot provide enough heat for a structure alone. They are most often supplemented by electric heating coils that operate when the heating capacity of the heat pump is insufficient to keep up with the heat loss. The heating elements also provide a backup source of heat if the heat pump fails. They are also called upon to warm the air when the heat pump is operating in the defrost mode. Supplemental heat for heat pumps may be required by local code.

A heat pump may have a heavier duty compressor than a typical air conditioner because of its year-round duty. However, today's compressors are often designed to handle both applications. An indoor heat-pump coil generally has a greater internal volume because it serves as the condensing coil in the heating mode. Heat pumps usually require an accumulator, also called a suction line accumulator, to trap liquid refrigerant before it can enter the compressor. The portion of the suction line outside of split-system units

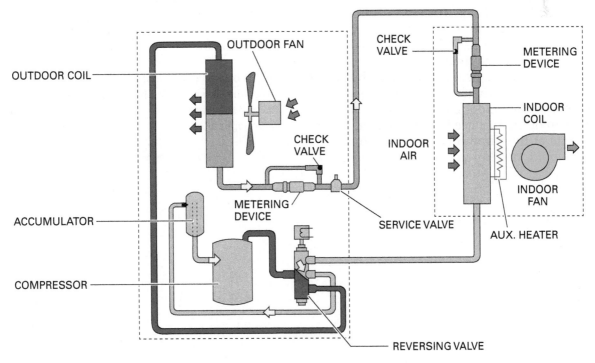

Figure 41 Schematic diagram of a heat pump system.

serves as the hot-gas line during the heating cycle.

Water-source heat pumps transfer heat using water instead of air. They are more effective and provide more consistent heating and cooling capacity than air-source units. For a standard water-source unit, the source of water may be a pond or lake. There are, however, inherent problems with using dirty water from these sources. A geothermal heat pump also uses water as a means of heat transfer, but not from a surface source. Groundwater from a well is a better option. The water is generally returned to the ground through a second well some distance away. Coils or long lengths of tubing can also be placed in the ground or along the bottom of a lake or pond (*Figure 42*), and water is circulated through the tubing to absorb or reject heat. This results in a closed-loop system, where the same water is recirculated through the loop. This significantly reduces problems with contaminants in the water.

Water-source heat pumps often have very high efficiency values, since the water source is stable in temperature. In both the heating and cooling modes, they can provide very efficient operation along with consistent capacity, regardless of outdoor air temperatures.

In commercial installations, water-source heat pumps may be connected to a boiler and cooling tower to heat and cool the water in a loop serving many heat pump units.

Dual-Fuel Systems – Heat pumps are sometimes combined with fossil-fuel furnaces. This type of system is referred to as a dual-fuel system. However, the furnace and the heat pump cannot operate at the same time. Electric heaters can operate along with the heat pump, since the heating elements are downstream of the indoor coil. With furnaces, the indoor coil is typically located downstream of the furnace heat exchanger. If the two operate at the same time, the heat from the furnace would quickly increase the operating pressure of the heat pump beyond a safe level. As a result, an outdoor thermostat is used to enable the furnace and disable the heat pump when the outdoor temperature falls below the point where the heat pump cannot keep up with the heating load. This temperature point, called the balance point, can be determined using a graph of the structure's heat loss. It is then compared to a graph of the heat-pump heating capacity at various temperatures. Refer to *Figure 43*. At the balance point, the heat pump's heating capacity matches the heat loss of the structure. If the temperature drops below the balance point, the heat pump cannot keep up on its own. In a dual-fuel system, the heat pump is shut down and the furnace assumes heating responsibilities. At or above the balance point, the heat pump can maintain the indoor setpoint alone, although it must run continuously to do so at or near the balance point. It is important to note that this applies primarily to air-to-air heat pumps. Remember that the capacity of heat pumps that extract heat from water or geothermal sources are not significantly affected by the outdoor air temperature.

Absorption Heat Pumps

Absorption heat pumps are driven by a heat source, rather than electricity. The most common heat source is natural gas, so absorption heat pumps are sometimes called gas-fired heat pumps. Solar energy, geothermal energy, and propane are also used as heat sources. The absorption heat pump uses an ammonia-water cycle. Ammonia is the refrigerant. The ammonia is absorbed into water, and a pump is used instead of a compressor to raise the pressure. The purpose of the heat source is to boil the water out of the refrigerant so the cycle can begin again.

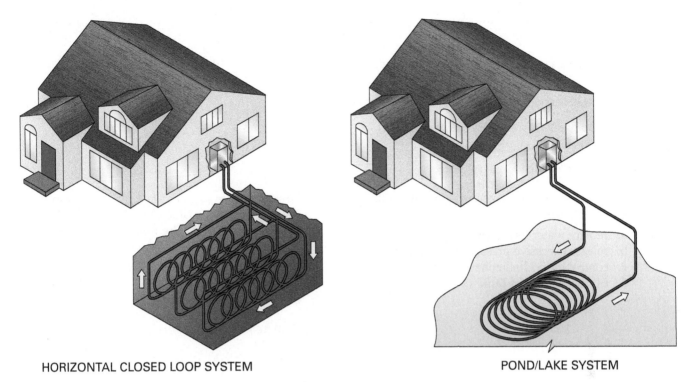

HORIZONTAL CLOSED LOOP SYSTEM   POND/LAKE SYSTEM

Figure 42 Water-source heat pump closed-loop examples.

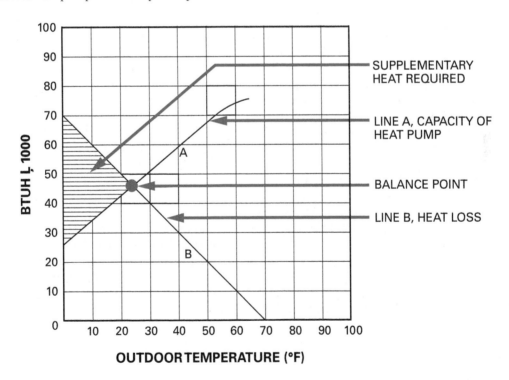

Figure 43 Balance point calculations.

Additional Resources

Refrigeration and Air Conditioning: An Introduction to HVAC, Larry Jeffus. 4th edition. New York, NY. Pearson.

NCCER Module 03203, *Introduction to Hydronic Heating.*

NCCER Module 03211, *Heat Pumps.*

3.0.0 Section Review

1. The make-up water valve in a hydronic heating system is used to _____.

   a. replace water lost from the system
   b. release excessive pressure in the boiler
   c. control individual zone temperatures
   d. fill the expansion tank

2. Refer to *Figure 43* in the text, A heat pump has heating capacities of 40,000 Btuh at 17°F and 70,000 Btuh at 47°F. The home has no heat loss at 70°F and a heat loss of 80,000 Btuh at 0°F. What is the balance point temperature?

   a. 20°F
   b. 26°F
   c. c.31°F
   d. d.34°F

SUMMARY

Gas-fired warm air furnaces and boilers provide heat for many of the buildings in the United States. Wood and coal are also used as heat sources, but their application is very limited because they are less convenient, messy, and their combustion products are more damaging to the environment. Gas heating systems operate on the principle of transferring the heat generated by fuel combustion to air or water, which is then circulated through the conditioned space. Furnaces (whether gas or oil) contain a fuel control, a combustion chamber, and heat exchangers. They must also have a means of moving the heat through the conditioned space, as well as a system for venting flue gases to the outside.

Older gas furnaces were inefficient since as much as 35 percent of the heat they produced was lost with the venting of the flue gases. Today's condensing furnaces have efficiency levels ranging from 90 to 97 percent. They use less fuel and are cheaper to operate than older furnaces. Hydronic heating systems that use fossil-fuel boilers to heat water or steam have also become far more efficient.

Air-to-air heat pumps are the primary choice for all-electric heating systems. Heat pumps are able to extract heat from the outside air using the mechanical refrigeration cycle. Electric furnaces that use resistance heating elements are also available. Resistance heating elements are also used to supplement and serve as a backup to heat pumps. Electric furnaces provide clean heat, but are usually more expensive to operate than natural gas. As a result, their use is limited primarily to areas where heating is a lower priority.

1. Placing an iron on clothing represents heat transfer by _____.
   a. radiation
   b. convection
   c. conduction
   d. capillary action

2. When you become warm while sitting some distance from a fire, heat is being transferred by _____.
   a. radiation
   b. convection
   c. conduction
   d. combustion

3. Humidifiers are used to _____.
   a. reduce condensation
   b. add moisture to the air
   c. keep furniture dry
   d. lower the room temperature

4. A temperature of 80°F is equal to _____.
   a. 12°C
   b. 26.7°C
   c. 176°C
   d. 202°C

5. Carbon monoxide (CO) is produced when _____.
   a. combustion is incomplete
   b. oil and oxygen are mixed
   c. there is water vapor in the flue gas
   d. there is excess air in the combustion chamber

6. A gas burner flame should be _____.
   a. yellow with a blue tip
   b. completely blue
   c. completely yellow
   d. blue with a yellow/orange tip

7. Of the following gases, the one with the highest heating value per cubic foot is _____.
   a. propane
   b. butane
   c. natural gas
   d. manufactured

8. In a forced-air furnace, air to warm the space _____.
   a. flows through the inside of the heat exchangers
   b. is heated directly by the burners
   c. flows over the outside of the heat exchangers
   d. is circulated through the furnace by convection

9. In a condensing furnace, the secondary heat exchanger _____.
   a. is used only in the cooling mode for dehumidification
   b. is used to reheat the flue gases before they are vented
   c. extracts heat by condensing moisture from the flue gas that flows through it
   d. extracts heat by condensing moisture from the conditioned air that passes over it

10. With few exceptions, the type of fan used to circulate conditioned air through a furnace is a(n) _____.
    a. backward-inclined centrifugal blower
    b. forward-inclined centrifugal blower
    c. propeller fan
    d. axial fan

11. Direct-venting of flue gases with PVC pipe may be used _____.
    a. only with high-efficiency condensing furnaces
    b. for all fan-assisted and high-efficiency furnaces
    c. only with natural-draft furnaces
    d. for any furnace when a masonry chimney is not used

12. A manometer may be used to measure _____.
    a. how much water is in the flue gas
    b. the gas pressure in the manifold
    c. the flame rectification current
    d. the percentage of primary air used in combustion

13. Sectional heat exchangers in a boiler are usually constructed of _____.

    a. steel
    b. cast iron
    c. stainless steel
    d. finned copper tubes

14. The high-temperature safety switch in an electric furnace is usually set to _____.

    a. close if the temperature exceeds 160°F
    b. open if the fuse fails to blow
    c. open if the induced-draft fan shuts down
    d. open if the temperature exceeds 160°F

15. In an electric furnace, the heating element replaces which of the following item(s) found in a gas furnace?

    a. Burner assembly and heat exchangers
    b. Burner assembly and blower
    c. Induced-draft motor and blower
    d. Condensing coil

Trade Terms Quiz

Fill in the blank with the correct trade term that you learned from your study of this module.

1. A device that is used to transfer heat from a warm surface to a cooler surface is called a(n) _____.

2. Furnaces with an electric spark ignition often use _____ to prove the pilot flame has been established.

3. The chemical reaction between fuel, heat, and oxygen is known as _____.

4. A furnace that uses a fan to draw in and exhaust combustion air is a(n) _____.

5. In a confined space or tightly sealed building, there may not be enough _____ to provide make-up air to the furnace.

6. When the thermostat calls for heat from a gas furnace, the main burner may be lit by a current flowing through the _____, causing it to become extremely hot.

7. Older furnaces typically have a(n) _____, which always remains lit.

8. The current industry standard for defining the overall efficiency of a furnace is the _____.

9. One instrument used to measure pressure is a(n) _____.

10. If a standing pilot flame goes out, a loss of voltage from the _____ will signal the gas valve control circuit to turn off the gas.

11. The air that is mixed in during the combustion process and helps push the byproducts of combustion out through the flue vent is known as _____.

12. Precisely drilled holes that control the flow of gas to burners are called _____.

13. Most modern furnaces have two independent gas valves in series, which together are known as a(n) _____.

14. A(n) _____ can reach efficiencies of 95 percent and more.

15. A(n) _____ has an AFUE of less than 78 percent.

16. The gas burner will have a blue flame when approximately 50 percent of the _____ is mixed with gas before ignition.

17. A number stamped on the _____ shows the orifice size.

18. A furnace that can be configured for upflow, counterflow, or horizontal installation is a(n) _____.

19. A hydronic heating unit that consists of similar portions and is usually assembled at the installation site is called a(n) _____.

20. When the outdoor temperature falls below the _____, a heat pump can no longer maintain the setpoint of a structure without supplemental heat.

21. Before fuel oil is burned in an oil furnace, the oil must be _____.

22. Heat that is transferred without any physical contact between substances is the result of _____.

23. A(n) _____ can open an electrical circuit but does not respond to current.

24. When a heat pump is paired with a fossil-fuel furnace, it is referred to as a(n) _____.

25. Warm, rising air that is displaced by cooler, falling air results in heat transfer by _____.

26. A heating/cooling system that uses the earth as both a heat source and a heat sink is called a(n) _____.

27. For variable-speed blower performance, a system is equipped with a(n) _____.

28. Two surfaces that are in contact with each other and have different temperatures can transfer heat through _____.

Trade Terms

Annual Fuel Utilization Efficiency (AFUE)
Atomized
Balance point
Combustion
Condensing furnace
Conduction
Convection

Dual-fuel system
Electronically commutated motor (ECM)
Flame rectification
Fusible link
Geothermal heat pump
Heat exchanger
Hot surface igniter (HSI)

Induced-draft furnace
Infiltration
Manometer
Multipoise furnace
Natural-draft furnace
Orifices
Primary air
Radiation

Redundant gas valve
Secondary air
Sectional boiler
Spud
Standing pilot
Thermocouple

Trade Terms Introduced in This Module

Annual Fuel Utilization Efficiency (AFUE): HVAC industry standard for defining furnace efficiency, expressed as a percentage of the total heat available from a fuel. The AFUE goes beyond the specific thermal efficiency of a unit in that it accounts for both the peak measurement of fuel-to-heat conversion efficiency and the losses of efficiency that occur during startup, shutdown, etc.

Atomized: Broken into tiny pieces or fragments, such as a liquid being broken into tiny droplets to create a fine spray

Balance point: An outdoor temperature value that represents an exact match of a structure's heat loss with the capacity of a heat pump to produce heat. At the heating balance point, a heat pump would need to run continuously to maintain the indoor setpoint, without falling below or rising above it.

Combustion: The process by which a fuel is ignited in the presence of oxygen.

Condensing furnace: A furnace that contains a secondary heat exchanger that extracts latent heat by condensing exhaust (flue) gases.

Conduction: In the context of heat transfer, the process through which heat is directly transmitted from one substance to another when there is a difference of temperature between regions in contact.

Convection: The movement caused within a fluid (air or water, for example) by the tendency of hotter fluid to rise and colder, denser material to fall, resulting in heat being transferred.

Dual-fuel system: A heating system typically comprised of a heat pump and a fossil-fuel furnace. Heat pumps supplemented by electric heat are not considered dual-fuel.

Electronically commutated motor (ECM): A DC-powered motor able to operate at variable speeds based on the programming of its electronic control module, which is AC-powered and converts AC to three-phase DC. ECMs also operate more efficiently than standard motors.

Flame rectification: A method of proving the existence of a pilot flame by applying an AC current to a flame rod, which is then rectified to a DC current as it flows back to a ground source. Monitoring the DC current flow provides the means of proving a pilot flame has been established.

Fusible link: An electrical safety device that melts to open a circuit but does not respond to current like a common fuse.

Geothermal heat pump: A heat pump that transfers heat to or from the ground using the earth as a source of heat in the winter and a heat sink in the summer. The system makes use of water as the heat-transfer medium between the earth and the refrigerant circuit.

Heat exchanger: A device that is used to transfer heat from a warm surface or substance to a cooler surface or substance.

Hot surface igniter (HSI): A ceramic device that glows when an electrical current flows through it. Used to ignite the fuel/air mixture in a gas furnace.

Induced-draft furnace: A furnace in which a motor-driven fan draws air from the surrounding area or from outdoors to support combustion and create a draft in the heat exchanger.

Infiltration: Air that unintentionally and naturally enters a building through doors, windows, and cracks in the structure.

Manometer: An instrument that measures air or gas pressure by the displacement of a column of liquid.

Multipoise furnace: A furnace that can be configured for upflow, counterflow, or horizontal installation.

Natural-draft furnace: A furnace in which the natural flow of air from around the furnace provides the air to support combustion and venting of combustion byproducts.

Orifices: Precisely drilled holes that control the flow of gas to the burners.

Primary air: Air that is pulled or propelled into the initial combustion process along with the fuel.

Radiation: In the context of heat transfer, the direct transmission of heat through electromagnetic waves through space or another medium. No contact between the two substances is required.

Redundant gas valve: A gas control containing two gas valves in series. If one fails, the other is available to shut off the gas when needed.

Secondary air: Air that is added to the mix of fuel and primary air during combustion.

Sectional boiler: A boiler consisting of two or more similar sections that contain water, with each section usually having an equal internal volume and surface area. Sectional boilers are often shipped in pieces and assembled at the installation site.

Spud: A threaded metal device that screws into the gas manifold. It contains the orifice that meters gas to the burners.

Standing pilot: A gas pilot that remains lit continuously.

Thermocouple: A device comprised of two different metals that generates electricity when there is a difference in temperature from one end to the other.

Additional Resources

This module presents thorough resources for task training. The following resource material is suggested for further study.

Refrigeration and Air Conditioning: An Introduction to HVAC, Larry Jeffus. 4th edition. New York, NY: Pearson.

National Fuel Gas Code, NFPA 54. Boston, MA: National Fire Protection Association. Available at **http://www.nfpa.org**

Proper Venting of Condensing Gas Furnaces. Indianapolis, IN: Bryant Heating and Cooling Systems. Available at **http://dms.hvacpartners.com/docs/1010/Public/05/01-8110-656-25-032008.pdf**

NCCER Module 03203, *Introduction to Hydronic Heating*.

NCCER Module 03211, *Heat Pumps*.

NCCER Module 03202, *Chimneys, Vents, and Flues*.

NCCER Module 03209, *Troubleshooting Gas Heating*.

Figure Credits

Answer	Section Reference	Objective
Section One		
1. c	1.1.3	1a
2. d	1.2.5	1b
Section Two		
1. d	2.1.0; Figure 3	2a
2. b	2.2.5	2b
3. d	2.3.1	2c
Section Three		
1. a	3.1.3	3a
2. b	3.2.2	3b

NCCER CURRICULA — USER UPDATE

NCCER makes every effort to keep its textbooks up-to-date and free of technical errors. We appreciate your help in this process. If you find an error, a typographical mistake, or an inaccuracy in NCCER's curricula, please fill out this form (or a photocopy), or complete the online form at **www.nccer.org/olf**. Be sure to include the exact module ID number, page number, a detailed description, and your recommended correction. Your input will be brought to the attention of the Authoring Team. Thank you for your assistance.

Instructors – If you have an idea for improving this textbook, or have found that additional materials were necessary to teach this module effectively, please let us know so that we may present your suggestions to the Authoring Team.

NCCER Product Development and Revision

13614 Progress Blvd., Alachua, FL 32615

Email: curriculum@nccer.org
Online: www.nccer.org/olf

❏ Trainee Guide   ❏ Lesson Plans   ❏ Exam   ❏ PowerPoints   Other _____

Craft / Level: _____   Copyright Date: _____

Module ID Number / Title: _____

Section Number(s): _____

Description: _____

_____

_____

_____

_____

Recommended Correction: _____

_____

_____

_____

_____

Your Name: _____

Address: _____

_____

Email: _____   Phone: _____

Introduction to Cooling

OVERVIEW

To service cooling equipment, you must have a clear understanding of the refrigeration cycle. You must also understand the function of the primary refrigeration-circuit components. Technicians can apply this knowledge to all refrigeration circuits. Despite the differences in the many refrigerants and their boiling points, the basic principles presented here apply to all direct-expansion refrigerant circuits.

Module 03107

Trainees with successful module completions may be eligible for credentialing through the NCCER Registry. To learn more, go to **www.nccer.org** or contact us at 1.888.622.3720. Our website has information on the latest product releases and training, as well as online versions of our *Cornerstone* magazine and Pearson's product catalog.

Your feedback is welcome. You may email your comments to **curriculum@nccer.org**, send general comments and inquiries to **info@nccer.org**, or fill in the User Update form at the back of this module.

This information is general in nature and intended for training purposes only. Actual performance of activities described in this manual requires compliance with all applicable operating, service, maintenance, and safety procedures under the direction of qualified personnel. References in this manual to patented or proprietary devices do not constitute a recommendation of their use.

03107 V5

INTRODUCTION TO COOLING

Objectives

When you have completed this module, you will be able to do the following:

1. Explain the fundamental concepts of the refrigeration cycle.
   a. Describe how heat affects the state of substances.
   b. Explain how heat is transferred from one substance to another.
   c. Describe pressure-temperature relationships.
   d. Describe basic refrigerant flow and the changes of state occurring in the refrigeration cycle.
   e. Identify common instruments used to measure pressure and temperature.
2. Identify common refrigerants and their identifying characteristics.
   a. Identify fluorocarbon refrigerants.
   b. Describe the use of ammonia as a refrigerant.
   c. Identify various refrigerant containers and their safe handling requirements.
3. Identify the major components of cooling systems and how they function.
   a. Identify various types of compressors.
   b. Identify different types of condensers.
   c. Identify different types of evaporators.
   d. Describe the devices used to meter refrigerant flow.
   e. Discuss basic refrigerant piping concepts.
   f. Identify various accessories used in refrigerant circuits.
4. Identify common controls used in cooling systems and how they function.
   a. Identify common primary controls.
   b. Identify common secondary controls.

Performance Tasks

Under the supervision of the instructor, you should be able to do the following:

1. Measure temperatures in an operating cooling system.
2. Calibrate a set of refrigerant gauges and thermometers.
3. Connect a refrigerant gauge manifold and properly calculate subcooling and superheat on an operating system using a temperature probe.
4. Identify refrigerants using cylinder color codes.
5. Identify compressors, condensers, evaporators, metering devices, controls, and accessories.

Trade Terms

Absolute pressure	Conduction	Hygroscopic	Pressure	Slug
Atmospheric pressure	Conductors	Insulators	Pump-down control	Specific heat
	Convection	Latent heat		Subcooling
Barometer	Desiccant	Latent heat of condensation	Radiation	Superheat
Brine	Enthalpy		Refrigerant floodback	Surge chamber
British thermal unit (Btu)	Fluorocarbon	Latent heat of fusion		Thermistor
	Gauge pressure		Rupture disk	Thermocouple
Comfort cooling	Halocarbon	Latent heat of vaporization	Saturation temperature	Ton of refrigeration
Compound	Halogens			
Condensation	Hydrocarbons	Mixture	Sensible heat	Total heat

Industry-Recognized Credentials

If you are training through an NCCER-accredited sponsor, you may be eligible for credentials from NCCER's Registry. The ID number for this module is 03107. Note that this module may have been used in other NCCER curricula and may apply to other level completions. Contact NCCER's Registry at 888.622.3720 or go to **www.nccer.org** for more information.

Contents

1.0.0 Refrigeration Cycle Fundamentals...3
  1.1.0 Heat and the States of Substances 4
    1.1.1 Temperature 4
    1.1.2 Heat Content 4
    1.1.3 Sensible and Latent Heat 6
    1.1.4 Specific Heat Capacity 9
  1.2.0 Heat Transfer...9
    1.2.1 Conduction 9
    1.2.2 Convection...9
    1.2.3 Radiation..9
    1.2.4 Conductors and Insulators 10
    1.2.5 Rate of Heat Transfer 10
  1.3.0 Pressure-Temperature Relationships 10
    1.3.1 Pressure 11
    1.3.2 Absolute and Gauge Pressure...11
    1.3.3 Pressure/Temperature Relationships...................................13
  1.4.0 The Refrigeration Cycle 14
    1.4.1 Basic Component Functions 16
    1.4.2 The Refrigeration Cycle in a Typical Air Conditioning System.....17
  1.5.0 Pressure and Temperature Measurement 20
    1.5.1 Thermometers 20
    1.5.2 Gauge Manifolds and Other Pressure Measurement Devices 22
2.0.0 Refrigerants...27
  2.1.0 Fluorocarbon Refrigerants...27
    2.1.1 CFCs, HCFCs, and HFCs...27
    2.1.2 Refrigerant Blends 29
  2.2.0 Ammonia (NH3)..29
  2.3.0 Refrigerant Containers and Handling...31
    2.3.1 Disposable Cylinders...31
    2.3.2 Returnable Cylinders 31
    2.3.3 Recovery Cylinders..33
    2.3.4 Identifying Refrigerants..33
    2.3.5 Refrigerant Safety Precautions 33
3.0.0 Cooling System Components...35
  3.1.0 Compressors 35
    3.1.1 Reciprocating Compressors...37
    3.1.2 Rotary Compressors 37
    3.1.3 Scroll Compressors 39

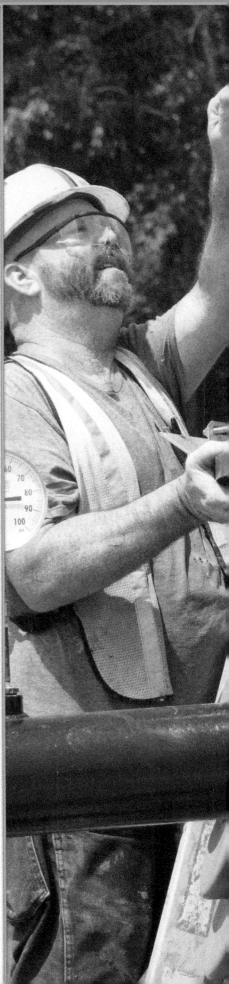

3.1.4   Screw Compressors 39

3.1.5   Centrifugal Compressors 40

3.2.0   Condensers 41

3.2.1   Air-Cooled Condensers 42

3.2.2   Water-Cooled Condensers 43

3.2.3   Cooling Towers 45

3.2.4   Evaporative Condensers 45

3.3.0   Evaporators 46

3.3.1   Direct-Expansion (DX) Evaporators 47

3.3.2   Flooded Evaporators 48

3.3.3   Evaporator Construction 49

3.3.4   Chilled-Water System Evaporators 50

3.4.0   Refrigerant Metering Devices 51

3.4.1   Fixed Metering Devices 52

3.4.2   Adjustable Metering Devices 53

3.5.0   Refrigerant Piping 55

3.5.1   Basic Principles 55

3.5.2   Suction Line 55

3.5.3   Hot-Gas Line 57

3.5.4   Liquid Line 58

3.5.5   Insulation 59

3.6.0   Refrigerant Circuit Accessories 59

3.6.1   Filter-Drier 59

3.6.2   Sight Glass and Moisture Indicator 59

3.6.3   Suction Line Accumulators 61

3.6.4   Crankcase Heaters 61

3.6.5   Oil Separators 62

3.6.6   Heat Exchangers 63

3.6.7   Receiver 64

3.6.8   Service Valves 64

3.6.9   Compressor Muffler 65

4.0.0   Cooling System Control 67

4.1.0   Primary Controls 67

4.1.1   Thermostats 67

4.1.2   Pressure Switches 69

4.1.3   Humidistat 70

4.1.4   Time Clock 70

4.2.0   Secondary Controls 70

4.2.1   Condenser Water Valves 71

4.2.2   Evaporator-Pressure Regulator 71

4.2.3   Check Valves 71

4.2.4   Pressure-Relief Devices 72

4.2.5   Oil-Pressure Safety Switches 72

4.2.6   Flow Switches 72

Figures and Tables

Figure 1   Mechanical refrigeration systems transfer heat from one location to another...5

Figure 2   The Fahrenheit and Celsius temperature scales...........................5

Figure 3   Changing the state of water. 6

Figure 4   Changing the state of water—metric values. 7

Figure 5   Terminology related to changing states. 8

Figure 6   The methods of heat transfer. 10

Figure 7   One ton of refrigeration illustrated. 11

Figure 8   Pressure is exerted in various directions, depending upon the state of a substance. 12

Figure 9   Absolute, gauge, and barometric pressure readings compared. 12

Figure 10 Pressure/temperature relationship for water. 14

Figure 11 Basic function of a window air conditioning unit. 15

Figure 12 Typical refrigeration cycle for a system using HFC-410A refrigerant. 19

Figure 13 Electronic thermometer. 20

Figure 14 Dial and electronic pocket thermometers. 21

Figure 15 An infrared, non-contact thermometer..21

Figure 16 The D:S ratio illustrated. 22

Figure 17 Calibrating a pocket thermometer. 22

Figure 18 Gauge manifold set...23

Figure 19 Digital gauge manifold. 23

Figure 20 Calibrating a refrigerant gauge...25

Figure 21 Digital and fluid-filled manometers..25

Figure 22 Refrigerant containers. 32

Figure 23 Rupture disk of a disposable refrigerant cylinder.......................32

Figure 24 Open, direct-drive compressors used in a large refrigeration application. 36

Figure 25 A hermetic scroll compressor. 37

Figure 26 A semi-hermetic compressor. 37

Figure 27 A cut-away showing some of the field-serviceable components of a semi-hermetic compressor...37

Figure 28 Reciprocating compressor operation. 38

Figure 29 Cut-away view of a hermetic reciprocating compressor.............38

Figure 30 Rotary-vane compressor operation. 38

Figure 31 Scroll compressor operation..39

Figure 32 Screw compressor operation...40

Figure 33 A cut-away of a small screw compressor...................................40

Figure 34 A centrifugal chiller. 40

Figure 35 Centrifugal compressor operation..41

Figure 36 Typical air- and water-cooled condensers. 42

Figure 37 Air-cooled condenser operation...43

Figure 38 A variety of water-cooled condenser types................................44

Figure 39 Natural-draft cooling tower. 47

Figure 40 A forced-draft, counterflow cooling tower. 47

Figure 41 An induced-draft cooling tower...48

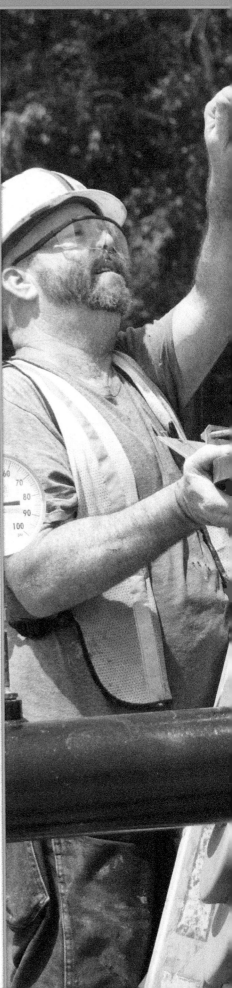

Figure 42   A typical evaporative condenser..48
Figure 43   An evaporator coil for commercial walk-in refrigerators.............48
Figure 44   The operation of direct-expansion and flooded evaporators......48
Figure 45   A cased direct-expansion evaporator coil....................................49
Figure 46   Evaporator construction. 49
Figure 47   Plate-freezing equipment for the food processing industry. 50
Figure 48   A large chiller with a centrifugal compressor. 52
Figure 49   Examples of common fixed and adjustable metering devices. ..53
Figure 50   The internal construction of a typical TXV..................................55
Figure 51   Major refrigerant lines. 56
Figure 52   A suction line riser..57
Figure 53   A reduced suction-line riser..57
Figure 54   A double suction riser. 57
Figure 55   Routing the suction line above the top of the evaporator coil. ...58
Figure 56   Hot-gas line...58
Figure 57   Common accessories used in the refrigeration circuit. 60
Figure 58   Filter-drier and moisture-indicating sight glass. 60
Figure 59   A cut-away of a typical filter-drier, showing the filter media and
            desiccant beads that trap water. 60
Figure 60   Suction line accumulator. 61
Figure 61   Crankcase heaters. 62
Figure 62   Oil separator. 63
Figure 63   Refrigerant-to-domestic water preheater. 64
Figure 64   Service valves. 64
Figure 65   A 4-in-1 valve-core tool...65
Figure 66   Compressor muffler. 68
Figure 67   Snap-action thermostat operation..68
Figure 68   A remote-bulb thermostat. 68
Figure 69   A programmable electronic thermostat. 69
Figure 70   Thermostat control using an app...69
Figure 71   A low-pressure switch (opens on a fall). 69
Figure 72   Non-adjustable pressure switch..70
Figure 73   Wall-mounted humidistat. 70
Figure 74   Electronic and mechanical time clocks. 71
Figure 75   Condenser water valve. 71
Figure 76   Evaporator-pressure regulating (EPR) valve. 71
Figure 77   Oil-pressure safety switch. 72
Figure 78   A water flow switch, with paddle attached.....................................72

Table 1   Specific Heat Values 9
Table 2   P/T Chart for HFC-410A and HCFC-22, in 5°F Increments...........13
Table 3   Examples of Old and New Refrigerant Names 28
Table 4   Color Codes for Common Refrigerants 33

1.0.0 REFRIGERATION CYCLE FUNDAMENTALS

Objective

Explain the fundamental concepts of the refrigeration cycle.

a. Describe how heat affects the state of substances.
b. Explain how heat is transferred from one substance to another.
c. Describe pressure-temperature relationships.
d. Describe basic refrigerant flow and the changes of state occurring in the refrigeration cycle.
e. Identify common instruments used to measure pressure and temperature.

Performance Tasks

1. Measure temperatures in an operating cooling system.
2. Calibrate a set of refrigerant gauges and thermometers.
3. Connect a refrigerant gauge manifold and properly calculate subcooling and superheat on an operating system using a temperature probe.

Trade Terms

Absolute pressure: Positive pressure measurements that start at zero (no atmospheric pressure). Gauge pressure plus the pressure of the atmosphere (14.7 psi at sea level at 70°F) equals absolute pressure. Absolute pressure is expressed in pounds per square inch absolute (psia). Absolute pressure = gauge pressure + atmospheric pressure.

Atmospheric pressure: The pressure exerted on all things on Earth's surface due to the weight of the atmosphere. It is roughly 14.7 psi at sea level at 70°F; the standard pressure exerted on Earth's surface. Atmospheric pressure may also be expressed as 29.92 inches of mercury.

Barometer: An instrument used to measure atmospheric pressure, typically in units of inches of mercury (in. Hg).

British thermal unit (Btu): The amount of heat needed to raise the temperature of one pound of water one degree Fahrenheit.

Comfort cooling: Cooling related to the comfort of humans in buildings and residences. The temperature range for comfort cooling is typically considered to be 60°F (15.5°C) to 80°F (26.6°C).

Condensation: The process related to the change of state of a substance, such as water, from the vapor or gaseous state to a liquid state.

Conduction: In the context of heat transfer, the process through which heat is directly transmitted from one substance to another when there is a difference of temperature between regions in contact.

Conductors: Relevant to heat transfer, materials that readily transfer heat by conduction.

Convection: The movement caused within a fluid (air or water, for example) by the tendency of hotter fluid to rise and colder, denser material to fall, resulting in heat being transferred.

Enthalpy: The total heat content (sensible and latent) of a refrigerant or other substance.

Gauge pressure: The pressure measured on a gauge, expressed as pounds per square inch gauge (psig) or inches of mercury vacuum (in. Hg vac.). Pressure measurements that are made without including atmospheric pressure.

Insulators: Materials that resist heat transfer by conduction.

Latent heat: The heat energy absorbed or rejected when a substance is in the process of changing state (solid to liquid, liquid to gas, or the reverse of either change).

Latent heat of condensation: The heat given up or removed from a gas in changing back to a liquid state (steam to water).

Latent heat of fusion: The heat gained or lost in changing to or from a solid (ice to water or water to ice).

Latent heat of vaporization: The heat gained in changing from a liquid to a gas (water to steam).

Pressure: The force exerted by a substance against its area of containment; mathematically defined as force per unit of area.

Radiation: The movement of heat in the form of invisible rays or waves, similar to light.

Saturation temperature: The boiling temperature of a substance at a given pressure. In the saturated condition, both liquid and vapor is likely present in the same space.

Sensible heat: Heat energy that can be measured by a thermometer or sensed by touch.

Specific heat: The amount of heat required to raise the temperature of one pound of a substance one degree Fahrenheit. Expressed as Btu/lb/°F. The specific heat of water (H_2O) is 1.0, thus providing the basis for defining one Btu.

Subcooling: The reduction in temperature of a refrigerant liquid after it has completed the change in state from a vapor. Only after the phase change is complete can the liquid begin to decrease in temperature as heat is removed.

Superheat: The additional, increase in temperature of a refrigerant vapor after it has completed the change of state from a liquid to a vapor. Only after the phase change is complete can the vapor begin to increase in temperature as additional heat is applied.

Thermistor: A semiconductor device that changes resistance with a change in temperature.

Thermocouple: A device made up of two different metals that generates electricity when there is a difference in temperature from one end to the other.

Ton of refrigeration: Large unit for measuring the rate of heat transfer. One ton is defined as 12,000 Btus per hour, or 12,000 Btuh.

Total heat: Sensible heat plus latent heat.

In nature, there is nothing from which heat or temperature is totally absent. Saying that something is cold simply means that it has less heat energy than something else does.

Refrigeration is a term that describes the intentional transfer of heat from one place, where it is not wanted, to another place where the additional heat is not important. Air conditioning units and refrigerators do not "pump" cold into a space. They simply transfer heat out of a space or object and move it elsewhere (*Figure 1*). In most cases, the heat is transferred outdoors. A fluid known as a refrigerant circulates through a refrigeration or air conditioning system. The refrigerant absorbs heat from the refrigerated space, and then carries it to a location outside the space where the heat is rejected. Various refrigerants are covered later in this module. Even water can be considered a refrigerant when it is used to transfer heat.

1.1.0 Heat and the States of Substances

To understand the refrigeration cycle, you must understand heat. Like light, electricity, and magnetism, heat is a form of energy. Like other forms of energy, heat can do work, but it cannot be destroyed. Heat can be measured and controlled, however. Its ability to do work depends on two characteristics: the temperature (intensity) and the heat content (quantity).

1.1.1 Temperature

Temperature provides a way to compare the level of heat contained in an object or substance. The intensity of heat is measured in degrees with a thermometer. Two of several different temperature scales are commonly used for measuring temperature. The temperature scale most often used in the United States is the Fahrenheit scale (*Figure 2*). The other scale commonly used worldwide, and sometimes in the United States, is the Celsius, or Centigrade, scale.

On the Fahrenheit scale, water boils at 212°F and freezes at 32°F. The distance between the two points is divided into 180 equal parts. On the Celsius scale, water boils at 100°C and freezes at 0°C. The distance between the two points is divided into 100 equal parts. The absolute zero point shown on the temperature scale in *Figure 2* is the theoretical point where all molecular motion stops, resulting in zero heat content. Note that the boiling points are based on the water being exposed to the standard atmospheric pressure for sea level.

The following formulas are used to convert temperatures from one scale to another:

$$C = \tfrac{5}{9} \times (F - 32°)$$
$$F = (\tfrac{9}{5} \times C) + 32°$$

For added convenience, a Fahrenheit-to-Celsius conversion table has been provided in *Appendix A*. Note_ that the formulas are used to convert from one measured temperature to the other. However, consider the following statement: *There is usually a 3°F difference between the first and second stage of cooling.* This is not a measured temperature, but a degree count. To convert this to °C, simple multiply it by $\tfrac{5}{9}$ (0.55). The same type of value in °C is converted to °F by multiplying by $\tfrac{9}{5}$ (1.8).

1.1.2 Heat Content

Heat content, or the quantity of heat, is the amount of heat energy contained in a substance. Heat content in the HVACR industry is measured by the British thermal unit (Btu) in the English system. In the metric system, the joule is the measure of heat content. One Btu is the amount of heat needed to raise the temperature of one lb of water by 1°F. For example, to heat 1 lb of water from 40°F to 50°F (a difference of 10°F), 10 Btus of heat must be added to the water. When the process is reversed, the same math holds true. If an lb of water is cooled by 10°F, 10 Btus are transferred out of the water.

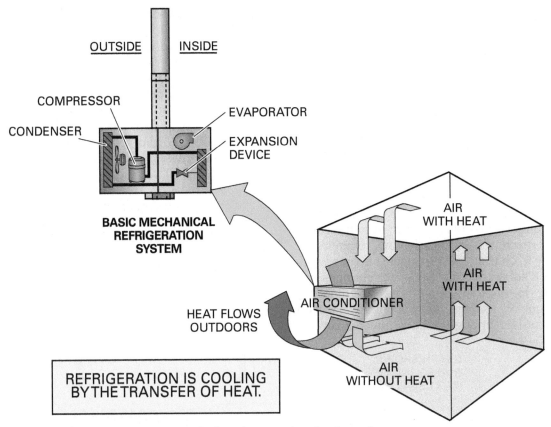

Figure 1 Mechanical refrigeration systems transfer heat from one location to another.

Pressure Affects Boiling Point

As altitude increases, atmospheric pressure decreases. Because temperature and pressure are directly related, a liquid will boil at lower temperatures as the altitude increases. In a high-altitude location like Denver, Colorado, it takes longer to cook most foods in water than it does in locations closer to sea level. The opposite condition occurs in a pressure cooker. The increased pressure inside the cooker causes boiling to occur at a higher temperature, so the contents cook faster than it would at atmospheric pressure.

Most man-made refrigerants are designed to boil at relatively low temperatures, so that heat passing over the evaporator at normal room temperature causes the refrigerant in the evaporator to boil. In an HFC-410A cooling system, for example, the low-side pressure is about 118 psig (814 kPa). This corresponds to a saturated refrigerant temperature of 40°F (4°C). Because of this, the heat absorbed from the air in the conditioned space is warm enough to cause the refrigerant to boil.

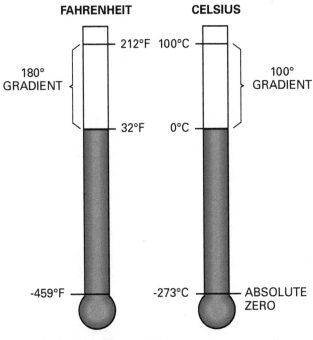

Figure 2 The Fahrenheit and Celsius temperature scales.

1.1.3 Sensible and Latent Heat

Depending on their heat content and temperature, materials or substances can exist in three states: solid, liquid, and gas. Using water as an example, the three states are ice (solid), water (liquid), and steam or vapor (gas).

In a refrigeration system, the refrigerant inside exists in either a liquid or a gaseous state, depending on its heat content. For this reason, these two states are the ones you will be most concerned with in your study of refrigeration. Since solids cannot flow through a system, this state is avoided in the refrigerant circuit. In truth, the vast majority of refrigeration systems cannot even approach the extremely low temperatures necessary to change most refrigerants into a solid.

When a substance such as water changes from one state to another, something peculiar happens (*Figure 3*). Assume there is one lb of ice, and the ice has had enough heat removed from it to reach a temperature of 0°F (–18°C). If heat is then added to the ice, a thermometer would show a rise in temperature until the reading reaches 32°F (0°C). At this point, the ice begins the process of changing to its liquid state (water). If heat is continually added, the thermometer reading will remain fixed at 32°F instead of continuing to rise as expected. In fact, it continues to read 32°F until the entire pound of ice has melted.

The increase in temperature from 0°F to 32°F (–18°C to 0°C) registered by the thermometer is called sensible heat. It is considered sensible because it can be sensed by a thermometer. Sensible heat is heat that is added to, or removed from, a substance without any change of state occurring.

But what about the heat that was added to the ice after the thermometer reached 32°F, where a change in state to a liquid began? The ice continually absorbed heat, without any sensible change in temperature, until all of the ice became water.

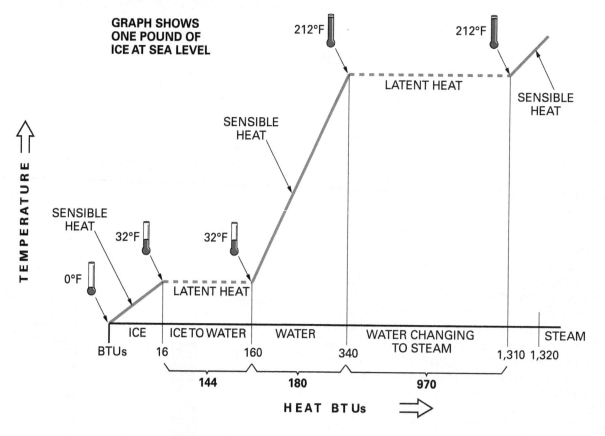

SENSIBLE HEAT:

HEAT THAT CAUSES A TEMPERATURE CHANGE AND CAN BE SENSED BY A THERMOMETER OR BY TOUCH.

LATENT HEAT:

HEAT THAT PRODUCES A CHANGE IN STATE WITHOUT A CHANGE IN TEMPERATURE.

Figure 3 Changing the state of water.

Put another way, a significant amount of heat was transferred to the ice during this period, but the thermometer did not show any change in temperature. This heat is called latent heat. Latent heat is heat energy that is added to, or removed from, a substance during a change-of-state process. While the substance is actively changing states, no measurable change in temperature occurs. All the heat energy transferred during that time is devoted solely to the change of state, and a measurable temperature change does not occur. The combination of sensible heat and latent heat together represent the total heat, or enthalpy, of a substance.

If heat is continually added to the pound of water after all the ice has melted, the thermometer will once again show an increase in temperature (sensible heat) until the temperature reaches 212°F (100°C). At this point, the water begins to boil. Once boiling begins, the water starts to change to its gaseous state. For water, this state is commonly known as steam. As additional heat is added, the thermometer reading remains at 212°F until all the water has turned into steam. During this time, latent heat is being absorbed by the water. Once all the water has been converted to steam, the thermometer will once again register sensible heat. *Figure 3* provides a graphical analysis of the process described here. To see the process in metric values, refer to *Figure 4*. The preceding discussion has been confined to one temperature scale to maintain clarity.

It takes a great deal more heat to cause a change in state than is needed for a sensible change in temperature. For water, it requires 144 Btus of latent heat to change one pound of ice at 32°F (0°C) into one pound of water at 32°F. Consider that this is 144 times as much heat as is needed to raise the temperature of water one degree. The change from water to steam requires an even greater amount of latent heat. It takes 970 Btus to change one pound of water at 212°F (100°C) to steam at 212°F.

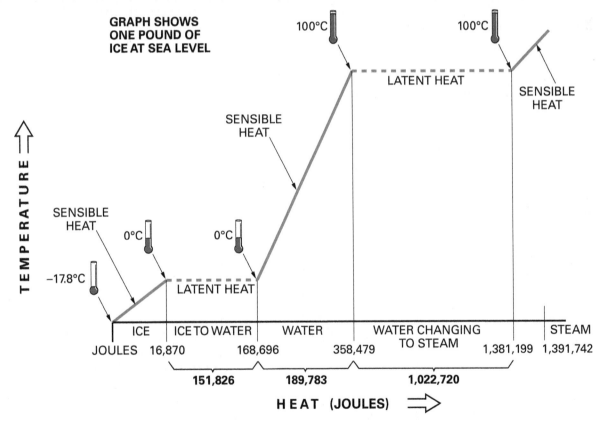

Figure 4 Changing the state of water—metric values.

Figure 5 graphically illustrates some important terminology related to the changing of states. Although water is still being used as an example, the terminology applies to all similar substances. Note first that the latent heat related to the change of state has several special names.

- Latent heat of fusion – The heat gained or lost in changing to or from a solid state (ice to water or water to ice).
- Latent heat of vaporization – The heat gained in changing from a liquid to a vapor (water to steam). The temperature at which this process begins is the boiling point. In refrigeration work, the boiling point is also known as the saturation temperature.
- Latent heat of condensation – The heat given up or removed from a vapor in changing back to a liquid state (steam to water).

Two additional terms shown in *Figure 5* are very important for HVACR technicians to understand. Superheat is the sensible (measurable) heat added to a vapor after the change of state from a liquid is complete. For example, water boils at 212°F (100°C) and remains at this temperature throughout the change of state. As the last bit of liquid water is converting to steam, the temperature remains the same at 212°F. However, once the last of the liquid water has become steam and then more heat is added, the temperature of the steam begins to rise. Once the temperature of the steam increases beyond 212°F, it is described as superheated. If the steam temperature rises 5°F more to 217°F (103°C), then it is said to have 5°F (2.8°C) of superheat.

Subcooling is the reverse of superheat. It is the sensible heat lost from a liquid (or solid) after a change to the next lower state is complete. For example, if water is cooled 5°F (2.8°C) below its boiling point to a temperature of 207°F (97.2°C), it has been subcooled 5°F (2.8°C). When ice is cooled 5°F below the freezing point to 27°F, it too is said to have 5°F of subcooling.

The importance of these two terms and their meaning cannot be overstressed. Understanding these terms and the related concepts is essential to understanding how the refrigeration cycle transfers heat. HVACR technicians who are engaged in troubleshooting and repairing cooling systems often work with these terms and concepts on a daily basis.

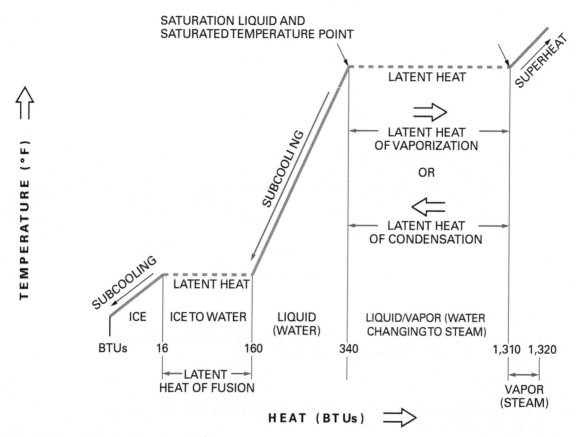

Figure 5 Terminology related to changing states.

1.1.4 Specific Heat Capacity

Specific heat is the amount of heat required to raise the temperature of one pound of a substance one degree Fahrenheit. Thus far, water has been used to demonstrate the process of changing states. It was shown that water in the liquid state has a specific heat of one Btu per pound, per each degree Fahrenheit of temperature change. This would be written as 1 Btu/lb/°F. When the discussion is clearly related to specific heat, it may be written simply as 1.00.

The specific heat of each substance is different. The specific heat of water at 1.00 provides the baseline for all other substances. *Table 1* shows the specific heat values for some common substances. Specific heat values vary from substance to substance, but also from one state of a given substance to another. As shown in *Table 1*, liquid water has a specific heat of 1.00. Ice, however, has a specific heat of 0.50. The result is that it takes twice as many Btus of heat to raise one pound of water one degree as it does to raise one pound of ice one degree. If one Btu of heat is added to one pound of ice, the temperature increases 2°F. Note that this fact was illustrated in *Figure 3* and *Figure 4*. Because the specific heat of ice is 0.50, it took only 16 Btus of heat to raise the temperature of the ice from 0°F to 32°F. Water with a specific heat of 1.00 required 180 Btus to raise the temperature from 32°F to 212°F.

1.2.0 Heat Transfer

Heat transfer is the movement of heat from one place to another, either within a substance or between substances. Heat always flows from a warmer location to a cooler location, like water running downhill. It is important to remember that the transfer of heat requires a difference in temperature. The three methods of heat transfer are conduction, convection, and radiation (*Figure 6*).

1.2.1 Conduction

Conduction is a means of heat transfer in which heat is moved from molecule to molecule within a substance, or between substances in contact with each other. When the molecules are heated, they become more active and move about, colliding with one another. These collisions continue in a direction toward the cooler part of the material, causing the movement of heat in the same direction. For example, when copper tubing is heated by a torch, the molecules in the tubing nearest the torch get heated first and begin to move and collide with the nearby molecules in the tubing. These

Table 1 Specific Heat Values

Water	1.00	Iron	0.10
Ice	0.50	Mercury	0.03
Air (dry)	0.24	Copper	0.09
Steam	0.48	Alcohol	0.60
Aluminum	0.22	Kerosene	0.50
Brass	0.09	Olive oil	0.47
Lead	0.03	Pine	0.67

molecules then collide with other molecules, causing the tubing to become heated. The result is that the heat is carried by conduction from the heated end toward the cold end.

1.2.2 Convection

Convection is the transfer of heat through the flow of a liquid or gas caused by a temperature differential. As shown in *Figure 6*, air near the fireplace is heated by conduction and becomes warmer than the air in the rest of the room. Since warm air rises, the heated air moves toward the ceiling. It gives up its heat to the structure and room air as it goes upward, and then settles back down to the floor as it cools. The cooler air at the floor level moves toward the fireplace to replace the rising warm air. It too will warm, rise, give off heat, and settle back down. This circulation of air is accomplished via convection. Convection can be either natural or forced. The fireplace in *Figure 6* depends on natural convection for the smoke to rise from the chimney. But it also heats indoor air, creating convection currents in the room. Forced convection uses fans or pumps, such as those found in home heating and air conditioning systems, to enhance the circulation process.

1.2.3 Radiation

Radiation is the movement of heat in the form of invisible rays or waves, similar to light. Like light, it needs no medium on which to travel. Even air is unnecessary. Radiation takes place free of convection. It travels in straight lines from the heat source to the point where it is absorbed without heating the space in between. Heat from the sun traveling through space and warming our homes is a good example of heat transfer by radiation. The solar radiation comes through the windows in a building, strikes the walls, floors, furniture, and people, and is absorbed by them. Infrared heating equipment is based on this method of heat transfer, where solid objects are heated by the unit, but the air in between is not. Note that the fireplace in *Figure 6* is also a source of radiant heat.

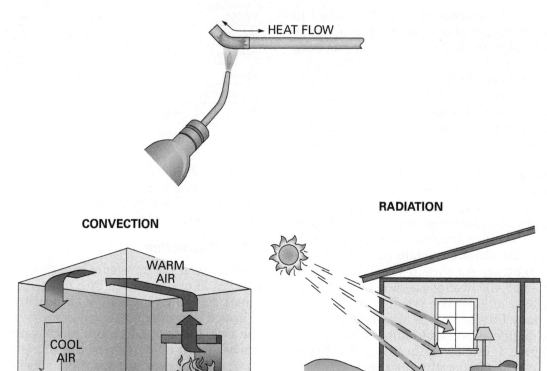

CONDUCTION

HEAT FLOW

CONVECTION

WARM AIR

COOL AIR

COLD AIR

RADIATION

Figure 6 The methods of heat transfer.

1.2.4 Conductors and Insulators

The rate of heat conduction varies for different substances. Some support the transfer of heat, while others restrict it. Materials in which the transfer of heat by conduction occurs readily and quickly are called conductors.Insulators are materials that resist conductive heat transfer. Cork, fiberglass, and polyurethane foam are examples of insulators. Most metals are good conductors of heat. Copper and aluminum are used in refrigeration systems because of their good heat-conduction ability. Good conductors of heat are typically good conductors of electricity as well.

1.2.5 Rate of Heat Transfer

The rate of heat transfer describes how fast heat can be added to or removed from an object or between objects. It is usually expressed in one of two ways in the HVACR industry. One way is in Btus per hour (Btuh). Another way is by the ton of refrigeration, which is a much larger unit of measure. The ton is commonly used in refrigeration work to describe the heat or cooling load for

a space, or the capacity of a piece of equipment or system.

One ton of refrigeration is defined as 12,000 Btus per hour (12,000 Btuh). See *Figure 7*. The ton is based on the amount of heat required to melt one ton of ice in a 24-hour period. As you learned earlier, one pound of ice at 32°F absorbs 144 Btus of heat while melting. Assuming that it takes one hour to melt, the rate of heat transfer is 144 Btus per hour, or 144 Btuh. Since a ton of ice contains 2,000 lb, the following math shows it requires 12,000 Btuh to melt a ton of ice in 24 hours:

144 Btus per pound × 2,000 pounds
= 288,000 Btus
288,000 Btus ÷ 24 hours
= 12,000 Btus per hour (Btuh)

1.3.0 Pressure-Temperature Relationships

An understanding of pressure-temperature relationships is essential for HVACR technicians. Before presenting details related to this relationship, the concept of pressure and its measurement must be examined.

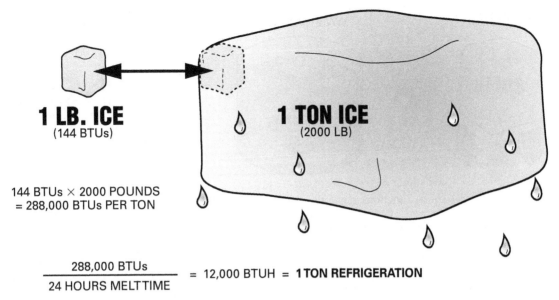

1 LB. ICE
(144 BTUs)

1 TON ICE
(2000 LB)

144 BTUs × 2000 POUNDS
= 288,000 BTUs PER TON

$$\frac{288,000 \text{ BTUs}}{24 \text{ HOURS MELT TIME}} = 12,000 \text{ BTUH} = \textbf{1 TON REFRIGERATION}$$

Figure 7 One ton of refrigeration illustrated.

1.3.1 Pressure

Pressure is defined as force per unit area. This is normally expressed in pounds per square inch (psi) in the United States. In HVACR work, the kilopascal (kPa) is the common unit of pressure measurement. The pascal is the official unit of metric pressure measurement. However, since the pascal is such a small unit of measure, the kilopascal (kPa) is the more common metric unit in HVACR work. One kilopascal is equal to 1,000 pascals. For convenience, a chart to convert various units of pressure measurement is provided in *Appendix B*. A variety of units for pressure measurement are encountered in the HVACR industry. For example, pressures below atmospheric pressure are often measured in inches of mercury (in. Hg). Technicians must be able to accurately convert pressure values between systems.

Depending on the state of a substance, pressure may be exerted in one direction, several directions, or all directions (*Figure 8*). Using the three states of water as an example, ice (a solid) exerts pressure only in a downward direction as the result of gravity. The same is true for all solid materials. As a liquid, water exerts pressure against all sides of the container in contact with it. As a gas (water vapor), it exerts pressure equally on all the surfaces of the container.

1.3.2 Absolute and Gauge Pressure

Earth is surrounded by a blanket of air called the *atmosphere*. Air is matter primarily consisting of oxygen, nitrogen, and water vapor. It has weight, and it exerts a force called *atmospheric pressure* on all things on Earth's surface. Atmospheric pressure can be measured with a barometer. For this reason, an atmospheric pressure measurement is often referred to as the barometric pressure.

Figure 9 shows a simple mercury tube barometer. The top of the barometer tube is sealed and in a vacuum, while the open end at the bottom rests in a container of mercury. Air pressure pushing down on the mercury in the container causes the column of mercury in the tube to rise. The extent of the rise is determined by the amount of pressure applied to the mercury.

Visualize a column of air with a cross-sectional area of one square inch and extending from Earth's surface at sea level to the farthest edge of the atmosphere. Also, assume that the temperature at sea level is 70°F (21°C). If this column of air is applied to the mercury tube barometer, the height of the mercury in the tube will be 29.92" (76 cm). The air column's weight will be 14.7 lb (6.67 kg). This means that every square inch of any surface at sea level has 14.7 lb of air pressure pushing down on it. These values of 29.92 in. Hg and 14.7 psi at sea level at 70°F are standards that are used frequently in refrigeration work.

ICE (SOLID)   WATER (LIQUID)   WATER VAPOR (GAS)

Figure 8 Pressure is exerted in various directions, depending upon the state of a substance.

There are two scales based on the unit of pounds per square inch. One pressure scale, called the absolute pressure scale, is based on the barometer measurements just described. On this scale, pressures are expressed as pounds per square inch absolute (psia), starting from zero, which represents a complete absence of pressure. Using the absolute pressure scale, standard atmospheric pressure is correctly expressed as 14.7 psia.

Another scale, called the gauge pressure scale, is normally used for refrigeration work. Gauge pressure scales use the atmospheric pressure as the 0-pressure starting point. Positive gauge pressures are expressed in pounds per square inch gauge, or psig. Negative pressures, those below 0 psig, are expressed in inches of mercury vacuum, or in. Hg vacuum. Note that any confined space that is below atmospheric pressure is said to be in some level of vacuum. A deep vacuum would generally describe a space that is at, or approaching, 0 psia.

Gauge pressures can easily be converted to absolute pressures by adding 14.7 to the gauge-pressure value. For example, a gauge pressure of 10 psig equals an absolute pressure of 24.7 psia, as shown here:

10 psig measured
+ 14.7 psi atmospheric pressure
———————————————
24.7 psia

A comparison of the gauge and absolute pressure scales is shown in *Figure 9*.

Refrigerant Gauge Pressure

Refrigerant gauges should always indicate a pressure of 0 psig when they are disconnected and open to the atmosphere. Any pressure reading of 0 and above on the gauges is always expressed in psig.

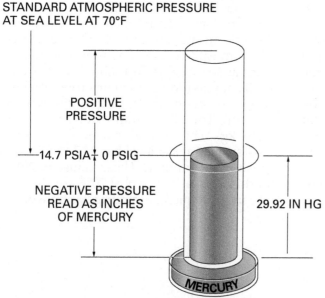

STANDARD ATMOSPHERIC PRESSURE
AT SEA LEVEL AT 70°F

POSITIVE PRESSURE

14.7 PSIA ‡ 0 PSIG

NEGATIVE PRESSURE READ AS INCHES OF MERCURY

29.92 IN HG

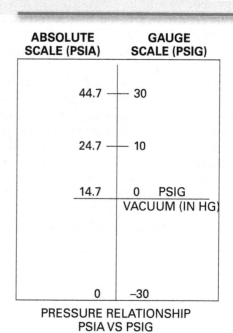

ABSOLUTE SCALE (PSIA)	GAUGE SCALE (PSIG)
44.7	30
24.7	10
14.7	0 PSIG
	VACUUM (IN HG)
0	−30

PRESSURE RELATIONSHIP
PSIA VS PSIG

Figure 9 Absolute, gauge, and barometric pressure readings compared.

1.3.3 Pressure/Temperature Relationships

Pressure and temperature have a special relationship. Two things are important to remember. First, the temperature at which a liquid or gas changes state depends on the pressure. Second, the boiling temperature of a liquid decreases as the pressure on it decreases. It will also increase as the pressure increases. Using our water example, water boils at 212°F (100°C) at standard atmospheric pressure (14.7 psia, 0 psig, or 101 kPa). With a lower atmospheric pressure of about 11.6 psia that exists at 5,000' above sea level, the same water would boil at the lower temperature of about 203°F (95°C). At the higher pressure of about 29.7 psia (15 psig + 14.7 psia atmospheric pressure), which can be reached in a pressure cooker, water boils at the higher temperature of about 250°F (121°C). *Figure 10* shows the pressure/temperature relationship for water.

There is an important lesson to be learned from this discussion that is directly related to the refrigeration cycle. The temperature at which liquid refrigerants change to a vapor, and the temperature at which refrigerant vapors change to a liquid (the process of condensation), can be manipulated by changing the pressure. Also remember that increasing the pressure of a refrigerant raises the temperature at which it boils. The reverse is also true; reducing the pressure lowers the boiling point.

Each refrigerant used in a mechanical refrigeration cycle has its own temperature/pressure relationship. A typical pressure/temperature (P/T) chart that covers several different common refrigerants is shown in *Table 2*. Note that this chart is provided in 5°F increments. For most purposes, a chart like this one will suffice; technicians can interpolate the pressures in between with limited effort. However, for more detailed work, a P/T chart with 1°F increments may be preferred.

Red figures (in Hg) vacuumNote that for each temperature provided, each refrigerant has only one pressure associated with it. Each combination of temperature and pressure represents the point of saturation for each refrigerant. For example, HFC-410A at 118 psig (814 kPa) has a corresponding temperature of 40°F (4.4°C). At this set of conditions, the refrigerant is in saturation, and both liquid and vapor coexist in some proportion. If any heat is added, the liquid will begin to vaporize (without a measurable change in temperature). If any heat is removed, the vapor will begin to condense to a liquid.

Table 2 P/T Chart for HFC-410A and HCFC-22, in 5°F Increments

Temperature		HCFC-22	HFC-410A
°F	°C		
–40	–40.0	0.5	11.6
–35	–37.2	2.6	14.9
–30	–34.4	4.9	18.5
–25	–31.7	7.4	22.5
–20	–28.9	10.1	26.9
–15	–26.1	13.2	31.7
–10	–23.3	16.5	36.8
–5	–20.6	20.1	42.5
0	–17.8	24.0	48.6
5	–15.0	28.2	55.2
10	–12.2	32.8	62.3
15	–9.4	37.7	70.0
20	–6.7	43.0	78.3
25	–3.9	48.8	87.3
30	–1.1	54.9	96.8
35	1.7	61.5	107
40	4.4	68.5	118
45	7.2	76.0	130
50	10.0	84.0	142
55	12.8	92.6	155
60	15.6	102	170
65	18.3	111	185
70	21.1	121	201
75	23.9	132	217
80	26.7	144	235
85	29.4	156	254
90	32.2	168	274
95	35.0	182	295
100	37.8	196	317
105	40.6	211	340
110	43.3	226	365
115	46.1	243	391
120	48.9	260	418
125	51.7	278	446
130	54.4	297	476
135	57.2	317	507
140	60.0	337	539
145	62.8	359	573
150	65.6	382	608

Vapor pressures – psig
Red figures (in Hg) vacuum a

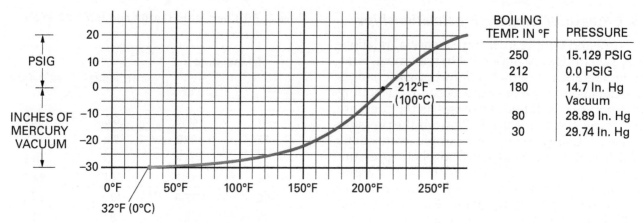

BOILING TEMP. IN °F	PRESSURE
250	15.129 PSIG
212	0.0 PSIG
180	14.7 In. Hg Vacuum
80	28.89 In. Hg
30	29.74 In. Hg

Figure 10 Pressure/temperature relationship for water.

The P/T chart helps technicians determine the state of a refrigerant inside the system, and to determine how much superheat or subcooling is present in different sections of the system. If a gauge is connected to a system and a thermometer is attached to the same point, the readings can be compared to the appropriate P/T chart. The previous example from the chart, using HFC-410A at 118 psig and 40°F, is used to demonstrate this.

- If the gauge and thermometer read the same values as shown on the P/T chart, then both liquid and vapor are present in the space. However, the refrigerant could be in transition from one state to another. Since there is no measurable temperature change during the process, it cannot be confirmed if a transition is in progress or not. This pressure and temperature combination can also indicate that the refrigerant is completely stable and no change of state is actively occurring.

- If the temperature reads 50°F (10°C) instead of 40°F (4.4°C), then two things are known immediately. First, there is no liquid refrigerant present at this point in the system; all of the refrigerant is in its vapor state. Further, the refrigerant vapor has been superheated 10°F (5.6°C). When superheat exists, liquid refrigerant cannot be present.

- If the temperature reads 30°F (–1.1°C) instead of 40°F (4.4°C), then two different facts are known. There is no vapor present in the system at this point; all of the refrigerant has changed to its liquid state. In addition, the refrigerant has been subcooled 10°F (5.5°C). When subcooling exists, refrigerant vapor cannot be present.

Trainees should not expect to fully understand these concepts without some practice with their application. The next section provides an overview of the refrigeration cycle in action, helping to further demonstrate the concepts learned thus far and how they apply to an operating system.

1.4.0 The Refrigeration Cycle

There are many variations of cooling systems used for personal comfort, food preservation, and industrial processes. Each of these systems uses a mechanical refrigeration system. *Figure 11* shows a very basic and common system with the following components identified:

- *Compressor* – Creates the pressure and temperature differences in the system needed to make the refrigerant flow and the refrigeration cycle work.
- *Condenser* – A heat exchanger that transfers heat absorbed by the refrigerant to the outdoor air or to another substance, such as water.
- *Metering, or expansion, device* – Creates pressure drop that lowers the boiling point of the refrigerant just before it enters the evaporator.
- *Evaporator* – A heat exchanger that transfers heat from the area or item being cooled to the refrigerant circulating within.

Also shown in *Figure 11* is the basic piping, referred to as the refrigerant lines. Together, the components and lines form a closed refrigeration circuit. The basic types of lines are:

- *Suction line* – The tubing that carries refrigerant gas from the evaporator to the compressor. The suction line is where the lowest pressures are found, and only refrigerant vapor is typically present here (by design).
- *Hot-gas line (also called the discharge line)* – The tubing that carries hot refrigerant gas from the compressor to the condenser. This line is where the highest pressures are found, and is another place where only refrigerant vapor is found.
- *Liquid line* – The tubing that carries liquid refrigerant from the condenser to the metering device and evaporator.

The arrows in *Figure 11* show the direction of flow through the system. The purpose of the refrigerant is to transport heat from one place to another. It is the medium through which heat can be moved into or out of a space or substance. Technically, a refrigerant is any liquid or gas that picks up heat by evaporating at a low temperature and pressure, and gives up heat by condensing at a higher temperature and pressure. Refrigerants often boil at extremely low temperatures. For example, HFC-410A boils at –60°F (–51°C) at atmospheric pressures.

Remember that the operation of a mechanical refrigeration system is basically the same for all systems. Only the type of refrigerant used, the size and style of the components, and the installed locations of the components and the lines change from system to system. Other devices, called *accessories*, are used in many systems to gain the desired cooling effect and to perform special functions. The events that take place within the system happen again and again in the same order. This repetitive series of events makes up the refrigeration cycle.

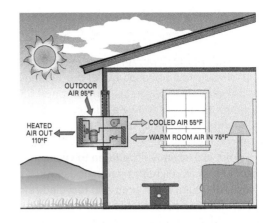

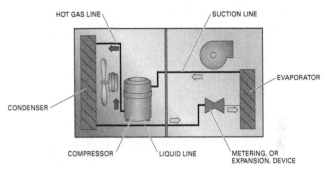

Figure 11 Basic function of a window air conditioning unit.

Predicting Weather

A barometer measures air pressure, and air pressure is used to show existing weather conditions and to predict changes in weather conditions. As a generalization, high pressure indicates that the weather is pleasant and low pressure indicates that the weather is cloudy and rainy. Changes in barometric pressure are used to predict changes in weather conditions. For example, if it is raining outside, but the barometer is rising, it indicates that better weather is on the way.

Figure Credit: Courtesy of Chelsea Clock Company

The Future of HCFC-22

As a result of the Clean Air Act, many refrigerants have already been phased out of production or the phaseout is in progress. Production of CFC refrigerants such as R-12 was halted in 1996. Production of HCFCs is scheduled to phase out over a period of years. All production of HCFC-22, for example, is presently scheduled to cease in 2020. Note that this does not mean their use must come to a sudden halt. Recovered and recycled refrigerant will continue to be available until the supply is exhausted. As the supply goes down however, the cost rises. The cost of HCFC-22 has risen sharply from its pricing prior to the lowering of production levels.

HCFC-22 is considered a Class II ozone-depleting substance, all of which have an ozone-depletion potential less than 0.2. Although the use of HCFCs in refrigeration systems is the topic here, there are many other HCFCs that are used in foam blowing, as aerosol propellants, and as fire-suppression agents. The import and production of all HCFCs must end by 2030, according to current legislation. Note that there are many new refrigerant blends that use HCFC-22 as a component. Their production must cease as well, in spite of their reduced impact on the ozone layer.

1.4.1 Basic Component Functions

The refrigeration cycle is based on two main principles that have been presented:

- As a liquid refrigerant changes state to a vapor, it absorbs large quantities of heat used to complete the latent process.
- The boiling point of a liquid can be changed by altering the pressure exerted on it.

Again referring to *Figure 11*, the refrigerant flows through the system in the direction indicated by the arrows. The compressor will serve as the starting point. It receives low-temperature, low-pressure vapor and compresses it. The compressor creates a low pressure in the suction line and a high pressure in the discharge line. This forces the refrigerant to travel in towards the lower pressure. Gases and liquids always flow from a point of higher pressure toward a lower pressure. Following the compression cycle, the refrigerant exits the compressor as a high-temperature, high-pressure vapor. The refrigerant vapor is now significantly hotter than the air outdoors. It travels to the condenser via the hot-gas, or discharge, line. The compression cycle also adds heat to the vapor, in addition to the heat it absorbed in the evaporator.

The condenser is a series of tubing coils through which the refrigerant flows. The condenser must reject any other heat that entered the system, including the heat of compression. As air cooler than the refrigerant vapor moves across the condenser coil, the hot vapor transfers its heat to the air. As it continues to give up heat to the outside air, it cools to its condensing temperature and begins to change from a vapor into a liquid. As more cooling takes place, all the refrigerant eventually changes to liquid. Once it has all changed into a liquid state, any additional cooling of the liquid is considered. The high-pressure, high-temperature, subcooled liquid now travels through the liquid line to the inlet of the metering device.

The metering device regulates the flow of refrigerant to the evaporator. It also decreases its pressure significantly. Think of it as little more than a restriction, like placing your thumb over the end of a garden hose. It is a small hole, or orifice, that reduces the pressure of the liquid refrigerant from the condenser. At the reduced pressure, the refrigerant is transformed into a cool or cold liquid at its saturation point, ready to boil when heat from the evaporator is added. As it leaves the metering device, it is primarily liquid but also contains some vapor. That will be discussed further at another time.

The evaporator receives low-temperature, low-pressure liquid refrigerant (with some vapor generally present) from the metering device. The evaporator is a series of tubing coils that exposes the cool liquid refrigerant to the warmer air passing over. Heat from the warm air is transferred through the tubing to the cooler refrigerant. This causes the refrigerant to boil and vaporize. It is important to realize that even though it has just boiled, it is still not hot, since refrigerants boil at such low temperatures. By the time it leaves the evaporator, it should have fully completed the process of changing into its vapor state. In addition, enough heat should be absorbed that some amount of superheat should be present. As it enters the suction line, it should be a low-temperature, low-pressure refrigerant vapor that travels to the compressor.

1.4.2 The Refrigeration Cycle in a Typical Air Conditioning System

Figure 12 shows a typical air conditioning system. The system is divided into two sections based on pressure. The high-pressure side includes all the components in which the pressure of the refrigerant is at or above the condensing pressure. This is often referred to as the head pressure, discharge pressure, or high-side pressure. The low-pressure side includes all the components in which the pressure of the refrigerant is at or below the evaporating pressure. This is often called the *suction pressure* or *low-side pressure*. The dividing line between the sections cuts through the compressor and the metering device. These two locations are where the pressures change from high-to-low, or low-to-high.

The refrigeration cycle can now be discussed in more detail. This system is typical of an air conditioner for comfort cooling that uses HFC-410A as the refrigerant. Note the P/T chart provided at the bottom of the figure. HFC-410A reaches saturation at 40°F (4.4°C) when at a pressure of 118 psig (814 kPa). This temperature is the typical target for evaporators used for comfort cooling.

The following example demonstrates the concepts and pressure/temperature relationships covered thus far. Assume an air temperature of 75°F (24°C) for the room being cooled and an outdoor air temperature of 95°F (35°C). These values, of course, vary from hour-to-hour and day-to-day in reality. The numbers in the discussion that follows correspond to the numbers shown in *Figure 12*. Note that, although the pressures shown in the figure are provided to one decimal point of precision, most refrigerant gauge sets cannot display this level of accuracy. Follow along on the figure as this system is described.

1. After compression, the now highly superheated gas from the compressor flows through the hot-gas line to the condenser. The gas is typically close to 180°F at 475 psig. Since the saturated temperature corresponding to 475 psig is 130°F (see the chart in *Figure 12*), the refrigerant vapor has now gained about 50°F (180°F − 130°F), or 28°C, of total superheat. This superheat must be removed before the vapor can be condensed into a liquid. The 180°F refrigerant in the hot-gas line will give up some of its superheat to the surrounding 95°F air before it ever reaches the condenser. The hot-gas line is normally not insulated and the tubing is a good conductor of heat.

2. Because the refrigerant in the condenser is significantly hotter than the outside air blowing over the condenser, it easily gives up the remaining superheat in the condenser. This drops its temperature to 130°F. As heat continues to be transferred from the vapor to the cooler outside air, the vapor begins to condense into a liquid. Remember that the refrigerant will remain at 130°F in the condenser until the change of state to a liquid is complete. No sensible change in temperature can occur during this latent process. After the refrigerant has traveled most of the way through the condenser, all the refrigerant has condensed into a liquid. Once this happens, subcooling of the liquid occurs as more heat is removed from the refrigerant.

3. During its remaining travel through the condenser, the liquid refrigerant drops in temperature. In this example, the temperature is lowered 15°F to 115°F. In other words, the liquid refrigerant is subcooled 15°F (130°F condensing temperature – 115°F actual temperature).

4. Subcooled liquid refrigerant from the condenser travels through the liquid line to the metering device. The liquid line is usually not insulated and may be long. Thus, the 115°F liquid refrigerant may be further subcooled as it gives up more heat to the cooler outside air. This drop could increase the subcooling by another 5°F, lowering the temperature of the liquid refrigerant to 110°F. As is the case in the suction line, this drop is completely dependent upon the environment in which the liquid line is routed.

5. The metering device controls the flow of liquid refrigerant to the evaporator. Here, the subcooled liquid from the condenser enters at a temperature of 110°F and a pressure of 475 psig. The pressure drop that occurs at the metering device results in a new pressure of 118 psig at its exit. The next event may be one of the most difficult parts of the cycle to understand. The pressure drop at the metering device creates a very unstable condition for the refrigerant, as it departs from the carved-in-stone conditions of the P/T chart for this refrigerant. At this new low pressure, the refrigerant must rapidly cool to the 40°F value shown on the chart. Since there is no other way to transfer its heat out, the refrigerant literally cools itself by instantly flashing some of the refrigerant

from a liquid to a vapor. As some of the liquid refrigerant changes state rapidly, latent heat is absorbed from the remaining liquid to make it happen. It was noted earlier at Point 1 that the refrigerant here is roughly 80 percent liquid and 20 percent vapor. This vapor is commonly known as flash gas, and it is the result of instantaneous cooling taking place in the refrigerant. It now conforms to the laws of physics and its own P/T chart. It enters the evaporator at a temperature of 40°F and a pressure of 118 psig, in a mostly liquid form and boiling at the slightest addition of heat. In the evaporator, heat is absorbed by the air and boiling continues until only vapor remains.

6. A mixture (roughly 80 percent liquid, 20 percent vapor) of HFC-410A is supplied from the metering device to the evaporator. This mixture is at a pressure of 118 psig, which corresponds to the 40°F boiling point of HFC-410A refrigerant. (See the chart in *Figure 12*.) The 40°F boiling point is used here because it is typical of the temperatures normally used for evaporators in air conditioning systems for comfort cooling.

7. Because the refrigerant flowing through the evaporator is cooler (40°F) than the warmer inside room air (75°F), it absorbs heat, causing the liquid refrigerant to boil and turn into a vapor. After traveling about 90 percent of the way through the evaporator tubing, all of the refrigerant has boiled into a vapor state. The amount of vapor consistently increased during the journey through the evaporator, while the amount of liquid decreased. As long as both liquid and vapor are present, the refrigerant is in saturation. Once all the liquid is gone, only vapor remains and the refrigerant is no longer saturated. Remember that the refrigerant remains at 40°F throughout this latent process of changing

states. The temperature of the refrigerant does not change until all of it has completely vaporized. Since the temperature shown at this point is 45°F, it is certain that no liquid remains.

8. During the remaining 10 percent of travel through the evaporator, the vapor continues to absorb heat from the warmer air, thus raising its temperature to 50°F. The saturated vapor is now superheated a total of 10°F (50°F – 40°F). This superheated vapor flows through the suction line and is drawn into the low-pressure side of the compressor. The cooled inside room air is recirculated by the evaporator fan back into the room at a temperature of about 55°F. The air temperature has been reduced 20°F by transferring its heat to the refrigerant.

Note that the temperature of the vapor at the compressor inlet is 5°F warmer than it was when it left the evaporator. The vapor may pick up an additional 3°F to 5°F (1.7°C to 2.8°C) of superheat because the vapor in the suction line absorbs more heat from the warmer surrounding air as it travels from the evaporator to the compressor. However, it is important to note that this completely depends on the environment in which the suction line is routed, and the efficiency of any insulation that has been applied to it. There may be no change at all, especially when the line is short. In a packaged unit, for example, the suction line may only be a couple of feet in length.

The refrigerant has now completed its cycle and is ready to start over again. Note that the above discussion is theoretical and does not take into account pressure drops that naturally occur as the refrigerant flows through components and within the piping in an actual system. In addition, in most operating systems, one or more conditions are changing constantly.

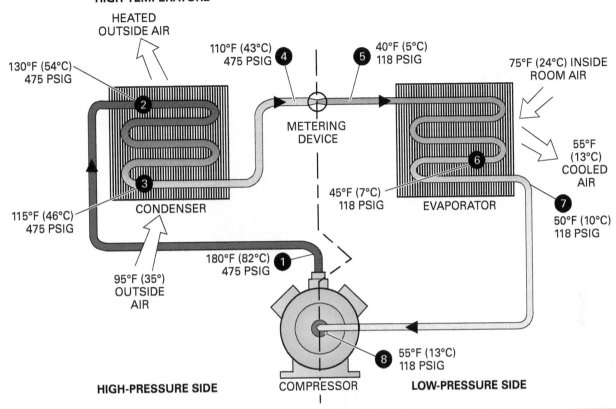

TEMPERATURE		REFRIGERANT/PRESSURE	
°F	°C	PSIG	kPa
30	−1	97	667
31	−1	99	681
32	0	101	696
33	1	103	709
34	1	105	724
35	2	107	738
36	2	109	753
37	3	111	768
38	3	114	783
39	4	116	798
40	4	118	814
41	5	120	829
42	6	123	845
43	6	125	862
44	7	127	878
45	7	130	895
46	8	132	911
47	8	135	929
48	9	137	946
49	9	140	963
50	10	142	980

TEMPERATURE		REFRIGERANT/PRESSURE	
°F	°C	PSIG	kPa
120	49	417	2,874
121	49	423	2,913
122	50	428	2,952
123	51	434	2,992
124	51	440	3,031
125	52	445	3,071
126	52	451	3,112
127	53	457	3,153
128	53	463	3,194
129	54	469	3,235
130	54	475	3,278
131	55	482	3,321
132	56	488	3,363
133	56	494	3,407
134	57	501	3,451
135	57	507	3,495
136	58	513	3,540
137	58	520	3,585
138	59	527	3,631
139	59	533	3,677
140	60	540	3,724

HFC–410A PRESSURE / TEMPERATURE CHART

	HIGH-TEMP, HIGH-PRESSURE VAPOR
	HIGH-TEMP, HIGH-PRESSURE LIQUID
	LOW-TEMP, LOW-PRESSURE MIXTURE
	LOW-TEMP, LOW-PRESSURE VAPOR

Figure 12 Typical refrigeration cycle for a system using HFC-410A refrigerant.

Several different instruments are used to make temperature and pressure measurements. Such measurements are crucial to ensuring that a system is operating properly, and to the troubleshooting process when problems occur.

1.5.1 Thermometers

Electronic, or digital, thermometers (*Figure 13*) now dominate temperature measurement in the HVACR industry. Temperature measurement features have also been built in to many electrical multimeters, further increasing their versatility. Electronic thermometers display temperature on either an analog meter or a digital liquid crystal display (LCD). They generally use either a thermocouple or thermistor type of temperature probe, or both, to sense the heat and generate a temperature indication. Thermocouple probes use a sensing device made of two dissimilar wires welded together at one end called a junction. When the junction is heated, it generates a low-level DC voltage that produces a temperature reading on the electronic thermometer indicator. Thermocouple probes tend to be rugged and less expensive. Thermistor probes use a semiconductor electronic element in which resistance changes with a change in temperature.

Often, several different probes are used with the same electronic thermometer to allow temperature measurement in different applications. For example, the long, slender probe in *Figure 13* is designed for use in air or for immersion in liquids. The other probe shown is designed specifically for use with pipes and tubes, as it has a spring-loaded clamp built in to firmly connect the probe to the line. This type is ideal for measuring the temperature of refrigerant lines. Refrigerant line temperature readings must be accurate, and accurate measurement requires excellent and consistent contact between the thermistor and the line.

Many electronic thermometers can also accommodate two or more probes at the same time, so that simultaneous measurements can be made at several locations. Most thermometers of this type can calculate and display the difference in temperature between the locations being measured.

Electronic thermometers are precise measuring instruments. Be sure to read and follow the manufacturer's instructions for operating electronic thermometers. Also, be sure to follow the manufacturer's instructions for calibration of the instrument.

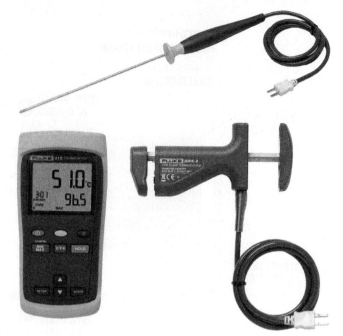

Figure 13 Electronic thermometer.

Dial thermometers are available in various forms (*Figure 14*). They are popular because they are rugged, small, and typically inexpensive. They come in many stem lengths and dial sizes. Because pocket-style dial thermometers have smaller calibrated scales, it is sometimes difficult to get accurate readings. A digital version of the pocket thermometer, such as the one shown in *Figure 14*, works better when a bit more accuracy is needed. Remember that most thermometers, regardless of the type, are generally more accurate in the mid-section of their range. When reading temperatures that are close to the minimum and maximum values in their range, they are less accurate.

Handheld infrared thermometers (*Figure 15*) are also very popular among technicians in the HVACR industry. Infrared thermometers allow the user to measure temperature without making contact with the surface of an object. For this reason, they are also called *non-contact thermometers*. Non-contact thermometers work by detecting infrared energy emitted by the device at which they are pointing. Laser-sighting devices allow the technician to aim the thermometer like a pistol at the surface to be measured. The temperature is displayed instantly. This type of thermometer is useful for taking temperature readings in hard-to-reach areas. However, it may not be effective on highly reflective surfaces. The unit may have an adjustable setting to allow the measurement of reflective surfaces. A thermometer with no such adjustment can still be used to measure the temperature of a shiny surface by applying a matte paint or non-reflective tape to the surface.

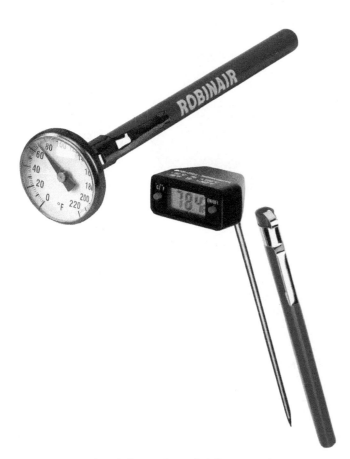

Figure 14 Dial and electronic pocket thermometers.

Figure 15 An infrared, non-contact thermometer.

It is important to understand the distance-to-spot ratio (D:S) of infrared thermometers. The D:S is the ratio of the distance to the object being measured compared to the diameter of the temperature-measurement area. For example, if the D:S ratio is 12:1, measurement of an object 12" (30.5 cm) away will average the temperature over a 1" (25 mm) diameter spot (*Figure 16*). The greater the distance, the larger the averaged spot becomes. When using an infrared unit with a 12:1 D:S from 12' (3.7 m) away, the spot then becomes 1' (0.3 m) in diameter. As a result, the temperature measurement may be inaccurate, unless the surface being measured is at least as large as the spot. Otherwise, the device will average the temperature of two or more unrelated surfaces.

Many thermometers can be calibrated when the readings become inaccurate. The calibration should be checked periodically to ensure accuracy. Dial-type pocket thermometers are often equipped with a hex nut to turn the dial in relation to the pointer for calibration. Refer to *Figure 17.*

To check or calibrate a thermometer at a relatively high-temperature point, prepare a pot of boiling water. Place the thermometer in the pot at least 2" (5 cm) deep, but be sure the pot is easily deep enough that the tip does not touch or get too close to the bottom. With the water at a full boil, check the temperature reading after the temperature stabilizes; it should be 212°F (100°C). If not, remove it from the pot and use the hex nut to adjust it to the proper reading. After making an adjustment, retest it.

> **WARNING!**
>
> Burners and boiling water can cause serious burns. Use the appropriate personal protective equipment (PPE) and handle the thermometer very carefully when it is removed from the water. Ensure that no flammable materials are near the heating surface.

To check or calibrate a thermometer at a low temperature, prepare a mixture of crushed ice and only enough cool water to bring it to the top of the ice. Make sure that the ice isn't floating off the bottom. Place the thermometer in the ice, again with at least 2" (5 cm) of the stem in the mixture and the tip off the bottom. Wait until the temperature stabilizes and then check the reading. Adjust the reading using the hex nut if it is not correct.

All thermometers that are capable of measuring temperatures in the range where water freezes and boils can be checked this way. However, not all thermometers can be calibrated in the field. Some may need to be returned to the manufacturer. Check the documentation that comes with the thermometer to determine how the calibration is done, or where to return it for service.

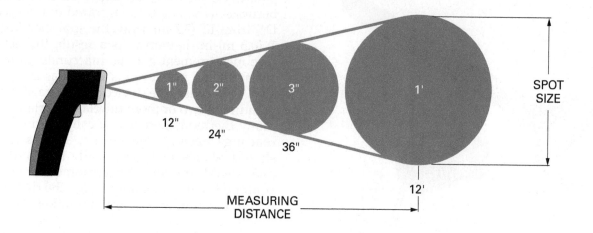

Figure 16 The D:S ratio illustrated.

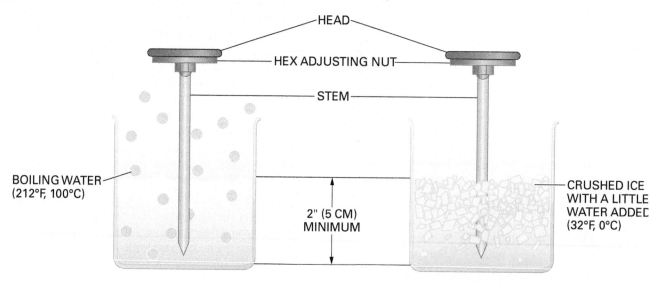

Figure 17 Calibrating a pocket thermometer.

1.5.2 Gauge Manifolds and Other Pressure Measurement Devices

The gauge manifold set is probably the most frequently used HVACR service instrument. It is used to install and check refrigerant charges and to measure low-side and high-side pressures in an operating system. The gauge manifold set is regularly used to route and control the flow of refrigerant to and from the closed circuit in support of servicing tasks.

Figure 18 shows a standard two-valve gauge manifold set. It consists of two pressure gauges mounted on a manifold assembly. A compound gauge is mounted on the left side of the manifold and a high-pressure gauge is mounted on the right. A compound gauge allows measurement of pressures both above and below atmospheric pressure. It is used to monitor and measure low-side, or suction, pressures. Most compound gauges used with the gauge manifold set today can measure pressures above atmospheric pressure in a range from 0 to 250 psig. However, compound gauges are often equipped with a retard range up to 500 psig. The retard range provides an extra safety cushion for the gauge, but there is no level of accuracy provided between 250 and 500 psig. Below atmospheric pressure they can measure from 0 to 30 in. Hg vac. The compound gauge is used to measure system low-side (suction) pressures, including any vacuum that exists in a system.

Note that a standard compound gauge is not an accurate way of measuring vacuum levels. Other instruments, called *micron gauges*, are needed to accurately read vacuum levels. The gauge manifold can at least identify the presence of a vacuum. Micron gauges will be presented elsewhere in the program.

The high-pressure gauge on the right side of the manifold is used to measure system high-side, or discharge, pressures. Most high-pressure gauges today can measure system pressures in the range of 0 to 800 psig. The range of both the low- and high-pressure gauges accommodates today's newer, environmentally friendly refrigerants such as HFC-410A that operate at higher pressures than their predecessors do. The hoses must also be rated for 800 psig. Digital versions of these gauges are also available (*Figure 19*). Many have the capability of reading the pressures and displaying the corresponding saturation temperature for the refrigerant in use.

A standard two-valve gauge manifold set has two hand valves and three hose ports. An additional port on the low-side and high-side is also quite common. The hand valves are adjusted to control the flow of refrigerant to and from the system during servicing activities. To read the system pressures, the valves remain closed. The gauge manifold hoses are connected to the system being serviced and/or other service equipment through a set of high-pressure service hoses. These hoses must be equipped with fast, self-sealing fittings that immediately trap refrigerant in the hoses when disconnected. Use of these fittings helps to meet the requirements of the Clean Air Act by minimizing the release of refrigerants into the atmosphere.

Many gauge manifold sets and service hoses are color-coded. The low-pressure compound gauge, hand valve, and low-pressure hose port are blue. A blue service hose is used to connect the manifold low-pressure hose port to the equipment suction service valve. Red is the color used to mark the high-pressure gauge, hand valve, and hose port. A red service hose is used to connect the high-pressure hose port to the equipment discharge service valve or liquid line valve. The center hose port is the utility port. This port is normally connected through a yellow service hose to other service instruments or devices. When not in use, the utility port should be capped. Note that the hoses are not different in any other way except color.

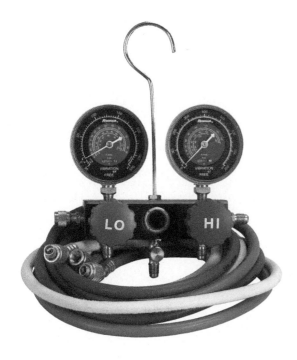

Figure 18 Gauge manifold set.

Figure 19 Digital gauge manifold.

Gauge manifold sets are also available with four hand valves and related hose ports. Use of this type of manifold can reduce service time by eliminating the need to switch a single utility hose between different service devices. Four-valve manifolds and related service hose sets are color-coded as follows: blue (low-pressure), red (high-pressure), yellow (charging), and black (vacuum).

The gauge manifold set is a precise measuring instrument. Its accuracy is critical for correct servicing. The technician must ensure that the gauge manifold set is always handled with care both during use and in transport. The calibration of the gauge manifold set should be checked regularly. Without any pressure applied and open to the atmosphere, the gauges should read 0 psig. To check their accuracy, look straight ahead at the needle. It should split the 0 psig mark on center.

Standard analog gauges are easily calibrated as follows:

Step 1 Ensure that there is no pressure inside the hoses. They should be open to the atmosphere, without any residual refrigerant or refrigerant/oil mixture inside. Alternatively, the hoses can be completely removed from the gauge manifold.

Step 2 Remove any protective boots and unscrew the protective lens.

Step 3 The calibration screw is typically a small screw located on the face of the gauge. The needle position can be adjusted by turning this screw in one direction or the other with a small common screwdriver until the needle accurately splits the 0 psig mark (*Figure 20*).

Step 4 Lightly tap the gauge on the back to ensure that the needle is not under any tension that causes it to move away from the desired position.

Step 5 Reinstall the protective lens covers. Do not overtighten.

Digital gauges can often be field-calibrated as well. For some models, it is as simple as turning the gauge on, and then quickly tapping the On button two times. Other models may require a different procedure. Check the manufacturer's documentation for the model in use to ensure the proper procedure is used.

Another instrument that is used to measure pressure is the manometer. Both electronic and non-electronic manometers (*Figure 21*) are in common use. Manometers are designed to measure very low pressures, often in inches of water column or similar units. They are often used to measure the gas pressure in heating systems and the pressure in air distribution ducts. You will learn more about manometers and their uses later in your training.

Figure 20 Calibrating a refrigerant gauge.

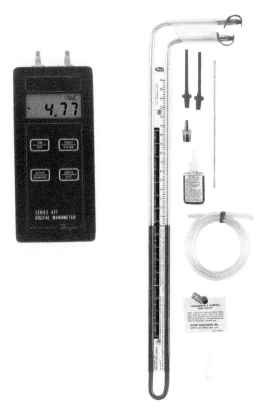

Figure 21 Digital and fluid-filled manometers.

Additional Resources

Refrigeration and Air Conditioning: An Introduction to HVACR. Latest Edition. New York, NY: Pearson.

1.0.0 Section Review

1. The Fahrenheit scale is based on the temperature water boils at sea level, which is _____.

   a. 100°F
   b. 180°F
   c. 212°F
   d. 459°F

2. The transfer of heat from one object to another by direct contact is called _____.

   a. radiation
   b. change of state
   c. convection
   d. conduction

3. The boiling temperature of a liquid is decreased _____.

   a. as the pressure is increased
   b. as the pressure is decreased
   c. in its solid state
   d. when it is subcooled

4. In a refrigerant circuit, the low-pressure, low-temperature liquid is converted into a low-pressure, low-temperature vapor in the _____.

   a. compressor
   b. evaporator
   c. expansion (metering) device
   d. condenser

5. A thermistor responds to temperature changes by _____.

   a. changing its resistance to electrical flow
   b. producing more or less current
   c. producing more or less voltage
   d. modulating its signal amplitude

2.0.0 REFRIGERANTS

Objective

Identify common refrigerants and their identifying characteristics.

   a. Identify fluorocarbon refrigerants.
   b. Describe the use of ammonia as a refrigerant.
   c. Identify various refrigerant containers and their safe handling requirements.

Performance Tasks

   4. Identify refrigerants using cylinder color codes.

Trade Terms

Compound: As related to refrigerants, a substance formed by a union of two or more elements in definite proportions by weight; only one molecule is present.

Fluorocarbon: A compound formed by replacing one or more hydrogen atoms in a hydrocarbon with fluorine atoms.

Halocarbon: Hydrocarbons, like methane and ethane, that have most or all their hydrogen atoms replaced with the elements fluorine, chlorine, bromine, astatine, or iodine.

Halogens: Substances containing chlorine, fluorine, bromine, astatine, or iodine.

Hydrocarbons: Compounds containing only hydrogen and carbon atoms in various combinations.

Mixture: As related to refrigerants, a blend of two or more components that do not have a fixed proportion to one another and that, however thoroughly blended, are conceived of as retaining a separate existence; more than one type of molecule remains present.

Rupture disk: A pressure-relief device that protects a vessel or other container from damage if pressures exceed a safe level. A rupture disk typically consists of a specific material at a precise thickness that will break or fracture when the pressure limit is reached, creating a controlled weakness, thereby protecting the rest of the container from damage.

Refrigerants are used in cooling systems to move heat into or out of a space or substance. This section briefly describes refrigerants, with the focus on their impact on the environment. You will learn more about their characteristics and specific uses later in your training. This section limits the discussion to fluorocarbon and ammonia refrigerants, since for all practical purposes they are the only ones in common use today.

2.1.0 Fluorocarbon Refrigerants

Most manmade, or synthetic, refrigerants in popular use today are a fluorocarbon compound or a mixture of fluorocarbon refrigerants. Included in this group are refrigerants such as HCFC-123 (R-123), HCFC-22 (R-22), HFC-410A (R-410A), and HFC-134a (R-134a). These refrigerants all originate from one of two base molecules: methane and ethane.

Methane and ethane are called hydrocarbons because they are organic compounds that contain only hydrogen and carbon atoms. When most or all of the hydrogen atoms in the methane or ethane molecule are replaced with elements such as chlorine, fluorine, or bromine, the changed molecule is called a halocarbon, short for halogenated hydrocarbon. Chlorine, fluorine, and bromine are chemically related elements called halogens. To halogenate a substance means to cause some other element to combine with a halogen. When all the hydrogen atoms in a hydrocarbon molecule are replaced with chlorine or fluorine, the molecule is said to be fully halogenated. Halocarbons in which at least one or more of the hydrogen atoms have been replaced with fluorine are called *fluorocarbons*. Fluorocarbon refrigerants fall into three groups: CFCs, HCFCs, and HFCs, based on their chemical structure. Note that fluorocarbons must contain some fluorine. Halocarbons may contain fluorine as well as other halogens, or they may not contain any fluorine at all.

2.1.1 CFCs, HCFCs, and HFCs

Traditionally, each refrigerant had a trade name or an R (refrigerant) name. Names like R-22, R-12, R-502, etc., were assigned by the American Society of Heating, Refrigerating, and Air-Conditioning Engineers (ASHRAE). These names were substituted for the true chemical names. For example,

Refrigerant and the Law

It is unlawful to knowingly release most refrigerants into the atmosphere, per Sections 608 and 609 of the US Environmental Protection Agency's Clean Air Act. The punishment for knowingly and willfully releasing refrigerants into the atmosphere can be significant.

Substantial releases, even those that are accidental, must be reported. Because these substances have been linked to both global warming and the destruction of Earth's ozone layer, new refrigerants that are more environmentally friendly have emerged. However, they too must be carefully managed and cannot be released into the atmosphere.

R-22 describes the refrigerant with the chemical name of chlorodifluoromethane. Craft professionals certainly agree that numbers are much better than having to call them by their chemical names.

However, the way refrigerants are named has changed. ASHRAE now uses acronyms ahead of the number, such as CFC, HCFC, and HFC. These acronyms describe the way the refrigerants are chemically structured and help to sort the chlorine-bearing products from those that do not contain chlorine. The number previously assigned to a given refrigerant by ASHRAE has been retained. Both old and new names are currently being used in the trade. Some examples of changed names for commonly used refrigerants are shown in *Table 3*.

CFCs, or chlorofluorocarbons, contain chlorine, fluorine, and methane. These refrigerants were very popular for many years. Well-known CFCs include CFC-12, previously used in automobiles and refrigerators, and CFC-11, a low-pressure refrigerant used in centrifugal chillers.

HCFCs, or hydrochlorofluorocarbons, contain hydrogen as well as the other substances, as the name implies. They have far less chlorine in their chemical structure. Popular HCFCs include HCFC-22, which is still in use in many comfort cooling systems but its production is being phased out.

Table 3 Examples of Old and New Refrigerant Names

Old Name	New Name
R-22	HCFC-22
R-123	HCFC-123
R-134a	HFC-134a
R-407C	HFC-407C
R-410A	HFC-410A

Refrigerant Compound Numbering

Each refrigerant compound has a specific chemical composition, which can be shown graphically. For example, the composition of the HCFC refrigerant R-22 appears like this:

$$H - C - F$$

with F above the carbon and Cl below the carbon.

The name of the refrigerant is derived from its components. You can see from this graphic that R-22 is an HCFC—H for hydrogen, C for chlorine, F for fluorine, and C for carbon. The numbers—22 in this case—relate to the number of atoms of each component in a single molecule of the refrigerant. Refrigerant blends have a much different numbering system.

HFCs, or hydrofluorocarbons, do not contain any chlorine. A number of the refrigerants used today are HFCs, such as HFC-410A and HFC-134a.

There is evidence that the ozone layer surrounding Earth has been damaged by various chemicals, most notably chlorine. The ozone layer filters out harmful radiation from the sun that would otherwise reach the Earth's surface and damage life. The chlorine in CFC and HCFC refrigerants is now known to contribute to this damage. Since the passage of the Clean Air Act in 1990, refrigerants bearing chlorine have come under increasing government regulation and control. The US Environmental Protection Agency (EPA) is responsible for making and enforcing laws pertaining to the use of these refrigerants.

It is important to note that CFCs as a group have had the greatest impact on the ozone layer because they contain the greatest volume of chlorine. Since this class of refrigerants poses the greatest threat to the environment, the production of new CFCs has been banned. Many new refrigerants have been developed to take their place. HCFC refrigerants are also being phased out, since they too contain chlorine in lesser amounts.

Most refrigerants in use today that are considered long-term products are HFCs that contain no chlorine at all. They are believed to be harmless to the ozone layer. Although this may help save the ozone layer from damage, it is now believed that

Refrigerant Certification

The US Environmental Protection Agency (EPA) requires that all persons who install, service, repair, or dispose of equipment containing a refrigerant possess a certification card. The certification card is obtained by passing a test for one or more categories of work as identified by the EPA. The categories are:

- *Type I*—Small appliances containing less than 5 lb of refrigerant, such as refrigerators and small air conditioners.
- *Type II*—Appliances that use high-pressure refrigerants such as HCFC-22 and HFC-410A.
- *Type III*—Appliances such as centrifugal chillers that use low-pressure refrigerants.
- *Type IV*—This is a universal certification that covers all the above categories.

HFCs contribute to the problem of global warming when released to the atmosphere. As a result, their intentional release is also unlawful. Global warming and the role of refrigerants in the process is creating additional concern in the scientific community. As research continues, HFC-based products may also be replaced with other refrigerants.

2.1.2 Refrigerant Blends

Many new refrigerants have been developed by blending two or more different refrigerants together. Blends can be made from combining HCFCs or HFCs together, or by combining the two types. Since HCFCs are being eliminated, any blend that contains an HCFC will eventually be banned for use in the United States. If a refrigerant blend contains an HCFC, then its name must also begin with that acronym.

A compound is a substance formed by a union of two or more elements in definite proportions by weight. When a compound is created, only one molecule is present. HCFC-22 is a good example of a pure compound. Compounds exhibit very predictable behavior in the refrigeration circuit.

However, many of today's new refrigerants that are created by blending are not pure compounds. When two or more substances are blended, they do not form a new compound. In fact, the individual molecules from the blend components remain intact and are still present in the blend. A blend that retains the properties of its components is simply a refrigerant mixture.

The behavior of a mixture is reasonably predictable, but there are differences in how they behave.

For example, when a leak develops in a system, it is possible that one refrigerant in the mixture will escape in a larger proportion than another refrigerant. The location of the leak and the system operating status determine whether this occurs or not. In addition, the superheat and subcooling values must be calculated differently. Consider a mixture that contains hypothetical Refrigerant 1 (R-1) and Refrigerant 2 (R-2). These two have different boiling points and have their own separate P/T charts. When they are mixed, they both can retain their previous characteristics. Therefore, while the new refrigerant mixture 1/2 is at work in the refrigeration circuit, R-1 may be in the process of boiling and vaporizing while R-2 has not yet absorbed enough heat to begin the process. This makes it more difficult to accurately determine the superheat or subcooling values for the mixture. Although each mixture comes with a new P/T chart, the saturation temperature becomes a range of temperatures on the chart, rather than one specific temperature. Chemists try to work with blend components that do not differ dramatically, but some difference in the behavior of the two components does often occur.

A more detailed review of all refrigerants is provided in a separate module of this program. However, it is important to understand that all refrigerants do not behave precisely as described earlier. Like most rules, there are exceptions.

2.2.0 Ammonia (NH3)

Anhydrous ammonia (R-717) has excellent heat-transfer qualities and is used mainly in large medium- and low-temperature applications. These include ice plants, ice skating rinks, food storage, and large food-processing plants. Since it is not in the same family as the other refrigerants discussed, the ASHRAE-assigned designation does not contain one of the CFC, HCFC, or HFC acronyms. The term *anhydrous* means that it is without water. Ammonia is highly soluble in water. When ammonia must be discharged from a system, it is frequently discharged in water to reduce the local odor.

Though not classified as poisonous, ammonia vapors have a harsh effect on the respiratory system. Only small quantities of the vapor can be safely inhaled. Exposure for 5 minutes to 50 parts per million (ppm) is the maximum exposure allowed by the Occupational Safety and Health Administration (OSHA). Ammonia is hazardous to life at 5,000 ppm and is flammable at 150,000 to 270,000 ppm. Ammonia has a distinct odor that can be detected at 3 to 5 ppm. This odor becomes very irritating at 15 ppm. Anyone working on an ammonia system needs to be specifically trained for that purpose. As far as the environment is concerned, ammonia is considered quite safe.

Ammonia has been used for many years in industrial refrigeration systems, and remains in

use today. Since it does not represent an environmental problem and is reasonable priced in comparison to synthetic refrigerants, it remains very popular. Ammonia has four distinct advantages over the halogenated refrigerants:

- Ammonia is roughly 3 to 10 percent more efficient in transferring heat than the halocarbon refrigerants. This means that less ammonia must be circulated through the system to transfer an equal amount of heat. As a result, the operating cost for ammonia-based systems can be significantly lower.
- Since less ammonia needs to flow through a system, pipe sizes can be smaller.
- Ammonia is substantially less expensive than CFCs or HCFCs.
- Ammonia is safe for the environment, since it has no potential for ozone destruction and does not contribute to global warming.

To be fair, there are also some disadvantages:

- Ammonia is poisonous in very high concentrations. However, ammonia's obvious smell is easy to detect at levels far below those that represent a hazard. Personnel who suspect a problem typically have plenty of time to react. Ammonia vapors also rise quickly, since they are lighter than air, helping to eliminate the hazard.
- Ammonia and copper are not compatible. This means that all components that ammonia encounters in an operating system must not

The Use of Ammonia

Ammonia has been used in refrigeration systems in the United States since the late 1800s. It was first used as a refrigerant to produce ice. The first patents for ammonia refrigeration machinery were filed in the 1870s. Practically everything we eat, as well as many beverages, is processed or handled in at least one facility that uses ammonia as a refrigerant before it reaches our hands.

Technicians working with ammonia systems must use refrigerant gauge manifolds and gauges constructed of steel. Gauge manifolds for fluorocarbon refrigerants are typically made of brass.

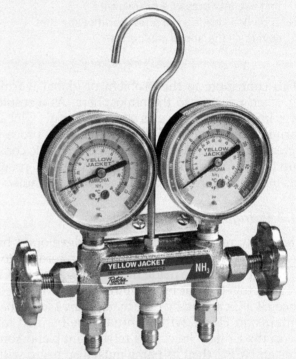

Figure Credit: Courtesy of Ritchie Engineering Company, Inc., YELLOW JACKET Products Division

contain copper. Piping systems for ammonia are typically made exclusively of steel. Heat transfer components such as evaporators or condensers usually contain steel or aluminum tubing.

Ammonia will likely remain a very popular refrigerant in the food-processing industry for years to come. Although it is not particularly cost-effective for use in small systems, it is an excellent choice for large systems.

Refrigerant Safety

All refrigerants and their containers must be handled with care. Refrigerants can cause injuries several different ways.

Any refrigerant container that holds refrigerant under pressure can potentially burst or leak. A container of refrigerant with a low boiling point that bursts will send liquid refrigerant that is rapidly absorbing heat and flashing from the liquid state to the vapor state in all directions. If the refrigerant liquid contacts human flesh, severe frostbite injury is likely. Whenever handling refrigerants, technicians should wear gloves and safety glasses at a minimum.

Although most refrigerants used are non-toxic, a great deal of refrigerant vapor released in a confined space can displace the oxygen and make breathing difficult or impossible. It is always a good idea to be aware of the volume of refrigerant contained in a room or similar space where work is going on. Consider what steps need to be taken quickly if there is a sudden and uncontrolled release of refrigerant.

2.3.0 Refrigerant Containers and Handling

Refrigerants come in disposable, returnable, or refillable metal containers that vary in shape and size (*Figure 22*). Low-pressure refrigerants such as HCFC-123 come in standard steel drums or cylinders. They have boiling points close to, or slightly above, common room temperatures. The pressure they exert on the container is much less than that of medium and high-pressure refrigerants, such as HCFC-22, HFC-134a, and HFC-410A. These refrigerants are liquefied compressed gases. If improperly handled, the pressurized containers can burst or leak, causing damage, injury, or even death under the right conditions.

2.3.1 Disposable Cylinders

Disposable cylinders are not manufactured for repeated use. These cylinders should be stored in dry locations to prevent rusting. They should also be transported carefully to prevent abrasion of their painted surfaces. Over time, rough handling or excessive heat could cause them to explode, especially if weakened by rust or corrosion. For added protection, keep disposable cylinders of refrigerant in their original cartons when practical.

For many years, the vast majority of disposable cylinders were either 30 lb or 50 lb models. However, many new refrigerants come in 24 lb, 25 lb, or 27 lb sizes.

> **WARNING!**
>
> Disposable cylinders must never be refilled. Not only is it dangerous, it is also against the law. Violators can be fined up to $25,000 and can face up to five years in jail.

Disposable cylinders should be recycled as scrap metal once they are empty. Be sure that the cylinder pressure is reduced to 0 psig, and then render the cylinder useless by puncturing the rupture disk (*Figure 23*).

2.3.2 Returnable Cylinders

Returnable cylinders go back to the manufacturer for reuse and refilling. They are not intended to be refilled in the field or to be used as a refrigerant recovery tank. However, you may see them used for that purpose when large volumes of refrigerant are recovered. These containers are not filled with more than 80 percent liquid. Excess liquid can result in an explosion when the contents is heated above the temperature at which they were filled and begins to expand. The pressure increases rapidly with even slight changes in temperature if the liquid has no room for expansion. Be aware that the term *full* means a vessel that is filled to 80 percent of its capacity. The remaining 20 percent of the volume must be available for expansion.

Returnable cylinders and tanks are normally stamped with various weight values. These values are shown in *Figure 22* and include the following:

- *Tare weight*—The empty weight of the vessel.
- *Gross weight*—The combined weight of the vessel (tare weight) plus the weight of the refrigerant when the vessel is full.
- *Net weight*—The weight of the contents in the vessel. For example, when ordering 50 lb of refrigerant from a supplier, it is the net weight that is being identified. Manufacturers design vessels so that when the full net weight is reached, 20 percent of the volume remains for liquid expansion.

Recovery cylinders, discussed further in the next section, also use these weight values.

The date when a cylinder was last tested is stamped on the shoulder or collar of returnable cylinders. Returnable and reusable cylinders must typically be retested every five years.

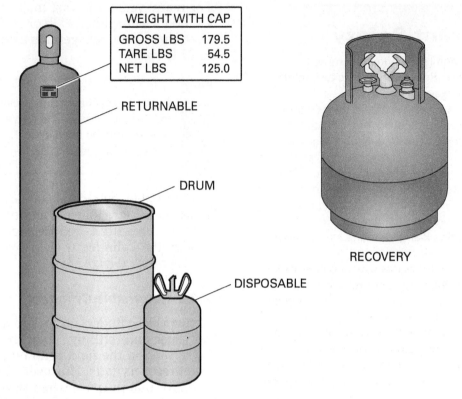

WEIGHT WITH CAP	
GROSS LBS	179.5
TARE LBS	54.5
NET LBS	125.0

RETURNABLE

DRUM

RECOVERY

DISPOSABLE

Figure 22 Refrigerant containers.

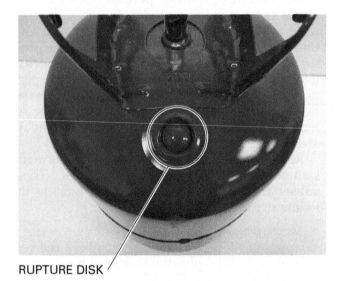

RUPTURE DISK

Figure 23 Rupture disk of a disposable refrigerant cylinder.

2.3.3 Recovery Cylinders

Recovery cylinders are typically used with refrigerant recovery/recycle units. Cylinders with a 50-pound capacity are commonly used for recovery. These cylinders are very strong and can be reused many times. According to EPA regulations, all cylinders used for the recovery and storage of used refrigerants are painted gray with the cylinder shoulders painted yellow. The label on the cylinder must be marked to properly identify the type of refrigerant it contains. A returnable cyl-inder should never be substituted for a recovery cylinder. If the cylinder does not have the proper markings of a recovery cylinder, do not use it for this purpose. Recovery cylinders must never be filled to more than 80 percent of capacity.

The US Department of Transportation (DOT) actually governs the construction and labeling of cylinders used for refrigerant storage and recovery. Cylinders used to recover HFC-410A must be specifically manufactured and labeled for use with this refrigerant due to its higher pressures.

2.3.4 Identifying Refrigerants

Refrigerant containers are color-coded and marked with labels to identify the type of refrigerant they contain. These labels also include important health information about the contents of the container. **Table 4** lists the color codes for some common refrigerant containers. The color codes help technicians readily identify refrigerants from a distance.

If the type of refrigerant used in a system is unknown, it can usually be identified by using one of the following methods:

- Check the manufacturer's service literature for the equipment.
- Check the nameplate on the equipment. If a refrigerant conversion has taken place, the unit should be clearly labeled.
- Check the data marked on the metering device.

Table 4 Color Codes for Common Refrigerants

HCFC-22	Light green	HCFC-123	Light blue-grey
HCFC-124	Dark green	HCFC-408A	Medium purple
HCFC-409A	Medium brown	HFC-404A	Orange
HFC-410A	Rose	HFC-407C	Brown
HFC-422B	Dark blue	HFC-134a	Light blue

Refrigerant recovery cylinders—Grey with yellow top

2.3.5 Refrigerant Safety Precautions

Butyl-lined gloves and safety glasses must be worn to avoid getting refrigerant on your skin or in your eyes. When accidentally released at atmospheric pressure, refrigerant can cause frostbite or burn the skin.

Refrigerants can cause suffocation if the amount and time of exposure is great enough. Although common refrigerants are not toxic, heavy concentrations displace oxygen, depriving the body of breathable air. Always maintain ample ventilation. Refrigerant vapor is invisible, has little or no odor, and is heavier than air. Be especially careful of low places where it might accumulate. Equipment rooms and other areas with systems holding large amounts of refrigerant must have alarm systems that detect low oxygen levels and sound an alarm. A self-contained respirator must be available near equipment rooms or other areas containing large volumes of refrigerant. Use the respirator if you must enter a contaminated area. Some equipment rooms have a mechanical ventilation system that can be used to clear contaminated air from the room.

In addition to wearing protective clothing and equipment, follow these rules when handling and using refrigerants:

- Always double-check to be sure you are using the proper refrigerant. The containers are color-coded and labeled to identify their contents. Container labels also include product, safety, and warning information.

- Refer to technical bulletins and Safety Data Sheets (SDSs) available from the manufacturers for information important to your health. They describe the flammability, toxicity, reactance, and health problems that could be caused by a particular refrigerant if spilled or incorrectly used.

> **NOTE**
>
> As of June 1, 2015, OSHA required new labeling methods for hazardous chemicals. The new labels are intended to make it easier for people to recognize and identify the chemical and its associated hazards. The new labels will include: a product identifier; a pictogram representing the nature of the hazard; a signal word such as Danger; precautionary statements; and identification of the supplier. A letter designation shows the minimum level of PPE required for a given exposure. This is all part of the Globally Harmonized System of Classification and Labeling of Chemicals, known as GHS. The GHS program also introduced a new Safety Data Sheet (SDS) format containing 16 sections that replaces the previous MSDS format.

- Do not drop, dent, or abuse refrigerant containers. Do not tamper with safety devices.
- When some refrigerants are exposed to an open flame, a toxic gas may be formed. Avoid situations that place refrigerant vapors and an open flame, such as a brazing torch, in the same space.
- Always use a proper valve wrench to open and close the cylinder valves.
- Replace the valve cap and hood cap to protect the cylinder valve when not in use or empty.
- Secure containers in place to prevent them from becoming damaged when moved (especially in a van or truck). Secure containers with straps or chains in an upright position for transport.
- Do not store containers where the temperature can cause the pressure to exceed the cylinder relief device settings.

The US Clean Air Act. **http://www.epa.gov/air/caa/**

International Institute of Ammonia Refrigeration (IIAR). **www.iiar.org**

2.0.0   Section Review

1. The difference between a halocarbon and a fluorocarbon refrigerant is that _____.

   a. the fluorocarbon always has fluorine in it, but not all halocarbons do
   b. fluorocarbon refrigerant has fluorine in it, while the halocarbon refrigerant still has carbon
   c. a halocarbon refrigerant never has chlorine in it while a fluorocarbon refrigerant does
   d. halocarbon refrigerants are highly toxic

2. OSHA establishes the maximum exposure to ammonia vapors at 5 minutes to a concentration of how many parts per million (ppm)?

   a. 25 ppm
   b. 50 ppm
   c. 75 ppm
   d. 100 ppm

3. HFC-410A refrigerant comes in a disposable container that is painted _____.

   a. dark brown
   b. light green
   c. dark green
   d. rose

Section Three

3.0.0 Cooling System Components

Objective

Identify the major components of cooling systems and how they function.

  a. Identify various types of compressors.
  b. Identify different types of condensers.
  c. Identify different types of evaporators.
  d. Describe the devices used to meter refrigerant flow.
  e. Discuss basic refrigerant piping concepts.
  f. Identify various accessories used in refrigerant circuits.

Performance Tasks

  5. Identify compressors, condensers, evaporators, metering devices, controls, and accessories.

Trade Terms

Brine: Water that is saturated with (contains) large amounts of salt. In the HVACR industry, the term typically describes any water-based mixture that contains substances such as salt or glycols that lower its freezing point.

Desiccant: A material or substance, such as silica gel or calcium chloride, that seeks to absorb and hold water from any adjacent source, including the atmosphere.

Pump-down control: A control scheme that includes a liquid refrigerant solenoid valve that closes to initiate the system off cycle. Once the valve closes, the compressor continues to operate, pumping most or all of the remaining refrigerant out of the evaporator coil. Pump-down control eliminates the possibility of excessive liquid refrigerant forming in the evaporator coil during the off cycle, and is primarily used in refrigeration applications. A thermostat typically controls the liquid solenoid valve, while a low-side pressure switch and/or a timer controls the compressor.

Refrigerant floodback: A significant amount of liquid refrigerant that returns to the compressor through the suction line during operation. Refrigerant floodback can have several causes, such as a metering device that overfeeds refrigerant or a failed evaporator fan motor.

Slug: Traditionally refers to a significant volume of oil returning to the compressor at once, primarily at startup. For example, a trap in a suction line may fill with oil during the off cycle, and then leave the trap as a slug at start-up. However, *slugging* is also a term incorrectly used by some to describe refrigerant floodback.

Surge chamber: A vessel or container designed to hold both liquid and vapor refrigerant. The liquid is generally fed out of the bottom into an evaporator, while vapor is drawn from the top of the container by the compressor to maintain refrigerant temperature through pressure.

Primary components required for the operation of a mechanical refrigeration system include compressors condensers, evaporators, metering devices, and piping. These components are reviewed in detail in this section.

3.1.0 Compressors

The compressor is the foundation of the refrigeration circuit. It creates the pressure difference that causes refrigerant to flow. Differences in pressure cause the flow of fluids. The direction of flow is always from a higher pressure to a lower pressure, much as heat moves from a higher temperature to a lower temperature. The result is that liquids and gases can be made to move by adjusting or changing the pressure around them. In refrigeration, a compressor is used to create the pressure differential that causes refrigerant vapors to flow in a system. Note that liquids cannot be compressed, and substantial amounts of liquid flowing through a compressor can damage or destroy it.

As the compressor ingests refrigerant vapor at a low pressure and compresses it to a higher pressure, additional heat is generated. The refrigerant vapor also absorbs some of the heat from the compressor motor in models where the motor is an integral part of the compressor structure.

Compressors are usually driven by an electric motor. Very large compressors can be driven by internal combustion engines or steam turbines, but this is rare. Compressors are divided into the following three groups based on the way they are joined to their motors or engines:

- *Open-drive compressor* – This type of compressor is physically separated from its motor (*Figure 24*). The drive shaft of the compressor extends outside the case. A mechanical seal is used with the rotating shaft to prevent leakage

Figure 24 Open, direct-drive compressors used in a large refrigeration application.

of the refrigerant. The compressor motor can turn the compressor using pulleys and belts. This is referred to as a *belt drive*. The motor can also be connected directly to the drive shaft for a direct-drive arrangement. Belt-driven arrangements allow the motor to run at one speed while the compressor can run at another. The proper combination of pulleys produces the desired speed of the compressor. Direct-drive units turn the compressor at precisely the same speed as the motor. Open-drive compressors can be opened for service and repair.

- *Hermetic (welded or sealed) compressor* – The compressor and motor share a common drive shaft. They are sealed in a welded steel enclosure or shell. Hermetic compressors (*Figure 25*) are more compact, less noisy, and require less maintenance than open-drive compressors because they have no belts or couplings to break or wear out. Because they are sealed, each entire unit must be replaced when it fails. However, open-drive compressors are available in much larger capacities than hermetic models. Hermetic compressors also cannot be serviced in the field. When internal problems occur, the compressor must be replaced in its entirety. The failed compressor is scrapped or returned to the manufacturer for evaluation or warranty credit. Hermetic models with capacities below 10 tons are most popular, but the use of compressors in higher capacities is increasing rapidly due to improved manufacturing techniques and the demand for larger compressors at a lower initial cost.

- *Semi-hermetic (serviceable hermetic) compressor* – Similar to the hermetic compressor, the compressor and motor share the same housing and the drive shafts are directly connected (*Figure 26*). These compressors offer some compromise between open-drive and hermetic models. They can be opened for service and repair of the intake and exhaust valves and some other components by removing the heads and/or the bottom and end of the housing (*Figure 27*). However, if the motor fails, the entire compressor is replaced. Failures related to the crankshaft, pistons, and piston rods usually result in complete replacement as well. These compressors are rarely destroyed when they fail. Instead, they are disassembled and remanufactured or rebuilt at shops designed for this purpose. They can then be resold.

Figure 25 A hermetic scroll compressor.

The following five types of compressors are commonly used in mechanical refrigeration systems:

- Reciprocating
- Rotary
- Scroll
- Screw
- Centrifugal

3.1.1 Reciprocating Compressors

Reciprocating compressors remain very common, although scroll compressors have taken their place

Figure 26 A semi-hermetic compressor.

Figure 27 A cut-away showing some of the field-serviceable components of a semi-hermetic compressor.

in many applications. They use one or more pistons moving back and forth within a cylinder or cylinders (*Figure 28*). The pistons compress the refrigerant vapors, like the pistons of an automotive engine. The suction and discharge valves are synchronized with the piston action. These valves control the intake and discharge of the refrigerant vapors.

Reciprocating compressors are typically used in domestic and commercial refrigerators, air conditioners, and commercial processing equipment. *Figure 29* shows a cut-away view of a hermetic reciprocating compressor. Semi-hermetic reciprocating compressors are used in commercial air conditioning and heat pumps above 10 tons. Open-drive reciprocating compressors are used mostly for refrigeration work and to support large industrial and commercial refrigeration systems anywhere in the 5- to 150-ton range.

3.1.2 Rotary Compressors

Rotary compressors are usually hermetic compressors. They are frequently used on appliances, room air conditioners, and central air conditioning systems below five tons. There are two basic types of rotary compressors: stationary vane and rotary vane. The rotary-vane type is used in HVACR applications.

Rotary-vane compressors (*Figure 30*) have a rotor centered on the drive shaft. However, the drive shaft is positioned off-center in the cylinder. Mounted on the rotor are two or more vanes that slide in and out to follow the shape of the cylinder. As the rotor turns, these vanes trap low-pressure suction gas and compress it against the cylinder wall, then force it out the discharge opening. The vanes also keep the compressed gas from mixing with the incoming low-pressure gas.

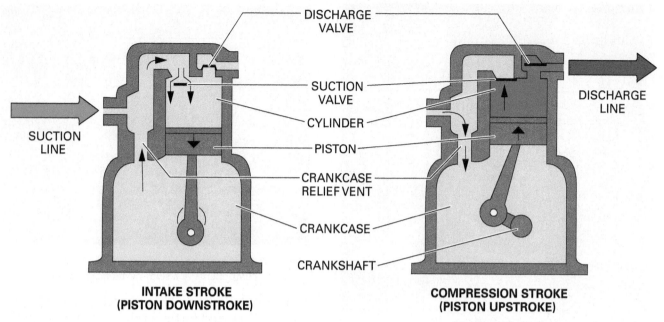

Figure 28 Reciprocating compressor operation.

INTAKE STROKE
(PISTON DOWNSTROKE)

COMPRESSION STROKE
(PISTON UPSTROKE)

DISCHARGE VALVE
SUCTION VALVE
CYLINDER
PISTON
CRANKCASE RELIEF VENT
CRANKCASE
CRANKSHAFT
SUCTION LINE
DISCHARGE LINE

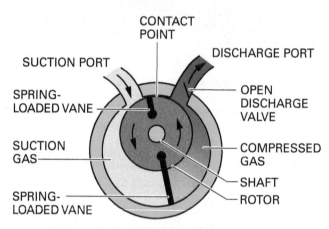

Figure 30 Rotary-vane compressor operation.

CONTACT POINT
SUCTION PORT
DISCHARGE PORT
SPRING-LOADED VANE
OPEN DISCHARGE VALVE
SUCTION GAS
COMPRESSED GAS
SHAFT
ROTOR
SPRING-LOADED VANE

Figure 29 Cut-away view of a hermetic reciprocating compressor.

Scroll Compressors

This cutaway view shows the interior of a scroll compressor. Notice the scrolls near the top, cut in half vertically. The other view shows what the two scrolls actually look like.

Scroll Compressor Sounds

Scroll compressors often produce unusual sounds when they start up and shut down. If you have never heard these sounds, you might think the compressor is defective. When you have the opportunity during your training, listen closely to these sounds so you can recognize them in the field. Some manufacturers of compressors and air conditioning equipment have training programs available on video or CD-ROM to help you identify these sounds.

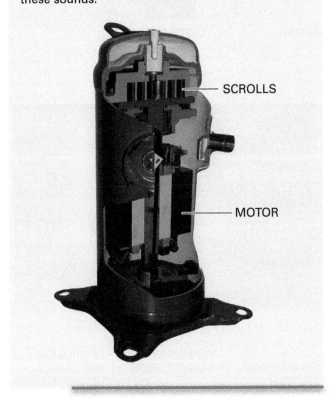

3.1.3 Scroll Compressors

Scroll compressors are always hermetic. Of all the compressor types, the scroll compressors have the fewest working parts. They operate efficiently even in applications that have larger pressure differences, such as commercial refrigeration and heat pumps. Since their development, their use has grown substantially. They are now used in most residential and light-commercial cooling systems that are built to provide high efficiency levels.

Unlike reciprocating compressors, there are no suction or discharge valves. Refer to *Figure 31*. However, a valve is used on the discharge side to help prevent reverse rotation. The scroll compressor achieves compression using two spiral-shaped parts called scrolls. One is fixed; the other is driven and moves in an orbiting action inside the fixed one. There is contact between the two. Refrigerant gas enters the suction port at the outer edge of the scroll, and after compression is squeezed out a separate discharge port at the center of the stationary scroll. The orbiting action draws gas into pockets between the two spirals. As this action continues, the gas opening is sealed off, and the gas is compressed and forced into smaller pockets as it progresses toward the center.

3.1.4 Screw Compressors

Screw compressors are used in large commercial and industrial applications requiring capacities from 20–750 tons. They are made in both open and hermetic styles. For comfort cooling, they are typically used to chill water.

Screw compressors use a matched set of screw-shaped rotors, one male and one female, enclosed within a cylinder (*Figure 32*). The compressor

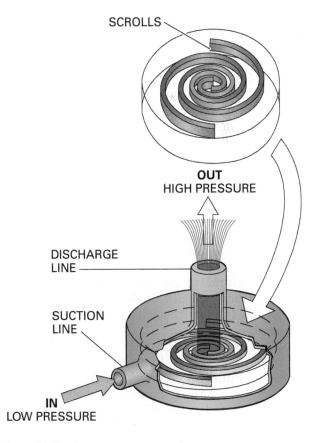

Figure 31 Scroll compressor operation.

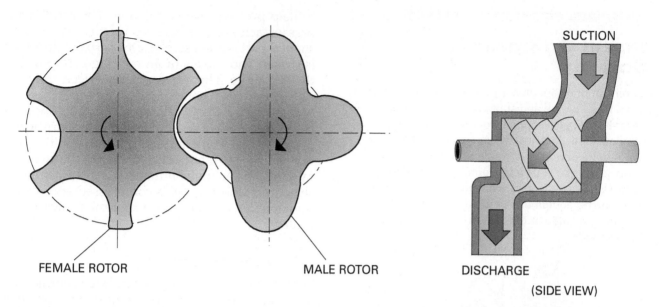

FEMALE ROTOR   MALE ROTOR   DISCHARGE

SUCTION

(SIDE VIEW)

Figure 32 Screw compressor operation.

motor drives the male rotor, which indirectly drives the female rotor. Normally the male rotor turns faster than the female rotor because it has fewer lobes. Typically, the male has four lobes and the female has six. As these rotors turn, they mesh with each other and compress the gas between them. The screw threads form the boundaries separating several compression chambers, which move down the compressor at the same time. In this way, the gas entering the compressor is moved through a series of progressively smaller compression stages until the gas exits at the compressor discharge in its fully compressed state. A cut-away view of the internal components and their arrangement is shown in *Figure 33*.

3.1.5 *Centrifugal Compressors*

Centrifugal compressors are made in open and semi-hermetic designs. They are typically used in commercial/industrial refrigeration and air conditioning systems with capacities larger than 100 tons. Standard models range up to 10,000 tons of capacity, with custom models exceeding 20,000 tons of capacity. In the centrifugal chiller shown in *Figure 34*, the compressor is located on top of the condenser and evaporator assemblies. The unit chills water to be circulated through a building for comfort cooling. It is also water-cooled, with water flowing through the condenser and then outdoors where the heat is rejected.

Figure 33 A cut-away of a small screw compressor.

COMPRESSOR DRIVE MOTOR HOUSING   CENTRIFUGAL COMPRESSOR   REFRIGERANT VAPOR INLET (SUCTION LINE)

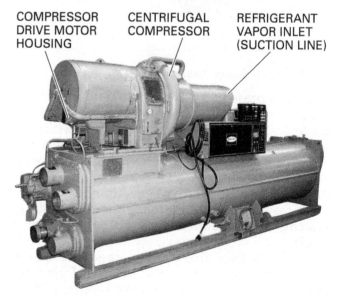

Figure 34 A centrifugal chiller.

Centrifugal compressors use a high-speed impeller with many blades that rotate in a spiral-shaped housing (*Figure 35*). The impeller is driven at high speeds (typically 10,000 rpm) inside the compressor housing. Refrigerant vapor is fed into the housing at the center of the impeller. The impeller throws this incoming vapor in a circular path outward from between the blades and into the compressor housing. This action, called *centrifugal force*, creates pressure on the high-velocity gas and forces it out the discharge port.

Often, several impellers are put in series to create a greater pressure difference and to pump a sufficient volume of vapor. A compressor that uses one impeller is considered a single stage unit, while one that uses two impellers is called a two-stage unit. If additional stages of compression are used, they are simply referred to as *multistage units*. When more than one stage is used, the discharge from the first stage is fed into the inlet of the next stage.

Prior to the elimination of CFC refrigerants, almost all centrifugal compressors operated using low-pressure refrigerants such as CFC-11. When low-pressure refrigerants are used, the low-pressure side of the refrigerant circuit usually operates in a vacuum. As a result, when leaks develop on the low side, air is drawn into the machine instead of refrigerant leaking out. Older centrifugal systems often leak, as low pressure units are not designed to withstand pressures over 15 psig. As a result, pressure testing can only be done at very low test pressures on low-pressure units.

As long as the leaks were controllable, the units were often left in that condition indefinitely. Prior to the 1990s, refrigerant was cheap and there was no legislation regarding refrigerant leaks. Today's centrifugal units offer much tighter construction. Although many still operate with low-pressure refrigerants like HCFC-123, others now use high-pressure refrigerants such as HFC-134a.

3.2.0 Condensers

Condensers (*Figure 36*) remove heat that was absorbed in other parts of the refrigeration cycle from the refrigerant. They take in high-pressure, high-temperature refrigerant gas from the compressor and change it into a high-temperature, high-pressure liquid. They do this by transferring the heat from the refrigerant to the air, to water, or possibly to both.

As the refrigerant progresses through the condenser, it first rejects any superheat and then fully condenses into a subcooled, high-temperature, high-pressure liquid. Condensers must reject the heat that was absorbed in the evaporator section, as well as the heat of compression. As a result, their design capacity for heat rejection must be greater than the evaporator's design capacity to absorb heat.

For the condenser to operate properly, the condensing medium must always be at a lower temperature than the refrigerant it is condensing. Condensers are classified according to the medium used to carry the heat away from the refrigerant vapor. The following are three types of condensers:

- Air-cooled
- Water-cooled
- Evaporative

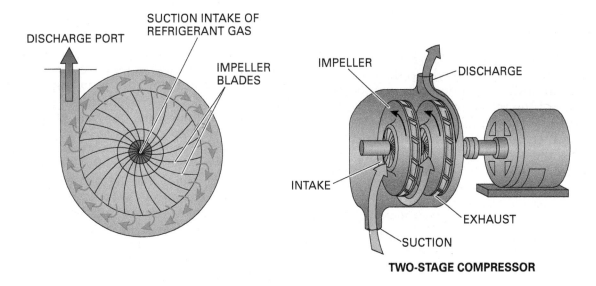

Figure 35 Centrifugal compressor operation.

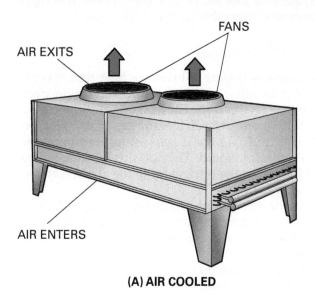

(A) AIR COOLED

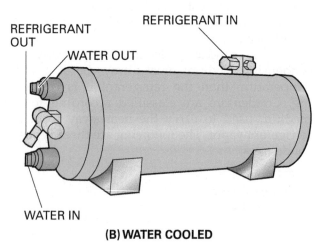

(B) WATER COOLED

Figure 36 Typical air- and water-cooled condensers.

3.2.1 Air-Cooled Condensers

Air-cooled condensers (*Figure 37*) reject the heat absorbed by the system directly to the outdoor air. At normal design conditions, the saturation temperature of the refrigerant flowing through the condenser is about 25°F to 35°F (14–19°C) warmer than the outside air to which it is rejected. This means that a saturation temperature of 120°F to 130°F (49–54°C) is typical in the condenser when the outdoor air is 95°F (35°C). The hot refrigerant vapor that initially enters the condenser may approach 200°F (93°C) under some conditions. Because the medium is outdoor air, the temperature difference needs to be greater than needed for water-cooled condensers.

Propeller (axial) fans are used with most air-cooled condensers to circulate air across the coil. Because air-cooled condensers require the circulation of air over their surfaces, their location and the temperature of the surrounding air are very important to efficient operation. The higher the temperature of the condensing air, the more work the system compressor must do to raise the refrigerant temperature to a level that allows efficient heat transfer. This causes the compressor to use more power.

Air-cooled condensers are typically used in residential and commercial air conditioning systems up to about 50 tons. However, much larger models up to several hundred tons of capacity are available when necessary. Air-cooled condensers are generally of two types: fin-and-tube condensers and plate condensers. Both are illustrated in *Figure 37*.

In the fin-and-tube condenser, the refrigerant vapor passes through the rows of tubing. The tubing is encased in metal ribs, or fins. The fins provide increased exposure to the surrounding air by increasing the surface area, thereby increasing heat transfer. Cooler air passing over the fins and tubing absorbs heat from the refrigerant. More fins generally result in better heat transfer and a smaller condenser coil to transfer the necessary heat. However, more fins also result in increased resistance to airflow and a tendency to collect more debris from the air stream. A compromise between size and fin spacing is often required.

Microchannel coils are a new way to get more from less. Coil manufacturers have developed coils that eliminate the standard copper or aluminum round tubing altogether. Instead, flat tubes that have multiple small passages inside are used. The idea is to increase the refrigerant surface area in contact with the heat-transfer surface, while also minimizing the refrigerant volume. Microchannel coils have increased efficiency while maintaining or reducing the size of the coils overall. Condenser coils were first fitted with microchannel coils, but they are also now being used for evaporator coils.

The plate condenser operates the same as the fin-and-tube condenser. Basically, it is formed by two sheets of metal that have been pressed or stamped into the correct shape and then welded together. The plates provide a larger surface area for the transfer of heat to the surrounding air. This approach is generally confined to very small domestic refrigeration units. Fin-and-tube condensers are the most popular air-cooled style by far.

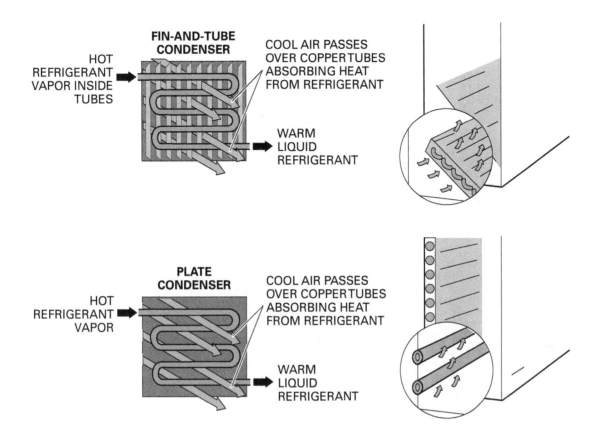

Figure 37 Air-cooled condenser operation.

3.2.2 Water-Cooled Condensers

Water-cooled condensers are more complex, more expensive, and require more maintenance than air-cooled condensers. However, they are more efficient and operate at much lower condensing temperatures (about 15°F [8°C] lower) than air-cooled condensers. This allows the system compressor to run at lower discharge pressures, requiring less power.

Depending on the type, the velocity of the water flowing in a water-cooled condenser should be between three and ten feet per second (one and three meters per second). If it flows too fast, the tubing may become pitted. If it flows too slowly, scaling from minerals in the water occurs. In areas where water is plentiful, the water that flows through the condenser may be used once and then drained into a waste system. This approach is very rare today.

Most often, the water portion of a water-cooled condenser is connected via piping to a cooling tower. In the cooling tower, the heat absorbed by the water in the condenser is rejected from the system by evaporation into the atmosphere. The cooled water is then returned to the system for reuse. There are four basic types of water-cooled condensers, shown in *Figure 38*:

- *Tube-in-tube* – In the tube-in-tube condenser, the water flows through one or more curved tubes. The high-temperature, high-pressure refrigerant supplied from the compressor flows in the opposite direction in a separate tube. In the condenser, the hot refrigerant gives up heat to the cooler water and condenses into a sub-cooled liquid.

- *Shell-and-tube (both vertical and horizontal configurations)* – Shell-and-tube condensers may be either horizontally or vertically mounted. Vertical condensers contain straight, vertical tubes encased in a metal shell. High-temperature, high-pressure refrigerant supplied from the compressor enters the top of the condenser metal shell that contains the tubes. The water also enters at the top and travels down through the vertical tubes where it absorbs heat from the surrounding refrigerant, then leaves at the bottom. As the refrigerant gas condenses on the outside of the water

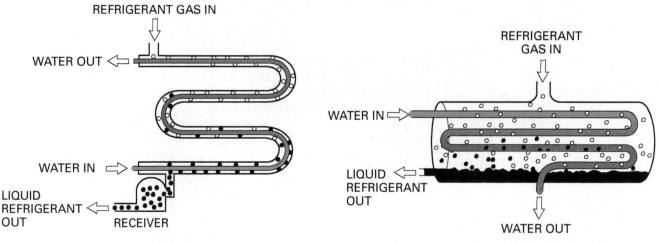

TUBE-IN-TUBE

REFRIGERANT GAS IN

WATER OUT

WATER IN

LIQUID REFRIGERANT OUT

RECEIVER

HORIZONTAL SHELL-AND-TUBE CONDENSER

REFRIGERANT GAS IN

WATER IN

LIQUID REFRIGERANT OUT

WATER OUT

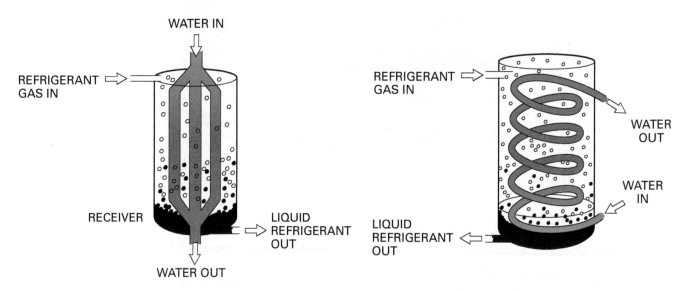

WATER IN

REFRIGERANT GAS IN

RECEIVER

LIQUID REFRIGERANT OUT

WATER OUT

VERTICAL SHELL-AND-TUBE CONDENSER

REFRIGERANT GAS IN

WATER OUT

WATER IN

LIQUID REFRIGERANT OUT

SHELL-AND-COIL CONDENSER

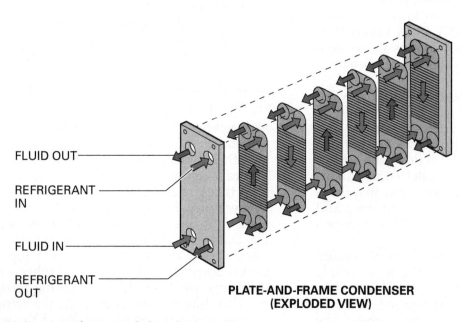

FLUID OUT

REFRIGERANT IN

FLUID IN

REFRIGERANT OUT

PLATE-AND-FRAME CONDENSER (EXPLODED VIEW)

Figure 38 A variety of water-cooled condenser types.

tubes, liquid refrigerant falls to the bottom of the shell where it collects and leaves the condenser at the outlet. The horizontal condenser design allows water flowing in the tubing to make several passes before leaving the condenser.

- *Shell-and-coil* – Construction of the shell-and-coil condenser is like that of the shell-and-tube condenser. The exception is that the tubing is wound around inside the shell, forming a coil, rather than being a straight length. As with other water-cooled condensers, the water flowing through the tubing cools the refrigerant in the shell surrounding it. As the refrigerant gas condenses on the cooler water tubes, liquid refrigerant falls to the bottom of the shell, where it collects and leaves the condenser at the outlet.

- *Plate-and-frame* – The plate-and-frame condenser, also known as a *plate heat exchanger*, consists of a series of metal plates held in place by a metal frame. Within the heat exchanger, the cooling medium flows through small channels, while the warm fluid flows through another. The plates are gasketed, welded, or brazed together to prevent the fluids from mixing. Plate-and-frame units are very efficient heat transfer devices. It is often surprising to see how much heat transfer performance can be packed into such a small area.

3.2.3 Cooling Towers

Cooling towers are not really condensers, since nothing is being condensed within the tower itself. Cooling towers support condensers in their function. In a cooling tower, the condenser water containing heat from the refrigerant is exposed to the outside air, which absorbs the heat from the water. There are many types of cooling towers used with water-cooled condensers.

Protecting Water-Cooled Condensers

Water-cooled condensers must be protected from freezing at all times. When a water-cooled condenser freezes, the ice can burst the tubes in one or more locations as it thaws. Water can then enter the refrigeration circuit in large volumes. Removing the water from the refrigeration circuit and returning the system to service is very expensive and time-consuming, and sometimes impossible.

One type is the natural-draft tower (*Figure 39*). These towers must be located where they can make use of natural air currents. They are made of a metal frame covering several layers or tiers of decks. Older towers used wood to construct the water decks. Since they use the natural air currents, no blowers are needed to move air through the tower. Water is piped up from the condenser located in the building below and is discharged in sprays over the decks. Spaces between the boards in the decks permit the water to drip or run from deck-to-deck, while being spread out and exposed to air breezes that enter the tower from the open sides. The cooled water is collected in a catch basin at the bottom of the tower where it is pumped back to the condenser for reuse. Natural-draft towers are no longer popular, as they must be much bigger than their forced-draft counterparts.

Forced-draft cooling towers are far more popular and widespread in the HVACR industry. The natural flow of air is replaced by air from propeller fans or blowers. Models like the one shown in *Figure 40* are called *counterflow towers*, since the air travels in the opposite direction through the tower (from bottom to top) than the water. Many cooling towers also have the added advantage of operating fans in stages or at various speeds. An induced-draft tower is a model that pulls the air through the tower rather than pushing it through (*Figure 41*). A propeller fan in the top of the tower draws air through the falling water droplets. The fan is usually belt-driven. Forced-draft cooling towers do not require nearly as much space as natural-draft models.

Since some water in the sump of the tower constantly evaporates as part of the cooling process, any water lost must be replaced to maintain the proper water level. This is generally done using a float that senses the water level in the catch basin and adds water as needed. Electronic water-level sensors are also used.

3.2.4 Evaporative Condensers

Evaporative condensers combine the functions of a water-cooled condenser and cooling tower. They first transfer heat from the refrigerant to the water, and then from the water to the outdoor air. The condenser water evaporates directly off the tubes of the condenser. Each pound of water that is evaporated removes about 1,000 Btus from the refrigerant flowing through the tubes.

As shown in *Figure 42*, air enters the bottom of the unit and flows upward over the condensing coil filled with refrigerant. At the same time, water is sprayed over the coil. Both the air and the

---

Multiple Cooling Towers

Cooling towers are often installed in groups, or cells, to gain the needed heat rejection capacity. In many cases, the individual tower cells can be isolated from each other so that maintenance can be performed on one cell while the system continues in operation.

Figure Credit: Courtesy of Baltimore Aircoil Company

water absorb heat from the refrigerant in the coil. Water eliminators, located above the water spray, remove water from the rising air, reducing the amount of water leaving the tower. The air is then moved out of the top of the unit using one or more fans. Cooled by both air and water, the refrigerant in the coil condenses into a subcooled liquid.

3.3.0 Evaporators

Evaporators (*Figure 43*) are used to extract heat from the conditioned space or substance. Low-temperature, low-pressure liquid refrigerant from the metering device enters the evaporator and immediately begins to absorb heat. As the refrigerant progresses through the evaporator, the heat in the warmer medium (air or water) causes it to boil

and change into a vapor state. During this latent process, a great deal of heat is absorbed.

It is very important to remember that the heat absorbed by the refrigerant during the change of state is exponentially greater than the heat it absorbs during a sensible change in temperature. If not enough liquid refrigerant is admitted to the evaporator, it changes to a vapor very quickly (flashes), and the majority of the evaporator coil contains only vapor. As a result, the capacity of the evaporator is seriously reduced, as vapor cannot do any substantial cooling work.

For evaporators to operate at peak efficiency, all the surface area inside the tubes must be exposed to flashing liquid. Ideally, the last remaining drop of liquid boils to a vapor as the refrigerant exits the coil. However, it is also essential that

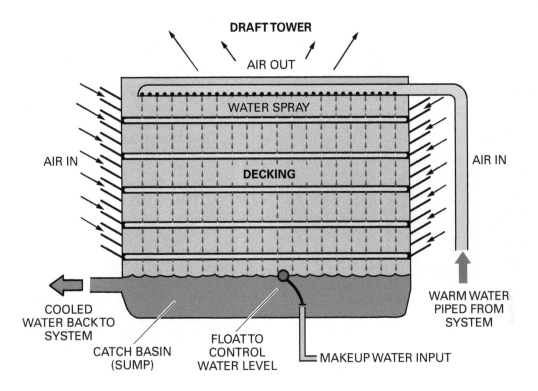

Figure 39 Natural-draft cooling tower.

liquid refrigerant does not return to the compressor. The challenge is to strike a balance between maximizing cooling capacity by coating as much of the evaporator surface as possible with flashing liquid, while ensuring that liquid refrigerant floodback does not occur.

Evaporators generally fall into one of two broad categories: direct-expansion type and flooded (*Figure 44*).

3.3.1 Direct-Expansion (DX) Evaporators

Direct-expansion (DX) evaporators are the most widely used, especially with fluorocarbon refrigerants. DX evaporators have one or more continuous tubes, or coils, through which the liquid refrigerant flows. The refrigerant, with a small amount of flash gas mixed in, enters the evaporator where it is gradually heated by the medium until it boils and becomes a vapor near the outlet.

The flow of refrigerant into the DX coil is controlled by the metering device at its input. It supplies just the right amount of refrigerant to the evaporator so that it is all transformed into vapor by the time it reaches the evaporator output.

Figure 40 A forced-draft, counterflow cooling tower.

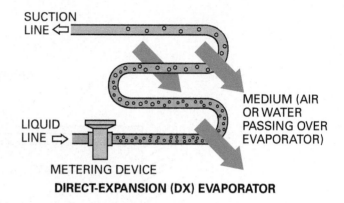

DIRECT-EXPANSION (DX) EVAPORATOR

Figure 41 An induced-draft cooling tower.

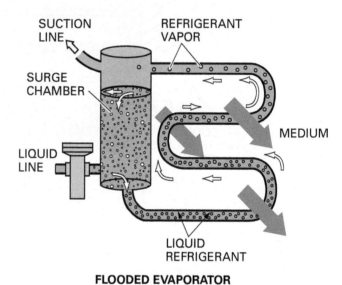

FLOODED EVAPORATOR

Figure 44 The operation of direct-expansion and flooded evaporators.

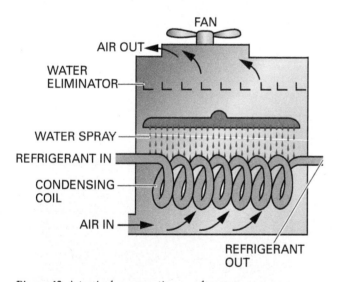

Figure 42 A typical evaporative condenser.

Figure 43 An evaporator coil for commercial walk-in refrigerators.

Movement of the medium (usually air or water) over the evaporator can be by natural draft, or it can be enhanced using fans or pumps.

Note the evaporator coil shown in *Figure 45*. Downstream of the expansion valve, a distributor is often found, with two or more smaller copper tubes feeding the refrigerant into the coil. This helps to ensure that liquid refrigerant is evenly distributed throughout the coil, improving its exposure to the medium being cooled. The distributor is designed to ensure that each branch receives an equal share of the refrigerant.

3.3.2 Flooded Evaporators

In a flooded evaporator, a special vessel known as a surge chamber is connected between the input and output of the evaporator tubing, as shown in *Figure 44*. Liquid refrigerant enters the surge chamber from the metering device. The metering device is designed to maintain a predetermined liquid level inside the chamber. Liquid then flows

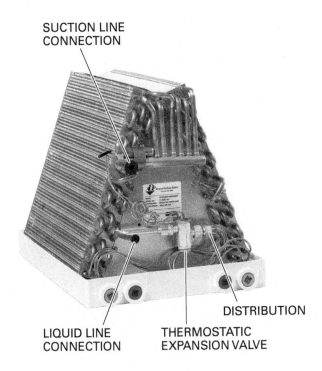

SUCTION LINE
CONNECTION

DISTRIBUTION

LIQUID LINE
CONNECTION

THERMOSTATIC
EXPANSION VALVE

Figure 45 A cased direct-expansion evaporator coil.

through the evaporator coil, where it boils and returns to the surge chamber. All the refrigerant vapor exits through the suction line on its way to the compressor. Any remaining liquid refrigerant falls into the surge chamber for recycling back through the evaporator. The surge chamber also protects the compressor from liquid refrigerant.

Flooded evaporators are typically used only in large food processing and other major refrigeration applications. They are not generally used with domestic refrigeration or comfort cooling evaporators, although some water chillers may use this approach.

3.3.3 Evaporator Construction

Evaporators can be classified by the way they are constructed. The primary classes are as follows (*Figure 46*):

- *Bare-tube evaporators* – There are two types of bare-tube evaporators: single-circuit (path) and multiple-path. Multiple-path evaporators are often used because they save space and reduce the number of metering devices needed. Each path is simply a copper, aluminum, or steel tube shaped in a way that best matches the job. The piping is the only surface used to transfer heat. Bare-tube evaporators have limited applications.

- *Finned-tube evaporators* – These are a variation of the bare-tube evaporator. Attached to the tubing are very thin spiral-wound or rectangular fins of aluminum or copper, like those used with the fin-and-tube condenser. These fins increase the amount of surface area exposed to the heated medium. This in turn increases the amount of heat that can be transferred to the refrigerant. This type of coil is by far the most popular in the HVACR industry. Coils with more closely spaced fins transfer heat more effectively. However, fins also create resistance to air movement. In addition, tight fin spacing also causes the coil to become fouled with airborne debris more easily. Since the evaporator coil is usually wet with condensate, dirt and other debris can collect on the fins very quickly. Evaporator fin spacing is determined for each application to provide the best balance between efficiency and maintenance issues. Fin spacing is measured in fins per inch (FPI).

- *Plate evaporators* – Similar to plate condensers, plate evaporators have a length of tubing woven along and firmly bonded to a metal plate. The plate provides greater surface area for heat transfer. This type of evaporator was typically used as a shelf in older domestic freezers. They are used heavily in food-processing applications. Foods such as fish can be placed in a single layer on each plate, or between two plates, where they can be frozen very quickly through maximum contact with the evaporator surface (*Figure 47*).

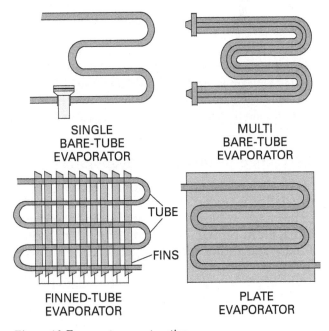

SINGLE
BARE-TUBE
EVAPORATOR

MULTI
BARE-TUBE
EVAPORATOR

TUBE

FINS

FINNED-TUBE
EVAPORATOR

PLATE
EVAPORATOR

Figure 46 Evaporator construction.

Figure 47 Plate-freezing equipment for the food processing industry.

Natural-Draft Evaporators

Natural-draft evaporators are quite rare today, but bare-pipe evaporator coils using only a natural draft to transfer heat have long been used in places like crab holding rooms, where the catch must retain a maximum amount of moisture. Forced air movement carries too much moisture away from the crabs. The bare coils are usually mounted on the walls around the perimeter of the storage area.

3.3.4 Chilled-Water System Evaporators

In many large system installations, cooling coils are installed a great distance apart. Because of the expense and problems associated with long runs of refrigerant piping, cooling is often done with chilled water rather than refrigerant. The chilled water is considered a secondary refrigerant. When the temperature of the water must be below freezing, brine is used in place of water. A refrigeration unit, called a *chiller*, is used to cool the water or brine. Technically, brine is defined as water saturated with salt. However, water that is mixed with alcohol, sodium chloride, calcium chloride, ethylene glycol, or propylene glycol is generally referred to as brine in the trade. It is used to lower the freezing point of the water.

The primary system generally uses shell-and-tube or shell-and-coil evaporators to absorb heat from the water. Like other evaporators, chiller evaporators can be of the direct-expansion or flooded type. Their construction is similar to that used for shell-and-tube or shell-and-coil condensers. In *Figure 48*, the black areas of the chiller are covered in insulation. The large insulated vessel beneath the compressor is the evaporator section. In a direct-expansion chiller, the colder liquid refrigerant runs through the tubing while the warmer secondary refrigerant (water or brine) circulates within the shell around the outside of the tubes. The heat contained in the water is transferred to the primary refrigerant flowing through the tubes. The refrigerant leaves the evaporator as a vapor on its way to the compressor inlet.

In a flooded chiller, the water is circulated through the tubing, which is located on the bottom of the evaporator shell. This tubing is submerged below the liquid refrigerant level within the shell, which is controlled by a float valve. The evaporator shell acts as a surge chamber. The top portion is left vacant so the refrigerant vapor can remain separated from the liquid refrigerant as it returns to the compressor.

3.4.0 Refrigerant Metering Devices

The metering device is located between the condenser outlet and the evaporator inlet. High-pressure, high-temperature liquid refrigerant from the condenser enters the metering device. It leaves as a low-pressure, low-temperature mixture of liquid and vapor. Regardless of type, the metering device performs two vital functions:

- It allows the liquid refrigerant to flow into the evaporator at a rate that matches the rate at which the evaporator boils liquid refrigerant into a vapor.
- It provides the pressure drop that lowers the boiling point of the refrigerant.

There are a number different metering device styles. They can be divided into two basic categories: fixed and adjustable. *Figure 49* shows some examples of both types. This section briefly describes the main types of fixed and adjustable metering devices. These devices are presented in greater detail in another module.

Fixed metering devices are used mainly in domestic refrigerators, freezers, and residential/commercial air conditioning units. Adjustable metering devices are used most often on systems with variable load requirements, and on systems that need to operate at higher efficiencies. Although fixed devices have been very popular for years in residential cooling systems, the demand for higher efficiency is making adjustable and modulating metering devices popular one again.

The performance of the metering device has a great deal to do with the condition of the liquid refrigerant it receives from the condenser or receiver. Indeed, the performance of the refrigeration circuit relies heavily on its condition. The condition being referred to here is the amount of subcooling.

Remember that some liquid refrigerant must flash to its vapor state as it passes through the metering device, providing the cooling effect for the remaining liquid. Enough liquid must flash to cool the liquid from its entering temperature (105°F, or 40°C, would be common) to the saturation temperature for the new pressure. In a comfort cooling system, 40°F (4 to 5°C) is considered normal. That results in a 65°F (36°C) temperature drop that must occur instantly. Any liquid refrigerant that flashes to vapor to provide this cooling effect is not available in the evaporator to transfer heat. Therefore, a lower liquid temperature (more subcooling) equals more cooling capacity in the evaporator coil. Consider how much cooling capacity is lost when the liquid leaving the condenser isn't subcooled at all. A clogged condenser coil can easily cause this condition, where the refrigerant vapor is not fully condensed. The result is even hotter liquid being fed to the metering device that must be cooled by flash gas, and less liquid refrigerant for it to work with.

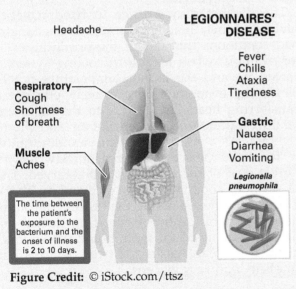

Figure 48 A large chiller with a centrifugal compressor.

3.4.1 Fixed Metering Devices

Fixed metering devices have a fixed opening, or orifice, size. The capillary tube, as shown in *Figure 49*, is one of the oldest metering devices. It is a fixed-length, small-diameter copper tube. Because of its high resistance to refrigerant flow, it restricts the flow of liquid refrigerant from the condenser to the evaporator. Longer lengths and smaller diameters produce the greatest pressure drop. Manufacturers determine the size and length of the capillary tube to provide a precise amount of refrigerant flow and pressure drop. Capillary tubes are often coiled to conserve space and protect them from damage.

Another type of fixed metering device similar to the capillary tube is simply known as a fixed restrictor. This device is a compact and rugged assembly that is installed at the evaporator inlet. The housing contains a piston. Pistons are made with different-sized orifices to match the capacities of different equipment. The smaller the orifice, the greater the resulting pressure drop. The easily replaced piston is installed in the housing at the time of system installation in order to match the metering device to the condensing unit. The amount of refrigerant flow through the fixed restrictor is controlled by the size of the orifice opening and the pressure applied. As the outdoor temperature increases, the pressure also increases, driving more refrigerant through the fixed metering device.

The distinct disadvantage to fixed metering devices is their inability to adjust to changing load conditions throughout the operating cycle. For example, when a comfort cooling system is started up and the indoor temperature is 90°F (32°C), the evaporator is extremely efficient at transferring heat. This is due to the high temperature difference between the air and the refrigerant. As a result, all of the liquid admitted by the fixed device vaporizes quickly, leaving a great deal of the evaporator coil starved. This causes poor cooling performance and an above-normal superheat, slowing the process of cooling the space significantly. A metering device that can sense the need for more liquid works much better. The advantages of the fixed device are its simplicity and very low cost.

THE METERING DEVICE CONTROLS THE AMOUNT OF REFRIGERANT SUPPLIED TO THE EVAPORATOR AND PROVIDES A PRESSURE DROP THAT LOWERS THE REFRIGERANT BOILING POINT.

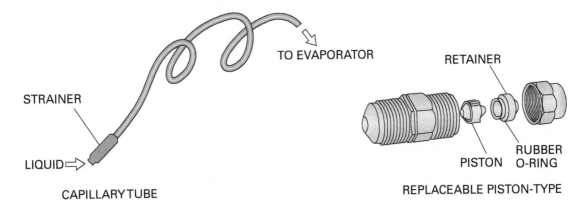

FIXED METERING DEVICES

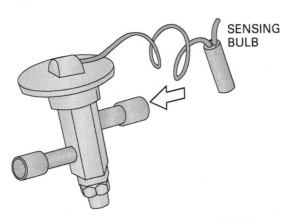

ADJUSTABLE METERING DEVICE (TXV)

Figure 49 Examples of common fixed and adjustable metering devices.

3.4.2 Adjustable Metering Devices

Adjustable metering devices differ significantly in their construction and how they are controlled. Adjustable metering devices work to regulate refrigerant flow so that the evaporator capacity matches the cooling load. There are six types of adjustable metering devices in common use:

- Hand-operated expansion valve
- Low-side float valve
- High-side float valve
- Automatic expansion valve
- Thermostatic expansion valve
- Electric and electronic expansion valves

Each of these devices is covered in detail in a later module. The first three on the list are generally associated with commercial/industrial refrigeration systems. However, the operation of a thermostatic expansion valve is briefly presented here.

The thermostatic expansion valve (TXV or TEV) controls the amount of refrigerant flowing through the evaporator by sensing the level of superheat in the suction line at the evaporator output. It is designed to maintain a constant superheat value. Refer to *Figure 50*.

A TXV has a diaphragm at the top, with a valve and seat below it. The top portion containing the diaphragm is called the power element. The pressure exerted on the underside of the power element diaphragm is a combination of the evaporator refrigerant pressure and the adjustment spring pressure. The spring is adjusted to exert a pressure that represents the desired level of evaporator superheat. Pressure on top of the diaphragm is applied via tubing from a remote sensing bulb used to monitor the superheat at the evaporator outlet. This bulb typically contains a refrigerant charge as the sensing fluid. However, it is important to note that the refrigerant inside the bulb and power element is not from the refrigerant circuit. It is a tiny amount of refrigerant permanently sealed inside by the manufacturer.

If the superheat increases or decreases as a result of changing load, a corresponding pressure change is transmitted from the bulb to the valve mechanism. This pressure change counteracts the combined evaporator and adjustment spring pressures to increase or decrease the valve opening, controlling the flow through the valve. Electronic expansion valves (EXVs) control the flow of refrigerant in much the same way as the thermostatic type. A small, pin-shaped valve that can fit tightly against a matching seat is positioned to allow more or less refrigerant to flow through the valve. However, the position of an EXV is controlled by a signal such as a 4–20mA current signal or a 0–10VDC voltage signal. An electronic circuit board transmits the signal. The signal intensity is based on the temperature reading of the refrigerant leaving the evaporator. The circuit board may also monitor refrigerant pressure, allowing it to consistently process and determine the current superheat value. The board then sends an electronic pulse to the EXV. Many operate in predetermined steps to position the valve for the needed amount of refrigerant.

Fixed Restrictors

Fixed restrictors replaced the capillary tube in many residential air conditioning and heat pump systems. The piston is less susceptible to plugging and is easy to service in the field. In addition, they greatly simplify the refrigerant circuitry required for heat pumps. The piston is free to move inside the housing. When pressure is applied from one direction, the piston is pushed against its seat and refrigerant can only flow through the orifice. When the direction of flow is reversed, the piston is pushed in the opposite direction, and refrigerant can flow both through and completely around it.

Capillary tubes are still used in room air conditioners. In some systems, several metering devices are built into the distribution lines that feed each evaporator circuit.

TXVs

This photo shows a thermostatic expansion valve installed in a system. Note that the sensing bulb is wrapped in insulation to make sure that it senses only the temperature of the refrigerant leaving the evaporator, and is not influenced by the external temperature. This TXV is externally equalized, as indicated by the connection from the bottom of the TXV to the suction line. This design, which is common in larger systems, allows the superheat to be accurately maintained regardless of the pressure drop through the evaporator.

Improper installation of a TXV can prevent the system from operating correctly. The sensing bulb must be securely fastened to a clean, straight, horizontal section of the suction line close to the evaporator outlet. In addition, the bulb must be thoroughly insulated with waterproof insulation to prevent it from being influenced by the surrounding air. The sensing bulb is attached to the suction line in various positions, depending upon the size of the line. As a general rule, the bulb should be fastened at the 4 or 8 o'clock position on line sizes of $\frac{7}{8}$" outside diameter (OD) and larger. On smaller suction lines, the position is not as critical, as long as it is not fastened at the very bottom. In this location, oil that remains at the bottom of the line may prevent an accurate temperature reading. Note that the outlet of the TXV has not been insulated; it will sweat when in operation if insulation is not added. Remember that the outlet side is at the same temperature and pressure as the evaporator coil. That short section should be insulated.

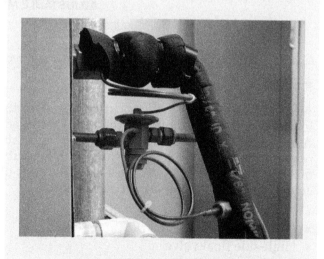

One advantage of the EXV is its increased accuracy. While a TXV is often set to maintain superheat values of 10°F–12°F (approximately 6°C) for comfort cooling applications, an EXV can reliably maintain superheat values below 5°F (3°C), without allowing refrigerant floodback. This allows the system to operate more efficiently, since more of the evaporator surface is bathed in boiling refrigerant.

3.5.0 Refrigerant Piping

Most piping used in refrigeration systems is constructed of ACR-grade copper tubing. Aluminum, steel, and stainless steel may also be used in some applications. The piping layout is usually determined by the system designer. The HVACR technician is interested in piping mainly from a servicing viewpoint.

There are four characteristics of a good piping layout:

- It provides refrigerant paths.
- It avoids excessive pressure drop.
- It returns oil to the compressor crankcase reliably.
- It protects compressors.

Figure 50 The internal construction of a typical TXV.

The purpose of the piping system is to provide a path for the flow of refrigerant from one component to another. Refrigerant flow must be accomplished without excessive pressure drops, such as those caused by friction, long risers, restrictions, and other piping conditions.

Some oil circulates in most refrigerant circuits. The piping layout, therefore, must return the oil to the compressor crankcase. A slug of oil or large amounts of liquid refrigerant entering the compressor through the suction line can seriously damage the compressor. Good piping practices help to minimize the potential for damage.

In *Figure 51*, the major refrigerant lines are identified. The suction line carries cold, low-pressure gas from the evaporator to the compressor. The hot-gas line carries hot, high-pressure gas from the compressor to the condenser. Where separate receivers are used, a condensate line is installed to drain refrigerant from the condenser to the receiver. The liquid line carries the liquid refrigerant from the receiver to the metering device. Note that some of the components shown here, such as the liquid-suction heat exchanger, are not necessary in every refrigerant circuit.

3.5.1 Basic Principles

Certain basic principles apply to all refrigerant lines. Keep them simple and pitch horizontal lines in the direction of flow. This helps to maintain the flow of oil in the right direction. The pressure drop in lines should be as low as practical, and the velocity of the refrigerant in the lines affects the pressure drop. The refrigerant velocity is also crucial to the movement of oil through the system, especially when it must travel up through a vertical riser. If the speed is too low here, the oil cannot make the trip to the top, and the compressor eventually starves for oil.

3.5.2 Suction Line

The suction line is the first line to be considered. It is the most critical and challenging from a layout and installation standpoint. The pressure drop at full load must be within practical limits. Larger line sizes provide lower pressure drop. However, oil movement and its return to the compressor must be maintained under minimum load conditions. Larger line sizes work against this requirement because the velocity slows down. The size of the suction line must represent a balance between these two needs. In some situations, the suction-line size must be different at various points in the system.

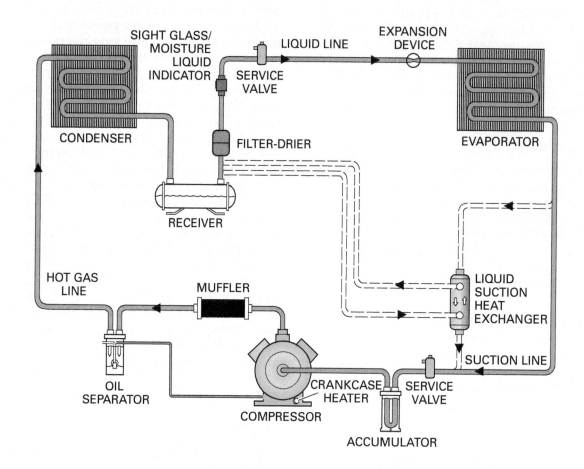

Figure 51 Major refrigerant lines.

The suction line must also prevent any liquid refrigerant and oil from draining directly to the compressor during shutdown. Although very small amounts of liquid returning to the compressor are tolerable, the return of oil in slugs and refrigerant floodback must be avoided.

The suction line is larger than the liquid line because it handles vapor. The same amount of refrigerant must flow through both the liquid and suction lines, but the vapor is in an expanded state. If the suction line is undersized, capacity is lost due to pressure drop. This causes the system to run longer to satisfy demand. The power consumption increases, resulting in higher energy costs to run the system. If the suction line is oversized, it will cause poor oil return. A larger line also increases the installation cost.

Unnecessary oil pockets should be avoided. Poor layout or complicated routing can result in pockets in which oil can collect. Also, allow room for expansion and vibration. All lines expand and contract with changes in temperature.

In suction risers, oil is carried up by the refrigerant gas, as shown in *Figure 52*. A minimum gas velocity must be maintained to keep the oil moving up. The trap at the bottom of the riser promotes free drainage of liquid refrigerant away from the TXV bulb, so that the bulb senses suction gas superheat instead of evaporating liquid temperature. In addition, the trap creates a point of turbulence in the line as the vapor turns the corners. This helps to mix, or entrain, the accumulated oil, and assists in lifting it to the top of the riser.

Where system capacity is varied through compressor capacity control or some other arrangement, a riser is usually sized smaller than the remainder of the suction line to ensure oil return up the riser (*Figure 53*). Where system capacity is variable over a wide range, it may not be possible to find a pipe size for a suction riser that ensures both oil movement at minimum capacity and an acceptable pressure drop at maximum capacity. In that case, a double suction riser (*Figure 54*) is

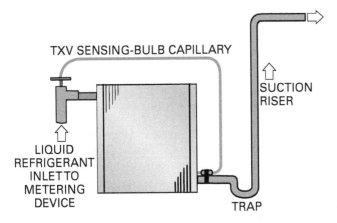

TXV SENSING-BULB CAPILLARY

LIQUID
REFRIGERANT
INLET TO
METERING
DEVICE

SUCTION
RISER

TRAP

Figure 52 A suction line riser.

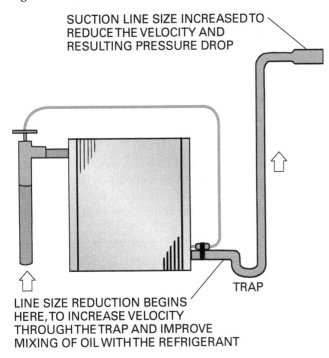

SUCTION LINE SIZE INCREASED TO
REDUCE THE VELOCITY AND
RESULTING PRESSURE DROP

TRAP

LINE SIZE REDUCTION BEGINS
HERE, TO INCREASE VELOCITY
THROUGH THE TRAP AND IMPROVE
MIXING OF OIL WITH THE REFRIGERANT

Figure 53 A reduced suction-line riser.

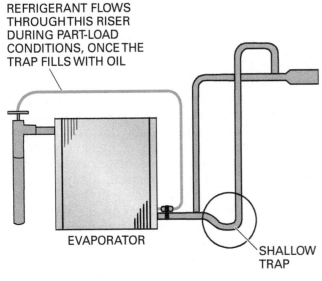

REFRIGERANT FLOWS
THROUGH THIS RISER
DURING PART-LOAD
CONDITIONS, ONCE THE
TRAP FILLS WITH OIL

EVAPORATOR

SHALLOW
TRAP

Figure 54 A double suction riser.

used. The design of the double suction riser is such that, under low demand, refrigerant flows up only one riser. Oil held in the trap due to insufficient velocity blocks the other riser. At full load, when the compressor is operating on all cylinders, refrigerant and oil travel up both risers. The trap for this purpose should be as shallow as possible, as shown in the figure.

When the compressor is on the same level or below the evaporator, a riser to at least the top of the evaporator coil should be placed in the suction line (*Figure 55*). This is to prevent liquid draining from the evaporator into the compressor during the off cycle. The suction line loop can be left out and the piping simplified, if the system is designed for pump-down control.

Pump-down control is accomplished by placing a solenoid valve in the liquid line, upstream of the metering device. A simple suction line, draining by gravity directly to the compressor and without traps, is allowed in this situation. With a pump-down control, the compressor does not shut down as soon as the thermostat is satisfied. Only the liquid-line solenoid valve closes. The compressor continues to run, vaporizing and pulling out any remaining refrigerant in the evaporator, until a pressure switch turns it off. Since the liquid flow has already stopped, any refrigerant pumped out is simply stored in the condenser or receiver.

It is important to prevent liquid refrigerant and oil from draining from the evaporator to the compressor during shutdown. However, it is equally important to avoid unnecessary traps in the suction line near the compressor. Such traps collect slugs of oil, which can be carried to the compressor during startup and cause serious damage. The part of the suction line near the compressor should be free draining to the compressor, and the admission of any significant liquid to this section should be prevented.

3.5.3 Hot-Gas Line

Considerations for the hot-gas lines are similar to those for suction lines. The pressure drop at full load must be maintained within reasonable limits; oil circulation must be maintained under minimum load conditions; and refrigerant and oil must be prevented from draining backwards to the head of the compressor during shutdown.

Figure 56 shows a hot-gas line in its simplest form. It should pitch down from the compressor to the condenser, unless the condenser is above the compressor. If a riser must be placed in the line, smaller pipe may be used to ensure oil movement at part-load conditions. Where the system load varies over a wide range, a double riser may be required.

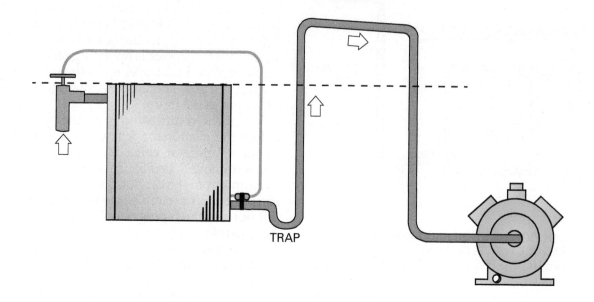

Figure 55 Routing the suction line above the top of the evaporator coil.

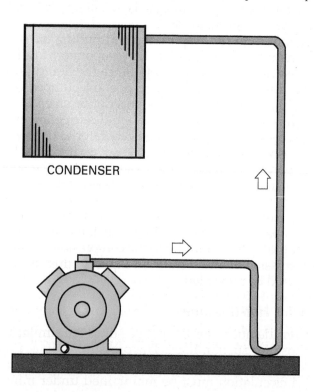

Figure 56 Hot-gas line.

Since the hot-gas line connects to the head of the compressor, provisions must be made to prevent oil or condensed refrigerant from flowing back through the line and into the compressor during the off cycle. A loop to the floor between the compressor and the riser (a deep trap) will normally provide an adequate reservoir to trap and hold the oil or condensed refrigerant that could come back down the riser during the off cycle.

3.5.4 Liquid Line

The layout of the liquid line is less critical but still very important. Oil mixes freely with many refrigerants when they are in liquid form. However, it is important to note that some newer refrigerants do not mix well with the oil, complicating line sizing and design. Therefore, it is not necessary to provide high velocities in liquid lines to ensure oil return. Traps in the liquid line do not create oil-return problems.

Large receivers are most likely found in commercial and industrial refrigeration systems. They are rarely needed in today's comfort cooling systems. The piping between the condenser and the receiver is sometimes referred to as the condensate line. However, it should not be confused with the condensate line that carries collected water away from an evaporator coil. The condensate line must provide for the drainage of condensed refrigerant to the receiver and the venting of any gas generated in the receiver back to the condenser. This line should be large enough to allow gas formed in the receiver to flow back to the condenser without restricting the drainage of liquid refrigerant from the condenser. Designers refer to this as sewer gas flow, since the principle is much the same. If this is impractical, a separate gas equalizer line from the receiver to the condenser is required. If gas that forms in the receiver is not provided with a way out, the receiver will become vapor-locked and cease to accept liquid.

It is desirable to have a subcooled liquid reach the metering device at a sufficiently high pressure for proper operation. An excessive pressure drop can result in the loss of capacity at the metering device; a pressure drop without sufficient subcooling causes some of the refrigerant to flash back into the vapor state. In fact, a liquid line that is too small becomes an undesirable metering device, much like a capillary tube. The pressure drop in liquid lines can be due to pipe friction, vertical rise, and/or accessories. Liquid line sizes are normally chosen such that a maximum allowable pressure drop is not exceeded.

Any vertical risers in the liquid line will experience a predictable pressure drop as the result of lifting the liquid. This is known as static loss, and it results from the weight of the liquid column. For common refrigerants, the pressure at the top of the riser will be lower by roughly 0.4 to 0.5 psig (2.7–3.4 kPa) for every foot of height. A 20' (6 m) riser will experience a pressure drop at the top of 8 to 10 psig (55–69 kPa) as a result. No line-sizing technique can eliminate this condition. Only a sufficient amount of subcooling in the liquid can prevent flash gas from occurring in a liquid-line riser. Accessories should be selected based on their pressure drop at a given cooling or flow capacity first, before the line size is considered. At times, the line size may be larger than that of the main liquid line as a result, but a smaller line size is rarely chosen.

3.5.5 Insulation

Liquid lines are not normally insulated, except where the surrounding temperature is significantly higher than the liquid refrigerant. Hot-gas lines are generally well above the surrounding temperature and need only be insulated for personnel protection. Suction lines are always insulated to prevent condensation on the outside of the line, and to reduce the amount of heat picked up from the environment; exceptions are rare. Some heat absorption may be desirable to help boil off very small amounts of liquid refrigerant that entered the line, but excessive heat gain by the suction gas must be avoided. Suction line insulation can be covered with a vapor barrier and weatherproofed when outdoors. However, vapor barriers and weatherproofing are rarely applied on residential and light commercial systems.

Refrigerant Line Sets

When installing refrigerant line sets, most technicians install the suction line before the liquid line to make sure there is adequate space to achieve the necessary pitch. The suction line with insulation is considerably larger.

3.6.0 Refrigerant Circuit Accessories

Additional components can be added to the basic refrigeration system to improve safety, endurance, efficiency, or servicing (*Figure 57*). Some of these components are factory-installed, while others may be installed in the field. This section briefly describes the most commonly used components. They include the following:

- Filter-drier
- Sight glass/moisture indicator
- Suction line accumulator
- Crankcase heater
- Oil separator
- Heat exchangers
- Receiver
- Service valves
- Compressor muffler

3.6.1 Filter-Drier

The filter-drier or strainer-drier combines the functions of a filter and a drier in one device (*Figure 58*). The filter protects metering devices and the compressor from foreign matter such as dirt, scale, or rust. The drier removes moisture from the system and traps it where it can do no harm. The drier portion is made of a desiccant material that has the ability to absorb water directly from the refrigerant (*Figure 59*).

Filter-driers are best installed in the liquid line just ahead of the metering device. They are typically replaced any time the system is opened for a lengthy period to make a system repair.

3.6.2 Sight Glass and Moisture Indicator

The sight glass is like a window that allows the technician to view the condition of the system refrigerant. It is often used when checking the refrigerant charge. The sight glass shows whether the refrigerant is fully condensed into a liquid or contains vapor. One common location for a sight glass is at the condenser outlet to view the condition and flow of refrigerant leaving the condenser. A more important location is near the inlet of the metering device, since this is where the refrigerant condition is most important.

> **NOTE**
>
> With most refrigerant blends, due to the differing characteristics of blend components, a sight glass may not provide a reliable indication of the refrigerant condition. One or more components may cause bubbles when subcooling is sufficient and present.

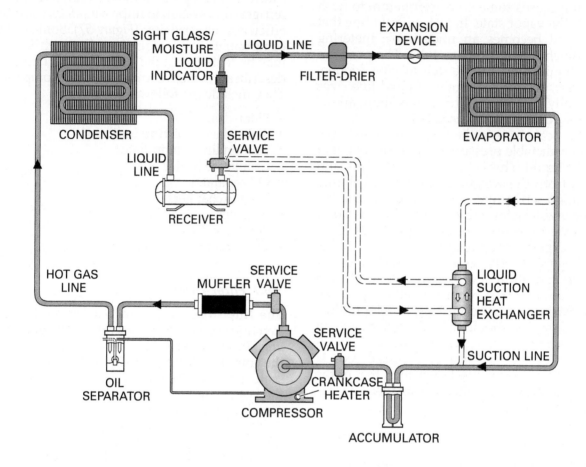

Figure 57 Common accessories used in the refrigeration circuit.

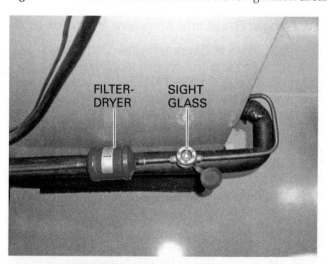

Figure 58 Filter-drier and moisture-indicating sight glass.

A moisture indicator is a small water-sensitive spot installed in the viewing port. It is exposed to the passing refrigerant and changes color depending on the amount of water in the refriger-

Figure 59 A cut-away of a typical filter-drier, showing the filter media and desiccant beads that trap water.

ant. Even a small amount of water in the circuit dilutes the oil, increases the chance of internal corrosion, and can create a restriction by freezing at the outlet of the metering device.

In heat pump systems, refrigerant flows in both directions. Standard filter-driers are designed for flow in one direction only. Bi-directional filter-driers are available for heat pump applications when necessary.

3.6.3 Suction Line Accumulators

An accumulator (*Figure 60*) is basically a trap designed to prevent refrigerant floodback to the compressor. The compressor cannot compress liquid. If a significant amount of liquid refrigerant enters the compressor, compressor damage or failure may result. Traces of liquid that return are not a problem, but a significant volume returning at once can damage a compressor. Accumulators are only found in the suction line, between the evaporator outlet and the compressor inlet.

Large accumulators may have heaters that help to vaporize refrigerant liquid that accumulates. Refrigerant vapor is drawn from the top of the accumulator to be returned to the compressor. A small orifice at the bottom of the internal U-tube allows tiny amounts of both liquid refrigerant and oil to return to the compressor. In larger systems, trapped oil is piped and metered back to the compressor. The small orifice shown ensures that oil can get back to the compressor, and that any liquid refrigerant that collects is controlled.

3.6.4 Crankcase Heaters

Crankcase heaters are installed on compressors to prevent refrigerant from condensing in the crankcase during the off cycle and causing damage. These heaters work by warming the oil to keep the refrigerant above its condensing temperature. They can be fastened to the bottom of the crankcase, inserted directly into the compressor crankcase (immersion type), or inserted into a well in the crankcase (insertion type). Wrap-around, or bellyband, heaters that encircle the outside shell of hermetic compressors are very common (*Figure 61*).

Filter-Driers

Filter-driers are most often installed in the liquid line. They are usually installed or replaced whenever the refrigerant circuit is opened to the air, because air contains moisture and other particulate contaminants that can damage the compressor.

If a severe compressor burnout has occurred, a special liquid line filter-drier designed for one-time use may be installed, then replaced with a clean one after the refrigerant has circulated for a while. Filter-driers used for this purpose are designed to capture acids that result from the compressor motor-winding sealants melting during an internal electrical failure. A suction line filter-drier may also be installed, for added protection and acid absorption. Once they have done their job and the system is deemed clean, suction line filter-driers should be removed from the circuit and not replaced. Note the gauge ports on the ends of the filter-drier. They allow a technician to check the pressure drop through the filter and determine if it is becoming fouled.

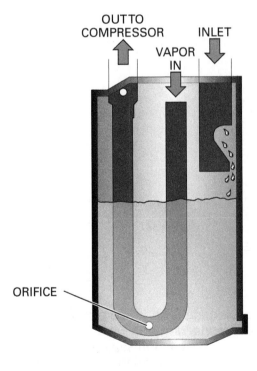

Figure 60 Suction line accumulator.

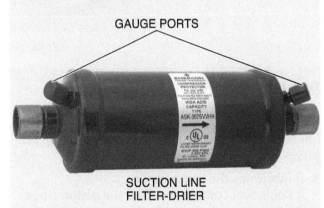

Figure Credit: Courtesy of Emerson Climate Technologies

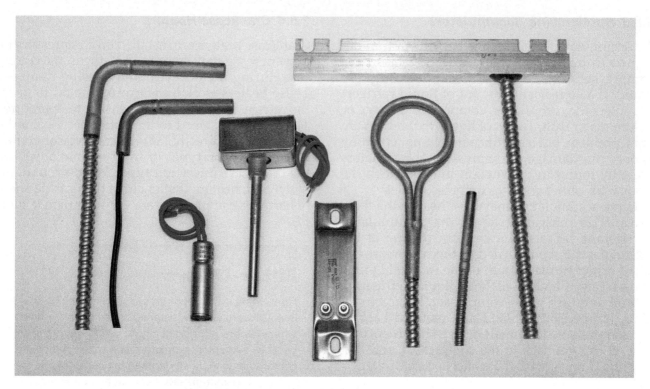

(A) CLAMP-ON, IMMERSION, AND INSERTION HEATERS

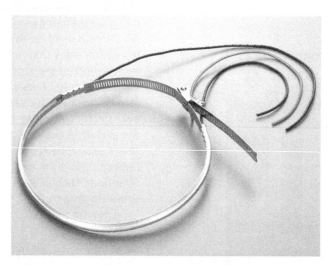

(B) BELLY-BAND HEATER

Figure 61 Crankcase heaters.

3.6.5 Oil Separators

Oil is used in a refrigeration system for four purposes:

- It lubricates the compressor.
- It helps seal components like the piston rings.
- It dampens compressor noise.
- It acts as a coolant for the compressor and motor.

Because there is oil in the compressor, it mixes with the refrigerant and travels with it to other areas of the system. Oil separators minimize the amount of oil that circulates through the system. Oil coats the inside of every component through which it passes. It reduces the heat-transfer ability and efficiency of the evaporator and condenser. Another reason for the oil separator is to slow down the accumulation of oil in places from which oil return is difficult. However, oil separators on residential and commercial comfort

Suction Line Accumulators on Heat Pumps

Suction line accumulators are widely used in residential heat pumps. During heating (or reverse-cycle) operation, the accumulator may become covered with frost. This is normal operation under these circumstances.

Keep in mind that the orifice allowing the liquid to return to the compressor in small amounts can be clogged by contaminants and debris in the piping. Compressor failures and service work can often create such contaminants. Once the orifice is plugged, the accumulator must completely fill before anything escapes. When it does, it will likely do so as a slug.

cooling systems are virtually non-existent and unnecessary. They are used primarily on large refrigeration systems.

Oil separators (*Figure 62*) are typically installed in the hot-gas line as close to the compressor discharge port as practical. Separators have a reservoir, or sump, to collect the trapped oil. A float valve in the sump maintains a seal between the high-pressure and low-pressure sides of the system. This valve automatically returns the oil to the compressor through an orifice when the oil level builds to a sufficient level.

The oil separator is charged with oil before the system is started up, to ensure that an adequate supply is available from the beginning.

Figure 62 Oil separator.

Otherwise, it may require the entire oil charge from the compressor to fill it up, leaving the compressor empty. A small filter-drier and a sight glass may be installed in the oil return line to the compressor. The sight glass simply helps the technician determine that oil is present and moving.

3.6.6 Heat Exchangers

Several types of heat exchangers can be used with refrigeration systems. The liquid-to-suction type and the refrigerant-to-water preheater are two possibilities.

The liquid-to-suction heat exchanger transfers some of the heat from the warm liquid refrigerant leaving the condenser to the cool suction gas leaving the evaporator. This increases efficiency by further subcooling the liquid refrigerant. In addition, it serves to help evaporate any small amount of liquid refrigerant that could return to the compressor from the evaporator. In refrigeration applications, the target superheat is generally set lower than it is for a comfort cooling application. As a result, there is slightly more potential for liquid refrigerant to return to the compressor. A liquid-to-suction heat exchanger provides a measure of protection.

Operation of the liquid-to-suction heat exchanger is quite simple. The liquid refrigerant leaving the condenser and the cool suction gas leaving the evaporator flow in opposite directions through the heat exchanger. The amount of heat that can be exchanged between the gas and liquid is determined by the temperature difference between the two, the amount of surface area, and how much time there is for the heat exchange to occur. It is important to note that these devices can improve system operation somewhat. However, in reality, there is no thermodynamic benefit, since heat is being exchanged within the system. No heat is actually being removed from the system in the process.

The refrigerant water preheater (*Figure 63*) is used to preheat the water supplied to the inlet of a water heater. In this heat exchanger, heat is transferred from the compressor hot-gas line to the water. This reduces energy consumption in both processes. The water de-superheats the refrigerant as it leaves the compressor. This allows the condenser to produce better subcooling. The water is preheated before it enters the water heater, reducing its energy consumption as well. Shell-and-coil and tube-in-tube type heat exchangers are typically used as water pre-heaters. Some units are available as an accessory through HVACR equipment manufacturers, while after-market manufacturers also market them.

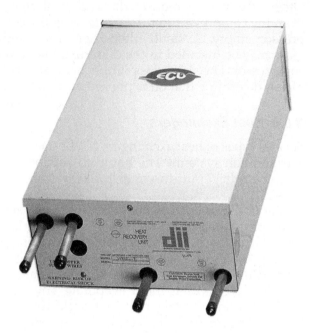

Figure 63 Refrigerant-to-domestic water preheater.

3.6.7 Receiver

The receiver is a tank or container used to store liquid refrigerant in the system. This storage is needed in some systems to accommodate changes during operation, to freely drain the condenser of refrigerant, and to provide a place to store the system charge during system service procedures or prolonged shutdowns. The receiver is installed in the liquid line between the condenser and the metering device. Receivers are used to store the excess refrigerant created by varying cooling loads in many systems that use self-adjusting metering devices. Note that most residential and commercial air conditioning systems store this excess refrigerant in the condenser coil. Receivers are not generally needed on these systems.

3.6.8 Service Valves

Service valves are access ports that installers and service technicians can use to measure system pressures and perform servicing procedures such as charging, evacuation, and dehydration. There are several types of service valves. The most basic type of service valve, known as a piercing valve or line tap, can be installed on tubing to test pressures. The piercing valve is clamped to the tubing and a needle-like point pierces the tubing as the valve is tightened. However, these valves often leak and provide poor service. They can be used to access a system that is otherwise completely sealed, and then later be removed and replaced

with a better device. But they are not reliable enough for permanent service.

Most air conditioning systems have factory-installed service valves like the ones shown in *Figure 64*. Note that there are two service valves, one for the system high side and the other for the low side. In this type of valve, a wrench is used to turn the valve stem and change the position of the valve. The valve has three positions. With the stem turned all the way in, or front-seated, the service port is closed off from the system. This position is used when connecting pressure gauges or other service equipment to the port. With the stem fully backed out, or back-seated, the refrigerant can readily flow to the service port. This position is used when charging, evacuating, or recovering refrigerant. When the stem is slightly open, or cracked, the service port is open to the system. This position is used when making pressure readings. When the valve is back-seated, the service port is inactive.

Note that not all service valves are designed to close off the access port when it is back-seated. Pressure is available at the access port in all valve positions.

Schrader valves are similar to the valves on automobile tires, but they are not identical or interchangeable. The Schrader valve opens when

Figure 64 Service valves.

the stem is depressed. Manifold gauge hoses are equipped with fittings designed to depress the stem as the fitting is tightened. When the Schrader valve is used for charging, evacuation, recovery, etc., the core is usually removed using a special tool designed for this purpose to decrease the pressure drop. The tool shown in *Figure 65* can remove and replace a defective core, or remove and hold the core internally while a system is being evacuated. A built-in ball valve provides another level of control.

Schrader valves are used for quick connection to a system with a gauge manifold. Without this type of valve, a more elaborate service valve is required. Although service valves are more reliable than Schrader valves, service valves increase the cost of a unit and require more time to access. Schrader valves do tend to leak small amounts of refrigerant, so it is very important to always keep them tightly capped when not in use.

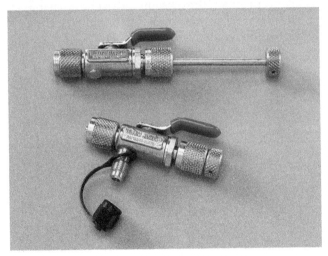

Figure 65 A 4-in-1 valve-core tool.

Schrader Valves

Schrader valves are used in a number of different applications. Almost every automobile tire worldwide is equipped with this style of valve. The valve is named for the founder of the company that first developed it. American August Schrader founded the Schrader Company in 1844 and patented the valve design in 1893.

3.6.9 Compressor Muffler

Mufflers are used in systems with open or semi-hermetic reciprocating compressors. Reciprocating compressors generate sound pulsations that can be transmitted along the piping. A muffler installed in the discharge line, as near the compressor as practical, is used to remove or dampen these pulsations. The muffler lowers the system noise and prevents possible damage from vibration. Smaller mufflers look just like filter-driers from the outside (*Figure 66*).

Figure 66 Compressor muffler.

3.0.0 Section Review

1. Reciprocating compressors can be found in the any of the following configurations *except* _____.

   a. hermetic
   b. semi-hermetic
   c. open-drive
   d. centrifugal

2. Forced-draft cooling towers in which the air-flow is forced up from the bottom as water trickles down are called _____.

   a. counterflow towers
   b. blast towers
   c. downflow towers
   d. natural towers

3. The metering device for a flooded evaporator is designed to _____.

   a. maintain a specific superheat setting
   b. maintain a predetermined liquid level
   c. maintain a specific subcooling value
   d. ensure that there is no vapor in the surge chamber

4. The amount of refrigerant flow through a fixed-orifice metering device is dependent upon the _____.

   a. superheat at the outlet of the coil
   b. amount of subcooling in the liquid line
   c. size of the opening and the pressure applied
   d. size of the opening and the temperature of the liquid

5. Which primary refrigerant line is considered the *least* critical in its layout and installation?

   a. Liquid line
   b. Suction line
   c. Hot-gas line
   d. Condensate line

6. Which refrigerant circuit accessory provides a place to store the system refrigerant charge during service procedures or prolonged shut-downs on larger systems?

   a. Accumulator
   b. Receiver
   c. Muffler
   d. Heat exchanger

SECTION FOUR

4.0.0 COOLING SYSTEM CONTROL

Objective

Identify the common controls used in cooling systems and how they function.
- a. Identify common primary controls.
- b. Identify common secondary controls.

Performance Tasks

5. Identify compressors, condensers, evaporators, metering devices, controls, and accessories.

Trade Terms

Hygroscopic: Describes a material that readily, and sometimes aggressively, absorbs water from the atmosphere or other adjacent material.

Controls are the devices used to start, stop, regulate, and/or protect the components of the mechanical refrigeration system. They can be divided into two groups: primary and secondary. Primary controls start or stop the refrigeration cycle either directly or indirectly by sensing temperature, humidity, or pressure, or by measuring time. Secondary controls regulate and protect the cycle and its components. This section introduces you to the most common control devices. You will study these devices again in greater detail later in your training.

4.1.0 Primary Controls

Primary controls start or stop the refrigeration cycle either directly or indirectly based on temperature, humidity, pressure, or time. Primary controls include, but aren't necessarily limited to, the following:

- Thermostat
- Pressure switches
- Humidistat
- Time clock

4.1.1 Thermostats

All thermostats sense and respond to the temperature in a conditioned space. They switch the system on or off at a preset temperature by opening a set of contacts in the system control circuit. This may be done in several different ways.

If a thermostat makes on a temperature rise and breaks on a temperature drop, it is considered a cooling thermostat. If it makes on a temperature drop and breaks on a temperature rise, it is considered a heating thermostat. Of course, many thermostats provide both functions.

Some are activated by the warping of a bi-metal strip (*Figure 67*). A bimetal device operates on the principle that different metals expand or contract at different rates when heated or cooled. The electrical contact is mounted directly to the bimetal strip. The mating side of the contact is magnetized. As the two become close, the magnetic energy overcomes the spring tension and the contacts snap together. These thermostats are known as snap-action thermostats.

Another type of thermostat is activated by pressure applied from a bellows attached to a chemical-filled sensing bulb with a capillary tube (*Figure 68*). When filled with a refrigerant (both liquid and vapor are present when the bulb is sealed), the bulb pressure increases or decreases as the temperature rises or drops. An increase in bulb pressure causes the bellows to expand, mechanically closing or opening the electrical contacts, depending on whether it is a heating or cooling thermostat. The action is the opposite for a pressure drop. This type of thermostat is called a *remote-bulb thermostat* because the sensing bulb can be located away from the thermostat.

WARNING!

Due to the health hazards associated with mercury, thermostats and other control devices that contain mercury have largely been banned from the marketplace. The regulations regarding their use and disposal requirements vary by state. Check the regulations in your state to determine the proper course of action for disposing of mercury thermostats and controls.

Mercury and all its compounds are toxic. Exposure can cause severe brain and kidney damage. Mercury can either be ingested or absorbed through the skin. Metallic mercury slowly evaporates when exposed to air. As a result, dangerous levels of mercury vapor can even develop in a room from a broken thermometer; mercury vapor inhalation can also lead to serious and chronic health problems.

Handle devices containing mercury with extreme caution; do not handle mercury with bare hands. For clean-up guidelines and related precautions, contact your state agency responsible for hazardous waste. Mercury spill kits are available and highly recommended for HVACR technicians.

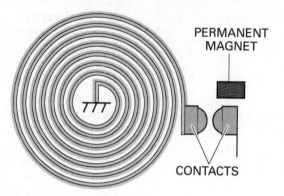

Figure 67 Snap-action thermostat operation.

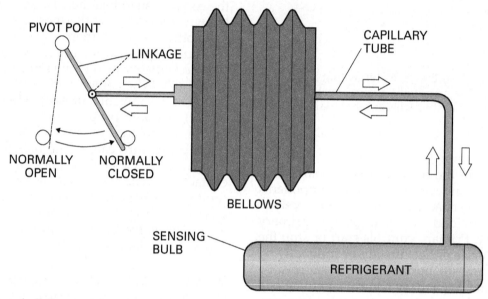

Figure 68 A remote-bulb thermostat.

Electronic, or digital, thermostats (*Figure 69*) use electronic components to sense temperature changes and perform switching functions. These thermostats generally use either a thermocouple or thermistor to sense the temperature. A thermocouple sensor is made of two dissimilar wires welded together at one end called a *junction*. When the junction is heated, it generates a low-level DC voltage that is applied to the switch circuits in the thermostat. A thermistor is a semiconductor device in which the resistance changes with a change in temperature.

Electronic thermostats are generally more accurate and reliable than other types. They contain microprocessor chips that allow them to be programmed for automatic startup, shutdown, and daily setpoint changes. Many of today's thermostats also incorporate humidity control, eliminating the need for a separate humidistat.

Wireless models can be connected to the local network and accessed through various apps (*Figure 70*). These thermostats allow a user to change the setpoint without even being in the building.

4.1.2 Pressure Switches

Pressure switches (*Figure 71*) control based on variations in pressure. They work on both the high-pressure and low-pressure sides of the system. Pressure switches can directly control the operation of a compressor when the low-side pressure has reached a preset value. This simple arrangement has often been used instead of a thermostat to control small refrigeration units.

A low-pressure switch is also used to control the compressor in a system designed for pump-down control. They also act as a safety device to open the operating circuit if a system pressure exceeds a safe level. As a result, pressure

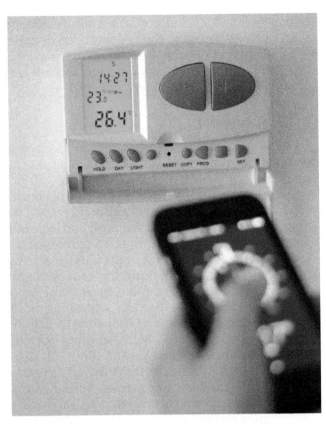

Figure 70 Thermostat control using an app.

switches are commonly used as both primary and secondary controls. They are often applied as primary controls for small refrigeration units, but act as secondary or safety controls in all types of systems.

A bellows pressure switch, shown in *Figure 71*, is directly connected to the refrigeration system through a capillary tube. As the pressure within the system changes, the pressure in the bellows also changes. The bellows expand with a pressure increase and contract with a pressure decrease. The movement in and out of the bellows

Figure 69 A programmable electronic thermostat.

Figure 71 A low-pressure switch (opens on a fall).

makes or breaks an electrical circuit as contacts mechanically linked to the bellows open or close.

Small, non-adjustable (fixed) pressure switches (*Figure 72*) have a snap-action design and are generally used as pressure safety switches.

4.1.3 Humidistat

Humidistats sense moisture in the air. Electromechanical humidistats use materials such as human hair or nylon that expand when they absorb moisture from the air. Such materials are considered hygroscopic. As the humidity increases, the hygroscopic material becomes moist and expands. This movement, applied through a spring-loaded lever, opens or closes the humidistat electrical contacts. A reduction in humidity causes the hygroscopic material to dry and contract.

Humidistats may be used to control humidification equipment that adds moisture to dry air. They are also used to control cooling systems or dehumidification equipment. Both of these systems remove moisture from the air. In some applications, maintaining precise humidity and temperature setpoints is critical. If so, the cooling system may be operated to remove moisture from the air while heat is added to the space to prevent overcooling.

Electronic humidistats (*Figure 73*) use electronic sensing elements with hygroscopic properties. Another type uses carbon particles embedded in a hygroscopic material. These sensing elements work like a thermistor. Changes in humidity affect the resistance of the material and thereby change the current in an electronic circuit. Many of today's sophisticated electronic thermostats also serve as humidistats. This provides comprehensive control.

4.1.4 Time Clock

Time clocks (*Figure 74*) are often used to start and stop a refrigeration system or selected components within a system. When a time clock is used, the system thermostat, pressure switch, or humidistat acts as the primary control during normal operation. However, when a timed event occurs, such as startup or shutdown, the time clock circuit overrides the other controls. Time clocks are also used to activate defrost cycles in heat pumps and medium/low-temperature refrigeration systems.

4.2.0 Secondary Controls

Secondary controls regulate and/or protect the refrigeration system while it is in operation. They are often referred to as operating controls because they keep the cycle adjusted and running properly. Operating controls include the following:

- Relays, contactors, and starters
- Condenser water valves
- Refrigerant solenoid valves
- Evaporator pressure regulators
- Check valves
- Timed devices

Safety controls include the following:

- Electrical overloads
- Current/temperature devices
- High- and low-temperature switches
- High- and low-pressure switches
- Relief valves, fusible plugs, and rupture discs
- Oil-pressure safety switches
- Flow switches

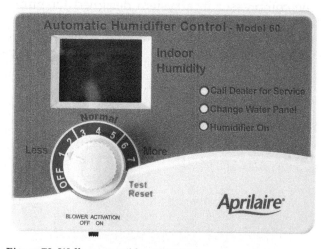

Figure 73 Wall-mounted humidistat.

Figure 72 Non-adjustable pressure switch.

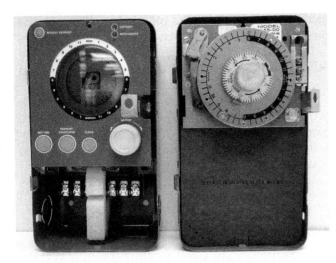

Figure 74 Electronic and mechanical time clocks.

Some of the basic electrical controls listed above are introduced in a separate training module, or have been discussed in this module. The following sections introduce those devices not discussed previously.

4.2.1 Condenser Water Valves

The condenser water valve (*Figure 75*) is used in systems with water-cooled condensers. It regulates the head pressure by controlling the flow of water through a condenser. It is usually a self-contained, pressure-actuated bellows valve. The desired head pressure setpoint is controlled through an adjusting stem at the top of the valve. The stem adjusts the spring pressure on the valve. As pressure increases in the bellows, the spring is pushed up and the valve opens a bit further. A re-

duction in the pressure applied to the bellows allows the spring to push down and close the valve slightly. These valves can provide a very steady head pressure as long as the water supply is relatively stable.

4.2.2 Evaporator-Pressure Regulator

An evaporator-pressure regulator (EPR), as shown in *Figure 76*, maintains a constant pressure and, therefore, a constant saturation temperature in the evaporator. It does this regardless of changes in pressure elsewhere in the system. It is typically a self-contained, pressure-actuated bellows valve. Its operating pressure is adjusted using a spring-tension screw. Evaporator pressure regulators are typically used on large refrigeration systems where a number of evaporators must operate at different temperatures, such as a food-processing facility. The evaporators that must operate at the lowest temperature (and pressure) do not generally need an EPR. However, all evaporators operating at higher temperatures require an EPR to keep them from becoming too cold.

4.2.3 Check Valves

Check valves are used to ensure single-direction flow. They prevent reverse flow. They use either a movable flapper or a ball that moves away from its seat to allow flow in the desired direction and seals on the seat to stop the flow in the wrong direction. Check valves are used in heat pump refrigeration circuits and various water circuits.

Figure 75 Condenser water valve.

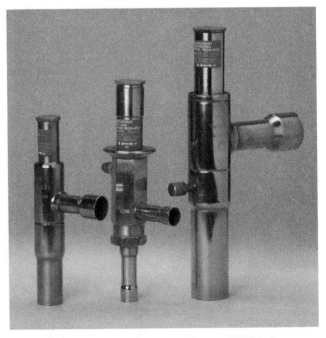

Figure 76 Evaporator-pressure regulating (EPR) valve.

4.2.4 Pressure-Relief Devices

Relief valves, fusible plugs, and rupture discs protect the refrigeration system or specific components from damage caused by an over-pressure condition. A relief valve is spring-loaded, which opens the valve fully when a pressure exceeding the setpoint is encountered.

A fusible plug has a soft metal core with a low melting point. In the case of extreme temperature or fire, the elevated temperature melts the core. This allows the gas to escape before hazardous pressures build up and an explosion occurs.

The rupture disc is a thin graphite disc that ruptures at pressures above the maximum desired pressure. They are generally used for protection at very low pressures, where spring sensitivity is questionable. Low-pressure centrifugal chillers are often fitted with rupture discs.

While pressure-relief valves typically close themselves once the pressure has been relieved, fusible plugs and rupture discs do not. Once they have been melted or ruptured, they must be replaced with a new one.

4.2.5 Oil-Pressure Safety Switches

An oil-pressure safety switch (*Figure 77*) is a pressure-actuated, electrical safety control used to protect the compressor against damage caused by a loss of oil pressure. Most operate on a differential pressure scheme. Since the oil pump is located on the low-pressure side of the compressor, the switch monitors both the suction pressure

in the crankcase and the oil pressure leaving the pump. The difference between the two pressures is the net oil pressure. When the net oil pressure falls below the setpoint, the contacts open, and the compressor shuts down.

To accommodate an oil pressure switch, compressors must have accessible service ports to connect the two required pressure inputs. Only a few very large hermetic compressors may be equipped with these fittings. They are used primarily on semi-hermetic compressors, as well as open-drive models.

4.2.6 Flow Switches

Flow switches are used to shut down a system when either air or water flow to the evaporator or condenser is inadequate. Flow switches also prevent the system from starting with an inadequate amount of flow. They can also be used to control devices such as electronic air cleaners, providing power to them once airflow has been established. The type used to sense air movement is referred to as a sail switch.

A water flow switch is an integral part of ensuring that a chiller has an adequate amount of water flow through the evaporator before startup. The accessory paddle shown on the flow switch in *Figure 78* can be cut to fit the application and the depth of the pipe where it is installed. However, the length of the paddle does affect the leverage of the water.

Figure 77 Oil-pressure safety switch.

Figure 78 A water flow switch, with paddle attached.

1. A substance that attracts and absorbs moisture from the air is referred to as being _____.

   a. hydraulic
   b. hygroscopic
   c. hydrogenated
   d. hyperdry

2. One clear example of a secondary control is a _____.

   a. pressure switch
   b. humidistat
   c. time clock
   d. pressure-relief valve

SUMMARY

Air conditioning is considered a collection of processes that heat, cool, clean, circulate, and control the moisture content of air. However, when the term *air conditioning* is used, it generally refers to a cooling process rather than heating.

The cooling portion of an air conditioning system depends on the mechanical refrigeration cycle. Refrigeration also depends on the same basic principles and components for the preservation of food and to conduct many industrial processes. The refrigeration cycle is based on the following concepts:

- Cold is merely the absence of heat. Heat is always present to some degree.

- Heat always flows from a warmer substance or location to a cooler substance or location.

- Heat must be added to or removed from a substance before a change in state can occur.

- The flow of a gas or liquid is always from an area of higher pressure to a lower pressure area.

- The temperature at which a liquid or gas changes state is dependent on the pressure applied.

The primary components included in a typical mechanical refrigeration cycle are the compressor, condenser, metering device, and evaporator. The compressor provides the pressure differentials, while the condenser and evaporator are the heat-exchange surfaces. The metering device controls the flow of refrigerant into the evaporator coil.

The layout and installation of the refrigerant piping system also plays a key role in the operation of the system. A number of accessories and controls are incorporated to make a complete and reliable operating system.

1. 5 lb of water heated to raise the temperature 2°F requires _____.

   a. 5 Btus
   b. 10 Btus
   c. 25 Btus
   d. 100 Btus

2. Sensible heat is heat that _____.

   a. produces a change in state without a change in temperature
   b. is gained when changing a liquid to a vapor
   c. is gained when changing a vapor to a liquid
   d. can be measured by a thermometer

3. Superheat can be gained _____.

   a. after all of a solid changes to a liquid
   b. in changing vapor to liquid
   c. after all liquid has been changed to vapor
   d. in changing liquid to a vapor

4. When the temperature of a liquid refrigerant is below its condensing temperature, the liquid is _____.

   a. superheated
   b. saturated
   c. subcooled
   d. boiling

5. If a gas is superheated 15°F (8.3°C) to a temperature of 87°F (31°C), we can determine that its _____.

   a. boiling point is 72°F (22°C)
   b. freezing point is 102°F (39°C)
   c. saturation point is 102°F (39°C)
   d. latent heat of fusion is 72°F (22°C)

6. Which type of heating unit or system is based on radiation?

   a. Warm-air furnaces
   b. Steam heating system with radiators
   c. Electric under-floor heating system
   d. Infrared heating units

7. Zero gauge pressure (0 psig) corresponds on the absolute scale to _____.

   a. 10.4 psia
   b. 14.7 psia
   c. 27.4 psia
   d. 44.7 psia

8. A gas or liquid always flows _____.

   a. from a lower pressure to a higher pressure
   b. from a higher temperature to a lower temperature
   c. from a higher pressure to a lower pressure
   d. in a straight line

9. In a refrigeration system, the low-pressure, low-temperature vapor is converted into a high-pressure, high-temperature vapor by the _____.

   a. compressor
   b. evaporator
   c. expansion (metering) device
   d. condenser

10. The low side of the refrigeration system includes the _____.

    a. inlet of the compressor
    b. muffler
    c. receiver
    d. condenser

11. The fluorocarbon refrigerants that are least harmful to the ozone layer are the _____.

    a. CFHs
    b. CFCs
    c. HCFCs
    d. HFCs

12. Which of the following statements is true about ammonia?

   a. Ammonia vapors are heavier than air.
   b. By the time a technician detects the odor of ammonia, the concentration is already high enough to cause injury.
   c. Ammonia is slightly more efficient at transferring heat than the halocarbon refrigerants.
   d. Ammonia is only used in large chilled water systems.

13. What is the required color code for refrigerant recovery cylinders?

   a. Orange with gray shoulders
   b. Green with yellow shoulders
   c. Gray with yellow shoulders
   d. Yellow with gray shoulders

14. A compressor with a piston that travels back and forth in a cylinder is a _____.

   a. reciprocating compressor
   b. rotary compressor
   c. centrifugal compressor
   d. scroll compressor

15. The main purpose of a condenser is to _____.

   a. store liquid refrigerant
   b. remove heat from the refrigerant
   c. remove water from the refrigerant
   d. add heat to the refrigerant

16. The main purpose of an evaporator is to _____.

   a. store liquid refrigerant
   b. remove heat from the refrigerant
   c. remove water from the refrigerant
   d. transfer heat into the refrigerant

17. A TXV regulates the flow of refrigerant to the evaporator to maintain a consistent _____.

   a. subcooling
   b. superheat
   c. discharge pressure
   d. airflow

18. The operation of a double-suction riser depends on _____.

   a. a sufficient amount of oil to close off the trap at the base
   b. a low velocity to get the oil to flow
   c. the compressor always operating at or near 100 percent of its capacity
   d. liquid refrigerant mixing with the oil

19. The accessory found only in the refrigeration circuit suction line is the _____.

   a. receiver
   b. accumulator
   c. muffler
   d. oil separator

20. What is the difference between a fusible plug and rupture disk?

   a. A fusible plug responds to high pressures; a rupture disk responds to high temperatures.
   b. A rupture disk responds to high pressures; a fusible plug responds to high temperatures.
   c. A rupture disk responds to high pressures; a fusible plug responds to low pressures.
   d. A fusible plug responds to high temperatures; a rupture disk responds to low temperatures.

Trade Terms Quiz

Fill in the blank with the correct trade term that you learned from your study of this module.

1. A pressure measurement that considers the zero-pressure point to be standard atmospheric pressure is known as _____.

2. Compounds containing only carbon and hydrogen atoms are known as _____.

3. The movement of heat in the form of invisible rays is _____.

4. The total heat content of a refrigerant is referred to as its _____.

5. The physical process in which heat is moved from one material to another by direct contact is _____.

6. If heat moves freely in and through materials, the materials are known as _____.

7. The heat absorbed by a substance while changing from a liquid state to a gaseous state is called the _____.

8. The term used to define the atmospheric pressure at sea level is _____.

9. Heat energy that increases the measurable temperature of a substance is called _____.

10. Chlorine, fluorine, and bromine are chemically related elements called _____.

11. The amount of heat needed to raise the temperature of one pound of water 1°F is called a(n) _____.

12. The transfer of heat through the flow of a liquid or gas, caused by a temperature difference between the two substances, is called _____.

13. The pressure that is exerted on all things on Earth's surface as the result of the weight of the air above is called _____.

14. A substance that is water-based but contains salt or other substances that reduce its freezing point is called _____.

15. A halocarbon in which at least one of the hydrogen atoms has been replaced with fluorine atoms is referred to as a(n) _____.

16. A pressure-relief device that fractures or breaks open when the pressure limit is reached to protect a container or vessel is called a(n) _____.

17. A hydrocarbon in which most or all of the hydrogen atoms have been replaced with fluorine, chlorine, bromine, astatine, or iodine is called a(n) _____.

18. The application of the refrigeration cycle for the purpose of maintaining comfortable conditions for humans is an example of _____.

19. Heat absorbed by a substance that cannot be measured with a thermometer is called _____.

20. An electrical device made of two different metals that generates a predictable electrical potential in response to temperature differences is called a(n) _____.

21. A device that changes its resistance to electrical flow in response to a temperature change is called a(n) _____.

22. A vessel designed to hold both refrigerant liquid and vapor, where the liquid is fed to a process and the vapor is drawn back to the compressor, is called a(n) _____.

23. The amount of heat, expressed in Btu/lb/°F, needed to raise the temperature of one pound of a substance 1°F is called its _____.

24. The heat given up or removed from a gas in the process of changing to a liquid is called the _____.

25. The process that identifies the change of state of a substance, such as water, from a gaseous state to a liquid state is called _____.

26. Substances that are not good conductors of heat are referred to as _____.

27. 12,000 Btuh is the same amount of heat as one _____.

28. When the sensible heat of a substance is added to its heat, the sum would be its _____.

29. An instrument used to measure atmospheric pressure is called a(n) _____.

30. The term that represents force per unit of area is _____.

31. The continued cooling of a liquid below its condensing temperature is called _____.

32. When two or more substances are combined, but more than one type of molecule remains after they are brought together, the combination is called a(n) _____.

33. The heat gained or lost in changing to or from a solid is called the _____.

34. When a refrigerant reaches its boiling point for a given applied pressure, it has reached its _____.

35. The additional, measurable heat that is added to a substance after it has completed a change of state into a vapor is known as _____.

36. A substance that is formed from the union of two or more elements, where only one new molecule results, is called a(n) _____.

37. A material that readily absorbs water from the atmosphere or other adjacent source is considered to be _____.

38. A significant amount of oil that returns to the compressor all at once is called a(n) _____.

39. The material inside a filter-drier that absorbs any water from the refrigerant stream is a type of _____.

40. When the compressor continues running at the end of an operating cycle long enough to remove any remaining refrigerant from the evaporator coil, the operating scheme is referred to as _____.

41. When the liquid refrigerant in an evaporator coil is not fully vaporized during operation, and the liquid returns to the compressor, it is called _____.

Trade Terms

Absolute pressure
Atmospheric pressure
Barometer
Brine
British thermal unit (Btu)
Comfort cooling
Compound
Condensation
Conduction
Conductors
Convection
Desiccant
Enthalpy
Fluorocarbon

Gauge pressure
Halocarbon
Halogens
Hydrocarbons
Hygroscopic
Insulators
Latent heat
Latent heat of condensation
Latent heat of fusion
Latent heat of vaporization
Mixture
Pressure
Pump-down control
Radiation

Refrigerant floodback
Rupture disk
Saturation temperature
Sensible heat
Slug
Specific heat
Subcooling
Superheat
Surge chamber
Thermistor
Thermocouple
Ton of refrigeration
Total heat

Appendix A

Fahrenheit to Celsius Conversion Table

$$°C = 5/9 \, (°F - 32°)$$

$$°F = (9/5 \times °C) + 32°$$

°C	°F	°C	°F	°C	°F	°C	°F	°C	°F
−99	−146.2	−60	−76.0	−21	−5.8	18	64.4	57	134.6
−98	−144.4	−59	−74.2	−20	−4.0	19	66.2	58	136.4
−97	−142.6	−58	−72.4	−19	−2.2	20	68.0	59	138.2
−96	−140.8	−57	−70.6	−18	−0.4	21	69.8	60	140.0
−95	−139.0	−56	−68.8	−17	1.4	22	71.6	61	141.8
−94	−137.2	−55	−67.0	−16	3.2	23	73.4	62	143.6
−93	−135.4	−54	−65.2	−15	5.0	24	75.2	63	145.4
−92	−133.6	−53	−63.4	−14	6.8	25	77.0	64	147.2
−91	−131.8	−52	−61.6	−13	8.6	26	78.8	65	149.0
−90	−130.0	−51	−59.8	−12	10.4	27	80.6	66	150.8
−89	−128.2	−50	−58.0	−11	12.2	28	82.4	67	152.6
−88	−126.4	−49	−56.2	−10	14.0	29	84.2	68	154.4
−87	−124.6	−48	−54.4	−9	15.8	30	86.0	69	156.2
−86	−122.8	−47	−52.6	−8	17.6	31	87.8	70	158.0
−85	−121.0	−46	−50.8	−7	19.4	32	89.6	71	159.8
−84	−119.2	−45	−49.0	−6	21.2	33	91.4	72	161.6
−83	−117.4	−44	−47.2	−5	23.0	34	93.2	73	163.4
−82	−115.6	−43	−45.4	−4	24.8	35	95.0	74	165.2
−81	−113.8	−42	−43.6	−3	26.6	36	96.8	75	167
−80	−112.0	−41	−41.8	−2	28.4	37	98.6	76	168.8
−79	−110.2	−40	−40.0	−1	30.2	38	100.4	77	170.6
−78	−108.4	−39	−38.2	0	32.0	39	102.2	78	172.4
−77	−106.6	−38	−36.4	1	33.8	40	104.0	79	174.2
−76	−104.8	−37	−34.6	2	35.6	41	105.8	80	176
−75	−103.0	−36	−32.8	3	37.4	42	107.6	81	177.8
−74	−101.2	−35	−31.0	4	39.2	43	109.4	82	179.6
−73	−99.4	−34	−29.2	5	41.0	44	111.2	83	181.4
−72	−97.6	−33	−27.4	6	42.8	45	113.0	84	183.2
−71	−95.8	−32	−25.6	7	44.6	46	114.8	85	185.0
−70	−94.0	−31	−23.8	8	46.4	47	116.6	86	186.8
−69	−92.2	−30	−22.0	9	48.2	48	118.4	87	188.6
−68	−90.4	−29	−20.2	10	50.0	49	120.2	88	190.4
−67	−88.6	−28	−18.4	11	51.8	50	122.0	89	192.2
−66	−86.8	−27	−16.6	12	53.6	51	123.8	90	194.0
−65	−85.0	−26	−14.8	13	55.4	52	125.6	91	195.8
−64	−83.2	−25	−13.0	14	57.2	53	127.4	92	197.6
−63	−81.4	−24	−11.2	15	59.0	54	129.2	93	199.4
−62	−79.6	−23	−9.4	16	60.8	55	131.0	94	201.2
−61	−77.8	−22	−7.6	17	62.6	56	132.8	95	203.0

°C	°F	°C	°F	°C	°F	°C	°F	°C	°F
96	204.8	135	275.0	174	345.2	213	415.4	252	485.6
97	206.6	136	276.8	175	347	214	417.2	253	487.4
98	208.4	137	278.6	176	348.8	215	419.0	254	489.2
99	210.2	138	280.4	177	350.6	216	420.8	255	491.0
100	212.0	139	282.2	178	352.4	217	422.6	256	492.8
101	213.8	140	284.0	179	354.2	218	424.4	257	494.6
102	215.6	141	285.8	180	356	219	426.2	258	496.4
103	217.4	142	287.6	181	357.8	220	428.0	259	498.2
104	219.2	143	289.4	182	359.6	221	429.8	260	500.0
105	221.0	144	291.2	183	361.4	222	431.6	261	501.8
106	222.8	145	293.0	184	363.2	223	433.4	262	503.6
107	224.6	146	294.8	185	365.0	224	435.2	263	505.4
108	226.4	147	296.6	186	366.8	225	437.0	264	507.2
109	228.2	148	298.4	187	368.6	226	438.8	265	509.0
110	230.0	149	300.2	188	370.4	227	440.6	266	510.8
111	231.8	150	302.0	189	372.2	228	442.4	267	512.6
112	233.6	151	303.8	190	374.0	229	444.2	268	514.4
113	235.4	152	305.6	191	375.8	230	446.0	269	516.2
114	237.2	153	307.4	192	377.6	231	447.8	270	518.0
115	239.0	154	309.2	193	379.4	232	449.6	271	519.8
116	240.8	155	311.0	194	381.2	233	451.4	272	521.6
117	242.6	156	312.8	195	383.0	234	453.2	273	523.4
118	244.4	157	314.6	196	384.8	235	455.0	274	525.2
119	246.2	158	316.4	197	386.6	236	456.8	275	527
120	248.0	159	318.2	198	388.4	237	458.6	276	528.8
121	249.8	160	320.0	199	390.2	238	460.4	277	530.6
122	251.6	161	321.8	200	392.0	239	462.2	278	532.4
123	253.4	162	323.6	201	393.8	240	464.0	279	534.2
124	255.2	163	325.4	202	395.6	241	465.8	280	536
125	257.0	164	327.2	203	397.4	242	467.6	281	537.8
126	258.8	165	329.0	204	399.2	243	469.4	282	539.6
127	260.6	166	330.8	205	401.0	244	471.2	283	541.4
128	262.4	167	332.6	206	402.8	245	473.0	284	543.2
129	264.2	168	334.4	207	404.6	246	474.8	285	545.0
130	266.0	169	336.2	208	406.4	247	476.6	286	546.8
131	267.8	170	338.0	209	408.2	248	478.4	287	548.6
132	269.6	171	339.8	210	410.0	249	480.2	288	550.4
133	271.4	172	341.6	211	411.8	250	482.0	289	552.2
134	273.2	173	343.4	212	413.6	251	483.8	290	554.0

°C	°F	°C	°F	°C	°F	°C	°F	°C	°F
291	555.8	313	595.4	335	635.0	357	674.6	379	714.2
292	557.6	314	597.2	336	636.8	358	676.4	380	716
293	559.4	315	599.0	337	638.6	359	678.2	381	717.8
294	561.2	316	600.8	338	640.4	360	680.0	382	719.6
295	563.0	317	602.6	339	642.2	361	681.8	383	721.4
296	564.8	318	604.4	340	644.0	362	683.6	384	723.2
297	566.6	319	606.2	341	645.8	363	685.4	385	725.0
298	568.4	320	608.0	342	647.6	364	687.2	386	726.8
299	570.2	321	609.8	343	649.4	365	689.0	387	728.6
300	572.0	322	611.6	344	651.2	366	690.8	388	730.4
301	573.8	323	613.4	345	653.0	367	692.6	389	732.2
302	575.6	324	615.2	346	654.8	368	694.4	390	734.0
303	577.4	325	617.0	347	656.6	369	696.2	391	735.8
304	579.2	326	618.8	348	658.4	370	698.0	392	737.6
305	581.0	327	620.6	349	660.2	371	699.8	393	739.4
306	582.8	328	622.4	350	662.0	372	701.6	394	741.2
307	584.6	329	624.2	351	663.8	373	703.4	395	743.0
308	586.4	330	626.0	352	665.6	374	705.2	396	744.8
309	588.2	331	627.8	353	667.4	375	707	397	746.6
310	590.0	332	629.6	354	669.2	376	708.8	398	748.4
311	591.8	333	631.4	355	671.0	377	710.6	399	750.2
312	593.6	334	633.2	356	672.8	378	712.4	400	752.0

PRESSURE CONVERSION CHART

To Convert To	Atmospheres	lbs per sq in	kg per sq cm (metric atmosphere)	Inches of Mercury at +32°F (+0°C)	Meters of Mercury at +32°F (+0°C)	Meters of Water at +60°F (+16°C)	Feet of Water at +60°F (+16°C)	Inches of Water at +60°F (+16°C)	Bar
From						Multiply By			
Atmospheres	1.000	14.700	1.033	29.921	0.760	10.340	33.910	406.900	1.013
Pounds per sq in	0.068	1.000	0.070	2.036	0.052	0.704	2.307	27.680	0.069
Kgs per sq cm	0.968	14.220	1.000	28.960	0.736	10.010	32.840	394.100	0.981
Inches of Mercury	0.033	0.491	0.035	1.000	0.025	0.346	1.132	13.590	0.034
Meters of Mercury	1.316	19.340	1.360	39.370	1.000	13.610	44.640	535.700	1.333
Meters of Water	0.097	1.421	0.010	2.893	0.073	1.000	3.281	39.370	0.098
Feet of Water	0.029	0.433	0.030	0.882	0.022	0.305	1.000	12.000	0.030
Inches of Water	0.002	0.036	0.003	0.074	0.002	0.025	0.083	1.000	0.002
Bar	0.987	14.500	1.020	29.530	0.750	10.210	33.490	401.800	1.000

Trade Terms Introduced In This Module

Absolute pressure: Positive pressure measurements that start at zero (no atmospheric pressure). Gauge pressure plus the pressure of the atmosphere (14.7 psi at sea level at 70°F) equals absolute pressure. Absolute pressure is expressed in pounds per square inch absolute (psia). Absolute pressure = gauge pressure + atmospheric pressure.

Atmospheric pressure: The pressure exerted on all things on Earth's surface due to the weight of the atmosphere. It is roughly 14.7 psi at sea level at 70°F; the standard pressure exerted on Earth's surface. Atmospheric pressure may also be expressed as 29.92 inches of mercury.

Barometer: An instrument used to measure atmospheric pressure, typically in units of inches of mercury (in. Hg).

Brine: Water that is saturated with (contains) large amounts of salt. In the HVACR industry, the term typically describes any water-based mixture that contains substances such as salt or glycols that lower its freezing point.

British thermal unit (Btu): The amount of heat needed to raise the temperature of one pound of water one degree Fahrenheit.

Comfort cooling: Cooling related to the comfort of humans in buildings and residences. The temperature range for comfort cooling is typically considered to be 60°F (15.5°C) to 80°F (26.6°C).

Compound: As related to refrigerants, a substance formed by a union of two or more elements in definite proportions by weight; only one molecule is present.

Conduction: In the context of heat transfer, the process through which heat is directly transmitted from one substance to another when there is a difference of temperature between regions in contact.

Conductors: Relevant to heat transfer, materials that readily transfer heat by conduction.

Convection: The movement caused within a fluid (air or water, for example) by the tendency of hotter fluid to rise and colder, denser material to fall, resulting in heat being transferred.

Desiccant: A material or substance, such as silica gel or calcium chloride, that seeks to absorb and hold water from any adjacent source, including the atmosphere.

Enthalpy: The total heat content (sensible and latent) of a refrigerant or other substance.

Fluorocarbon: A compound formed by replacing one or more hydrogen atoms in a hydrocarbon with fluorine atoms.

Gauge pressure: The pressure measured on a gauge, expressed as pounds per square inch gauge (psig) or inches of mercury vacuum (in. Hg vac.). Pressure measurements that are made without including atmospheric pressure.

Halocarbon: Hydrocarbons, like methane and ethane, that have most or all their hydrogen atoms replaced with the elements fluorine, chlorine, bromine, astatine, or iodine.

Halogens: Substances containing chlorine, fluorine, bromine, astatine, or iodine.

Hydrocarbons: Compounds containing only hydrogen and carbon atoms in various combinations.

Hygroscopic: Describes a material that readily, and sometimes aggressively, absorbs water from the atmosphere or other adjacent material.

Insulators: Materials that resist heat transfer by conduction.

Latent heat of condensation: The heat given up or removed from a gas in changing back to a liquid state (steam to water).

Latent heat of fusion: The heat gained or lost in changing to or from a solid (ice to water or water to ice).

Latent heat of vaporization: The heat gained in changing from a liquid to a gas (water to steam).

Latent heat: The heat energy absorbed or rejected when a substance is in the process of changing state (solid to liquid, liquid to gas, or the reverse of either change).

Mixture: As related to refrigerants, a blend of two or more components that do not have a fixed proportion to one another and that, however thoroughly blended, are conceived of as retaining a separate existence; more than one type of molecule remains present.

Pressure: The force exerted by a substance against its area of containment; mathematically defined as force per unit of area.

Pump-down control: A control scheme that includes a liquid refrigerant solenoid valve that closes to initiate the system off cycle. Once the valve closes, the compressor continues to operate, pumping most or all of the remaining refrigerant out of the evaporator coil. Pump-down control eliminates the possibility of excessive liquid refrigerant forming in the evaporator coil during the off cycle, and is primarily used in refrigeration applications. A thermostat typically controls the liquid solenoid valve, while a low-side pressure switch and/or a timer controls the compressor.

Radiation: The movement of heat in the form of invisible rays or waves, similar to light.

Refrigerant floodback: A significant amount of liquid refrigerant that returns to the compressor through the suction line during operation. Refrigerant floodback can have several causes, such as a metering device that overfeeds refrigerant or a failed evaporator fan motor.

Rupture disk: A pressure-relief device that protects a vessel or other container from damage if pressures exceed a safe level. A rupture disk typically consists of a specific material at a precise thickness that will break or fracture when the pressure limit is reached, creating a controlled weakness, thereby protecting the rest of the container from damage.

Saturation temperature: The boiling temperature of a substance at a given pressure. In the saturated condition, both liquid and vapor is likely present in the same space.

Sensible heat: Heat energy that can be measured by a thermometer or sensed by touch.

Slug: Traditionally refers to a significant volume of oil returning to the compressor at once, primarily at startup. For example, a trap in a suction line may fill with oil during the off cycle, and then leave the trap as a slug at start-up. However, *slugging* is also a term incorrectly used by some to describe refrigerant floodback.

Specific heat: The amount of heat required to raise the temperature of one pound of a substance one degree Fahrenheit. Expressed as Btu/lb/°F. The specific heat of water (H_2O) is 1.0, thus providing the basis for defining one Btu.

Subcooling: The reduction in temperature of a refrigerant liquid after it has completed the change in state from a vapor. Only after the phase change is complete can the liquid begin to decrease in temperature as heat is removed.

Superheat: The additional, increase in temperature of a refrigerant vapor after it has completed the change of state from a liquid to a vapor. Only after the phase change is complete can the vapor begin to increase in temperature as additional heat is applied.

Surge chamber: A vessel or container designed to hold both liquid and vapor refrigerant. The liquid is generally fed out of the bottom into an evaporator, while vapor is drawn from the top of the container by the compressor to maintain refrigerant temperature through pressure.

Thermistor: A semiconductor device that changes resistance with a change in temperature.

Thermocouple: A device made up of two different metals that generates electricity when there is a difference in temperature from one end to the other.

Ton of refrigeration: Large unit for measuring the rate of heat transfer. One ton is defined as 12,000 Btus per hour, or 12,000 Btuh.

Total heat: Sensible heat plus latent heat.

Additional Resources

This module presents thorough resources for task training. The following reference material is recommended for further study.

Refrigeration and Air Conditioning: An Introduction to HVACR. Latest Edition. New York, NY: Pearson.
The US Clean Air Act. **http://www.epa.gov/air/caa/**
International Institute of Ammonia Refrigeration (IIAR). **www.iiar.org**.

Figure Credits

© Sean Pavone/Shutterstock.com, Module opener
Courtesy of Chelsea Clock Company, SA01
Courtesy of National Refrigerants, Inc., Table 2
Fluke Corporation, reproduced with permission, Figure 13
Courtesy of SPX Service Solutions, Figures 14, 18
Courtesy of Dwyer Instruments, Inc., Figure 21
Courtesy of Ritchie Engineering Company, Inc., YELLOW JACKET Products Division, SA03, Figure 65
Courtesy of Hill Phoenix, Figure 24
Courtesy of Evapco, Figures 40, 41
© iStock.com/ttsz, SA06
Courtesy of Baltimore Aircoil Company, SA07
Courtesy of RAE Corporation, Century Division, Figure 43
Advanced Distributor Products, Figure 45
Courtesy of Tucal S. L., Figure 47
Courtesy of Minnesota State University-Mankato, Figure 48
Courtesy Of Parker Hannifin, Sporlan Division, Figure 50
Courtesy of Emerson Climate Technologies, Figures 59, 66, SA09
Courtesy of HotWatt, Figure 61A
Backer Springfield, Figure 61B
Courtesy of Temprite USA, Figure 62
Courtesy of ECU, Figure 63
© iStock.com/lucadp, Figure 69
© iStock.com/adrian825, Figure 70
Courtesy of Sealed Unit Parts Co., Inc., Figure 72
Aprilaire, Figure 73
Photo courtesy of Johnson Controls, Inc., Figures 75, 78
Courtesy of Danfoss, Figures 76, 77
Courtesy of The Garlock Family of Companies, Appendix B

Answer	Section Reference	Objective
Section One		
1. c	1.1.1	1a
2. d	1.2.1	1b
3. b	1.3.3	1c
4. b	1.4.1	1d
5. a	1.5.1	1e
Section Two		
1. a	2.1.0	2a
2. b	2.2.0	2b
3. d	2.3.4	2c
Section Three		
1. d	3.1.1	3a
2. a	3.2.3	3b
3. b	3.3.2	3c
4. c	3.4.1	3d
5. a	3.5.4	3e
6. b	3.6.7	3f
Section Four		
1. b	4.1.3	4a
2. d	4.2.0	4b

NCCER CURRICULA — USER UPDATE

NCCER makes every effort to keep its textbooks up-to-date and free of technical errors. We appreciate your help in this process. If you find an error, a typographical mistake, or an inaccuracy in NCCER's curricula, please fill out this form (or a photocopy), or complete the online form at **www.nccer.org/olf**. Be sure to include the exact module ID number, page number, a detailed description, and your recommended correction. Your input will be brought to the attention of the Authoring Team. Thank you for your assistance.

Instructors – If you have an idea for improving this textbook, or have found that additional materials were necessary to teach this module effectively, please let us know so that we may present your suggestions to the Authoring Team.

NCCER Product Development and Revision
13614 Progress Blvd., Alachua, FL 32615

Email: curriculum@nccer.org
Online: www.nccer.org/olf

❏ Trainee Guide   ❏ Lesson Plans   ❏ Exam   ❏ PowerPoints   Other _____

Craft / Level: _____ Copyright Date: _____

Module ID Number / Title: _____

Section Number(s): _____

Description: _____

_____

_____

_____

Recommended Correction: _____

_____

_____

_____

_____

Your Name: _____

Address: _____

Email: _____ Phone: _____

Air Distribution Systems

OVERVIEW

Most heating and cooling systems use ductwork to deliver conditioned air to the spaces being cooled or heated. The ductwork may be made of sheet metal, fiberglass ductboard, fabric, or flexible duct. The performance of an HVAC system is closely linked to the quality of the air distribution system. The ductwork must be of the proper size and type, and must be correctly installed and sealed.

Module 03109

Trainees with successful module completions may be eligible for credentialing through the NCCER Registry. To learn more, go to **www.nccer.org** or contact us at 1.888.622.3720. Our website has information on the latest product releases and training, as well as online versions of our *Cornerstone* magazine and Pearson's product catalog.

Your feedback is welcome. You may email your comments to **curriculum@nccer.org**, send general comments and inquiries to **info@nccer.org**, or fill in the User Update form at the back of this module.

This information is general in nature and intended for training purposes only. Actual performance of activities described in this manual requires compliance with all applicable operating, service, maintenance, and safety procedures under the direction of qualified personnel. References in this manual to patented or proprietary devices do not constitute a recommendation of their use.

03109 V5

AIR DISTRIBUTION SYSTEMS

Objectives

When you have completed this module, you will be able to do the following:

1. Describe the factors related to air movement and its measurement in air distribution systems.
   a. Describe how pressure, velocity, and volume are interrelated in airflow.
   b. Describe air distribution in a typical residential system.
   c. Identify common air measurement instruments.
2. Describe the mechanical equipment and materials used to create air distribution systems.
   a. Describe various blower types and applications.
   b. Describe various fan designs and applications.
   c. Demonstrate an understanding of the fan laws.
   d. Describe common duct materials and fittings.
   e. Identify the characteristics of common grilles, registers, and dampers.
3. Identify the different approaches to air distribution system design and energy conservation.
   a. Identify various air distribution system layouts.
   b. Describe heating and cooling air movement resulting from various air distribution system designs.
   c. Explain how to maximize energy efficiency through the proper sealing and testing of air distribution systems.

Performance Tasks

Under the supervision of the instructor, you should be able to do the following:

1. Use a manometer to measure static pressure in a duct system.
2. Use a velometer to measure the velocity of airflow at the output of air system supply diffusers and registers.
3. Use a velometer to calculate system cfm.
4. Read and interpret equivalent length charts and required air volume/duct size charts.

Trade Terms

A_k factor
Blower door
Boots
Cubic feet per minute (cfm)
Dew point
Dry-bulb temperature
External static pressure (ESP)
Free air delivery

Pitot tube
Plenum
Psychrometric chart
R-value
Relative humidity (RH)
Revolutions per minute (rpm)
Static pressure (s.p.)
Stratify

Takeoffs
Total pressure
Transfer grille
Vapor barrier
Velocity
Velocity pressure
Venturi
Wet-bulb temperature

Industry-Recognized Credentials

If you are training through an NCCER-accredited sponsor, you may be eligible for credentials from NCCER's Registry. The ID number for this module is 03109. Note that this module may have been used in other NCCER curricula and may apply to other level completions. Contact NCCER's Registry at 888.622.3720 or go to **www.nccer.org** for more information.

Contents

1.0.0 Air Movement and Air Measurement 1
  1.1.0 Airflow and Pressure in Distribution System Ductwork 2
    1.1.1 Absolute Pressure (PSIA) vs. Gauge Pressure (PSIG) 6
  1.2.0 Residential Air Distribution Systems 7
  1.3.0 Air Measurement Instruments 9
    1.3.1 Temperature and Humidity Meters 9
    1.3.2 Pressure Measurement Instruments 10
    1.3.3 Air Velocity Measurement Instruments 13
    1.3.4 Measuring Rotational Speed 14
2.0.0 Air Distribution Equipment and Materials 16
  2.1.0 Blowers 16
    2.1.1 Belt- and Direct-Driven Blowers and Fans 16
    2.1.2 Centrifugal Blowers 16
  2.2.0 Fans 18
    2.2.1 Propeller Fans 18
    2.2.2 Duct Fans 19
  2.3.0 The Fan Laws 19
    2.3.1 Fan Curves 21
  2.4.0 Duct Materials and Fittings 21
    2.4.1 Galvanized Steel Ducts.. 22
    2.4.2 Fiberglass Ductboard 23
    2.4.3 Flexible Duct 25
    2.4.4 Fittings and Transitions.. 27
  2.5.0 Diffusers, Registers, Grilles, and Dampers 30
    2.5.1 Air Dampers 32
    2.5.2 Fire and Smoke Dampers 32
3.0.0 Air Distribution System Design and Energy Conservation 35
  3.1.0 Air Distribution System Layouts 35
    3.1.1 Perimeter-Loop Duct System 36
    3.1.2 Radial Duct System... 36
    3.1.3 Extended-Plenum Duct System 36
    3.1.4 Reducing-Trunk Duct System... 36
    3.1.5 Furred-In Duct Systems 38
  3.2.0 Heating and Cooling Room Airflow....................................... 39
  3.3.0 Energy Efficiency in Air Distribution Systems 41
    3.3.1 Insulation and Vapor Barriers 41
    3.3.2 Duct Sealing... 43

Figures and Tables

Figure 1   Basic forced-air distribution system 2
Figure 2   Pressures in an air distribution system 3
Figure 3   Some causes of friction loss in an air distribution system 4
Figure 4   Measuring duct pressure 5
Figure 5   Comparison of atmospheric pressure to water-column values 5
Figure 6   Absolute, gauge, and barometric pressure scales compared 6
Figure 7   Typical residential air distribution system 8
Figure 8   Air damper 9
Figure 9   Electronic thermometer 9
Figure 10 Sling psychrometer 10
Figure 11 Electronic psychrometer 10
Figure 12 Manometers 12
Figure 13 Portable differential-pressure gauge 12
Figure 14 Pitot tube and static pressure tips 12
Figure 15 A velometer and a rotating-vane anemometer 13
Figure 16 Hot-wire anemometer 14
Figure 17 Combination contact/noncontact tachometer 15
Figure 18 Belt-drive and direct-drive blowers 17
Figure 19 Industrial centrifugal blower 17
Figure 20 Forward-curved centrifugal blower wheel 17
Figure 21 Backward-inclined centrifugal blower wheels 18
Figure 22 Radial centrifugal blower wheels 18
Figure 23 Propeller fan 19
Figure 24 Tube-axial and vane-axial duct fans 20
Figure 25 Typical fan curve chart 21
Figure 26 Metal ducts 23
Figure 27 Common rectangular metal duct connection methods 26
Figure 28 Snap-lock duct connection 26
Figure 29 Duct flex connectors 27
Figure 30 Fiberglass ductboard joint 27
Figure 31 Flexible duct 28
Figure 32 Duct fittings 29
Figure 33 Example of equivalent length 29
Figure 34 Registers and diffusers 31
Figure 35 Typical air dampers 33
Figure 36 Fire damper fusible link 33
Figure 37 Combination fire and smoke damper 33
Figure 38 Upflow furnace in an attic 35
Figure 39 Two types of perimeter duct systems 37
Figure 40 Extended-plenum system 37
Figure 41 Reducing trunk system 37
Figure 42 Furred-in duct system 38
Figure 43 Fan coil unit designed for furred-in installation 38
Figure 44 Room air distribution patterns for a perimeter duct system 39
Figure 45 Room air distribution patterns for high sidewall outlets 40

Figures and Tables (continued)

Figure 46   Room air distribution patterns for ceiling diffusers 40
Figure 47   Condensation on a supply air duct 41
Figure 48   Duct liner 42
Figure 49   Insulated main duct and branches 42
Figure 50   Example of R-value calculation using
            ASHRAE Standard 90-80 42
Figure 51   A pressure pan in use 43
Figure 52   A blower door depressurizing a home 44
Figure 53   Duct sealing mastic 44
Figure 54   Duct sealing locations 45
Figure 55   Typical register boot 46
Figure 56   Fiberglass mesh tape 46

Table 1   Example Airflow Performance Chart 8
Table 2   Duct Materials and Their Applications 23
Table 3   Standard Sheet Metal Gauges (Inches) 24
Table 4   Typical Metal Duct Gauge Thickness and Aspect Ratios 25
Table 5   Performance Data for a Square Diffuser 31

1.0.0 AIR MOVEMENT AND AIR MEASUREMENT

Objective

Describe the factors related to air movement and its measurement in air distribution systems.

a. Describe how pressure, velocity, and volume are interrelated in airflow.
b. Describe air distribution in a typical residential system.
c. Identify common air measurement instruments.

Performance Tasks

1. Use a manometer to measure static pressure in a duct system.
2. Use a velometer to measure the velocity of airflow at the output of air system supply diffusers and registers.
3. Use a velometer to calculate system cfm.

Trade Terms

Blower door: An assembly containing a fan used to depressurize a building by drawing air out, to support duct leak testing. It is usually installed temporarily in place of an outside door.

Cubic feet per minute (cfm): A unit for the volume of air flowing past a point in one minute. Cubic feet per minute can be calculated by multiplying the velocity of air, in feet per minute (fpm), times the area it is moving through, in square feet (cfm = fpm × area). The metric value is cubic meters per hour (m3/h).

Dew point: The temperature at which air becomes saturated with water vapor, and the water starts to condense into droplets; a state of 100 percent relative humidity.

Dry-bulb temperature: The temperature measured using a standard thermometer. It represents the measure of sensible heat present.

External static pressure (ESP): The total resistance of all objects and ductwork in the air distribution system beyond the blower assembly itself.

Pitot tube: A tool used to capture pressure measurements in a moving air stream.

Psychrometric chart: A graphic method of showing the relationship of various air properties.

Relative humidity (RH): The ratio of the amount of moisture present in a given sample of air to the amount it can hold at saturation. Relative humidity is expressed as a percentage.

Revolutions per minute (rpm): The number of rotations made by a spinning object over the course of one minute.

Static pressure (s.p.): The pressure exerted uniformly in all directions within a duct system, usually measured in inches of water column (in. w.c.) or centimeters of water column (cm H2O).

Total pressure: The sum of the static pressure and the velocity pressure in an air duct.

Velocity: The speed at which air is moving. The rate of airflow usually measured in feet per minute.

Velocity pressure: The pressure in a duct due to the linear movement of the air. It is the difference between the total pressure and the static pressure.

Wet-bulb temperature: Temperature taken with a thermometer that has a wick wrapped around its sensing bulb, saturated with distilled water before taking a reading. The reading from a wet-bulb thermometer, through evaporation of the water, takes into account the moisture content of the air. It reflects the total heat content (sensible and latent) of the air.

An HVAC system performs no better than its air distribution system. Understanding air distribution is essential to evaluating system performance. An adequate air distribution system must do all of the following:

- Supply the right volume of air to each conditioned space
- Provide air to each space without causing discomfort
- Operate efficiently without noise
- Require minimum maintenance

Air distribution systems are forced-air systems. Natural air movement cannot provide the momentum required for a proper quantity of conditioned air to travel through a series of ducts to conditioned areas. The major components that make up a forced-air system are as follows:

- An air handling unit (AHU) or fan coil unit (FCU)
- An air distribution system for conditioned air
- An air distribution system for return air
- Grilles and registers to distribute conditioned air and collect return air

An air handling unit is also called an *air handler*. A smaller unit used in a residential or light commercial application that has a coil assembly in the same cabinet with the blower is commonly called a *fan coil unit* (FCU). Regardless of which name is used, the air-moving equipment must include a blower and is responsible for creating the air movement.

Figure 1 shows the basic components of a forced-air system. Each system may use a different type of air handling equipment, different duct materials, and different air terminal devices. The installation location of the terminals varies dramatically as well. Additional accessories or equipment are often added to the air distribution system to achieve the desired humidity, air volume, or air cleanliness.

Duct System Design

HVAC technicians are not generally expected to design or size the duct system. That is the job of the system designer. However, understanding the factors that go into good system design is very helpful when comfort or performance problems are encountered. Especially in the commercial environment, complaints related to the air distribution system are very common.

1.1.0 Airflow and Pressure in Distribution System Ductwork

Air can be moved by creating positive pressure (above atmospheric pressure) in one area and negative pressure (below atmospheric pressure) in another. Creating a pressure difference between two areas causes air to move from the area of highest pressure to the area where a lower pressure exists. The pressure difference does not need to be substantial for some movement to occur.

All fans and blowers produce these changes in pressure. The air inlet to a spinning blower wheel is below atmospheric pressure, while the outlet is above atmospheric pressure. The indoor blower of a system creates pressure differences in a duct system. As noted earlier, the air moves from the areas of higher pressure toward the areas of lower pressure in the ducts (*Figure 2*).

At the return air grille and in the duct, the air pressure is lower than atmospheric pressure in the room. Therefore, air moves into the duct. The pressure decreases to its lowest point at the blower-wheel inlet. The air pressure is increased to its highest level at the blower-wheel outlet. From there, the air continues its flow from the blower discharge into the conditioned space. The difference in the pressures, although small, causes air to move from the supply duct opening toward the return duct opening. As it does, room air is mixed with the supply air.

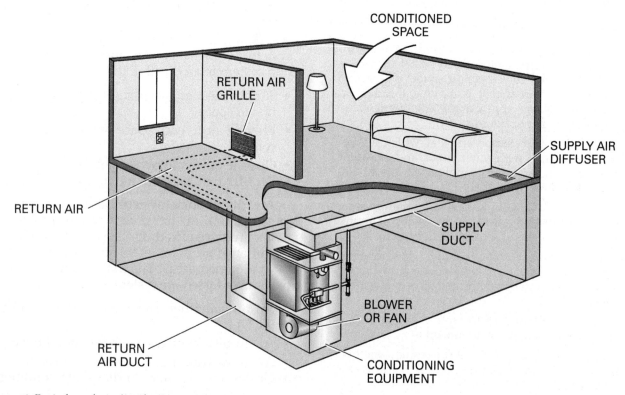

Figure 1 Basic forced-air distribution system.

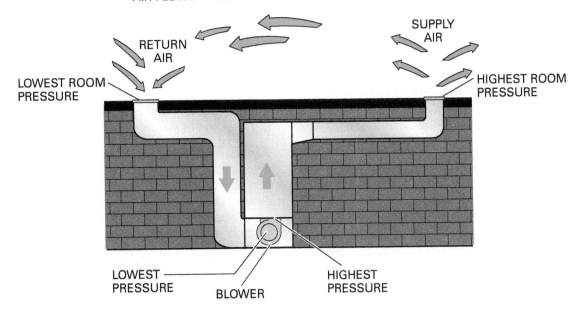

Figure 2 Pressures in an air distribution system.

The amount of pressure difference the blower creates affects the velocity of the air as well as the volume. The velocity, or speed, at which the air moves is measured in feet per minute (fpm). In the metric system, meters per second (m/s) are most often used. The volume is measured in cubic feet. However, the volume needs to be related to time to be of value. The volume of air movement in air distribution systems is commonly measured in cubic feet per minute (cfm). In the metric system, it is measured in cubic meters per hour (m³/h).

Air velocity and volume can be determined in air distribution systems using velometers and anemometers, which are discussed later in this module. The air volume in cubic feet per minute can be calculated by multiplying the velocity of air (in fpm), times the cross-sectional area that it is moving through (in square feet). The equations look like this:

$$\text{Volume flow rate (cfm)} = \text{duct area (ft}^2) \times \text{velocity (fmp)}$$

Or:

$$\text{Volume flow rate (m}^3/\text{h)} = \text{duct area (m}^2) \times \text{velocity (m/s} \times 3{,}600 \text{ s/h)}$$

Other variations of the formula used to find other values are as follows:

$$\text{Velocity (fpm)} = \frac{\text{Volume flow rate (cfm)}}{\text{Area (ft}^2)}$$

$$\text{Area (ft}^2) = \frac{\text{Volume flow rate (cfm)}}{\text{Velocity (fpm)}}$$

As an example, consider a duct that is 14" × 12" in size. The air is moving at a velocity of 900 fpm. The duct dimensions must be converted from inches to feet before using the formula. Each dimension is divided by 12 to convert from inches to feet:

$$14" = 1.166'$$
$$12" = 1'$$
$$\text{Duct size in feet} = 1.166' \times 1'$$

Now calculate the duct area in square feet:

$$1.166 \text{ ft} \times 1 \text{ ft} = 1.166 \text{ ft}^2$$

Now the air volume can be calculated using the area and velocity data provided:

$$\text{Volume flow rate} = \text{duct area} \times \text{velocity}$$
$$\text{Volume flow rate} = 1.166 \text{ ft}^2 \times 900 \text{ fmp}$$
$$\text{Volume flow rate} = 1{,}049 \text{ cfm}$$

To use the metric formula, the area of the cut must be determined in square meters. The velocity is converted to meters per second, and then multiplied by 3,600 seconds per hour to change to meters per hour.

Although ductwork is required to guide the air to its destination, the inside surface offers resistance to the flow of air. Therefore, the velocity of airflow across the duct is not really uniform. It varies from zero at the duct walls to a maximum velocity in the center of the duct. The velocity of the moving air decreases next to the duct walls due to friction. The rougher the duct surface, the greater the resistance to airflow. Also remember that resistance against airflow increases as the velocity increases. Some of the energy used by the blower is used in overcoming the resistance.

Additional friction and resistance can also be caused by changes in direction, various duct-lining materials, and anything protruding into the airstream. *Figure 3* shows some of the many fittings and changes in direction found in duct systems. The blower must work against this resistance as well. Fittings that add resistance decrease the quantity of air that could potentially flow through. The size and design of the registers and grilles also affects the airflow in a system. For example, a return air grille with fins spaced $\frac{1}{2}$" apart across its face resists airflow more than one with fins spaced 1" apart. The angle of the fins also makes a difference. These friction losses have a significant effect on sizing the blower and the ductwork. Friction losses are also referred to *static pressure loss*, *static pressure drop*, or *simple pressure drop*.

There are three pressure values that exist in an air distribution system: static pressure (s.p.), velocity pressure, and total pressure.

Static pressure is the pressure exerted uniformly in all directions within a duct system. In a supply air duct, think of it as the bursting or exploding pressure that acts on all internal surfaces. As shown in *Figure 4*, static pressure can be applied to a manometer for measurement via the static pressure openings in a pitot tube or static pressure tip connected to the manometer. Note that the static pressure openings in the pitot tube are perpendicular to the direction of airflow. The static pressure can also be measured by inserting a probe just through the wall of the duct, with the opening remaining perpendicular to the direction of flow. The use of this equipment is covered further in this module.

Velocity pressure is the pressure in a duct system caused by the forward movement of air.

Velocity pressure acts in the direction of airflow only. It is the difference between the total pressure and the static pressure. Total pressure is the sum of the static and the velocity pressures in a duct system. This relationship is shown mathematically as follows:

Total pressure = static pressure + velocity pressure
Static pressure = total pressure − velocity pressure
Velocity pressure = total pressure − static pressure

Figure 4 shows a manometer connected to a special pitot tube in the duct, configured to measure the velocity pressure. As shown, the total pressure is being measured through the opening pointing into the airstream. Static pressure is ported separately through the side of the tube to the opposite end of the manometer. The two pressures oppose each other, and the reading represents velocity pressure.

A manometer is the perfect tool to measure low pressures in inches of water column (in. w.c.), which express the height, in inches, to which the pressure exerted will lift a column of water. The atmosphere exerts a pressure of roughly 14.7 pounds per square inch (psi) at sea level with 70°F dry air. This atmospheric pressure of 14.7 psi will support a column of water 406.8" high, or 33.9' (*Figure 5*). For every pound per square inch of pressure, a column of water can be raised 27.74", or about 2.3'.

In the metric system, pressure is often measured in kilopascals (kPa). The normal pressure of the atmosphere at sea level is 101.3 kPa. To convert psi to kPa, multiplying psi by 6.895 provides sufficient accuracy. The right side of *Figure 5* shows the conversion of pressure in kPa to centimeters of water column (cm H_2O).

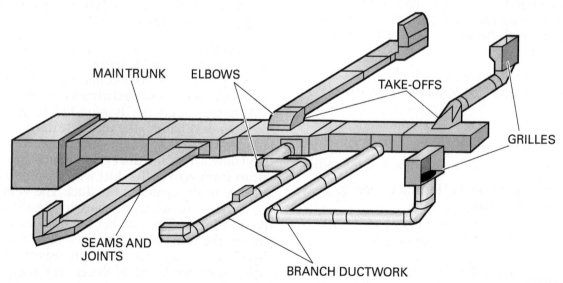

Figure 3 Some causes of friction loss in an air distribution system.

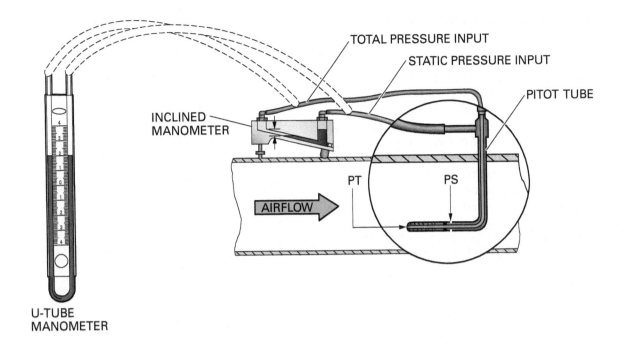

TOTAL PRESSURE INPUT
STATIC PRESSURE INPUT
PITOT TUBE
INCLINED MANOMETER
PT
PS
AIRFLOW
U-TUBE MANOMETER

PITOT TUBE SENSES TOTAL AND STATIC PRESSURES. MANOMETER MEASURES VELOCITY PRESSURE (DIFFERENCE BETWEEN TOTAL AND STATIC PRESSURES).

PT = TOTAL PRESSURE
PS = STATIC PRESSURE

Figure 4 Measuring duct pressure.

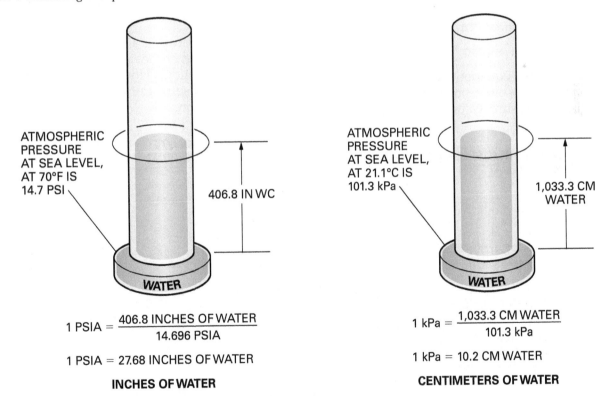

ATMOSPHERIC PRESSURE AT SEA LEVEL, AT 70°F IS 14.7 PSI

406.8 IN WC

$$1 \text{ PSIA} = \frac{406.8 \text{ INCHES OF WATER}}{14.696 \text{ PSIA}}$$

1 PSIA = 27.68 INCHES OF WATER

INCHES OF WATER

ATMOSPHERIC PRESSURE AT SEA LEVEL, AT 21.1°C IS 101.3 kPa

1,033.3 CM WATER

$$1 \text{ kPa} = \frac{1,033.3 \text{ CM WATER}}{101.3 \text{ kPa}}$$

1 kPa = 10.2 CM WATER

CENTIMETERS OF WATER

Figure 5 Comparison of atmospheric pressure to water-column values.

1.1.1 Absolute Pressure (PSIA) vs. Gauge Pressure (PSIG)

Every square inch of Earth's surface at sea level, when the air temperature is 70°F, has 14.7 pounds of air pressure pushing down on it. Variations in temperature will change the density of the air, and therefore the precise pressure. It has also been presented that the atmospheric pressure of 14.7 psi at sea level will support a column of water 406.8" high when measured with a manometer. When measured with a mercury-tube barometer like the ones used by meteorologists, the same atmospheric pressure of 14.7 psi will cause the mercury in the tube to rise to a height of 29.92". The values of 14.7 psi and 29.92" of mercury (in. Hg), at sea level and at 70°F, are important to understand in the HVACR craft.

Refer to *Figure 6*. The absolute pressure scale is based on the barometer measurements just described. On this scale, pressures are expressed in pounds per square inch absolute (psia) starting from 0 psi. Another scale, called the *gauge pressure scale*, is more frequently used to define pressure levels. The gauge pressure scale considers the atmospheric pressure of 14.7 psia, or 0 psig, as the starting point. The following mathematical expressions illustrate the relationship between gauge pressure and atmospheric pressure:

$$0 \text{ psig} = 14.7 \text{ psia}$$

$$\text{psia} = \text{psig} + 14.7$$

As shown in the equations, gauge pressure can be converted to absolute pressure by *adding* 14.7 to the gauge pressure value. Likewise, absolute pressure can be converted to gauge pressure by *subtracting* 14.7. Negative pressures (those below 0 psig) are expressed in inches of mercury vacuum (in. Hg vacuum). Therefore:

$$0 \text{ psia} = 29.92 \text{ in. Hg vacuum}$$

Conversion between absolute and gauge pressure scales is sometimes necessary when making pressure calculations. However, for most practical purposes, the use of the gauge pressure scale is far more common.

Duct Pressure Ranges

The Sheet Metal and Air Conditioning Contractors' National Association (SMACNA) assigns classifications to ducts according to their pressure range. The operating pressure range is a significant factor in the structural characteristics of the duct. Classifications are stated in inches of water column, or inches of water gauge. These terms are used interchangeably. The duct pressure classifications are as follows:

- ½"   Up to ½ in. w.c.
- 1"   ½–1 in. w.c.
- 2"   1–2 in. w.c.
- 3"   2–3 in. w.c.
- 4"   3–4 in. w.c.
- 6"   4–6 in. w.c.
- 10"   6–10 in. w.c.

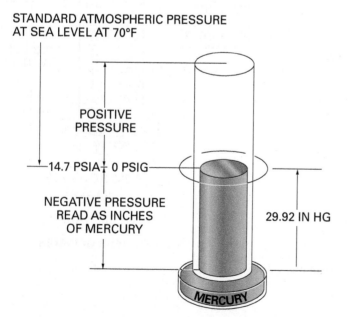

STANDARD ATMOSPHERIC PRESSURE
AT SEA LEVEL AT 70°F

POSITIVE PRESSURE

14.7 PSIA ‡ 0 PSIG

NEGATIVE PRESSURE
READ AS INCHES
OF MERCURY

29.92 IN HG

MERCURY

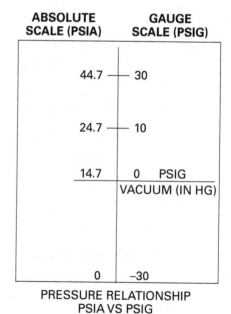

ABSOLUTE SCALE (PSIA)	GAUGE SCALE (PSIG)	
44.7	30	
24.7	10	
14.7	0	PSIG
	VACUUM (IN HG)	
0	−30	

PRESSURE RELATIONSHIP
PSIA VS PSIG

Figure 6 Absolute, gauge, and barometric pressure scales compared.

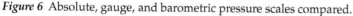

1.2.0 Residential Air Distribution Systems

Figure 7 shows an air distribution system for a typical single-story house. The discussion that follows demonstrates the concepts and pressure relationships that have been covered so far, along with some new ones.

In most cases, more airflow is needed for cooling than for heating. Therefore, the air handler (blower) and air distribution system must be able to supply the higher volume of air needed for the cooling mode. Air distribution systems, including the blower, must be designed to function properly for the highest volume of airflow to be handled.

In this residential example, assume that 3 tons of cooling capacity is required, based on heat load calculations. In HVACR, a rule of thumb is that comfort cooling systems require about 400 cfm of air per ton of cooling. Therefore, the blower in this system must be able to supply 1,200 cfm of air (3 tons × 400 cfm) or more. In *Figure 7*, a fan coil unit that can provide three tons of cooling capacity has been selected.

Manufacturers provide airflow performance charts for air handling equipment to show the capacity of the included blower. See *Table 1*. The selected unit can provide slightly more than the needed 1,200 cfm against a static pressure of 0.5 in. w.c. with high speed selected. At lower static pressures, more airflow is produced. Note that there is a very slight difference in blower performance based on the voltage provided to the motor.

Note the use of the term total external static pressure (ESP) at the top of the airflow performance chart. The external static pressure is the resistance of all things in the air distribution system beyond the blower assembly itself. This value usually does not include the static pressure loss of the indoor coil. Charts from the manufacturer provides the pressure loss associated with a variety of components, including the indoor coil, at a given air volume.

In the example shown in *Figure 7*, the static pressure loss of the evaporator coil is shown as 0.08 in. w.c. When designing a system, it is always necessary to confirm what components have or have not been accounted for in an airflow performance chart. If the cooling coil has not been considered in the manufacturer's chart, it can make a significant difference. Wet coils especially (as the result of water condensing on their surface) impose a considerable amount of resistance to airflow. Once an evaporator coil has been cooled by the refrigerant, it typically becomes wet very quickly.

The system shown in *Figure 7* has a total of 12 air supply outlets, selected for airflows between 50 and 150 cfm. The return air is taken into the system through two centrally located return air grilles.

While reviewing the drawing, remember that the home or other structure is part of the system too. The supply-air grilles are often located around the perimeter of a home, since the perimeter is where a great deal of the heat gain or loss occurs. The supply air leaves the registers and washes up the walls of the house. It then travels through the conditioned space within the house as it makes its way back to the return air grilles.

The discussion can begin at the return air grilles. The blower has created a slightly negative pressure at the grilles. As shown, the fan creates a vacuum of –0.02 in. w.c. just behind the face of each grille. This results in the air being motivated to enter the return air grilles. As the air flows down the return duct towards the blower, additional static pressure losses are encountered. At the inlet to the blower wheel, the air pressure is at its lowest point in the system. For this example, it is at –0.17 in. w.c. The return air moves through the blower wheel, where it is increased to the highest pressure in the duct system. In this example, it is 0.26 in. w.c. The total difference in static pressure between the inlet and outlet of the blower (the lowest and highest pressures in the system, respectively) is 0.43 in. w.c.

The air at the blower outlet moves out through the supply duct. After the air enters the duct, the tee used to change the direction causes some pressure drop. The friction of the duct walls adds to the static pressure, as do the transitions protruding into the supply air duct. As the air passes through the supply air grilles, some additional static-pressure loss results. In the example, it is shown to be 0.06 in. w.c.

The transitions in the supply duct size shown in *Figure 7* play an important role in system performance. Each initial duct section after the tee must handle 600 cfm of air. Two branch duct outlets, totaling 200 cfm of air, are supplied from these sections. This reduces the quantity of air supplied to the next sections of the trunk to 400 cfm for each side. The next sections also supply 200 cfm of air each to the conditioned space. This reduces the quantity of air supplied to the final sections of the supply duct to 200 cfm on each side.

At each point in the duct following a substantial change in the air volume, the duct is transitioned to a smaller size. The duct is resized to maintain a relatively equal velocity and static pressure through each section. This has

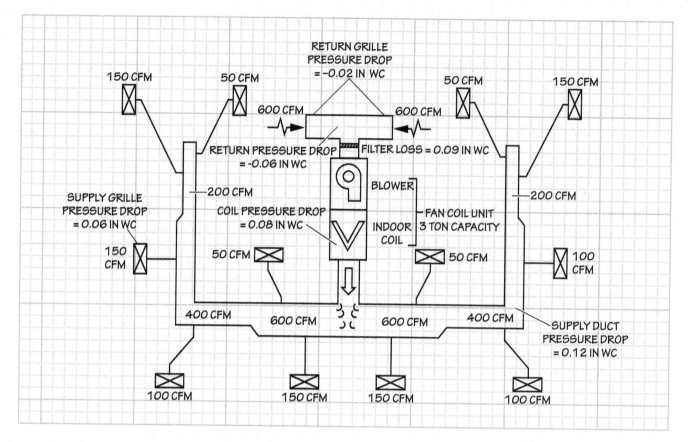

- SYSTEM COOLING CAPACITY = 3 TONS
- AT 400 CFM PER TON, TOTAL AIRFLOW REQUIRED = 1,200 CFM
- TOTAL RETURN DUCT STATIC PRESSURE AT 1,200 CFM = 0.19 IN WC
    - –0.06 IN WC RETURN DUCT STATIC PRESSURE LOSS
    - –0.09 IN WC FILTER STATIC PRESSURE LOSS
    - –0.04 IN WC TOTAL RETURN AIR GRILLE LOSS
- TOTAL SUPPLY DUCT SYSTEM STATIC PRESSURE AT 1,200 CFM = 0.29 IN WC
    - 0.12 IN WC SUPPLY DUCT STATIC PRESSURE LOSS
    - 0.17 IN WC TOTAL SUPPLY GRILLE STATIC PRESSURE LOSS
- TOTAL EXTERNAL STATIC PRESSURE = 0.48 IN WC

Figure 7 Typical residential air distribution system.

Table 1 Example Airflow Performance Chart

Model and Size	Motor Speed	Total External Static Pressure (in. w.c.)											
		0.10		0.20		0.30		0.40		0.50		0.60	
		208V	230V	208V	230V	208V	230V	208V	230V	208V	230V	208V	230V
FCU-12 Size 036	High	1495	1566	1428	1487	1365	1435	1311	1366	1232	1285	1150	1190
	Medium	1260	1388	1228	1344	1176	1281	1132	1235	1060	1166	996	1077
	Low	1080	1238	1034	1181	1004	144	955	1094	919	1028	862	963

a self-balancing effect on the system. If the size remained the same throughout, too much air would tend to travel down the full length of the duct and try to escape through the grilles closest to the end. The grilles closest to the fan coil unit would not deliver the desired airflow, while too much air would be delivered at the grilles beyond them.

Normally, dampers would be installed in each branch to provide a means of balancing the quantity of air supplied to each area. With a damper, such as the one shown in *Figure 8*, resistance can be added to a branch line, decreasing the amount of airflow into it. Although the design may call for a specific amount of airflow from each grille, the actual air volume needed can vary due to several factors. Some occupants, for example, may prefer a warmer or cooler room. Dampers provide a means to make minor adjustments in the airflow.

To finalize the example, refer back to *Table 1* and compare the total ESP in the example to the performance of the blower. This blower should have no problem moving the desired volume of air against an ESP of 0.43 in. w.c., when it is set to high speed. Depending upon the voltage provided to the unit, the medium speed may provide the needed airflow.

Note that this example does not represent a perfect design. It simply demonstrates the effects of static pressure loss and how all portions of the air distribution system contribute some level of resistance to airflow.

1.3.0 Air Measurement Instruments

Evaluating the performance of an air distribution system involves many considerations. These include the temperature and humidity of the air, the velocity at which it moves, and the pressure in the duct system. A variety of instruments are used to measure these values.

1.3.1 Temperature and Humidity Meters

Several instruments are used to measure the temperature and humidity of the conditioned air when working on air distribution systems. The following are three of the most common instruments:

- Electronic thermometers
- Psychrometers
- Hygrometers

Electronic Thermometers – Electronic, or digital, thermometers (*Figure 9*) are used to measure temperatures in air distribution systems and for many other purposes. Electronic thermometers with a digital display are commonly used for HVACR service work. They use either a thermocouple or thermistor-type temperature probe to sense the temperature. Many electronic thermometers have two or more probes so that measurements can be made at multiple locations at the same time. Most electronic thermometers of this type can calculate and display the difference in temperature between the locations as they are being measured.

Many digital multimeters (DMMs) can also be used to measure temperature. This feature requires the use of thermocouple or thermistor accessories that convert the DMM into an electronic thermometer. Some DMMs can also use a noncontact infrared accessory to measure temperature from a distance.

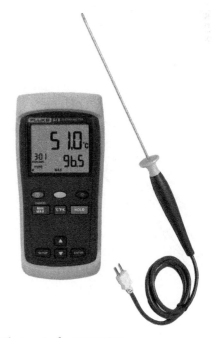

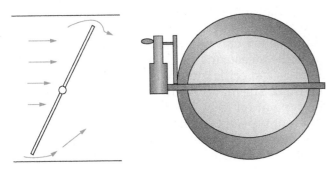

Figure 8 Air damper.

Figure 9 Electronic thermometer.

Electronic thermometers are precision instruments. Be sure to follow the manufacturer's instructions for operating and calibrating electronic thermometers to prevent inaccurate measurements.

Psychrometers – A sling psychrometer (*Figure 10*) is also used measure temperature. It has two thermometers, one to measure the dry-bulb temperature and the other to measure the wet-bulb temperature. The sensing bulb of the wet-bulb thermometer is covered with a wick that is saturated with water before taking a reading. To make sure the temperatures are accurate, the sling psychrometer is spun rapidly in the air. This enhances evaporation from the wet-bulb thermometer wick, giving it a lower temperature reading. The measured wet-bulb and dry-bulb temperatures can be used to find the percent of relative humidity (RH) in the air. This is done using either a built-in chart on the psychrometer or a separate psychrometric chart. Relative humidity, expressed as a percentage, compares the amount of moisture present in air to the amount it can hold when it is fully saturated.

Hygrometers – A hygrometer is used to measure and give direct readings of relative humidity. Many varieties of hygrometers are available, including both electronic and dial types.

Figure 11 shows an electronic psychrometer, which is a thermometer and hygrometer combined. It measures dry-bulb temperature, wet-bulb temperature, relative humidity, and dew point. The dew point is the temperature at which a quantity of air becomes saturated (100% RH).

Figure 11 Electronic psychrometer.

Any water that it is unable to hold falls out in the form of tiny droplets. As the outdoor air cools at night, it often falls below the dew point for the moisture it contains. The result is dew on the ground and other surfaces in the morning. As the air temperature rises the following day, the water is re-evaporated back into the air. Instruments like the one shown in *Figure 11* have an optional temperature probe that allows the user to measure the temperature differential between the air and a surface.

1.3.2 Pressure Measurement Instruments

Several instruments and accessories are used to measure the pressures in an air distribution system. Among these are three common devices:

- Manometers
- Differential pressure gauges
- Pitot tubes and static pressure tips

Manometers – Manometers are used to measure the low pressures found in air distribution systems. Manometers used for air-distribution service are calibrated in inches of water column. Manometers can use water or a special type of oil as the measuring fluid. Popular models use oil that has a specific gravity of 0.826 as the measuring fluid. The manufacturer of the gauge specifies the type of oil to be used, so substitution with a different type of oil is not recommended.

Manometers come in many styles, including U-tube, inclined, and combined U-inclined (*Figure 12*). Electronic manometers are also widely used. Pitot tubes or static pressure tips, described later in this section, are almost always used with manometers when measuring pressures in duct systems.

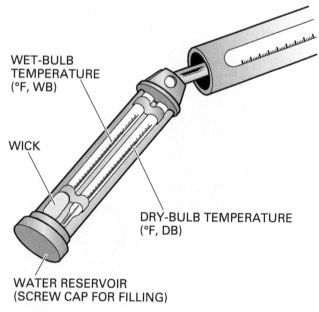

WET-BULB
TEMPERATURE
(°F, WB)

WICK

DRY-BULB TEMPERATURE
(°F, DB)

WATER RESERVOIR
(SCREW CAP FOR FILLING)

Figure 10 Sling psychrometer.

Psychrometrics

The atmosphere consists of a mixture of air (mostly nitrogen and oxygen) and water vapor. The study of air and its properties is called *psychrometrics*. In 1911, Dr. Willis Carrier presented his Rational Psychrometric Formula to the American Society of Mechanical Engineers (ASME). His formula led to the development of the psychrometric chart like the one shown here. It gives a graphical representation of the interrelationships that exist for all the properties of air. HVAC engineers, designers, and technicians typically use a psychrometric chart to predict the values for the various properties of air when designing an HVAC system, or before adjusting or modifying an existing HVAC system.

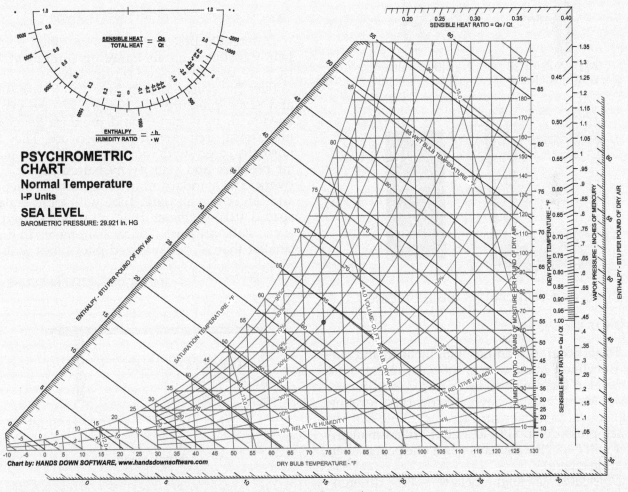

Figure Credit: Courtesy of Hands Down Software, www.handsdown software.com

To check the difference in pressure between two sources, each is attached to one of the manometer ports. The column moves in relation to the difference between the two pressures. However, in many cases, one port is left open to the atmosphere. This provides a direct reading of the pressure on the connected port.

Manometers work on the principle that pressure causes a change in the level of the liquid column. When a positive or negative pressure is applied to one side, the column of liquid moves until it reaches equilibrium again.

U-tube and inclined manometers are available in several low-pressure ranges. Inclined manometers are usually calibrated in the lowest pressure ranges and are more sensitive than U-tube manometers. U-inclined manometers combine the sensitivity of the inclined manometer with the higher-range capability of the U-tube manometer. Inclined-vertical manometers combine an inclined section for high accuracy with a vertical manometer section for extended range.

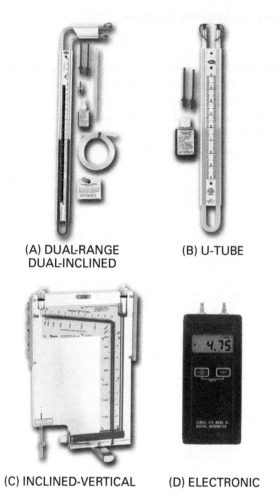

(A) DUAL-RANGE
DUAL-INCLINED

(B) U-TUBE

(C) INCLINED-VERTICAL

(D) ELECTRONIC

Figure 12 Manometers.

To get accurate readings with inclined and vertical-inclined manometers, the instrument must be level. A built-in spirit level is used for this purpose. Many models have a screw-type leveling adjustment to help ensure accuracy. Note that too much pressure applied to a liquid-filled manometer will expel the contents when one port is open to atmosphere.

Differential Pressure Gauges – A differential pressure gauge (*Figure 13*), also known as a *Magnahelic®* *gauge*, provides a direct reading of pressure. These gauges are typically used to measure blower pressures, filter resistance, and furnace flue draft values. With the appropriate accessories, they can also measure air velocity. Single-scale models are calibrated in either in. w.c. or psi. Dual-scale gauges are normally calibrated for pressure in in. w.c. and for air velocity in fpm. Several models are available covering pressure ranges from 0.0 in. w.c. to 10 in. w.c. These gauges are also available with metric units of measure, such as millimeters of water, centimeters of water, Pascals, and kilopascals.

Figure 13 Portable differential-pressure gauge.

Differential pressure gauges are often permanently installed on air handling equipment, but they are also carried as a service tool.

Pitot Tubes and Static Pressure Tips – Pitot tubes and static pressure tips (*Figure 14*) are probes used with manometers and pressure gauges to make measurements inside ductwork. They are normally used with portable models to make air pressure and velocity measurements in ductwork. The standard pitot tube used in ducts 8" and larger has an outer tube with eight equally spaced 0.04"-diameter (1.02 mm) holes used to sense static pressure. For measurements in ducts smaller than 8", pocket-sized pitot tubes with an

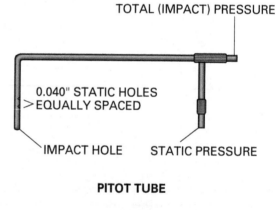

TOTAL (IMPACT) PRESSURE

0.040" STATIC HOLES
EQUALLY SPACED

IMPACT HOLE

STATIC PRESSURE

PITOT TUBE

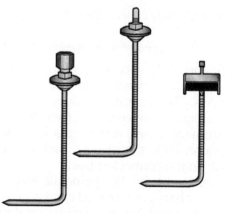

STATIC PRESSURE TIPS

Figure 14 Pitot tube and static pressure tips.

outer tube and four equally spaced static-pressure holes are best. Pitot tubes come in various lengths ranging from 6" to 60" (15 cm to 1.5 m), with graduation marks to show the depth of insertion in a duct.

A pitot tube consists of an impact tube fastened concentrically inside a larger tube. The inner tube receives the total pressure input. The outer tube receives static-pressure inputs from the radial sensing holes. The tubes transfer pressure from the sensing holes to the ports of the manometer.

When the total pressure tube is connected to the high-pressure port of the manometer alone, velocity pressure is indicated directly. The other manometer port remains open to the atmosphere for this use. To be sure of accurate velocity pressure readings, the pitot tube tip must be pointed directly into the duct air stream.

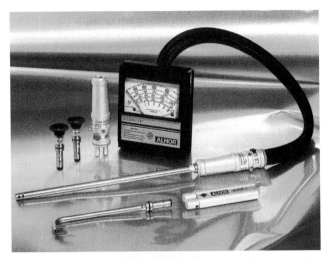

VELOMETER

Flow Hood

A flow hood, also known as a *balometer*, is a special type of velometer used in balancing air distribution in commercial systems. It is held over a diffuser and displays the air velocity on an integral readout. Air velocity can also be measured using a manometer, and then applying the appropriate formula.

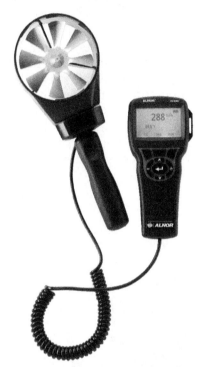

ROTATING VANE ANEMOMETER

Figure 15 A velometer and a rotating-vane anemometer.

Static pressure tips are used with manometers and differential pressure gauges to measure static pressure. They are typically L-shaped and have four 0.04" (1.02 mm) sensing holes.

1.3.3 Air Velocity Measurement Instruments

Velometers and anemometers, such as the ones shown in *Figure 15*, are used to measure the velocity of airflow. The measurement of air velocity is done to check the operation of an air distribution system. It is also done when balancing system airflow.

Most velometers give direct readings of air velocity in feet per minute or meters per second. Some can provide direct readings in cfm or m³/h with the input of duct size so that internal calculations can be made. Velometers with both analog scales and digital readouts are in common use.

Some anemometers use a rotating vane (propeller) or balanced-swing vane to sense the air movement. When the rotating-vane anemometer is positioned to make a measurement, the vane rotates in the air stream. The rotational speed is converted into a velocity reading for display. In the swinging-vane velometer, the air stream causes the vane to tilt at different angles in response to the measured air velocity. The position of the vane is converted into a velocity reading.

Another velocity-measuring device, known as a *hot-wire anemometer* or *thermal anemometer*, (*Figure 16*), gives direct readings of air velocity. This instrument uses a sensing probe that contains a small resistance-heating element. When the probe is held perpendicular to the air stream being measured, the temperature of the element changes due to the airflow. This causes its resistance to change, which alters the amount of current flow to the meter circuitry. Hot-wire anemometers are typically used for very-low velocity readings.

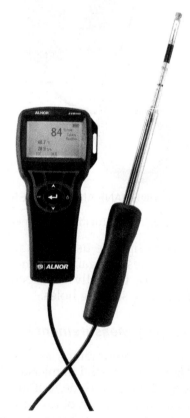

Figure 16 Hot-wire anemometer.

Depending on the probe or attachment used, velometers can often measure air velocities up to 10,000 fpm (50.8 m/s). Some electronic models can average up to 250 individual readings taken across a duct area to calculate an average velocity. Some velometers also include a micro-printer to record the readings, and can interface with a computer to download a series of recorded values for charting.

A velometer can be used to determine the average velocity in a duct. The average velocity must be calculated by taking a number of velocity readings that represent the velocity in each section. For rectangular ducts with dimensions less than 30" × 30" (76 cm × 76 cm), 25 readings are typically taken, equally spaced across the length and width of the duct. The readings are then averaged. The duct length and width is then measured to determine its cross-sectional area. Multiplying the area by the average velocity provides a reasonably accurate air volume value, in cubic feet per minute or cubic meters per hour.

1.3.4 *Measuring Rotational Speed*

The rotational speed of motors and blowers must also be measured for various purposes when working with air distribution systems. The speed at which the shaft of an air-moving device is rotating is measured in revolutions per minute (rpm). The easiest way to determine the fan rpm is to measure it directly with a tachometer (*Figure 17*).

There are two types of tachometers: contact and noncontact. Use of the noncontact type is safer and more convenient when the motor is located in a hard-to-reach place. Some manufacturers make a combination contact/noncontact model tachometer that can be used to make rpm measurements by either method, such as the one shown in *Figure 17*. To measure the rpm of a shaft with the contact-type tachometer, do the following:

Step 1 Turn the drive motor on.

Step 2 Contact the end of the motor or blower shaft with the tachometer sensor tip.

Step 3 Allow the reading to stabilize, and then read the rpm.

Noncontact tachometers are seldom used to measure the rpm of a shaft directly. They are much better suited for use with pulleys or blower wheels of a much greater diameter than a small drive shaft. To measure the rpm with the noncontact tachometer, use the following steps:

Step 1 Ensure the drive motor is off and properly disabled.

Step 2 Place a reflective mark or a piece of reflective tape on the device to be measured.

Step 3 Turn the motor back on.

Step 4 Point the tachometer light beam at the pulley or blower wheel and hold it steady. The optical eye of the tachometer counts the rotations of the reflective surface.

> **NOTE**
>
> If the entire surface of the pulley or wheel is reflective, the surface may need to be dulled to provide an accurate measurement.

Figure 17 Combination contact/noncontact tachometer.

Additional Resources

Air Distribution Systems: Introduction to Thermo-Fluids Systems Design, A. G. McDonald and H. L. Magande. 2012. Chichester, UK: John Wiley & Sons, Ltd.

1.0.0 Section Review

1. Multiplying the cross-sectional area of a duct by the air velocity is used to calculate _____.

   a. static pressure
   b. velocity pressure
   c. air volume
   d. air density

2. The supply-air grilles in a residential system are often placed around the perimeter of the structure because _____.

   a. they are easier to install there
   b. the perimeter is where most of the heat gain or loss occurs
   c. the return-air grilles are also located there
   d. it allows the air distribution ducts to be shorter in length

3. A manometer is used to measure _____.

   a. pressure
   b. velocity
   c. dry-bulb temperature
   d. relative humidity

2.0.0 AIR DISTRIBUTION EQUIPMENT AND MATERIALS

Objective

Describe the mechanical equipment and materials used to create air distribution systems.

a. Describe various blower types and applications.
b. Describe various fan designs and applications.
c. Demonstrate an understanding of the fan laws.
d. Describe common duct materials and fittings.
e. Identify the characteristics of common grilles, registers, and dampers.

Performance Tasks

4. Read and interpret equivalent length charts and required air volume/duct size charts.

Trade Terms

A_k factor: The area factor of registers, grilles, and diffusers that reflects the free area for airflow relative to a square foot.

Boots: Sheet metal fittings designed to transition from the branch duct to the receptacle for the grille, register, or diffuser to be installed.

Free air delivery: The condition that exists when there are no effective restrictions to airflow (no external static pressure) at the inlet or outlet of an air-moving device.

Plenum: A chamber at the inlet or outlet of an air handler. The air distribution system attaches to the plenum.

Stratify: To form or arrange into layers. In air distribution, layers of air at different temperatures will tend to stratify unless an outside influence forces them to move and mix. Warmer air will stratify on top of cooler air.

Takeoffs: Connection points installed on a trunk duct that allow the connection of a branch duct.

Vapor barrier: A barrier placed over insulation to stop water vapor from passing through the insulation and condensing on a cold surface.

Venturi: A ring or panel surrounding the blades on a propeller fan to improve fan performance.

An air distribution system consists of the fan or blower that moves the air, the ductwork, and the grilles or registers that are used as air entry and exit points. In addition, dampers are used in some systems to control airflow.

2.1.0 Blowers

The blower provides the pressure difference required to move the air through the duct system and into the conditioned space. It must overcome the pressure losses caused by both the supply and return ductwork, and any obstructions that may exist inside.

The terms *fan* and *blower* are sometimes used interchangeably, but the term used usually depends on the application. For example, a blower generally describes the device used to move air against the resistance of a duct system. A fan describes the device used to move air when there is little or no significant resistance to airflow.

2.1.1 Belt- and Direct-Driven Blowers and Fans

Two types of blowers are commonly used in air distribution systems: belt-driven and direct-drive (*Figure 18*). In belt-driven blowers, the blower motor is connected to the blower wheel by a belt and pulley. The blower speed is adjusted mechanically by a change in the pulley diameter. Today, belt-driven blowers are more commonly used in commercial HVAC products.

In direct-drive blowers, the blower wheel is mounted directly on the motor shaft. Most residential and light commercial equipment use multispeed or variable-speed motors with direct-drive blowers. The blower speed is adjusted electrically by changing the blower motor wiring connections, or by changing an ECMs program settings. This enables the speed of the motor to be adjusted easily to match the requirements of the individual system. It also allows the speed to be changed for heating and cooling seasons without human intervention.

2.1.2 Centrifugal Blowers

Centrifugal blowers (*Figure 19*) are used with forced-air systems because they work best against the resistance of the duct. They can be used in very large systems and those that are considered high-pressure systems. As is true of most blowers, the airflow is at a right angle (perpendicular) to the shaft on which the wheel is mounted. The

(A) BELT-DRIVE BLOWER

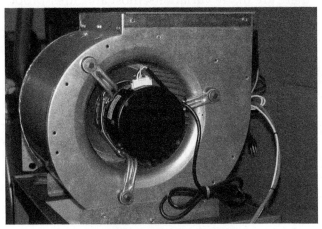

(B) DIRECT-DRIVE BLOWER

Figure 18 Belt-drive and direct-drive blowers.

Figure 19 Industrial centrifugal blower.

wheel is mounted in a scroll-shaped housing, which is necessary to develop the needed pressure changes. Centrifugal blowers are identified by the wheel blade position with respect to the direction of rotation. The basic types of centrifugal blowers include the following:

- *Forward-curved* – Forward-curved centrifugal blowers are extremely common in both residential and commercial systems. They are also used in ventilation systems. As shown in *Figure 20*, the tips of the blades in a forward-curved blower are inclined in the direction of rotation.

- *Backward-inclined* – The blades of the backward-inclined centrifugal blower are inclined away from the direction of rotation. Visually, to the uninformed, they appear to be installed backwards. Typically, these blowers are used in industrial systems that require heavy-duty blower construction and stable air delivery at a significant resistance. The individual blades are often welded in place. Backward-inclined blowers operate at a higher efficiency than forward-curved blowers. However, they also operate at higher speeds, and therefore tend to be noisier.

  Smaller backward-inclined wheels are usually supplied with flat blades, while larger wheels are supplied with airfoil-shaped blades to improve efficiency. Blowers using the airfoil blades generally run more quietly than other types. *Figure 21* shows a backward-inclined fan wheel and an airfoil fan wheel.

- *Radial* – Radial blower wheels (*Figure 22*) have straight blades that are, to a large extent, self-cleaning. This makes them more suitable for use in air systems that have particles or grease

Figure 20 Forward-curved centrifugal blower wheel.

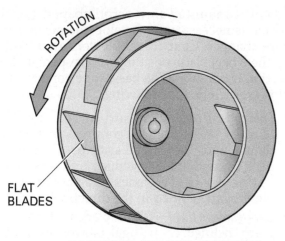

FLAT
BLADES

BACKWARD-INCLINED FAN WHEEL

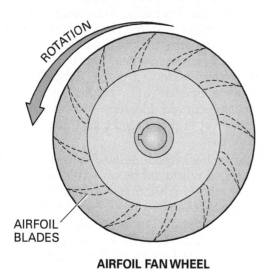

AIRFOIL
BLADES

AIRFOIL FAN WHEEL

Figure 21 Backward-inclined centrifugal blower wheels.

in the air stream. They can also be used in other applications such as pneumatic conveying systems. The wheels of radial blowers are simple in construction, with narrow blades that often resemble paddles. They can withstand the high speeds needed to operate at higher levels of resistance. However, for high static pressures and high-speed operation, the blades are welded in place.

2.2.0 Fans

Fans are typically used in applications where little resistance to airflow exists, such as with through-wall exhaust fans and condenser fans. Although there are many variations, most are either propeller fans (axial) or duct fans (tube-axial). In fans, the airflow is normally parallel to the drive shaft, as opposed to the perpendicular relationship found in blowers.

2.2.1 Propeller Fans

Propeller, or axial, fans (*Figure 23*) have good efficiency in free air delivery and are commonly used as condenser fans or through-wall exhaust fans in HVACR applications. Free air delivery is the condition that exists when there is little or no restriction to airflow. Propeller fans are usually mounted in a venturi to cause the air to flow in a straight line from one side of the fan to the other. A venturi is a ring surrounding the blades on a propeller fan. To achieve the best performance, the blade must be properly set in the venturi opening. If the position is other than that specified by

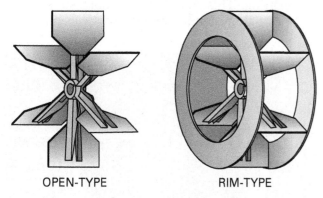
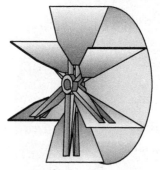
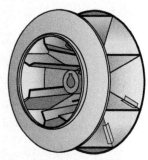

OPEN-TYPE   RIM-TYPE   ONE-SIDE CLOSED   BACKPLATE-TYPE

Figure 22 Radial centrifugal blower wheels.

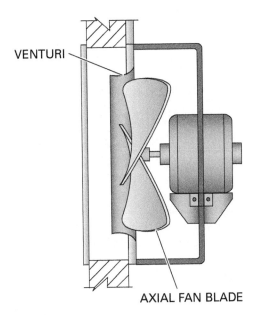

VENTURI

AXIAL FAN BLADE

Figure 23 Propeller fan.

the manufacturer, the performance will drop and the fan might be noisy. Propeller fans make more noise than centrifugal blowers or fans, so they are normally used where noise is not a factor.

2.2.2 Duct Fans

In duct fans, airflow is also parallel to the shaft on which the wheel is mounted. However, duct fans are housed in a cylindrical duct or tube. The tube acts as a long venturi. This design allows duct fans to operate at higher static pressures than propeller fans. Duct fans are commonly used in ducted exhaust systems that move a great deal of air against significant resistance. A fan is considered ducted if the attached duct length is more than the distance between the inlet and outlet of the fan blades.

The two types of ducted fans commonly used are tube-axial fans and vane-axial fans. A tube-axial fan discharges air in a helical or screw-like motion. A vane-axial fan has vanes on the discharge side of the propeller that cause the air to discharge in a straight line. This reduces the amount of turbulence, thereby improving efficiency and pressure capacity. *Figure 24* shows both tube-axial and vane-axial versions of duct fans.

2.3.0 The Fan Laws

The performance of all fans and blowers is governed by three rules commonly known as the *Fan Laws*. The volume of air movement, fan speed, external static pressure, and horsepower (hp) are all closely related. There are three fan laws:

- *Fan Law 1* – The volume of air delivered by a fan varies directly with the speed of the fan:

$$\text{New volume flow rate} = \frac{\text{New rpm} \times \text{existing volume flow rate}}{\text{Existing rpm}}$$

Or:

$$\text{New rpm} = \frac{\text{New volume flow rate} \times \text{existing rpm}}{\text{Existing volume flow rate}}$$

- *Fan Law 2* – The static pressure of a system varies directly with the square of the ratio of the fan speeds:

$$\text{New static pressure} = \text{Existing static pressure} \times \left(\frac{\text{New rpm}}{\text{Existing rpm}}\right)^2$$

- *Fan Law 3* – The horsepower varies directly with the cube of the ratio of the fan speeds:

$$\text{New hp} = \text{existing hp} \times \left(\frac{\text{new rpm}}{\text{existing rpm}}\right)^3$$

Example:

The existing conditions for an air-moving system are as follows:

- 5,000 cfm of air-volume flow rate
- 1,000 rpm blower speed
- 0.5 in. w.c. ESP
- 2 hp motor

If the airflow is to be increased to 6,000 cfm, what are the new rotational speed, static pressure, and horsepower values?

Solution:

First, use Fan Law 1 to calculate the new rpm:

$$\text{New rpm} = \frac{\text{New volume flow rate} \times \text{existing rpm}}{\text{Existing volume flow rate}} =$$

$$\frac{6,000 \text{ cfm} \times 1,000 \text{ rpm}}{5,000 \text{ cfm}} =$$

$$\frac{6,000,000}{5,000 \text{ cfm}} = 1,200 \text{ rpm}$$

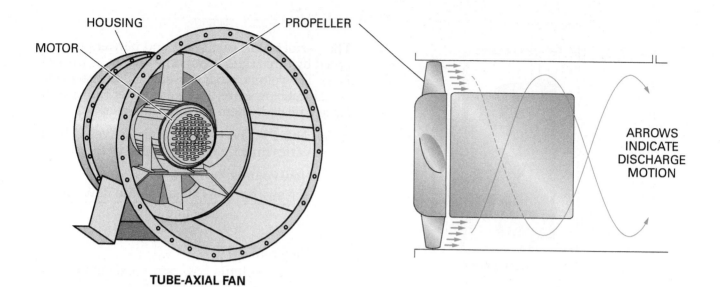

TUBE-AXIAL FAN

ARROWS INDICATE DISCHARGE MOTION

VANE-AXIAL FAN

DISCHARGE MOTION

Figure 24 Tube-axial and vane-axial duct fans.

Use Fan Law 2 to calculate the new static pressure:

New static pressure =

Existing static pressure $\times \left(\dfrac{\text{New rpm}}{\text{Existing rpm}} \right)^2 =$

0.5 in. w.c. $\times \left(\dfrac{1,200 \text{ rpm}}{1,000 \text{ rpm}} \right)^2 =$

0.5 in. w.c. $\times 1.2^2 =$

0.5 in. w.c. $\times 1.44 = 0.72$ in. w.c.

Use Fan Law 3 to calculate the new horsepower:

New hp = existing hp $\times \left(\dfrac{\text{new rpm}}{\text{existing rpm}} \right)^3$

New hp = 0.5 hp $\times \left(\dfrac{1,200 \text{ rpm}}{1,000 \text{ rpm}} \right)^3$

New hp = 0.5 hp $\times 1.2^3$

New hp = 0.5 hp $\times 1.728$

New hp = 0.864 hp

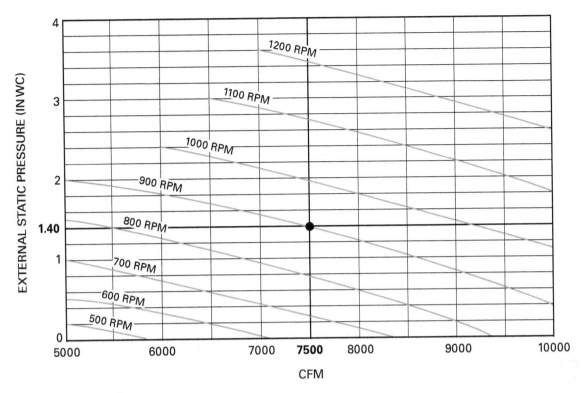

Figure 25 Typical fan curve chart.

Although these calculations aren't needed by an HVACR technician daily, there are times when they are needed to change the airflow characteristics of a system. It is important to understand how significantly one factor affects another. The knowledge alone will help when troubleshooting airflow problems.

2.3.1 Fan Curves

Manufacturer's fan curves are used to show the performance of a fan or blower at different conditions. Refer to *Figure 25*. If the values for any two of the three characteristics shown on the chart are known, the value for the other characteristic can easily be found.

For example, assume that the static pressure of a system is 1.4 in. w.c. and the blower is running at 900 rpm. The intersection of these two lines is represented by the black dot in *Figure 25*. To find the air volume, trace down vertically from the intersection, and read the value of 7,500 cfm. This means that this particular fan can produce 7,500 cfm at these conditions. As the external static pressure and speed change, so does the air volume.

2.4.0 Duct Materials and Fittings

Building code requirements are not completely standardized across the nation. Almost all localities have minimum standards or codes that determine the type of materials and methods that can be used in air distribution systems. The HVACR technician must become familiar with and follow the local codes that apply to each job.

Load calculations determine the heating and cooling loads for a building. From the calculated load, the needed air volume can be determined for area of the building. The size of trunk and branch ducts used in a system is then based on the air volume needed to satisfy the requirements. A variety of duct-sizing charts and calculators can be used to find the correct duct sizes. The needed components can then be selected and/or fabricated.

This section describes the primary components of residential air distribution systems and presents some of the application considerations. The basic components of a duct system include the following:

- Main trunk and branch ducts
- Fittings and transitions
- Air diffusers, registers, and grilles
- Dampers

A wide variety of materials are used in HVAC system duct designs. *Table 2* shows a few of the common materials used to construct duct systems, and where the material would be applied. As the table shows, the most common material used is galvanized steel. The other materials are used for more specific applications such as kitchen exhausts, moist air, and so on. Air shafts, typically found in multi-story buildings, are usually made of concrete or gypsum board.

Duct systems are installed in basements, crawl spaces, attics, and even within concrete floors. Ducts are most often made from metal, fiberglass ductboard, fabric, or plastic materials. Galvanized sheet metal, fiberglass, and flexible duct materials are typically used for comfort systems. When installed in a concrete slab, ducts are usually made of metal or plastic. Where weight is a factor, aluminum or fabric duct can be used.

2.4.1 Galvanized Steel Ducts

Galvanized steel or sheet metal duct can be round, oval, or rectangular (*Figure 26*). All three shapes may be used in the same duct system. Popular sizes of round and rectangular steel duct, along with an assortment of standard fittings, can be acquired from HVAC supply houses, with some

assembly required. Many HVAC shops are also capable of fabricating most of the components needed to construct a system.

Because sheet metal duct is rigid, the layout must be well planned, and all the pieces cut precisely. An appropriate thickness of metal must be used for a given duct size to maintain its structural integrity. The thickness of sheet metal is expressed as a gauge. Galvanized sheet metal is typically available from 8 to 31 gauge (ga.). Higher gauges indicate thinner metal, as shown in *Table 3*. Galvanized sheet metal heavier than 8 ga. is called out by its actual thickness in inches. Note that the thickness of metals such as stainless steel and aluminum are different from that of galvanized steel.

Larger ducts are made from heavier gauges than smaller ducts. This helps reduce "drumming" and making popping noises when the

Did You Know?

Fabric Duct

Ductwork made of heavy-duty fabric has become popular in open-ceiling environments such as factories, warehouses, restaurants, and sports arenas. This ductwork is available in round and half-round configurations. The latter is ideal for surface-mount installations. Duct sections can be zippered together to obtain the required length. Fittings such as tees and elbows are also available. The air is diffused either through the fabric itself, or through small openings placed along its length. In many cases, cleaning the duct consists of removing it from its hangers and throwing it into a commercial washing machine. They are particularly well-suited for areas where corrosive materials and excessive moisture represent a problem for metal duct systems, such as indoor swimming pools. Compared to metal duct systems, the installation is very simple and far less labor intensive.

Figure Credit: Courtesy of DuctSox Corporation

Table 2 Duct Materials and Their Applications

Material	Applications
Galvanized steel	Widely used for most HVAC systems
Aluminum	For systems with high moisture-laden air or special exhaust systems
Stainless steel	For kitchen exhaust, fume exhaust, or high moisture-laden air
Concrete	Underground ducts and air shafts
Rigid fibrous glass	Interior, low-pressure HVAC systems
Gypsum board	Ceiling plenums, corridor ducts, and air shafts

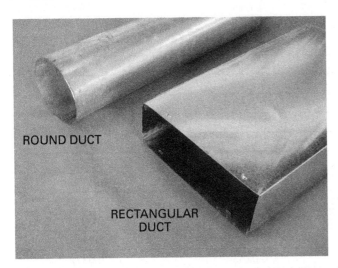

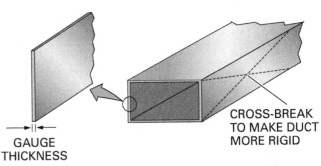

Figure 26 Metal ducts.

system blower starts and stops, in addition to improving the structural integrity. Lines or ridges, called *cross-breaks*, are also used on large sheet metal panels or ducts to make them more rigid and resistant to popping in and out. A cross-break is shown in *Figure 26*.

The aspect ratio of a duct is often used to classify duct sizes and estimate its cost. Aspect ratio is the ratio of the duct's width to its height. For example, if a duct is 18" wide and 6" high (46 cm × 15 cm), the aspect ratio is 18:6, or 3:1. *Table 4* shows some typical gauge thicknesses used for metal ducts. It also shows a tabulation of duct aspect ratios.

Sections of square or rectangular duct are assembled using any one of several methods. S-slip and drive connectors (*Figure 27*) are commonly used for smaller rectangular duct. Standing seams are quite waterproof, and standing-S seams add additional strength and rigidity to the joint. Some prefabricated duct may be assembled using a snap-lock approach (*Figure 28*). In addition to the connection methods shown here, there are other styles used for larger and heavier duct. Round duct sections are normally fastened together with self-tapping sheet metal screws. One end of the joint is crimped to allow it to slip inside the other piece. When additional sealing

is needed, the joint can be sealed with a flexible duct-sealing compound applied with a paint brush or caulking gun. Leaking duct joints reduce the amount of air available for delivery to the outlets and waste energy.

A ductwork system must be well supported so that it does not move significantly or fall. If it is not properly supported, movement can occur when the fan starts, which causes a surge of air through the system. Long runs of duct may create stress due to thermal expansion and contraction as it heats and cools. This type of movement can be accommodated by using flexible or fabric joints at different points in the system.

Metal duct systems can transmit vibrations from the air handling equipment. Vibration transmission can be prevented by using flexible connectors or fabric joints at the connections to the air handler. *Figure 29* shows how flex connectors are applied.

2.4.2 Fiberglass Ductboard

Fiberglass ductboard can be used instead of metal duct in some areas and applications. Its surface creates more friction loss than metal duct, but it is quieter because it absorbs more noise. It is also self-insulating, while metal duct must be insu-

Table 3 Standard Sheet Metal Gauges (Inches)

Gauge	Galvanized Steel	Aluminum	Stainless Steel
3	—	0.2294	0.2500
4	—	0.2043	0.2344
5	—	0.1819	0.2187
6	—	0.1620	0.2031
7	—	0.1443	0.1875
8	0.1680	0.1285	0.165
9	0.1532	0.1144	0.1562
10	0.1382	0.1019	0.1406
11	0.1233	0.0907	0.1250
12	0.1084	0.0808	0.1094
13	0.0934	0.0720	0.0937
14	0.0785	0.0641	0.0781
15	0.0710	0.0571	0.0703
16	0.0635	0.0508	0.0625
17	0.0575	0.0453	0.0562
18	0.0516	0.0403	0.0500
19	0.0456	0.0359	0.0437
20	0.0396	0.0320	0.0375
21	0.0366	0.0285	0.0344
22	0.0336	0.0253	0.0312
23	0.0306	0.0226	0.0281
24	0.0276	0.0201	0.0250
25	0.0247	0.0179	0.0219
26	0.0217	0.0159	0.0187
27	0.0202	0.0142	0.0172
28	0.0187	0.0126	0.0156
29	0.0172	0.0113	0.0141
30	0.0157	0.0100	0.0125
31	0.0142	0.0089	0.0109
32	0.0134	0.0080	0.0102
33	—	0.0071	0.0094
34	—	0.0063	0.0086
35	—	0.0056	0.0078
36	—		0.0070

Proprietary Connectors

Proprietary connectors are unique connection systems that have been invented and patented by their manufacturer. Such duct connection systems are often used with large ductwork and ductwork that requires a very tight seal. The illustrations show examples of two such duct connection systems that are very popular for commercial and industrial duct applications.

Table 4 Typical Metal Duct Gauge Thickness and Aspect Ratios

Rectangular Duct Width in Inches	Commercial		Residential
	Sheet Metal Galvanized	Aluminum	Sheet Metal Galvanized
UP TO 12	26	0.020	28
13–23	24	0.025	26
24–30	24	0.025	24
31–42	22	0.032	–
43–54	22	0.032	–
55–60	20	0.040	–
61–84	20	0.040	–
85–96	18	0.050	–
OVER 96	18	0.050	–

RECTANGULAR DUCT

Round Duct Diameter	Commercial Sheet Steel Galvanized Gauge	Residential Sheet Steel Galvanized Gauge
UP TO 12	26	30
13–18	24	28
19–28	22	–
27–36	20	–
35–53	18	–

ROUND DUCT

Duct Class (Aspect Ratio)	Width in Inches	Perimeter
1	6–8	24–72
2	12–24	36–72
3	26–40	70–106
4	24–88	60–220
5	48–90	116–216
6	90–145	210–336

ASPECT RATIO

lated when significant temperature differences are expected. Fiberglass ductboard may even be used to insulate metal duct in some applications. Fiberglass board is available in flat sheets for fabrication, or as prefabricated round sections. Fiberglass board is commonly sold in 1", 1.5", or 2" (25, 38, and 51 mm) thicknesses with a foil backing. The backing is reinforced with fibers to make it strong. The inside surface of the ductboard is coated to prevent the erosion of the duct fibers into the supply air. Fiberglass particles released into the air can be harmful to health.

Duct is fabricated from the sheets of fiberglass using special knives or programmable cutting machines. The material cuts very easily and neatly. The knives cut away the fiberglass while leaving the foil backing intact to form laps (*Figure 30*). When two pieces are fastened together, the lap can be stapled to the adjoining board using special staples. Special tape is then applied to make the joint airtight.

Once the ductboard has been assembled, it must be sealed using the appropriate closure system, in accordance with applicable codes and job specifications. Most tape available at the local hardware store is not acceptable for this use. Only closure systems that comply with *UL Standard 181A* are suitable for use with rigid fiberglass duct systems. These closure systems include the following:

- Pressure-sensitive aluminum foil tapes listed in *UL 181A, Part (P)*
- Heat-activated aluminum foil or scrim tapes listed in *UL 181A, Part (H)* (This is the preferred method.)
- Mastic and glass fabric tape closure systems listed in *UL 181A, Part (M)*

Round fiberglass duct is easy to install because it can be cut to length with a knife, but it is rarely used. Fiberglass duct systems must be properly supported or they will sag over long runs. Special hangers with a wide surface area must be used to avoid cutting into the duct. The requirements for hanger spacing are also different from those of sheet metal ducts.

One disadvantage of fiberglass duct is that it is not as sturdy as sheet metal duct. Therefore, it cannot be installed in areas where it might be subject to physical damage. It also cannot withstand higher pressures like metal duct.

2.4.3 Flexible Duct

Flexible round duct (*Figure 31*) comes in sizes up to 24" (61 cm) in diameter. It comes wrapped with insulation protected by a vapor barrier made of fiber-reinforced vinyl or foil backing for use in unconditioned areas.

Flexible duct is typically used in spaces where obstructions make rigid duct difficult or impossible to install. A common use is to connect a branch line to the main duct. Flexible duct is easy to route around corners and other bends. Duct runs should be kept as short and as straight as possible. Gradual bends should be used, since

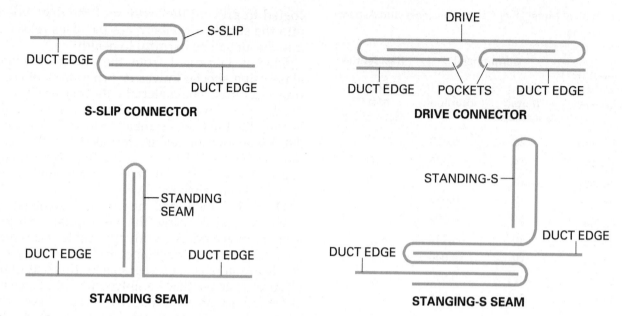

Figure 27 Common rectangular metal duct connection methods.

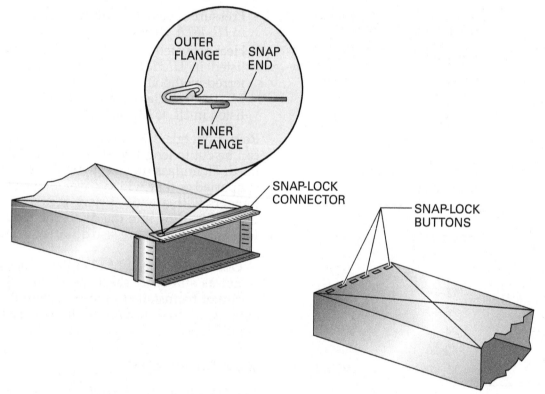

Figure 28 Snap-lock duct connection.

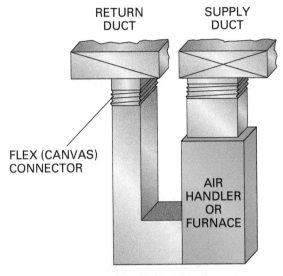

RETURN DUCT   SUPPLY DUCT

FLEX (CANVAS) CONNECTOR

AIR HANDLER OR FURNACE

VIBRATION/NOISE CONTROL AT AIR HANDLER

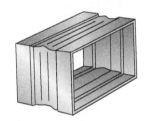

VIBRATION/NOISE/MOVEMENT CONTROL IN DUCT RUNS

Figure 29 Duct flex connectors.

tight turns can greatly reduce the airflow. If a connection to a ceiling diffuser needs an elbow, it is better to use an insulated metal elbow at the input to the diffuser than to bend flexible duct tightly to form the connection. This is because the elbow is smoother inside, and will not collapse.

Long runs of flexible duct are not recommended unless the friction loss is taken into account. Even when properly installed, most flex ducts cause at least two to four times as much resistance as an equal metal duct. The material should be stretched when installed in a straight run, but not too tightly. This helps reduce the internal friction loss. To avoid sags in the run, flexible duct should be amply supported with 1" (2.5 cm) wide or wider bands to keep the duct from collapsing and reducing the inside dimension. Some flexible duct comes with built-in eyelet holes in the seam for hanging.

Flexible duct is easy to install and provides good noise and vibration attenuation. Flex duct branches should never exceed 12' (3.7 m) in length; 6' (1.8 m) is a more desirable goal. Local codes often dictate the maximum length of each flexible-duct run.

2.4.4 Fittings and Transitions

Duct fittings such as elbows, takeoffs, and boots change the direction of airflow or change its velocity. Transitions are typically used to change

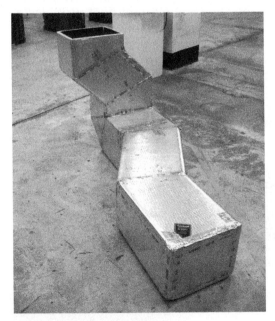

Figure 30 Fiberglass ductboard joint.

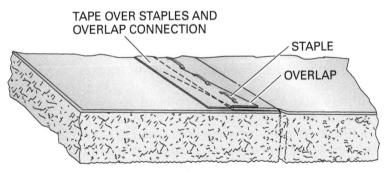

TAPE OVER STAPLES AND OVERLAP CONNECTION

STAPLE

OVERLAP

Health Issues

At one time, there were three major concerns about ductboard. The first was the potential for the porous material to absorb moisture and allow the growth of mold and fungus. This could cause severe health problems for building occupants. The second concern was the difficulty in cleaning ductboard, since its surface tends to collect and hold dust and dirt. The third concern was the potential for the glass fibers to loosen from the duct and be swept into the air stream. These concerns have been largely eliminated by biocides in the fiberglass to prevent the growth of microorganisms, and by the application of an interior coating.

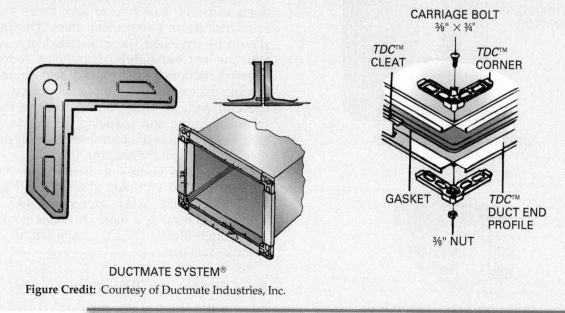

DUCTMATE SYSTEM®

Figure Credit: Courtesy of Ductmate Industries, Inc.

from one size duct to another. They are also used to change from one duct shape to another. *Figure 32* shows examples of these fittings.

Air moving in a duct has inertia that makes it tend to flow in a straight line. Changing the course of airflow increases the friction loss, since the air must flow against a duct surface to encourage the turn. Each fitting in a duct run adds friction. It takes energy to overcome the added resistance of the fittings. Because of this, the number of fittings used in a duct system must be minimized to reduce the total amount of friction in the system.

Fittings and transitions are related to some equivalent feet of duct length. This means that each fitting creates a pressure drop equal to a certain number of feet of straight duct of the same size. Therefore, adding fittings has the same effect as increasing its overall length. The use of unnecessary fittings must be avoided.

For each standard type of fitting, the friction loss has been converted to the equivalent feet of duct length. This information is available in a set of charts available in American Society of Heating, Refrigerating, and Air Conditioning Engineers (ASHRAE) and Sheet Metal and Air Conditioning Contractors' National Association (SMACNA) publications.

The total equivalent length for a duct run is calculated by adding all the equivalent lengths for fittings to the length of straight duct used. *Figure 33* shows several types of elbows as an example of how their equivalent lengths change. Also shown is an elbow with an equivalent length of 30' (9.1 m) added to two sections of duct totaling 100' (30.5 m) of straight length. The resulting pressure drop is the same as the pressure drop of a straight duct 130' (39.6 m) long.

Figure 31 Flexible duct.

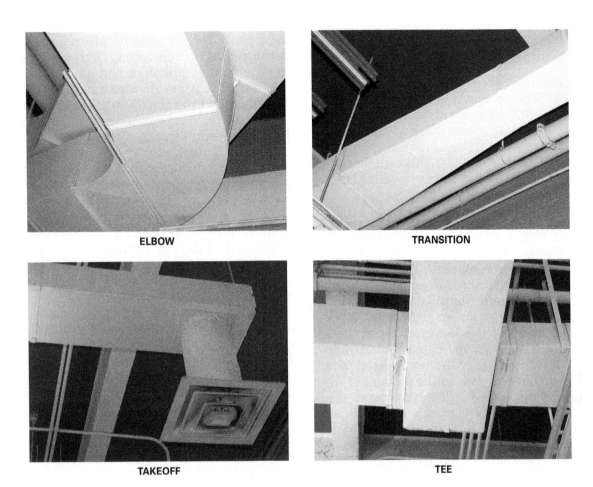

Figure 32 Duct fittings.

E = 5'   F = 10'   G = 30'   H = 15'   I = 30'

TURNING VANES

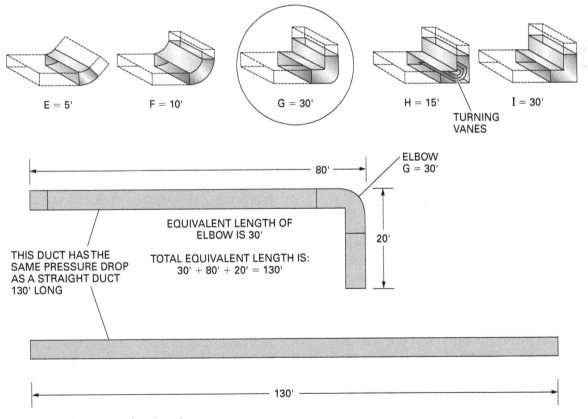

ELBOW
G = 30'

80'

20'

THIS DUCT HAS THE
SAME PRESSURE DROP
AS A STRAIGHT DUCT
130' LONG

EQUIVALENT LENGTH OF
ELBOW IS 30'

TOTAL EQUIVALENT LENGTH IS:
30' + 80' + 20' = 130'

130'

Figure 33 Example of equivalent length.

2.5.0 Diffusers, Registers, Grilles, and Dampers

Air outlets distribute the supply air into the conditioned space. When properly selected, they blend the supply air with the room air so that the room is comfortable without excess noise or drafts. The terms *diffuser, register,* and *grille* are used to describe different kinds of outlets. Although they are often used interchangeably in conversation, they each have their own specific definition:

- *Diffuser* – A diffuser is an outlet that discharges supply air into a room in a widespread fan-shaped pattern.
- *Register* – A register is an outlet that discharges supply air into a room in a more concentrated, non-spreading pattern. Many have one-way and two-way adjustable air stream deflectors.
- *Grille* – A grille is the louvered covering of an opening created for the passage of air into a room. It controls the distance, height, and spread of airflow, as well as the amount of air delivered to the space. Grilles have many different designs. Some are fixed and can direct air in one direction only. Others are adjustable and can be set to send air in different directions. Grilles with no adjustments are typically used as covers for the return air duct.

Figure 34 shows common registers and diffusers. As shown, the floor register is relatively long and narrow. It gives excellent performance for both heating and cooling when used in perimeter duct systems. A floor register is usually installed parallel to the room's outside wall, about 6" to 8" (15 to 20 cm) away from the wall. Floor registers are fed from below and discharge air upward. Typically, fixed vanes are installed crosswise at an angle to spread the air stream. Floor registers normally have a built-in shutoff and balancing damper.

Low-sidewall registers are excellent for heating when used in perimeter distribution systems. They also work well for cooling, if designed to discharge air upward. If not so designed, a plastic air scoop accessory may be added to redirect the supply air upward. Low sidewall registers are fed from the back and mounted flush with the wall just above the baseboard trim. Some are made to discharge air in two or three directions. Low sidewall registers normally include a built-in shutoff damper.

High-sidewall registers provide poor heating performance in cold climates because they tend to leave a cold layer of air in the lower half of the room. However, they are adequate for heating in warmer climates. They provide good performance in cooling when used with central returns. When used with room-by-room returns, the cooling performance is even better. High sidewall registers are mounted flush on the room's inside wall. The top edge is usually mounted 6" to 12" (15 to 30 cm) down from the ceiling. They typically discharge air horizontally toward the outside wall. High sidewall registers normally include a built-in shutoff damper.

Baseboard diffusers are long and narrow. They are mounted on the floor with their back against the wall. Supply air is fed from below and discharged upward close to the wall to cover a wide section of the outside wall and windows. Baseboard diffusers normally have a built-in shutoff damper. They perform well in both heating and cooling when used with perimeter duct systems.

Ceiling diffusers can be round, square, or rectangular. The vast majority of them are found in commercial buildings. Some mount flush to a hard ceiling, while others are surface-mounted. They are often cut into acoustic ceilings, with their weight supported from the T-bars. The supply air is fed from above. Ceiling diffusers can distribute supply air equally in all directions, or they can supply air in one, two, or three directions only. Accessory volume dampers are a common option. They perform very well for cooling and are adequate for heating. However, they do allow warm air to stratify in the upper part of a room, like most systems that deliver warm air from above.

Manufacturers provide tables that show the performance of their grilles, registers, and diffusers. *Table 5* shows an example of the performance data for a square ceiling diffuser. The table provides the designer with the pressure loss for a given velocity of air exiting the diffuser. This is known as the *face velocity*. In addition, the air volume is also provided, along with the throw. The term *throw* refers to the distance the air will travel when it exits the diffuser at the face velocity shown. An air velocity of 50 fpm (0.254 m/s) is considered the point where the air is no longer being thrown with any real force.

Designers usually reference the chart with a desired air volume in mind. For example, assume that a copier room requires an air supply of 400 cfm (680 m³/h). According to the chart (**Table 5**), a 14" (36 cm) neck size will provide 410 cfm at face velocity of 500 fpm, a pressure drop of 0.16 in. w.c., and a throw of 11.5'. The neck size refers to the size of the duct connection. A 16" (41 cm) diffuser can provide the same airflow at a face velocity of 400 fpm and a throw of 10'. In other words, the 16" diffuser will move the same amount of air a bit more softly.

Table 5 Performance Data for a Square Diffuser

Face Velocity (fpm)		300	400	500	600	700	800	900	1,000
Pressure Loss (in w.c.)		0.006	0.010	0.016	0.022	0.031	0.040	0.050	0.062
Neck Size 6"	CFM	50	65	85	100	115	130	150	165
A_k 0. 165	Throw	3.5	4.5	5.5	6.5	8.0	9.0	10.0	11.0
Neck Size 8"	CFM	85	110	140	170	195	225	250	280
A_k 0.280	Throw	4.5	5.5	7.0	8.5	10.0	11.0	12.0	14.0
Neck Size 10"	CFM	125	170	210	250	295	335	380	420
A_k 0.420	Throw	5.0	6.5	8.0	9.5	11.5	13.0	15.0	16.0
Neck Size 12"	CFM	180	240	300	355	415	475	535	595
A_k 0.595	Throw	6.0	8.0	10.0	11.5	13.5	15.5	17.5	19.0
Neck Size 14"	CFM	245	330	410	490	575	655	740	820
A_k 0.820	Throw	7.0	9.0	11.5	13.5	16.0	18.0	20.0	22.5
Neck Size 16"	CFM	310	410	515	620	720	825	925	1,030
A_k 1.030	Throw	7.5	10.0	12.5	15.0	18.0	20.0	22.0	25.0
Neck Size 18"	CFM	400	530	665	800	930	1,065	1,200	1,330
A_k 1.330	Throw	8.5	11.0	14.0	17.0	20.0	23.0	26.0	28.0
Neck Size 20"	CFM	480	640	800	960	1,120	1,280	1,440	1,600
A_k 1.600	Throw	9.5	12.0	16.0	18.0	22.0	25.0	28.0	31.0
Neck Size 22"	CFM	570	760	950	1,140	1,330	1,520	1,710	1,900
A_k 1.900	Throw	10.5	13.5	17.0	19.0	24.0	27.0	30.0	33.0
Neck Size 24"	CFM	690	920	1,150	1,380	1,610	1,840	2,070	2,300
A_k 2.30 0	Throw	11.0	14.5	18.5	22.0	26.0	30.0	33.0	36.0

Terminal velocity of 50 fpm.

FLOOR

CEILING

BASEBOARD

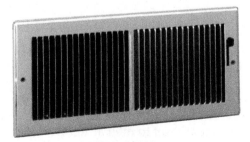

SIDEWALL

Figure 34 Registers and diffusers.

A face velocity of 500 fpm (2.54 m/s) from a ceiling diffuser is usually considered a good velocity for personal comfort. Higher velocities and longer throws allow for greater mixing of the room air, but with higher pressure drops and a risk of the room feeling drafty to occupants. The final selection is made based on the designer's knowledge of the room use and the desired effect. Note that each diffuser also has an A_k factor assigned. The A_k factor is used to determine the air volume from an outlet by multiplying it by the average velocity of the air leaving the outlet. The A_k factor represents the free area of the grille in square feet. Air balancing is presented in detail in another module.

2.5.1 Air Dampers

Dampers are used to control and balance airflow in duct systems. Without them, some rooms receive too much air while others do not receive enough. This is due to the common imperfections in system design and installation.

Some dampers are manually adjustable. Others, often used in zoned heating or cooling systems, are motorized, or have actuators that change their position in response to system controls. Sometimes dampers are used to influence the mixture of two airflows from different sources. Fire and smoke dampers help to protect building occupants in emergency situations.

A balancing damper should be installed in an accessible place in each branch supply duct. The closer the dampers are to the main duct or supply air plenum, the better. Branch take-offs are the ideal location. Balancing dampers should be tight fitting with minimum leakage possible around the blade. The built-in or accessory dampers on supply diffusers and registers should not be used to balance an air system. When partially closed, they disrupt the performance of the diffuser or register and may increase noise significantly. These dampers should only be used by room occupants to make minor adjustments in the airflow. *Figure 35* shows three types of dampers used in air distribution systems.

2.5.2 Fire and Smoke Dampers

Fire dampers are used to maintain the fire-resistance ratings of walls, partitions, and floors penetrated by HVAC ducts. Fire dampers also prevent the spread of fire. The dampers are normally held open with a fusible link (*Figure 36*) that is usually set to melt at 165°F (78°C). Dampers with ratings of 212°F (100°C) and 286°F (141°C)

are also available. If a fire occurs, the link melts and the dampers close with significant force by spring action.

Fire dampers usually have a fire resistance rating of either 1½ or 3 hours. Building partitions with a 3-hour fire rating require the use of a 3-hour rated damper while partitions with fire ratings of less than 3 hours would require the use of 1½ hour rated dampers. The rating for the fire damper must be at least 75 percent of the fire rating for the wall, floor, or partition. Therefore, a fire damper rated for 1½ hours can be used for a barrier rated for up to 2 hours—a common fire rating for walls. Likewise, a damper rated for 3 hours can be used in barriers with fire ratings up to 4 hours.

In addition to the fire-resistance rating, fire dampers have a static or dynamic air-closure rating. Static-rated dampers can only be used in HVAC systems where the HVAC equipment is automatically shut down in case of fire. In that case, no air should be flowing within the ducts. Dynamic-rated dampers are designed to close even if the HVAC system remains running and air is moving through the ducts. They are capable of positive closure despite the added air pressure.

Smoke can be as deadly as fire, so controlling the spread of smoke in a building is critical. Smoke dampers in the ducts perform this critical task. Smoke dampers can be passive in their function where they operate to simply shut off and isolate a section of duct. They also can be part of an engineered smoke control system that directs smoke outdoors once the fire has been extinguished.

Most smoke dampers are operated electrically and are controlled by a smoke or heat detector, fire alarm, or automated building control system. Smoke dampers are rated for leakage in accordance with *UL Standard 555S, Leakage Rated Dampers for Use in Smoke Control Systems*. Class 1 has the lowest leakage rating, and Classes 2 and 3 having higher leakage ratings. Smoke dampers have temperature, air velocity, and air pressure ratings that show how they will perform in various conditions.

Combination fire and smoke dampers (*Figure 37*) are also available. These dampers must conform to the rating agency requirements for both fire and smoke dampers. The use of fire and smoke dampers is governed by *NFPA 90, Installation of Air Conditioning and Ventilation Systems*.

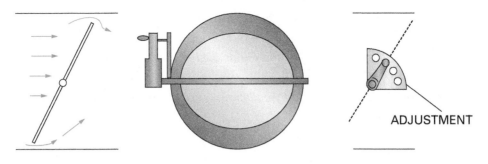

BUTTERFLY DAMPER

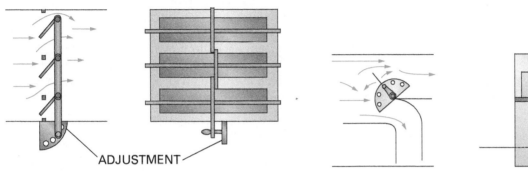

MULTIPLE-VANE DAMPER

SPLITTER DAMPER

Figure 35 Typical air dampers.

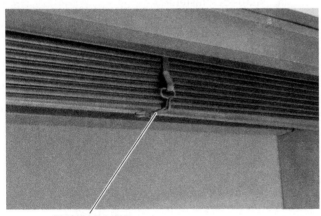

FUSIBLE LINK

Figure 36 Fire damper fusible link.

Figure 37 Combination fire and smoke damper.

Additional Resources

Air Distribution Systems: Introduction to Thermo-Fluids Systems Design, A. G. McDonald and H. L. Magande. 2012. Chichester, UK: John Wiley & Sons, Ltd.

NFPA 90, Installation of Air Conditioning and Ventilation Systems. Current edition. Quincy, MA: National Fire Protection Association.

2.0.0 Section Review

1. Which type of blower is commonly used in residential systems?

   a. Backward-inclined centrifugal
   b. Forward-curved centrifugal
   c. Radial
   d. Vane-axial

2. The type of fan usually housed in a short venturi is the _____.

   a. duct fan
   b. tube-axial fan
   c. vane-axial fan
   d. propeller fan

3. An air distribution system is equipped with a ½ hp motor and a belt-driven blower. The system is moving 1,200 cfm of air with the blower turning at 950 rpm. The system static pressure is 0.38 in. w.c. If the belt loosens and the blower speed slows to 790 rpm, what will the new air volume be, in cfm? (Round your answer to the nearest whole number.)

   a. 625 cfm
   b. 998 cfm
   c. 1,332 cfm
   d. 1,443 cfm

4. A homeowner wants to add air conditioning to an existing warm-air heating system. The existing system moves 870 cfm of air at a static pressure of 0.37 in. w.c., and has a ⅓ hp motor driving a blower turning at 1,050 rpm. If the system needs to move 1,000 cfm of air to support the new air conditioning system, what motor horsepower will be required? Round your answer to two decimal places.

   a. 0.29 hp
   b. 0.37 hp
   c. 0.51 hp
   d. 0.61 hp

5. Round metal duct sections are normally connected using _____.

   a. snap-lock fasteners
   b. duct tape
   c. sheet metal screws
   d. drive clips

6. An air outlet device that directs air in a concentrated non-spreading stream is a _____.

   a. diffuser
   b. register
   c. louver
   d. damper

SECTION THREE

3.0.0 AIR DISTRIBUTION SYSTEM DESIGN AND ENERGY CONSERVATION

Objective

Identify the different approaches to air distribution system design and energy conservation.

  a. Identify various air distribution system layouts.
  b. Describe heating and cooling air movement resulting from various air distribution system designs.
  c. Explain how to maximize energy efficiency through the proper sealing and testing of air distribution systems.

Trade Terms

R-value: A number, such as R-19, that is used to indicate the ability of insulation to resist the flow of heat. The higher the R-value, the better the insulating ability.

Transfer grille: A grille usually installed in walls or doors, with a grille of the same size mounted on each side, that allows air to pass freely in or out of an enclosed space.

Air distribution systems typically consist of a supply duct system and a return duct system. The supply duct is usually more complex in its layout. The supply duct system distributes the air in the conditioned space. The return duct system collects air from the conditioned space and returns it to the blower. Air distribution system design depends on the structure and its intended use. Large and specialized systems can be quite complex.

Since air distribution in residential applications is more uniform and relatively simple, they are used as the basis for learning. Except for the system layout and size of the components, the principles of operation and the types of parts used in duct systems are basically the same for many commercial applications.

To size the system, a designer begins with a heat load analysis. This determines the heat gain or loss for the structure and its individual areas. To create the analysis, the designer divides heat gain and heat loss factors into two categories: external and internal.

External factors include the outside air temperature range, the relative humidity, and the position of the building relative to the sun. Internal factors include lights, mechanical equipment, office machines, the number of people in the building, and the operating schedule. These factors vary based on the type and size of the building, how it is used, and where it is located.

Major factors affecting the design approach are building construction and climate. For practical reasons, building construction is usually the dominant factor. The designer must also consider other factors such as the applicable building codes and overall cost. The best system design is one that balances performance with cost while ensuring a safe and reliable installation.

3.1.0 Air Distribution System Layouts

The type of duct system used in a building is determined by the climate and the building construction features. Most homes use perimeter duct systems where construction features allow it. Perimeter systems have floor or ceiling diffusers near the building walls. They can be applied in homes with crawl spaces or attics, and even to homes built on concrete slabs. In multi-story homes, the perimeter outlets may be in the ceiling instead of the floor. Floor outlets provide a better combined heating and cooling performance in cold climates than ceiling outlets.

Air handlers and furnaces in the attic may be suspended from the rafters with vibration isolators to keep vibration from being transmitted to the structure. They may also be laid on the attic floor. The unit is usually placed in a horizontal position, but an attic with enough height may allow it to be installed in an upflow or counterflow position (*Figure 38*).

Figure 38 Upflow furnace in an attic.

The main ducts can be made from fiberglass ductboard or insulated sheet metal with a vapor barrier. Branch ducts are usually round flexible duct or round sheet metal covered with foil-faced insulation. Remember that the excessive use of flexible duct may result in poor airflow. Branch lines longer than 6' (1.9 m) should start at the main trunk as metal, and then transition to flexible duct within 6' of the terminal. The return grille is generally in a central hallway ceiling or wall, but larger installations often require more than one for best performance.

Perimeter systems can have various layouts. Four common layouts are as follows:

- Perimeter loop
- Radial
- Extended plenum
- Reducing trunk

3.1.1 Perimeter-Loop Duct System

Perimeter-loop duct systems (*Figure 39*) are used in structures built on concrete slabs. They offer little advantage for homes of any other construction. The perimeter loop is a continuous duct of consistent size embedded in the slab. It runs close to the outer walls, with the outlets located in the floor. The loop is fed by several branches from the plenum. When the furnace fan is running, the warm air in the ducts helps keep the entire slab warmer. Heat loss to the outside is reduced by insulation placed around the edges of the slab. The loop has a consistent pressure with equal pressure applied to all outlets.

3.1.2 Radial Duct System

In a radial duct system, also shown in *Figure 39*, each perimeter outlet is fed from directly from the supply air plenum. Also known as *spider systems* or *plenum systems*, radial duct systems are used in small homes or additions. The ductwork can be installed in a concrete slab, crawl space, or attic. The location of the ductwork determines the location of the supply outlets. Volume control dampers should be used in all branch duct runs to facilitate balancing the system airflow. This may not be possible if the duct is encased in a slab floor. A single return air grille is typically used. Radial duct systems are economical to install but not very common due to size limitations, noise, and questionable performance unless the structure is very small.

3.1.3 Extended-Plenum Duct System

The extended-plenum system (*Figure 40*) uses round or rectangular trunks for the main supply and return. The supply ducts remain the same size over their entire length. These systems can be installed in a crawl space or attic, but are not recommended for slab floors. This is another situation where a slab floor does not provide access to balancing dampers. Like radial systems, dampers for balancing are essential for acceptable performance.

Separate branch ducts run from the main trunk to each supply outlet. The extended-plenum system works best when the air handler is in the center of the main duct's length. However, the supply duct can also be routed in one direction only. A volume damper should be installed in each branch at the takeoff. This allows the airflow to be balanced since the trunk duct has no transitions in size to help balance the airflow. Without volume dampers, system performance is poor.

An extended-plenum system requires careful duct sizing and balancing dampers, but can provide reasonable performance in small systems. In a larger structure that requires a significant amount of duct to reach supply outlets, it is a poor choice. The following are some recommended practices for laying out an extended-plenum duct system:

- The supply and return ducts should extend no more than 24' (7.3 m) from the air handler.
- The first branch should be at least 18" (46 cm) from the beginning of the main duct.
- The main trunk should extend at least 12" (30 cm) past the last branch takeoff.
- Balancing dampers should be installed at each branch takeoff.

3.1.4 Reducing-Trunk Duct System

A reducing-trunk system (*Figure 41*) provides the best performance but at a slightly higher cost. It works well in larger buildings that require longer duct runs. Remember that extended-plenum systems are limited to 24' (7.3 m). It is a better choice for systems where the air handler is installed on one end of the main trunk duct rather than in the middle. Indeed, it is a better choice for most applications. When properly sized, the same pressure drop is maintained from one end of the duct system to the other. This allows all branches to operate at roughly the same pressure. When it is properly designed and installed, it requires less

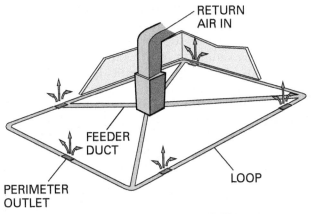

RETURN
AIR IN

FEEDER
DUCT

LOOP

PERIMETER
OUTLET

(A) PERIMETER LOOP SYSTEM

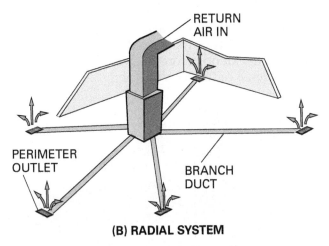

RETURN
AIR IN

PERIMETER
OUTLET

BRANCH
DUCT

(B) RADIAL SYSTEM

Figure 39 Two types of perimeter duct systems.

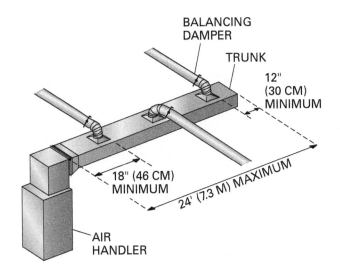

BALANCING
DAMPER

TRUNK

12"
(30 CM)
MINIMUM

18" (46 CM)
MINIMUM

24' (7.3 M) MAXIMUM

AIR
HANDLER

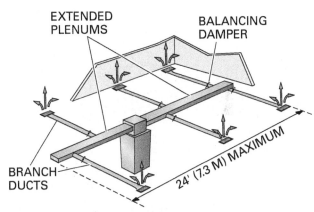

EXTENDED
PLENUMS

BALANCING
DAMPER

BRANCH
DUCTS

24' (7.3 M) MAXIMUM

Figure 40 Extended-plenum system.

effort to balance. The following are some recommended practices for laying out a reducing trunk duct system:

- The first main duct section should be no longer than 20'.
- The length of each reducing section should not exceed 24'.
- The first branch duct connection down from a single-taper transition should be at least 4' from the beginning of the transition fitting. This distance allows the air turbulence caused by the fitting to subside before the air is sent into the next branch duct. If the distance is less than 4', the branch ducts near the transition can be hard to balance because the air may tend to bypass them.
- The trunk duct should extend at least 12" from the last branch duct.

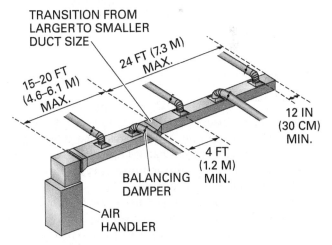

TRANSITION FROM
LARGER TO SMALLER
DUCT SIZE

24 FT (7.3 M)
MAX.

15–20 FT
(4.6–6.1 M)
MAX.

12 IN
(30 CM)
MIN.

4 FT
(1.2 M)
MIN.

BALANCING
DAMPER

AIR
HANDLER

Figure 41 Reducing trunk system.

3.1.5 Furred-In Duct Systems

Some residential structures, such as apartments, have neither a crawl space nor an attic. When independent systems are provided for each apartment, the air distribution system is often furred-in (*Figure 42*). To do so, the ceiling is lowered in the area where the duct is installed. This is usually done in a hallway. With the registers and diffusers in the high sidewall of the adjacent rooms, the ceiling in the other areas can remain at a normal height. This approach can be used with extended-plenum and reducing trunk systems. Note, however, that this type of system does not carry the supply air to the perimeter of the building, although the supply is generally blown in that direction.

The air handler for a furred-in system is often placed in a closet. However, small fan coil units that are capable of supporting small systems are made for installation in the same furred-in space (*Figure 43*). A return air grille is then installed immediately behind the unit. For these units, as well as air handling units installed in closets, the return duct may only be a few inches or centimeters long. Air handling units in closets are usually connected directly to a grille on the opposite side of a closet wall. They may also simply draw air through a ventilated closet door or non-ducted wall grille. In this case, the closet is acting as a return air plenum.

The registers are placed on the high sidewall of the adjoining interior walls. As a result, the branch line connections are very short, extending only to the walls of the furred area. If the supply duct length extends more than 24' (7.3 m) from the air handler, a reducing trunk should be used.

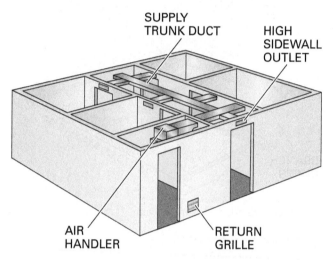

Figure 42 Furred-in duct system.

Figure 43 Fan coil unit designed for furred-in installation.

Radon

Radon is a radioactive gas that is colorless, odorless, and tasteless. These characteristics make it difficult to know when it is present. Without evidence, it is out of the minds of homeowners in many areas. Radon forms as uranium goes through its natural radioactive decay process. There are many approaches to dealing with a known radon problem. One basic approach is to capture gases from the soil beneath a home and exhaust them outdoors, before they enter the structure.

Homes with an air distribution system in a concrete slab floor are more exposed to the potential of radon gas entry, again without the occupants being aware of it. Before installing a system in a slab, radon testing should be done to determine if there is a problem that requires a solution. Although it is important that ducts in the slab be properly sealed, the presence of radon makes it crucial to the health of the occupants. The entry of radon gas into the return air ducts is most likely, since they operate at a reduced pressure that tends to draw soil gases in through leaks in the duct.

3.2.0 Heating and Cooling Room Airflow

The layouts discussed in the previous section produce different airflow in the areas they serve. Understanding how the airflow differs from system to system helps to determine which layout is best for an application. It also helps when a technician must investigate comfort-related complaints.

The airflow from a perimeter system sweeps the outside walls and windows. This helps offset the heat losses and gains associated with the walls, windows, and doors at the source. The return air grilles are located on the interior walls or ceilings in a central location. Central returns are the most common. For improved performance, individual returns can be installed in each room. However, individual room returns increase the installation cost of a system significantly, and are rarely chosen as a result.

The location of the return grilles in a central location helps to draw air from the perimeter toward the center of the building. To allow air to leave the rooms at the same rate as it enters, doors must remain open or be undercut. Alternatively, a transfer grille can be placed in the wall or door. However, transfer grilles have a negative effect on acoustics—voices can easily be carried from room to room. Without a way for room air to return to the system, the supply air volume will drop sharply once the room is pressurized.

Figure 44 shows the room airflow for a perimeter system with floor registers. During the heating mode, the heated air moves up the wall quickly to the ceiling. Because it is warmer and lighter than the room air, it spreads across the ceiling, and then down the inside wall as it mixes and cools. Some room air is induced from the floor into the flow of warm air and mixes with it. A stratified zone of cooler air is formed at the floor. This is improved upon by locating the return grille low.

During the cooling mode, the supply air travels up the wall, with some striking the ceiling. Because it is cooler and heavier than the room air, it travels only a short distance and then drops back down into the room as shown. The cool air mixes with the room air well, leaving only a small stratified layer of warm air near the ceiling. High return grilles minimize this problem, but doing so has a negative effect on heating performance.

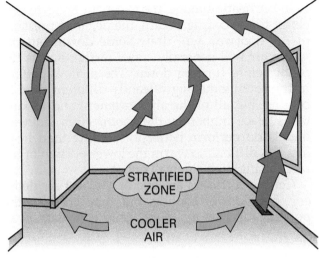

HEATING MODE

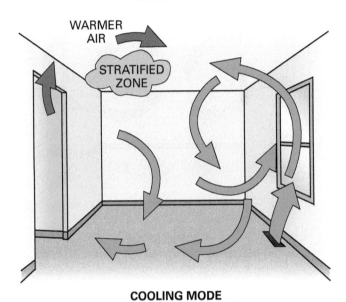

COOLING MODE

Figure 44 Room air distribution patterns for a perimeter duct system.

Figure 45 shows the room airflow with high sidewall outlets. This pattern is typical of furred-in duct systems. In the cooling mode, cool air moves across the ceiling and down the far wall. The room air mixes well with the supply air and almost no stratification occurs. In the middle of the room, the air has the tendency to tumble gently. Since the cool air falls, some is drawn down from the ceiling quickly as it exits through the door.

In the heating mode, the warm air tends to stay closer to the ceiling. Again, the negative pressure of the doorway will draw some down, but the warm air will generally travel farther across the room before turning down. The airflow pattern in the room is not significantly different than in cooling, but all of the air movement and mixing takes place higher in the room. High sidewall outlets do perform better for cooling than heating. It is difficult to warm the lower portion of the room using high sidewall outlets.

Ceiling diffusers are one of the best air delivery methods for cooling, but they are also the poorest for heating. High ceilings create even more problems for heating with ceiling diffusers. In the cooling mode, supply air mixes well with the room air (*Figure 46*). Air motion in the room is good with few stagnant areas. In the heating mode, a ceiling diffuser will perform poorly, especially if the return grille is also mounted in the ceiling. The warm air clings to the ceiling with little of it reaching the occupied space in the room until it cools. Using a return grille mounted low on the wall encourages better air movement and mixing to some degree. However, since air is being delivered to the room from above, return air often must be drawn from above as well for practical reasons.

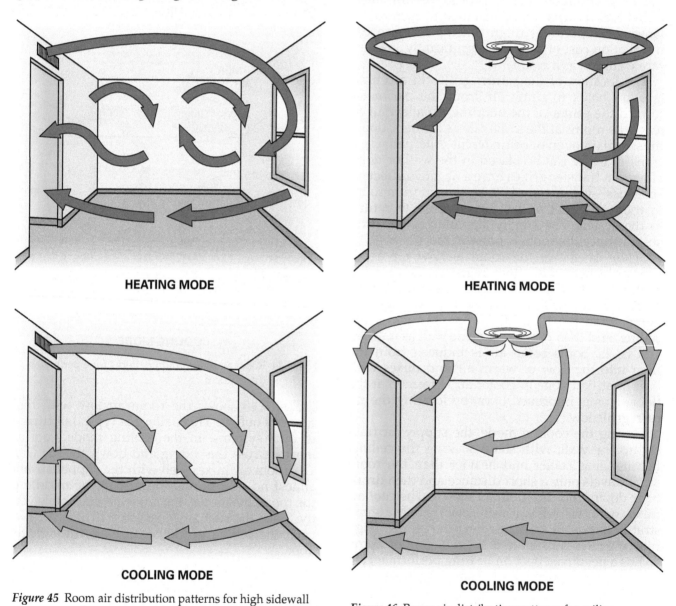

HEATING MODE

HEATING MODE

COOLING MODE

COOLING MODE

Figure 45 Room air distribution patterns for high sidewall outlets.

Figure 46 Room air distribution patterns for ceiling diffusers.

3.3.0 Energy Efficiency in Air Distribution Systems

Duct systems must be properly installed, sealed, and insulated to be energy efficient. Ductwork in unconditioned spaces must be insulated to prevent it from losing or gaining heat from the surrounding air. Poorly sealed duct joints are also known to be a significant cause of energy loss. Leaking supply ducts can lose conditioned air before it arrives at its destination. Leaking return ducts can draw unconditioned air from attics and crawl spaces. In the process, they can also draw in foul air.

Special pressure testing can determine if ductwork is properly sealed. This type of testing might be done as part of a whole-house energy efficiency analysis. Such testing is often performed by HVACR contractors who have incorporated building weatherization into their list of services. Duct testing is usually a specified part of the start-up and commissioning process for commercial buildings.

3.3.1 Insulation and Vapor Barriers

When ductwork passes through an unconditioned space, heat transfer takes place between the duct walls and the surrounding air. In the cooling mode, an astounding amount of water from the surrounding air can condense on supply-duct surfaces. Since the supply air temperature is commonly 55°F (13°C), the duct surface is below the dew point of air in unconditioned spaces. As a result, insulation with a reliable vapor barrier must be applied to the duct. When fiberglass ductboard is used, the material serves as both the insulation and the vapor barrier.

The importance of a good vapor barrier cannot be overstated. Condensation forms very quickly on the outside of a cold duct in an unconditioned space (*Figure 47*). This also happens in unseen and sometimes inaccessible locations. The results can be devastating. Allowing an area to stay consistently moist can cause mold and mildew growth, as well as structural failure from wood rot. Entire structures have been made uninhabitable as a result. Mold and mildew growth presents serious short- and long-term health concerns. It also wastes considerable energy, due to the significant latent heat load of condensing water.

Although the insulating material provides a thermal barrier, water vapor can easily find its way through and condense on the duct wall. The vapor barrier must provide a waterproof shield around the duct and insulation. Even a small flaw in the vapor barrier can allow water vapor

Figure 47 Condensation on a supply air duct.

to enter. If the insulating material becomes wet, both the insulation and vapor barrier are compromised and the problem becomes progressively worse.

Metal duct can be insulated in two ways. It can be insulated on the outside and/or on the inside. Insulation inside the duct, known as duct liner, is installed by the duct fabricator. It is either glued or fastened to tabs mounted on the inside duct wall as shown in *Figure 48*. Duct liner also helps to make the system quieter. However, if liner is to be installed, the designer must consider the liner thickness when sizing the duct. Since duct liner is commonly 1" (2.5 cm) thick, a 12" × 12" (30 cm × 30 cm) duct becomes a 10" × 10" (25 cm × 25 cm) duct once the liner is in place. For a duct that size, the loss of area is roughly 30 percent. Increasing the size of the duct to accommodate the liner increases the overall cost. Duct liner also has a higher friction loss than smooth sheet metal, reducing the airflow unless the fan is properly selected. Duct liner is most often applied in commercial installations to provide insulation and noise attenuation. On the inside of the duct, it also cannot provide any sort of vapor barrier

Insulation and a vapor barrier can be wrapped around the outside of the ductwork before or after it has been installed. The insulation is usually a foil or vinyl-backed fiberglass. The material is commonly referred to as duct wrap. It comes in several thicknesses, with 2" being typical. The insulation must be taut, but cannot be pulled too tightly. If the fiberglass is compressed, it loses some of its insulating value. The backing provides the vapor barrier. If the duct operates below the dew point of the surrounding air, the use of a vapor barrier is essential. Once applied, all seams in the insulation and vapor barrier must be properly sealed with an approved tape. To avoid

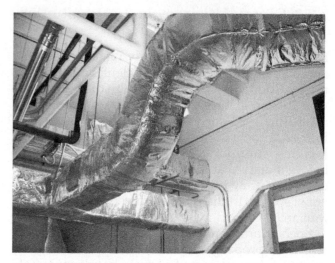

Figure 49 Insulated main duct and branches.

Even though return air duct has a lesser temperature difference with the surrounding air, it should also be insulated. Another common practice is to line return-air plenums with duct liner to reduce fan noise, especially if the return grille is close to the furnace.

Figure 48 Duct liner.

condensation, any punctures or slits in the vapor barrier must also be well sealed. When branch ducts are made of sheet metal, external insulation is installed over each branch duct as well. *Figure 49* shows a good example of an insulated duct system. In many cases, the taped areas are painted over with a sealant to enhance the vapor barrier.

As noted previously, the ductwork must be insulated to avoid heat gains or losses. *ASHRAE Standard 90-80* specifies the minimum acceptable R-value of insulation that must be used. *Figure 50* shows an example of the *ASHRAE Standard 90-80* method used to calculate the R-value for the insulation of cold duct. The R-value must also be calculated for heating operation in the same manner. The R-value of the insulation applied is determined by which operating mode shows the greatest need for insulation.

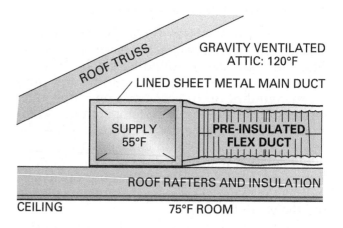

TD DUCT = 120 − 55   R = TD/15   1" = ABOUT R-4
TD DUCT = 65°F   R = 65/15
                                    R = 4.3

R = TD/15

WHERE:

R =   THERMAL RESISTANCE TO HEAT FLOW
       OF THE DUCT WITH INSULATION
       (R-VALUE)

TD =   TEMPERATURE DIFFERENCE BETWEEN
       INSIDE AND OUTSIDE DUCT

Figure 50 Example of R-value calculation using *ASHRAE Standard 90-80.*

3.3.2 Duct Sealing

It is important that all air distribution systems be free of air leaks. Leaks allow conditioned air to be lost, or air from outside the conditioned space to be drawn into the return duct. Leaks result in higher energy costs.

A pressure-pan accessory (*Figure 51*) can be used in conjunction with a blower door (*Figure 52*) to identify leaking ducts. To find air leaks, the blower door is used to depressurize the home. A variable-speed fan motor is used in the blower door. Its speed is adjusted to remove enough air from the house to obtain a 50-Pa (0.2 in. w.c.) pressure difference between the outside and the inside of the structure. The home is then at a negative pressure relative to the atmospheric pressure. The following preparations are made to perform the testing:

- The HVAC system is turned off.
- All grilles, registers, and dampers are opened fully.
- Any fresh air intakes are tightly closed.
- If the duct system is outside the conditioned space (attic or crawlspace, for example), ensure that the space is open to the outdoors.

With the home depressurized, the pressure pan is connected to a manometer and placed over each register or grille to obtain a pressure reading. Since the entire structure is under a negative pressure, and all grilles and registers are open, the duct system should be at the same negative pressure. Leaks that allow air to enter the duct result in the duct pressure being higher. A completely sealed duct should show no pressure difference. A substantial pressure difference indicates a large leak.

Any leaks must then be located. A flashlight and inspection mirror can be used to inspect hard-to-see areas. Since it is unlikely that metal duct will leak along the length of a section, joints are the focus of the inspection. Poorly supported and sagging duct sections tend to open joints. If ductwork is sagging, make sure it is properly supported before attempting to seal leaks. Repositioning the duct after sealing will likely reopen the leak. Listening for whistling noises when the system is running helps locate leaks, as well as visible dirty streaks near joints.

> **WARNING!**
> Always wear the proper personal protective equipment (PPE) including safety glasses and a dust mask or filter when inspecting duct systems.

> **WARNING!**
> Vermiculite was once widely used as loose attic insulation. The product looks like small light-brown or gold-flecked pebbles. During mining, some of this product was contaminated with asbestos. If you find vermiculite attic insulation, take care not to disturb it and use appropriate breathing protection around it.

> **GOING GREEN**
>
> **Sealing Duct Leaks**
>
> The US Department of Energy estimates that the energy lost each year in the United States through duct leakage is equivalent to the energy burned by 13 million cars in one year. For that reason, it is important to find and seal leaks in forced-air duct systems.

Figure 51 A pressure pan in use.

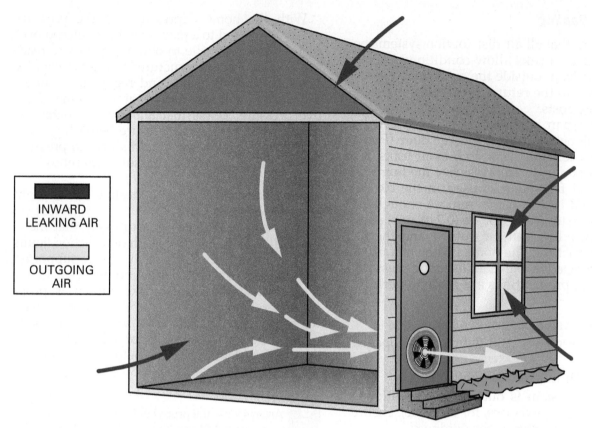

Figure 52 A blower door depressurizing a home.

It can generally be assumed that older sheet metal duct systems leak. Sheet metal duct in newer homes is less likely to leak because codes now require them to be properly sealed at the time of installation. When sealing a duct system, either of these two approaches can be used:

- Spend the time to find each leak and seal only the leaking areas.
- Seal all the accessible seams and joints in the system.

Finding individual leaks can take a lot of time. It is often more cost effective to simple seal all seams and joints.

Mastic duct sealant (*Figure 53*) is non-toxic and easy to apply. It comes in pails for brush application, or in tubes for use in a caulking gun.

Follow the manufacturer's instructions regarding the technique and the proper PPE when applying mastic. All joints and seams should be sealed (*Figure 54*). That includes seams on plenums, air handling unit connections, joints between duct sections, register boots (*Figure 55*), takeoffs, and branch line connections at the takeoffs. Longitudinal seams in snap-lock round and rectangular duct must also be sealed.

Before sealing a joint, remove any dust or dirt. If a joint has a gap less than ¼" (6–7 mm), use a brush to apply a thick coat of mastic that is 2" (5

Figure 53 Duct sealing mastic.

cm) wide. If the joint has a gap greater than ¼", embed fiberglass mesh in the first coat of mastic (*Figure 56*). Apply another coat of mastic to cover the mesh. Do not turn on the air handler or try to re-insulate the duct until the mastic has cured.

Minor slits, punctures, and gaps in ductboard can be repaired using a closure material such as fiberglass mesh tape embedded in mastic. The amount of mesh and mastic used is based on the size of the damage.

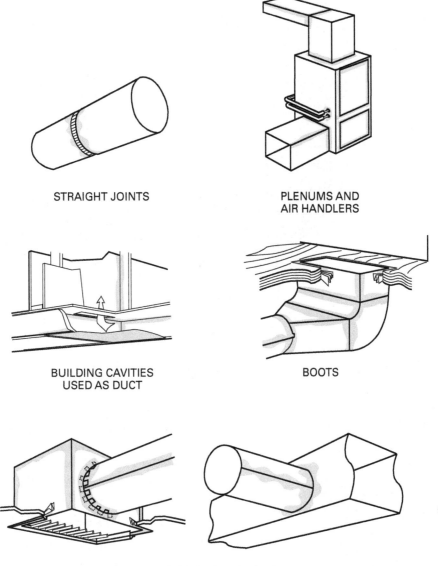

STRAIGHT JOINTS

PLENUMS AND
AIR HANDLERS

BUILDING CAVITIES
USED AS DUCT

BOOTS

TEES, WYES, AND ELLS

Figure 54 Duct sealing locations.

Sealing Leaks with Duct Tape

It would seem that duct tape would be a natural choice for sealing duct leaks. In reality, it is a poor choice, and temporary at best. Scientists at the Lawrence Berkeley National Laboratory tested over a dozen duct tape products. The tests simulated actual system operating conditions. They found that some duct tapes held up better than others, but all the products failed over time. The products used today to secure ducts is especially designed for that purpose.

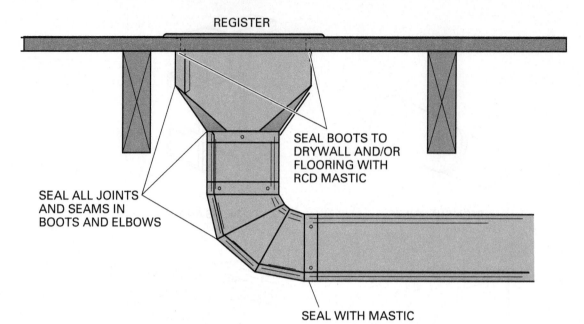

SEAL BOOTS TO
DRYWALL AND/OR
FLOORING WITH
RCD MASTIC

SEAL ALL JOINTS
AND SEAMS IN
BOOTS AND ELBOWS

REGISTER

SEAL WITH MASTIC

Figure 55 Typical register boot.

Figure 56 Fiberglass mesh tape.

Flexible ducts that have become crushed or punctured may be replaced or repaired. To repair a small puncture or a small crushed area, try the following procedure:

- Cut the vapor barrier and insulation from around the inner core at point of the damage.

High-Tech Duct Sealing

A process is now available that allows air duct leaks to be sealed from the inside. A special machine injects a mist containing the duct sealant into the duct. The sealant seeks out the leaks and plugs them from the inside. This high-tech procedure is expensive and requires special equipment. Workers also require special training.

- Use wire cutters to cut the wire reinforcement and remove the damaged portion. The duct section is now cut in two.
- Insert a sheet metal sleeve at least 4" (10 cm) long into the ends of each duct section to splice the ends together.
- Secure the inner portion of the flexible duct to the splice with nylon ties.
- Apply mastic and tape as necessary over the sliced area to ensure a reliable seal.
- Pull the insulation and vapor barriers of the two ducts together over the splice and seal them in the same manner.

Additional Resources

Air Distribution Systems: Introduction to Thermo-Fluids Systems Design, A. G. McDonald and H. L. Magande. 2012. Chichester, UK: John Wiley & Sons, Ltd.

Weatherization Technician, NCCER. 2010. New York, NY: Pearson.

Mechanical Insulating Level One, NCCER. 3rd Edition. 2018. New York, NY: Pearson.

Mechanical Insulating Level Two, NCCER. 3rd Edition. 2018. New York, NY: Pearson.

Mechanical Insulating Level Three, NCCER. 3rd Edition. 2018. New York, NY: Pearson.

3.0.0 Section Review

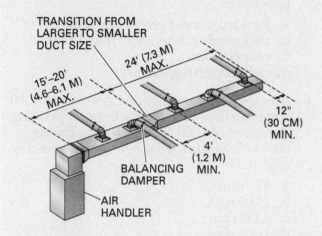

Figure SR01

1. What type of duct system is shown in *Figure SR01*?

   a. Extended plenum
   b. Perimeter loop
   c. Reducing trunk
   d. Radial

2. Which supply air diffuser or register location offers the best cooling performance?

   a. Floor
   b. Low sidewall
   c. Ceiling
   d. Baseboard

3. Which of the following statements about pressure-testing residential duct systems is *true*?

   a. The HVAC system must be running during the test.
   b. The pressure in the house must be reduced to roughly 10 in. w.c. below atmospheric pressure.
   c. A higher pressure measured in the duct than in the home during a test indicates a leak.
   d. All grilles and registers, except for the one being tested, must remain closed.

SUMMARY

Proper air distribution is necessary to provide comfort. It is also a critical factor in many manufacturing processes. An HVACR technician must evaluate the air distribution in a conditioned space and make decisions regarding its performance. An HVACR system can perform no better than its air distribution system allows.

A variety of instruments are used to measure pressure, velocity, humidity, and temperature of the air. It is important for HVACR technicians to understand how to use these instruments to test and balance air distribution systems. Further, a technician must know how to interpret and respond to measured values.

The layout of an air distribution system is normally determined by the construction of the building and the climate. Ideally, the system air handler will be in installed in an area that allows the duct length to be as short as possible and a minimal number of fittings.

Ductwork must be properly sealed and insulated to ensure energy-efficient operation. A lack of insulation and improperly sealed duct joints are major causes of energy loss. Sealing duct systems requires patience and attention to detail to ensure a reliable seal is created.

1. Within an air distribution system, the highest pressure is found at the _____.

   a. conditioned space
   b. inlet to the return duct
   c. inlet to the blower
   d. outlet of the blower

2. The static pressure, velocity pressure, and total pressure measured in a duct system are typically measured in _____.

   a. pounds per square inch or kPa
   b. inches of mercury or centimeters of mercury
   c. inches of water column or centimeters of water
   d. cubic feet per minute or cubic meters per hour

3. An air distribution system for a light commercial building is designed to work with a 5-ton cooling system. About how many cfm of airflow should the system blower be capable of supplying?

   a. 1,200 cfm
   b. 1,650 cfm
   c. 2,000 cfm
   d. 2,350 cfm

4. External static pressure loss is the _____.

   a. total pressure loss in the air handler and its component parts
   b. total pressure loss in an air distribution system, excluding the blower itself
   c. total pressure drop across the air handler
   d. difference between the total pressure and the static pressure

5. A psychrometer is used to measure _____.

   a. wet-bulb and dry-bulb temperature
   b. static pressure and velocity pressure
   c. wet-bulb temperature, dry-bulb temperature, and dew point
   d. wet-bulb temperature, dry-bulb temperature, and static pressure

6. Pitot tubes are used for measuring _____.

   a. air temperature
   b. air density
   c. relative humidity
   d. air pressure and velocity

7. The type of blower or fan that is best suited to applications with a high resistance to airflow is the _____.

   a. forward-curved centrifugal blower
   b. backward-inclined centrifugal blower
   c. propeller fan
   d. radial centrifugal blower

8. An existing air distribution system has 6,000 cfm of airflow created by a blower operating at a speed of 1,200 rpm. To reduce the airflow to 5,000 cfm, what new blower speed is required?

   a. 833 rpm
   b. 1,000 rpm
   c. 1,150 rpm
   d. 1,300 rpm

9. The duct material that has the least friction loss is _____.

   a. sheet metal
   b. fiberglass ductboard
   c. flexible duct
   d. internally insulated sheet metal

10. Fittings and transitions used in a duct system have _____.

    a. the same friction loss per foot as the same size straight duct
    b. less friction loss per foot as the same size straight duct
    c. a friction loss equal to some length of the same size straight duct
    d. twice the friction loss as the same size straight duct

11. Dampers built into supply diffusers and registers are best used to _____.

    a. add moisture to the air
    b. reduce duct noise
    c. make minor adjustments in airflow
    d. completely balance system airflow

12. The ideal location for a balancing damper in a branch duct is at the _____.

    a. takeoff
    b. boot
    c. transitions
    d. air handling unit

13. A fire damper will close when _____.

    a. the fire has burned for 3 hours
    b. the fusible link melts
    c. the HVAC equipment shuts off
    d. smoke is detected

14. A duct system installed in a concrete slab that helps keep the entire floor warm is the _____.

    a. perimeter-loop system
    b. reducing trunk system
    c. perimeter radial system
    d. overhead trunk system

15. Under what conditions would it be necessary to seal a gap in a duct with fiberglass tape embedded in mastic?

    a. It is always done that way.
    b. When the opening exceeds ¼" (6–7 mm).
    c. It is only done to seal fiberglass ductboard.
    d. When the opening is less than ¼" (6–7 mm).

Trade Terms Quiz

Fill in the blank with the correct trade term that you learned from your study of this module.

16. The volume of air flowing past a point in one minute is measured in _____.

17. The device containing a fan that is used to depressurize a building for duct leak-testing is called a _____.

18. The temperature measured using a standard thermometer is called the _____.

19. A graphic representation of the relationship between air properties is known as a _____.

20. The sum of the static pressure and the velocity pressure in an air duct provides the _____.

21. The ring or panel surrounding the blades of a propeller fan that increases the fan's performance is called a(n) _____.

22. The temperature at which air becomes saturated with water vapor, causing the water to condense into water droplets and fall from the air, is called the _____.

23. A tool that is inserted into a duct to capture pressure values is a(n) _____.

24. The effectiveness of insulation is reflected in its _____.

25. The ratio of the amount of moisture present in a given sample of air compared to the amount it can hold at saturation defines _____.

26. The pressure exerted uniformly in all directions within a duct system is called _____.

27. _____, usually measured in feet per minute, reflects how fast air is moving in a duct.

28. The pressure in a duct due to the forward movement of the air is referred to as the _____.

29. A fan moving air with no effective restrictions to airflow at its inlet or outlet is called _____.

30. The sealed enclosure at the inlet or outlet of an air handler is called a(n) _____.

31. Temperature taken with a thermometer that has a saturated wick wrapped around its sensing bulb before taking a reading is called the _____.

32. The material installed over insulation to prevent moisture from entering is referred to as a(n) _____.

33. Air masses at different temperatures will _____ unless an outside force encourages it to move and mix.

34. Adding up the total friction loss of the ducts and other objects in the air distribution system beyond the blower itself will result in the _____.

35. The speed at which the shaft of an air-moving device is rotating over a period of time is reported in _____.

36. The _____ reflects the free area of a grille, register, or diffuser in relation to one square foot.

37. Branch ducts need _____ so they can be connected to the main duct.

38. To transition from branch ducts to the grille and registers, _____ are typically required.

39. A _____ allows air to move freely out of an enclosed area.

Trade Terms

A_k factor	External static pressure (ESP)	Relative humidity (RH)	Transfer grille
Blower door	Free air delivery	Revolutions per minute (rpm)	Vapor barrier
Boots	Pitot tube	Static pressure (s.p.)	Velocity
Cubic feet per minute (cfm)	Plenum	Stratify	Velocity pressure
Dew point	Psychrometric chart	Takeoffs	Venturi
Dry-bulb temperature	R-value	Total pressure	Wet-bulb temperature

Tony Vazquez
President/CEO, Careersafety Center
Training Manager, Hubbard
www.careersafetycenter.com

How did you get started in the construction industry (i.e., took classes in school, a summer job, etc.)?

I come from a family of furniture makers that emigrated from Cuba to the United States. Influenced by my father, I entered the construction industry. It wasn't long before my father and I opened our own company, but the need to find good craftsmen led me to a career of training others. As the need for quality training became more apparent, I eventually opened my own construction training school (VisionQuest-Academy) in Ocala, Florida. As a Master Trainer and Subject Matter Expert (SME) with NCCER in the construction and green job trades, I enjoy sharing my love of building by encouraging others to excel in the industry that's made me successful.

Who or what inspired you to enter the industry (i.e., a family member, school counselor, etc.)? Why?

Here's my story of how the construction industry and its craft has impacted my life. I was raised in a second-generation construction/craftworking family. I was inspired as a boy by my father and grandfather back in Cuba where they hand-made furniture. Later, here in the United States, my father continued his craft as a construction worker. As a teenager, I often worked on the weekends with him. After high school, my father and I started a construction company. During my earlier years in the construction industry, I took all and any construction trade classes that were available in my county.

During this time, I also learned the construction crafts from my father and those that worked with us. After a few years, I realized that there had to be a better way to acquire these trade skills. In 1990, while in the pursuit of trained construction workers, I ended up at Miami Job Corps, and I was impressed about the training going on in their program. Later that year, I began volunteering my time to help their program and was asked to take over the program when the instructor retired few years later. With all the scattered construction courses I had taken and the field experiences I acquired, I had minimum qualification to teach. I continued to take more specific courses and returned to college and obtained an instructor/teacher certification in the industry.

Now, many years later, I am currently an NCCER-certified Master Trainer and Subject Matter Expert (SME), qualified to teach others the industry crafts. I am in my 21st year of teaching and I still enjoy passing the skills that I have learned in the trades to others as if it was my first day.

What do you enjoy most about your career (i.e., variety, environment, career satisfaction, etc.)?

I still love to build. However, today I get the most satisfaction from teaching others the art of craftsmanship. I have taught at both the secondary and post-secondary level for over 21 years. In past decades, the basics of construction have been taught at a few high schools (shop classes) and a few post-secondary training programs scattered throughout the country. However, as our industry now becomes more united in providing industry-driven curriculum and nationally recognized training, it is revitalizing the art of construction craftsmanship. It has come during a much-needed time as our older workforce is preparing to retire and we have the need of new blood and craftsmanship in our industry.

Why do you think training and education are important in construction? If so, why?

Yes, training and education is at its most influential time. Tough times today require tough and specialized training. In today's way of building, each trade has become a specialized trade, where those with the most training experience and credentials get the job. As our industry moves forward in creating nationally recognized credentials and certifications, specialized training centers and schools have also emerged.

Why do you think credentials are important in construction?

They are extremely important, in two distinct and different ways. First, by establishing standardized trainer credentials, the industry assures that training taking place across the country is similar and of the same quality. Second, completion points and credentials given to trainees are also similar and of the same quality across the country, thus making it possible for national certification that is valid anywhere.

NCCER has brought much-needed training that is nationally organized and accredited to our industry. On a personal level, the NCCER has given me the opportunity to connect all of my educational abilities to a single, widely recognized entity. Being able to acquire multiple credentials has allowed me to excel. I have reached the Master Trainer and Subject Matter Expert status with NCCER in the construction and green job trades. This in turn has given me the incredible opportunity to expand how I can teach and train others. It has given me the opportunity to create my own construction training academy.

How has training/construction impacted your life and career (i.e., advancement opportunities, better wages, etc.)?

Construction education and training have impacted my life since my early teen years both directly and indirectly. In the early years, it provided the technical background for the work I was performing. Years later, training in specialized areas took place in a more formal setting. Currently, the combination of field work and formal classroom training has provided me with a good salary and lifestyle. Furthermore, it has provided me the opportunity to open VisionQuest-Academy, a full service post-secondary educational and craft training academy in Ocala, Florida. The academy is dedicated to providing cutting-edge construction and green job training and is offered in English and Spanish. NCCER has been a key to this opportunity—otherwise, many of our county's adults and the working population would not have the opportunity to obtain national industry credentials.

Would you recommend construction as a career to others? Why?

Yes, I absolutely do. As I tell all of my students, it's never too late for formal or informal education to take place. However, a well planned training strategy is like a well planned trip that includes a road map to your destination. NCCER has that covered for the construction and green job industry along with the provision of pathways to success. Ultimately, the construction and green job industry is unique in that, at the end of the day, you have something tangible.

What does craftsmanship mean to you?

Craftsmanship can be defined in a multitude of ways. In the construction industry, craftsmanship is displayed in the final results of a job, task, or project completed within the quality standards set for that industry. However, I personally define it this way. It is the skills, expertise, ability, and technique used by a person to shape, mold, transform, and or convey top-quality products of value to others. A craftsman's level of work and performance can only be achieved by a combination of hard work, formal and informal training, time and dedication in the industry, and the love of a craft.

Trade Terms Introduced in This Module

A$_k$ factor: The area factor of registers, grilles, and diffusers that reflects the free area for airflow relative to a square foot.

Blower door: An assembly containing a fan used to depressurize a building by drawing air out, to support duct leak testing. It is usually installed temporarily in place of an outside door.

Boots: Sheet metal fittings designed to transition from the branch duct to the receptacle for the grille, register, or diffuser to be installed.

Cubic feet per minute (cfm): A unit for the volume of air flowing past a point in one minute. Cubic feet per minute can be calculated by multiplying the velocity of air, in feet per minute (fpm), times the area it is moving through, in square feet (cfm = fpm × area). The metric value is cubic meters per hour (m3/h).

Dew point: The temperature at which air becomes saturated with water vapor, and the water starts to condense into droplets; a state of 100% relative humidity.

Dry-bulb temperature: The temperature measured using a standard thermometer. It represents the measure of sensible heat present.

External static pressure (ESP): The total resistance of all objects and ductwork in the air distribution system beyond the blower assembly itself.

Free air delivery: The condition that exists when there are no effective restrictions to airflow (no external static pressure) at the inlet or outlet of an air-moving device.

Pitot tube: A tool used to capture pressure measurements in a moving air stream.

Plenum: A chamber at the inlet or outlet of an air handler. The air distribution system attaches to the plenum.

Psychrometric chart: A graphic method of showing the relationship of various air properties.

R-value: A number, such as R-19, that is used to indicate the ability of insulation to resist the flow of heat. The higher the R-value, the better the insulating ability.

Relative humidity (RH): The ratio of the amount of moisture present in a given sample of air to the amount it can hold at saturation. Relative humidity is expressed as a percentage.

Revolutions per minute (rpm): The number of rotations made by a spinning object over the course of one minute.

Static pressure (s.p.): The pressure exerted uniformly in all directions within a duct system, usually measured in inches of water column (in. w.c.) or centimeters of water column (cm H$_2$O).

Stratify: To form or arrange into layers. In air distribution, layers of air at different temperatures will tend to stratify unless an outside influence forces them to move and mix. Warmer air will stratify on top of cooler air.

Takeoffs: Connection points installed on a trunk duct that allow the connection of a branch duct.

Total pressure: The sum of the static pressure and the velocity pressure in an air duct.

Transfer grille: A grille usually installed in walls or doors, with a grille of the same size mounted on each side, that allows air to pass freely in or out of an enclosed space.

Vapor barrier: A barrier placed over insulation to stop water vapor from passing through the insulation and condensing on a cold surface.

Velocity: The speed at which air is moving. The rate of airflow usually measured in feet per minute.

Velocity pressure: The pressure in a duct due to the linear movement of the air. It is the difference between the total pressure and the static pressure.

Venturi: A ring or panel surrounding the blades on a propeller fan to improve fan performance.

Wet-bulb temperature: Temperature taken with a thermometer that has a wick wrapped around its sensing bulb, saturated with distilled water before taking a reading. The reading from a wet-bulb thermometer, through evaporation of the water, takes into account the moisture content of the air. It reflects the total heat content (sensible and latent) of the air.

Additional Resources

This module presents thorough resources for task training. The following resource material is suggested for further study.

Air Distribution Systems: Introduction to Thermo-Fluids Systems Design, A. G. McDonald and H. L. Magande. 2012. Chichester, UK: John Wiley & Sons, Ltd.

NFPA 90, Installation of Air Conditioning and Ventilation Systems. Current edition. Quincy, MA: National Fire Protection Association.

Weatherization Technician, NCCER. 2010. New York, NY: Pearson.

Mechanical Insulating Level One, NCCER. 3rd Edition. 2018. New York, NY: Pearson.

Mechanical Insulating Level Two, NCCER. 3rd Edition. 2018. New York, NY: Pearson.

Mechanical Insulating Level Three, NCCER. 3rd Edition. 2018. New York, NY: Pearson.

Figure Credits

© Branka Tasevski/Shutterstock.com, Module opener

Fluke Corporation, reproduced with permission, Figure 9

Courtesy of Extech Instruments, a FLIR Company, Figures 11, 17

Courtesy of Hands Down Software, www.handsdownsoftware.com, SA01

Courtesy of Dwyer Instruments, Inc., Figures 12, 13

Courtesy of TSI/Alnor Instruments, Figures 15, 16

Courtesy of Northern Blower Inc., Figure 19

Courtesy of DuctSox Corporation, SA03

Courtesy of Ductmate Industries, Inc., SA04

Hart & Cooley Inc., Figures 31, 34, Table 5

Courtesy of Nailor Industries, Figure 37

Carrier Corporation, Figure 43

Courtesy of American Basement Solutions, www.americanbasementsolutions.com, Figure 47

Courtesy of CertainTeed Insulation, Figure 48

Courtesy of G.E. Insulation, Figure 49

Courtesy of ShopThermalStar.com, Figure 51

Courtesy of RCD Corporation, Figures 53, 56

Answer	Section Reference	Objective
Section One		
1. c	1.1.0	1a
2. b	1.2.0	1b
3. a	1.3.2	1c
Section Two		
1. b	2.1.2	2a
2. d	2.2.1	2b
3. b	2.3.0	2c
4. c	2.3.0	2c
5. c	2.4.1	2d
6. b	2.5.0	2e
Section Three		
1. c	3.1.4; Figure 41	3a
2. c	3.2.0	3b
3. c	3.3.2	3c

2.0.0 SECTION REVIEW

Question 3

Use Fan Law 1 to determine the new volume flow rate:

$$\text{New volume flow rate} = \frac{\text{New rpm} \times \text{existing volume flow rate}}{\text{Existing rpm}}$$

$$\text{New volume flow rate} = \frac{790 \text{ rpm} \times 1{,}200 \text{ cfm}}{950 \text{ rpm}}$$

$$\text{New volume flow rate} = \frac{948{,}000 \text{ rpm-cfm}}{950 \text{ rpm}}$$

$$\text{New volume flow rate} = 998 \text{ cfm}$$

Question 4

First, use Fan Law 1 to determine the new rpm:

$$\text{New rpm} = \frac{\text{New volume flow rate} \times \text{existing rpm}}{\text{Existing volume flow rate}}$$

$$\text{New rpm} = \frac{1{,}000 \text{ cfm} \times 1{,}050 \text{ rpm}}{870 \text{ cfm}}$$

$$\text{New rpm} = \frac{1{,}050{,}000 \text{ cfm-rpm}}{870 \text{ cfm}}$$

$$\text{New rpm} = 1{,}207 \text{ rpm}$$

Then, use Fan Law 3 to determine the new horsepower:

$$\text{New hp} = \text{existing hp} \times \left(\frac{\text{new rpm}}{\text{existing rpm}}\right)^3$$

$$\text{New hp} = 0.333 \text{ hp} \times \left(\frac{1{,}207 \text{ rpm}}{1{,}050 \text{ rpm}}\right)^3$$

$$\text{New hp} = 0.333 \text{ hp} \times 1.15^3$$

$$\text{New hp} = 0.333 \text{ hp} \times 1.52$$

$$\text{New hp} = 0.51 \text{ hp}$$

The horsepower is **0.51 hp**.

NCCER CURRICULA — USER UPDATE

NCCER makes every effort to keep its textbooks up-to-date and free of technical errors. We appreciate your help in this process. If you find an error, a typographical mistake, or an inaccuracy in NCCER's curricula, please fill out this form (or a photocopy), or complete the online form at **www.nccer.org/olf**. Be sure to include the exact module ID number, page number, a detailed description, and your recommended correction. Your input will be brought to the attention of the Authoring Team. Thank you for your assistance.

Instructors – If you have an idea for improving this textbook, or have found that additional materials were necessary to teach this module effectively, please let us know so that we may present your suggestions to the Authoring Team.

NCCER Product Development and Revision

13614 Progress Blvd., Alachua, FL 32615

Email: curriculum@nccer.org
Online: www.nccer.org/olf

❏ Trainee Guide ❏ Lesson Plans ❏ Exam ❏ PowerPoints Other _____

Craft / Level: _____ Copyright Date: _____

Module ID Number / Title: _____

Section Number(s): _____

Description: _____

_____

_____

_____

_____

Recommended Correction: _____

_____

_____

_____

_____

Your Name: _____

Address: _____

_____

Email: _____ Phone: _____

Basic Copper and Plastic Piping Practices

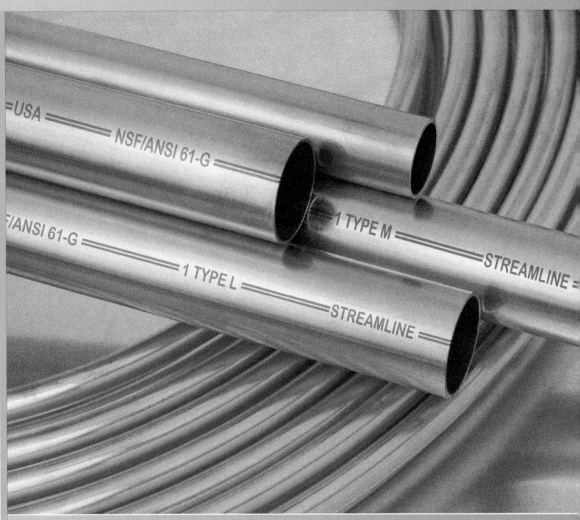

OVERVIEW

Copper tubing and piping is used extensively in HVAC work. Copper is the base material for virtually all common refrigerant piping. Plastic piping is used in heating and cooling systems for a variety of purposes. Plastic piping may be used to carry chilled water for large cooling systems, condenser water for water-cooled mechanical systems, or even to vent high-efficiency furnaces.

Module 03103

Trainees with successful module completions may be eligible for credentialing through the NCCER Registry. To learn more, go to **www.nccer.org** or contact us at **1.888.622.3720**. Our website has information on the latest product releases and training, as well as online versions of our *Cornerstone* newsletter and Pearson's product catalog.

Your feedback is welcome. You may email your comments to **curriculum@nccer.org**, send general comments and inquiries to **info@nccer.org**, or fill in the User Update form at the back of this module.

This information is general in nature and intended for training purposes only. Actual performance of activities described in this manual requires compliance with all applicable operating, service, maintenance, and safety procedures under the direction of qualified personnel. References in this manual to patented or proprietary devices do not constitute a recommendation of their use.

03103 V5

Objectives

When you have completed this module, you will be able to do the following:

1. Recognize and identify different types of copper tubing and their related fittings.
   a. Describe and identify copper tubing characteristics.
   b. Identify various copper fittings.
2. Describe and demonstrate how to join copper tubing mechanically.
   a. Measure, cut, and bend copper tubing to prepare it for joining.
   b. Describe and demonstrate the methods and tools used to join copper tubing.
   c. Describe common hangers and supports associated with copper tubing installations.
3. Recognize different types of plastic piping and show how it can be joined.
   a. Identify different types of plastic piping.
   b. Identify the tools and products needed and demonstrate how to join plastic piping.

Performance Tasks

Under the supervision of the instructor, you should be able to do the following:

1. Cut and bend copper tubing.
2. Safely join copper tubing using mechanical fittings.
   - Flare tubing and complete a flared connection.
   - Use a compression fitting and ferrule to make a connection
   - Use a swaging tool to swage a piece of tubing.
3. Cut and join lengths of plastic pipe.

Trade Terms

Annealed
Brazing
Chamfer
Conductivity
Copper-clad
Ferrule
Halocarbon refrigerants
Hard-drawn
Malleable

Oxidation
Plate
Pressure drop
Reamer
Soldering
Swaging
Witness mark
Work hardening

Industry Recognized Credentials

If you are training through an NCCER-accredited sponsor, you may be eligible for credentials from NCCER's Registry. The ID number for this module is 03103. Note that this module may have been used in other NCCER curricula and may apply to other level completions. Contact NCCER's Registry at 888.622.3720 or go to **www.nccer.org** for more information.

Contents

1.0.0 Copper Tubing and Fittings 1

1.1.0 Characteristics of Copper Tubing 1

1.1.1 Copper Tube Types 2

1.1.2 Identifying Markings 4

1.1.3 Copper Tube Sizing 4

1.2.0 Copper Tube Fittings 6

1.2.1 Flare Fittings 6

1.2.2 Compression Fittings 7

1.2.3 Sweat Fittings 8

1.2.4 Pressed Fittings 9

1.2.5 Swaged Joints 9

2.0.0 Joining Copper Tubing and Fittings 12

2.1.0 Measuring, Cutting, and Bending Copper Tubing 12

2.1.1 Measuring Tubing 12

2.1.2 Cutting Tubing 13

2.1.3 Bending Tubing 15

2.2.0 Joining Copper Tubing 15

2.2.1 Making Flared Connections 16

2.2.2 Making Compression Connections 18

2.2.3 Swaging a Tube 19

2.2.4 Pressure Testing 19

2.3.0 Hangers and Supports 20

3.0.0 Plastic Piping 26

3.1.0 Types of Plastic Tubing 26

3.1.1 ABS (Acrylonitrilebutadiene Styrene) Pipe 26

3.1.2 PE (Polyethylene) Tubing 26

3.1.3 PVC (Polyvinyl Chloride) Pipe 26

3.1.4 CPVC (Chlorinated Polyvinyl Chloride) Pipe 28

3.1.5 PEX (Cross-linked Polyethylene) Tubing 28

3.1.6 Plastic Piping Schedules 29

3.2.0 Joining Plastic Pipe 30

3.2.1 Solvent-Cementing Products 31

3.2.2 Solvent-Cementing Plastic Pipe 32

3.2.3 Plastic Pipe Support Spacing 34

Figures and Tables ————————

Figure 1 Annealed (soft) and hard-drawn copper tubing 2
Figure 2 Identifying markings on Type ACR tubing 4
Figure 3 Tubing dimensions for a nominal 1" copper tube size, for each type of copper tubing 5
Figure 4 Pressure control, designed to connect to flared tubing 6
Figure 5 Types of flare fittings... 7
Figure 6 Flare nut wrench 7
Figure 7 Common compression fittings.. 7
Figure 8 Compression ring, or ferrule 7
Figure 9 Sweat fittings 9
Figure 10 A reducing coupling and a reducing bushing............................. 9
Figure 11 Pressed fittings 10
Figure 12 Ridgid® press tool and jaws.. 11
Figure 13 Swaging tools 11
Figure 14 Swaged joint 11
Figure 15 Typical tubing measurements to accommodate sweat fittings.... 13
Figure 16 Compression joint cut-away 13
Figure 17 Handheld tubing cutters 13
Figure 18 Cutting with a handheld tubing cutter 14
Figure 19 Reaming tool. This style reams the inside of the tube as well as the outside.. 14
Figure 20 Using a bending spring 15
Figure 21 Handheld tubing bender 16
Figure 22 The components of a typical flared connection......................... 16
Figure 23 Flaring tool 17
Figure 24 Cross-section of a single thickness flare joint.......................... 17
Figure 25 Forming a double-thickness flare... 17
Figure 26 Nylon abrasive pads and sand cloth 17
Figure 27 Flaring copper tubing.. 18
Figure 28 Hammer-driven swaging tool 19
Figure 29 Swaging a tube with a hammer-driven tool.............................. 19
Figure 30 Flaring tool set with swaging adapters 19
Figure 31 Nitrogen tank and accessories for pressure testing.................... 21
Figure 32 Various beam clamps.. 22
Figure 33 Clevis and split ring hangers.. 22
Figure 34 Pipe straps 22
Figure 35 Riser clamp 22
Figure 36 Pipe straps and channel... 24
Figure 37 Pipe straps with plastic inserts... 25
Figure 38 Pipe straps designed to accommodate insulation....................... 25
Figure 39 A clevis hanger fitted with an insulation saddle......................... 25
Figure 40 Plastic piping used in a condensing furnace installation............. 27
Figure 41 ABS pipe and fittings 27
Figure 42 PE tubing... 27
Figure 43 PVC pipe and fittings.. 28

Figures and Tables (continued)

Figure 44 PVC pipe markings 28

Figure 45 CPVC pipe and fittings 29

Figure 46 PEX tubing 29

Figure 47 Plastic tubing and pipe cutter 31

Figure 48 PVC hand saw 31

Figure 49 Pipe wrap for measuring, marking straight cut lines,
and laying out complex cuts 29

Figure 50 PVC/CVPC clear pipe cleaner 29

Figure 51 PVC/CVPC primer/cleaner with a purple color 29

Figure 52 A variety of PVC, CPVC, and ABS cement products 29

Figure 53 Fabricating a solvent-cemented joint 31

Table 1   Copper Tubing Color Codes 9

Table 2   Dimensions and Physical Characteristics of
Type ACR Copper Tubing 10

Table 3   Minimum Bend Radius for Copper Tubing 17

Table 4   Recommended Hanger Spacing for Copper Tubing 22

Table 5   Schedule 40 Pipe Dimensions 27

Table 6   Schedule 80 Pipe Dimensions 27

Table 7   Average Handling and Set-Up Times for PVC/CVPC Cements 30

Table 8   Average Cure Times for PVC/CVPC Cements 30

Table 9   General Guidelines for Horizontal Support
Spacing – Plastic Pipe 32

SECTION ONE

1.0.0 COPPER TUBING AND FITTINGS

Objective

Recognize and identify different types of copper tubing and the related fittings.

a. Describe and identify copper tubing characteristics.
b. Identify various copper fittings.

Trade Terms

Annealed: Metal that has been heat-treated to soften it, making it formable. Hard-drawn copper tubing is annealed to make it soft and formable.

Brazing: A heat-bonding method of joining metals using another alloy with a melting point lower than the metal(s) being joined. The alloy is used as a filler metal to bond and fill any gaps between the pieces by capillary action. Brazing uses filler metals that have a melting point above 842°F (450°C).

Conductivity: The ease and rate in which energy passes through a material. Thermal conductivity refers to the ease and rate at which heat passes through a material; electrical conductivity refers to the rate at which electricity passes.

Ferrule: A ring or bushing placed around a tube that squeezes or bites into the tube beneath when compressed, forming a seal.

Halocarbon refrigerants: Any of a class of organic compounds containing carbon and one or more halogens, such as chlorine or fluorine. Refrigerants such as R-22 contain chlorine, which has been linked to the destruction of the ozone layer. As a result, refrigerants bearing chlorine are being phased out.

Hard-drawn: The process of heating copper and drawing it through dies to form its shape and size. Each die it is drawn through is progressively smaller until the desired size is reached.

Malleable: A characteristic of metal that allows it to be pressed or formed to some degree without breaking or cracking. Copper is malleable, while cast iron is not.

Oxidation: The process of combining with oxygen at a molecular level. Copper and oxygen join to form copper oxides, appearing as darkened deposits on the copper surface. Common rust is an iron oxide.

Pressure drop: The reduction in pressure between one point in a pipe or tube to another point. Pressure drop results from friction in piping systems, which robs energy from the flowing fluid or vapor. Pressure drops increase as flow velocity (speed) increases.

Soldering: A heat-bonding method of joining metals using another alloy with a melting point lower than the metal(s) being joined. The alloy is used as a filler metal to bond and fill any gaps between the pieces by capillary action. Soldering uses filler metals that have a melting point below 842°F (450°C).

Swaging: The process of using a tool to shape metal. In context, it describes the process of forming a socket in the end of a copper tube that is the correct size to accept another piece of tubing or a component.

An HVAC technician must be able to work with a variety of piping and tubing materials. Copper tubing is used to transport refrigerant in virtually all residential and commercial air conditioning systems. Steel is also used with a few refrigerants, such as ammonia. Steel and plastic piping are both used to transport water in HVAC systems that use it as a heat exchange medium. Steel, copper, and a few other special types of tubing are used to provide fuel gases to heating systems. The final choice of material for many applications, including fuel gases, is often controlled by building codes. This module focuses on handling, cutting, bending, and mechanically joining copper tubing and plastic piping. Mastering these skills is essential to an HVAC technician's success.

1.1.0 Characteristics of Copper Tubing

HVAC technicians and installers are in contact with copper tubing and fittings frequently. It is the primary material used for refrigerant piping both on the inside and on the outside of air conditioning and refrigeration equipment. Therefore, it is critical that good copper-tube-working skills be developed.

Air conditioning and refrigeration systems using halocarbon (halogenated hydrocarbon) refrigerants normally use copper tubing. Some aluminum is used to construct coils, but its use is limited due to issues of durability and workability. Steel pipe and fittings, as well as aluminum, are often used with non-halocarbon refrigerants such as ammonia. In spite of the higher cost of copper, it significantly outperforms all other materials in many HVAC applications, including refrigerant circuits.

Copper tubing of all types must meet the specifications set forth by the American Society for Testing and Materials (ASTM). It is constructed of 99.9 percent copper. Copper is relatively soft and malleable, with high thermal and electrical conductivity. The surface molecules of copper tend to combine with oxygen in the air, causing its natural, bright reddish-orange or yellow-orange color to change to a dull brown or even black color. The copper reacts with the oxygen in the air to form copper oxides, commonly known as tarnish. The process is referred to as oxidation. In the presence of heat, such as that from a torch when soldering or brazing, the oxidation process accelerates dramatically. During the brazing process, the copper oxides form so quickly that they often form loose, black flakes that readily fall away from the tubing. The surface of the tubing also remains discolored.

In the trades, there are two primary categories of copper tubing. The two categories are based on the temper, or hardness of the material. Annealed copper tubing is often referred to as soft copper. It is generally sold in manageable rolls of various lengths. Both 25-foot and 50-foot rolls are common. Annealed tubing is very easy to bend and form. As long as it is done with the proper tools and technique, it can be formed without suffering significant damage. Hard-drawn tubing, commonly known in the trade as hard copper, is fabricated by drawing, or pulling, the copper through a series of dies to decrease its size to the desired diameter. This process tends to harden the tubing, making it more rigid. Both annealed and hard-drawn tubing are shown together in *Figure 1*. As a general rule, the working pressure rating of hard copper is about 60 percent higher than that of soft copper. Both types however, are quite capable of handling common refrigerant circuit pressures. Hard copper tubing is sold in straight lengths, usually 20 feet long. It is not recommended that hard-drawn tubing be formed under field conditions. Both soft and hard copper can be joined and installed using a variety of methods and fittings.

Soft copper tubing is available in very tiny sizes for use as metering devices. Soft copper tubing is most often used for smaller refrigerant lines. Although 1⅜" and 1⅝" OD sizes are available, it is rare for sizes larger than 1⅛" OD to be used. Hard copper tubing is commonly available from ¼" OD to 6⅛" OD.

1.1.1 Copper Tube Types

Different types of copper tubing are identified with one or more letters. The primary difference in each type is the wall thickness. A copper tubing size of 1" will be used in this section to demonstrate the variance in wall thickness.

Type M copper tubing has the thinnest wall of the commonly used products. It is also the lightest in weight and lowest in cost as a result. Type M is typically used in domestic plumbing systems for water, and is suitable for many other low pressure and drainage applications. A 1" Type M copper tube has a wall thickness of 0.035".

Type L copper tubing has a thicker wall and can be used in higher pressure applications than Type M. A 1" Type L copper tube has a wall thickness of 0.050".

Before proceeding, it is important to note that, as the wall thickness of copper tubing increases from type to type, the outside diameter for a given

Copper

The cost of copper has sky rocketed in recent years due to excessive demand across the world. Regardless of its cost and value, copper is 100-percent recyclable and should never be disposed of in a landfill or other permanent disposal site. If a material can be recycled, it should be recycled. Collect copper scrap and waste from piping installations and demolished facilities, and then turn it over to a responsible party for recycling. To deter copper thieves and prevent them from selling their ill-gotten gains, documentation of the exchange is now required in many areas.

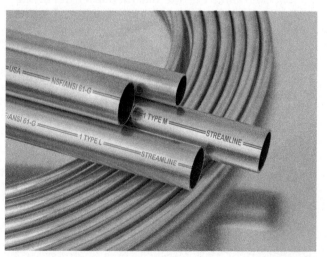

03103-12_F01.EPS

Figure 1 Annealed (soft) and hard-drawn copper tubing.

size does not change. In other words, 1" Type M and 1" Type L copper tubing have the same outside diameter. As the wall thickness increases, the inside diameter of the tubing decreases.

Type K tubing has the thickest wall and, of course, is the heaviest in weight per foot. Type K is typically used in the most challenging and critical applications. It is rarely needed or specified in the HVAC industry for field use, but may be chosen for underground use. It is important to note, though, that a heavier tube type can always be used in place of a lighter weight. Specifications usually identify the minimum type and weight to be used, but not a maximum value. However, the cost of heavier copper tubing is much higher. Therefore, a heavier weight is almost never substituted for a lighter one by contractors.

Type DWV has thin walls. The acronym *DWV* stands for drain, waste, and vent. It is not designed for pressurized applications. Since it is primarily designed for DWV service, the smallest size available is 1¼". This size has a wall thickness of 0.040". Due to the cost of the material compared to plastic options, it is typically only used in DWV applications where an attractive appearance is desired. There is little incentive to use it where the installation is hidden from public view. HVAC personnel very rarely need to handle Type DWV.

Type ACR tubing is somewhat unusual when compared to the others. ACR is an acronym for air conditioning and refrigeration. This tubing was specifically developed for refrigerant piping use. However, it can be used in almost any application where Type L has been specified. Type ACR is available in both hard and soft versions. This type is the same in construction as Type L, with the same dimensions. What sets it apart is the fact that it is purged of all air, and then charged with nitrogen after it is manufactured. It is then tightly capped to maintain the nitrogen atmosphere inside at a very low pressure. This eliminates the presence of oxygen inside the tube, thereby eliminating copper oxides. When a new length of Type ACR is first opened, it will be extremely clean, bright, and shiny inside. This level of cleanliness is very important to the refrigeration circuit. The caps or plugs should always be replaced on the unused portion of ACR tubing once it is cut. Although any nitrogen is now gone and air has entered, replacing the plugs prevents the constant entry of fresh oxygen and debris.

One additional feature of importance regarding Type ACR is that it is called out by size differently. While a length of Type L tubing would be called out as 1" (its nominal diameter), Type ACR of the same identical size would be called 1⅛" tubing (the actual outside diameter). This is a key point to remember. In conversation and in purchasing situations, HVAC personnel refer to copper tube sizes by their actual outside diameter (OD), because they are generally working with Type ACR. Plumbers, on the other hand, call out copper tubing by the nominal size, since they are generally working with types other than ACR. This can cause some confusion between craftworkers and material providers. Craftworkers must be specific about the copper tubing type and size they need.

Tubing? Pipe? Which Is It?

The terms *pipe* and *tube* are often used interchangeably. Some say that there is one characteristic that separates the two. Pipe is typically sized by its inside diameter, while tubing is sized by its outside diameter. However, this is really not the case.

For example, a common 1" steel pipe would be expected to have an inside diameter of 1". However, the actual inside diameter is roughly 0.050" greater. The copper industry refers to its tubular products as tubing. So a 1" copper tube would be expected to have an outside diameter of 1". Again, this is not the case. In fact, most copper tubing products have an outside diameter ⅛" larger than their spoken size. Therefore, a 1" copper tube, regardless of its wall thickness, has an outside diameter of 1⅛". The size of any pipe or tube, as spoken, is usually based on the approximate, or nominal, size. That specific dimension is rarely found on the material, inside or outside. Further, it is usually based on the inside diameter, except for one particular type of copper tubing.

Another important fact does nothing to clarify the difference. Both pipes and tubes come in different wall thicknesses. As the wall thickness increases, the outside diameter does not change. The added wall thickness shrinks the inside diameter, rather than increasing the outside diameter. This is true for both pipes and tubes. However, the nominal size remains the same.

The issue is undoubtedly confusing. Regardless of the many discrepancies and valid arguments about the differences in pipes and tubes, it is the tendency of experienced HVAC technicians and installers to consider tubes to be relatively thin-walled materials, often with some level of flexibility, while the walls of a pipe are relatively thick.

There are also medical gas types of copper tubing. They are typically identified as Medical Gas Type K and Medical Gas Type L.

1.1.2 Identifying Markings

The labeling on copper tubing contains very important information. Manufacturers must permanently etch or stamp Types K, L, M, DWV, and medical gas copper tubing to show the tube type, the name or trademark of the manufacturer, and the country of origin. This same information must be printed on the hard-drawn tubes in a specific color that corresponds to the tube type. Type ACR does not require these same permanent markings, but they may be present. Hard ACR tubing is best identified by the printed markings (*Figure 2*). Soft ACR tubing may not have printed or permanent markings at all. *Table 1* shows five types of copper tubing and the corresponding color code for labeling. Note that the Type ACR color code is the same as Type L.

1.1.3 Copper Tube Sizing

The sizing of most copper tubing is based on the nominal, or standard, size of the inside diameter. This is true for Types K, L, M, and DWV. However, remember that each of these types have different wall thicknesses, and the inside diameter (ID) of the tubing changes as the wall thickness changes. The OD remains the same. This means that the OD of Type M, Type L, and Type K tubes are the same for any given size. The ID of each one, however, is different.

Figure 3 provides a comparison of dimensions for each type of copper tubing, based on a nominal size of 1". Remember that Type ACR is referred to by its actual OD. In the example shown then, all types would be identified as 1" except for the Type ACR. It is identified as 1⅛" tubing. The dimensions shown are based on the ASTM standards. *Table 2* provides some detailed information about Type ACR tubing. Charts for all tubing types are readily available from a number of sources, such as the Copper Development Association.

Table 1 Copper Tubing Color Codes

Type	Color Code
K	Green
L	Blue
M	Red
DWV	Yellow
ACR	Blue

Did You Know?

Copper is one of the most plentiful metals. It has been in use for thousands of years. In fact, a piece of copper pipe used by the Egyptians more than 5,000 years ago is still in good condition. When the first Europeans arrived in the new world, they found the Native Americans using copper for jewelry and decoration. Much of this copper was from the region around Lake Superior, which separates northern Michigan and Canada.

One of the first copper mines in North America was established in 1664 in Massachusetts. During the mid-1800s, new deposits of copper were found in Michigan. Later, when miners went west in search of gold, they uncovered some of the richest veins of copper in the United States.

03103-12_F02.EPS

Figure 2 Identifying markings on Type ACR tubing.

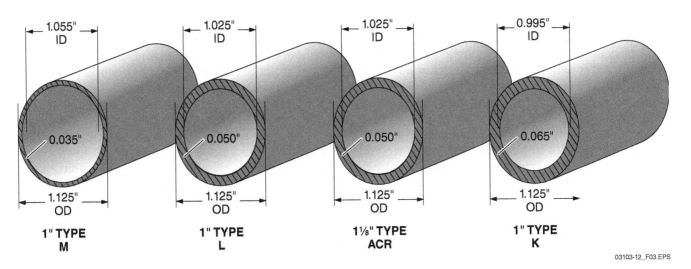

Figure 3 Tubing dimensions for a nominal 1" copper tube size, for each type of copper tubing.

Table 2 Dimensions and Physical Characteristics of Type ACR Copper Tubing

Nominal or Standard Size, inches		Nominal Dimensions, inches			Calculated Values (based on nominal dimensions)				
		Outside Diameter	Inside Diameter	Wall Thickness	Cross Sectional Area of Bore, sq inches	External Surface, sq ft per linear ft	Internal Surface, sq ft per linear ft	Weight of Tube Only, pounds per linear ft	Contents of Tube, cu ft per linear ft
⅛	A	.125	.065	.030	.00332	.0327	.0170	.0347	.00002
³⁄₁₆	A	.187	.128	.030	.0129	.0492	.0335	.0575	.00009
¼	A	.250	.190	.030	.0284	.0655	.0497	.0804	.00020
⁵⁄₁₆	A	.312	.248	.032	.0483	.0817	.0649	.109	.00034
⅜	A	.375	.311	.032	.076	.0982	.0814	.134	.00053
	D	.375	.315	.030	.078	.0982	.0821	.126	.00054
½	A	.500	.436	.032	.149	.131	.114	.182	.00103
	D	.500	.430	.035	.145	.131	.113	.198	.00101
⅝	A	.625	.555	.035	.242	.164	.145	.251	.00168
	D	.625	.545	.040	.233	.164	.143	.285	.00162
¾	A	.750	.680	.035	.363	.196	.178	.305	.00252
	A	.750	.666	.042	.348	.196	.174	.362	.00242
	D	.750	.666	.042	.348	.196	.174	.362	.00242
⅞	A	.875	.785	.045	.484	.229	.206	.455	.00336
	D	.875	.785	.045	.484	.229	.206	.455	.00336
1⅛	A	1.125	1.025	.050	.825	.294	.268	.655	.00573
	D	1.125	1.025	.050	.825	.294	.268	.655	.00573
1⅜	A	1.375	1.265	.055	1.26	.360	.331	.884	.00875
	D	1.375	1.265	.055	1.26	.360	.331	.884	.00875
1⅝	A	1.625	1.505	.060	1.78	.425	.394	1.14	.0124
	D	1.625	1.505	.060	1.78	.425	.394	1.14	.0124
2⅛	D	2.125	1.985	.070	3.09	.556	.520	1.75	.0215
2⅝	D	2.625	2.465	.080	4.77	.687	.645	2.48	.0331
3⅛	D	3.125	2.945	.090	6.81	.818	.771	3.33	.0473
3⅝	D	3.625	3.425	.100	9.21	.949	.897	4.29	.0640
4⅛	D	4.125	3.905	.110	12.0	1.08	1.02	5.38	.0833

Thankfully, one feature of copper tube sizes holds true. Tubing sizes stated as nominal have an actual OD that is ⅛" larger than the stated size. For example, 1" Types K, L, and M all have an OD of 1⅛". Type ACR of the same size would be called 1⅛" OD. As a result, the same fittings can be used on, and will properly fit, any of the four types of copper tube.

It should be noted that copper tubing is made to metric standards as well as imperial standards.

ASTM Standard B88 covers imperial copper tubing standards. *ASTM Standard B88M* is a companion publication covering the metric sizes. Metric copper tubing has several different types as well, identified as Types A, B, and C. Type A has the thickest wall, while Type C has the thinnest wall. There are quite a few other standards to which copper tubing is manufactured across the world.

It is important to remember that the metric and imperial tubing sizes and fittings are not interchangeable. Unlike metric hardware, such as nuts and bolts, metric copper tubing is virtually nonexistent in North American products or installations. You should not expect to encounter it in the foreseeable future. If it is encountered, it would be handled in the same manner as the imperial sizes. Remember that metric fittings would need to be matched with the metric tubing.

HVAC technicians should also be aware that the copper tubing used by manufacturers to construct finned coils for air conditioning and refrigeration systems do not meet the same specifications as commercially available copper tubing. The wall thickness is typically very thin, allowing for maximum heat transfer. This also makes them more prone to both physical and environmental damage. All other copper tubing in a typical HVAC unit, other than that contained within the coil assembly, is usually Type ACR or Type L.

1.2.0 Copper Tube Fittings

There are many types of copper tube fittings. Each type is used for a specific joining method. The different fittings will be reviewed based on the joining process used for them. The joining processes include flare joints, compression fittings, soldering, and brazing.

Regardless of the type of fitting used, virtually any flaw in the wall of a pipe or tube creates turbulence and pressure drop. All fittings create some level of obstacle or friction that increases pressure drop. In addition, abrupt or excessive changes in the direction of flow also create friction and increase pressure drop. For this reason, it is important to plan the route of tubing and piping systems carefully, and to minimize the number of fittings and changes in direction. More fittings in any piping system also increase the possibility of leaks.

1.2.1 Flare Fittings

The flared joint is a common method of joining copper tubing. Although it is rarely used today to connect tubing sizes greater than ⅝" OD, it is very common for smaller tubing sizes. The most common application for flare fittings is to connect re-

frigerant lines on small refrigeration circuits, and to connect various pressure controls to all sizes of systems (*Figure 4*). Flare fittings are not as reliable as heat-bonded joints on copper tubing. However, mechanical joints like flare fittings can be taken apart and reassembled much more easily when component service is required.

The flare fittings used in refrigeration and air conditioning consist of a variety of elbows, tees, unions, adapters, and the flare nuts that match the size and angle of the flare (*Figure 5*). They are drop-forged brass and are accurately machined to form the 45-degree flare face. Although flare fittings used in other industries may have different face angles, the 45-degree flare angle is used in the HVAC industry.

Fitting sizes are based on the size of tubing they fit. Flare nuts are hexagon-shaped, like most nuts. Couplers and adapters also have a hexagon-shaped area to accommodate a wrench. Tees and elbows typically offer only two flat sides for a

03103-12_F04.EPS

Figure 4 Pressure control, designed to connect to flared tubing.

Tubing Sets

Interconnecting tubing packages are available to connect the evaporator and condensing units of split-system air conditioners and heat pumps for residential and light commercial use. The copper tubing in most packages is Type ACR, factory-charged with nitrogen and fitted with plugs. Others may even be pre-charged with refrigerant. The suction, or low-pressure refrigerant vapor, line is usually pre-insulated as well. When using a tubing package, leave the plugs in place until just before making connections. This will minimize the entry of air and contaminants.

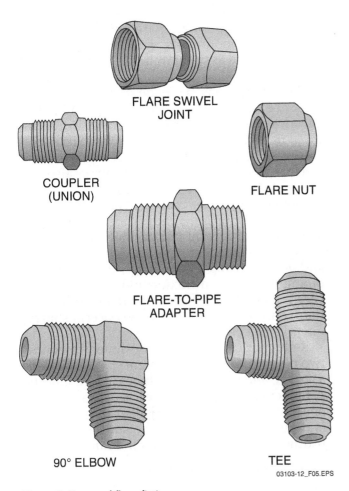

FLARE SWIVEL
JOINT

COUPLER
(UNION)

FLARE NUT

FLARE-TO-PIPE
ADAPTER

90° ELBOW

TEE

03103-12_F05.EPS

Figure 5 Types of flare fittings.

wrench. A flare nut wrench (*Figure 6*) is best used with these fittings when possible. However, the correct size of open-end wrench, as well as an adjustable wrench when used properly, is acceptable. A flare nut wrench is not designed to hold fittings such as the tee and the elbow.

Flare-to-pipe adapters are often needed to connect a copper line to a component such as a compressor. Since the point of connection on the component is often a pipe thread, it must be adapted to allow for the connection of a copper line. The flare-to-pipe adapter, usually called a flare adapter, accomplishes this task. Such adapt-

ers are also available as elbows and tees, and in both male and female forms.

1.2.2 Compression Fittings

Compression fittings (*Figure 7*) are sometimes used to connect tubing to various components. This method takes less time than making flared or heat-bonded connections. A seal is obtained by connecting the tubing to the threaded fitting with a coupling nut and a compression ring. The compression ring is known as a ferrule (*Figure 8*). It is also referred to as an olive in the United Kingdom. The joint is formed by compressing the ferrule between the nut and the threaded fitting, forming a tight seal. Compression couplings are available in types similar to those for flare fittings and in tubing sizes from ¼" to 1" nominal. Adapters are available to join these couplings to other fittings. Compression joints can be created faster than flare joints.

Ferrules can come in several different shapes and materials depending on the intended use. Compression fittings work with plastic tubing as well as copper. Ferrules used with plastic tubing may be plastic or copper. Some types of compression fittings may also have a metal or plastic sleeve that slides on the tubing and passes through the compression nut.

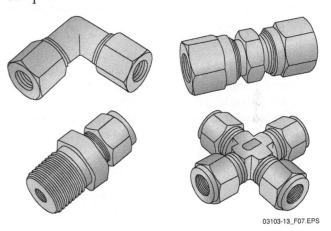

03103-13_F07.EPS

Figure 7 Common compression fittings.

03103-13_F08.EPS

Figure 8 Compression ring, or ferrule.

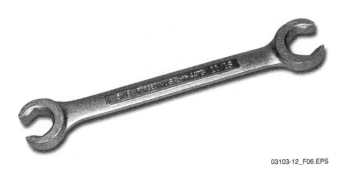

03103-12_F06.EPS

Figure 6 Flare nut wrench.

Common compression fittings are not used on refrigerant tubing in the HVAC field. Most HVAC experts prefer flare fittings to compression fittings in most situations due to reliability issues. However, there are a number of specialized compression fitting systems on the market that are used for refrigerant piping. Several different styles with higher levels of reliability have been developed by manufacturers to make connections to small systems easy and as foolproof as possible. When these fittings are encountered in the field, follow the manufacturer's instructions for assembly carefully.

Common compression fittings are often used on small water lines to connect humidification equipment, grease lines for large bearings, and a few other noncritical applications. However, the use of common compression fittings in the HVAC industry as a whole is somewhat limited.

1.2.3 Sweat Fittings

Copper tubing is often joined by soldering or brazing techniques in which a soft metal is melted in the joint between the walls of the two parts. A separate training module is devoted to this joining process.

Soldering and brazing are both heat bonding techniques. They are often referred to as sweating. Fittings known as sweat fittings (*Figure 9*) are made for making soldered and brazed connections. Sweat fittings may be made of copper or brass. They are made slightly larger than the tubes to be joined, leaving enough room for solder to flow into the joint. Some sweat fittings, like the male adapter shown in *Figure 9*, allow copper tubing to be joined to a threaded connection. Sweat adapters also come in female form, with the threads on the inside.

Elbows, or ells as they are commonly called, are available in long radius and short radius forms. For refrigerant piping, long radius ells are always preferred. Long radius ells reduce friction and pressure drop in the refrigerant flow. Smooth piping with limited friction and pressure drop is very important in refrigerant circuits. Where friction and pressure are less of a concern, short radius ells can be used.

Reducing couplings are designed to connect two pieces of tubing of different sizes. Each end is a socket, and slips over the tubes to be connected. A reducing bushing however, is designed so that one end slips over tubing while the opposite end is designed to slip inside the socket of another fitting. The bushing end has the same OD as an

Tee Fittings

The orientation of tee fittings in a system where significant flow occurs can be very important. Tee fittings have branch and run connections, as shown here. Improperly connecting a tee can lead to a condition known as bull-heading. Bull-heading results in turbulence, greatly adding to the pressure drop. Bull-heading occurs when tees are installed in such a position that the flow input is applied to the branch instead of the run. Tees should be installed so that the pressure input is applied to the run. If more than one tee is installed in the line, a straight piece of pipe at least 10 tube-diameters in length between tees is recommended to avoid additional turbulence whenever possible.

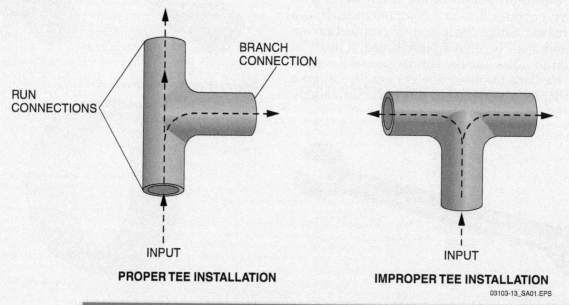

PROPER TEE INSTALLATION

IMPROPER TEE INSTALLATION

03103-13_SA01.EPS

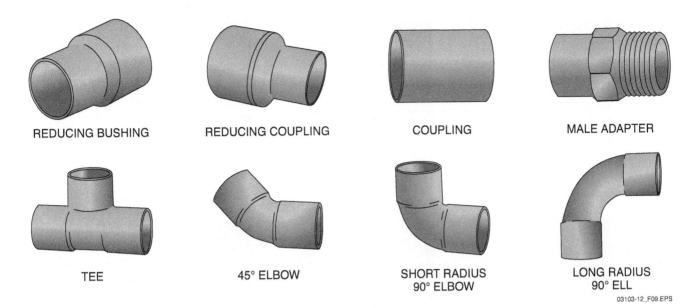

REDUCING BUSHING   REDUCING COUPLING   COUPLING   MALE ADAPTER

TEE   45° ELBOW   SHORT RADIUS 90° ELBOW   LONG RADIUS 90° ELL

03103-12_F09.EPS

Figure 9 Sweat fittings.

equally sized piece of tubing. *Figure 10* shows the two types of fittings together for clarity.

1.2.4 Pressed Fittings

Pressed fittings (*Figure 11*) are designed to allow installers to connect pipe and tube without flame using innovative press tools. Press fitting systems were introduced to North America by Viega LLC, which subsequently designed press fittings for use with not only copper but also stainless steel, black iron pipe, and PEX tubing. With this press technology, connections can be made in as few as four seconds with greater consistency than soldering or brazing. Pressed fittings are available for a wide variety of applications, ranging from potable water to natural gas and corrosive chemicals.

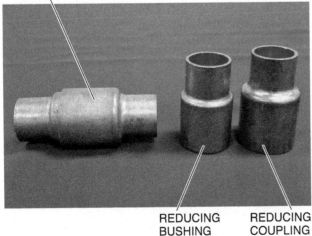

THE BUSHING CAN
SLIP INSIDE THE COUPLING

REDUCING BUSHING   REDUCING COUPLING

03103-13_F10.EPS

Figure 10 A reducing coupling and a reducing bushing.

The use of pressed fittings for refrigerant piping has been somewhat limited, but they are used in numerous other systems. Special pressing tools (*Figure 12*) that use different pressing jaws for each size of fitting are required. Fittings to accommodate tubing sizes up to 4" are available.

The assembly of pressed fittings is not covered in this module. However, it is important to ensure that the chosen fittings are indeed compatible with the tool in use. The manufacturers of the pressing tools often do not manufacture the fittings.

1.2.5 Swaged Joints

Swaged joints are made without an additional fitting. Technically, swaging describes any process of using a tool to shape metal. In the HVAC trade, it refers to the use of a tool to expand the end of a piece of tubing. The process forms a female socket that is the proper size to accept the end of another piece of tubing.

This type of joint is created with a swaging tool (*Figure 13*). The tool is used to expand the end of a piece of tubing to match the size of a fitting socket (*Figure 14*). When properly done and heat-bonded by soldering or brazing, swaging produces very secure joints. However, it does take more time than using a sweat coupling. Swaged joints are almost exclusively used to join two lengths of tubing, so their use on the job site is very limited. Tubing is often swaged in the factory environment using automated equipment to reduce manufacturing costs. Soft copper is fairly easy to swage; hard copper should not be swaged on site, although it can be done with the proper equipment.

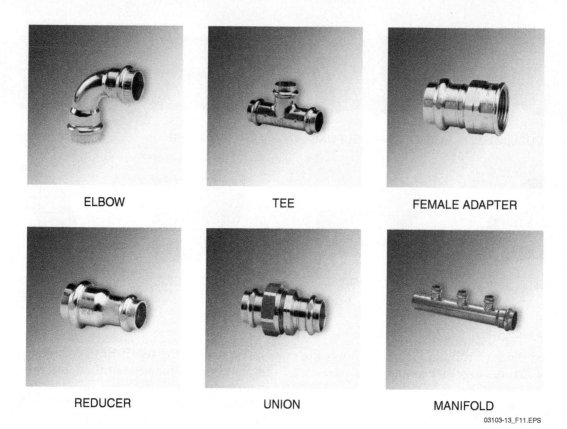

ELBOW   TEE   FEMALE ADAPTER

REDUCER   UNION   MANIFOLD

03103-13_F11.EPS

Figure 11 Pressed fittings.

03103-13_F13.EPS

Figure 13 Swaging tools.

03103-13_F12.EPS

Figure 12 RIDGID® press tool and jaws.

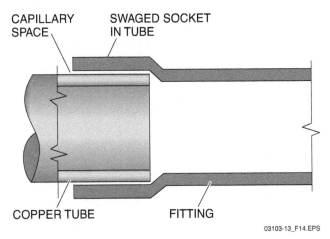

CAPILLARY SPACE

SWAGED SOCKET IN TUBE

COPPER TUBE

FITTING

03103-13_F14.EPS

Figure 14 Swaged joint.

Additional Resources

The Copper Tube Handbook. A publication of the Copper Development Association. **http://www. copper.org/publications/pub_list/pdf/copper_ tube_handbook.pdf**

1.0.0 Section Review

1. The copper tubing listed here with the thickest wall is _____.

   a. Type K
   b. Type L
   c. Type M
   d. Type ACR

2. Common compression fitting joints _____.

   a. take more time to create than flare joints
   b. are commonly used for refrigerant lines
   c. are more reliable than flare joints
   d. are not used for refrigerant lines

2.0.0 Joining Copper Tubing and Fittings

Objective

Describe and demonstrate how to join copper tubing mechanically.

   a. Measure, cut, and bend copper tubing to prepare it for joining.

   b. Describe and demonstrate the methods and tools used to join copper tubing.

   c. Describe common hangers and supports associated with copper tubing installations.

Performance Tasks

1. Cut and bend copper tubing.
2. Safely join copper tubing using mechanical fittings.
   - Flare tubing and complete a flared connection.
   - Use a compression fitting and ferrule to make a connection
   - Use a swaging tool to swage a piece of tubing.

Trade Terms

Copper-clad: Components that have been coated or covered with a thin copper layer.

Plate: To apply a layer of a metal on the surface of another metal. Other, less-precious metals are often plated with gold to reduce the value or expense.

Reamer: A tool designed to remove the sharp lip and burrs left inside a pipe or tube after cutting.

Work hardening: The repetitive reforming (such as bending or flexing) of a material, causing a permanent change in its crystalline structure and an increase in its strength and hardness.

With the primary materials now introduced, the methods and steps to join copper tubing can be presented in detail. Soldering and brazing are skills that are presented in another training module. The process of joining copper tubing using flare and compression fittings is the focus of this module. Swaging copper tubing is also presented. Some of the basic skills required, such as cutting and bending, are also needed for soldering and brazing.

2.1.0 Measuring, Cutting, and Bending Copper Tubing

Before a copper tubing joint can be made, the material must be cut to the proper length. Of course, if a very long run of tubing is being installed, many of the joints will not require cutting. Full lengths of copper tubing can be joined, until changes in direction or component connections are required. Soft copper tubing can also be bent and formed, eliminating the need for some fittings.

2.1.1 Measuring Tubing

It is extremely important to measure tubing carefully. It is relatively easy to measure between two known points and cut a piece of tubing to that length. However, when fittings are involved, the dimensions of the fitting must be taken into account. The tubing must be cut to a shorter dimension in order to allow room for the fittings between obstacles. If it is too long, the assembly will not fit the specified location. If it is too short, then the assembly will be too short to mount, or the tubing will not extend to the bottom of a sweat-fitting socket. To arrive at the proper cut length, the additional length added by the fitting must be taken into consideration. This is true of each type of fitting, both mechanical and sweat.

To establish the length of tubing needed to connect two fittings, two of three dimensions must be considered (see *Figure 15*). For sweat fittings, a measurement can be made between the faces of the two fittings. If so, then the socket depth of the fitting, multiplied by two, must be added to arrive at the proper length. Alternatively, a measurement can be made from the centers of the fittings, or to the backs of the fittings. In these cases, the length that the fittings add must be subtracted to arrive at the proper cut length. The installation situation typically determines the simplest and most accurate approach.

Compression fittings have a socket like sweat fittings, but it is generally very shallow (*Figure 16*). However, the socket depth must be accounted for with compression fittings as well. For flare fittings, the tube must be long enough to reach the top of the beveled edge, after the flare is applied. The flaring process shortens the length of the tube, although the difference is very slight. For a precision fit, the loss of length in making the flare must also be considered. However, when small tubing is being used, and when there are bends formed in the tubing, the cut length of the tubing need not be quite so precise to terminate it.

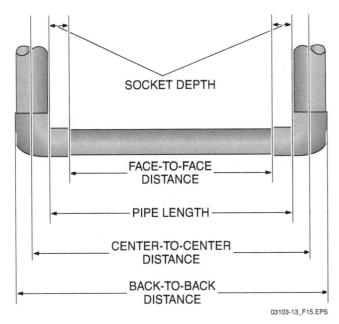

Figure 15 Typical tubing measurements to accommodate sweat fittings.

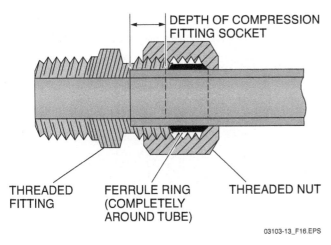

Figure 16 Compression joint cut-away.

2.1.2 Cutting Tubing

Copper tubing is typically cut with either a hand-held tubing cutter (*Figure 17*) or a hacksaw. The tubing cutter is always preferred because it produces a square end and creates no saw cuttings. Particles from the cutting action of a saw can be very harmful in refrigerant lines and other piping systems. The only time a hacksaw would be chosen is when there are significant obstructions to the swing of a tubing cutter. Even in those unusual situations, the tubing cutter can often be used to cut through the majority of the tube, and then the cut finished with a hacksaw to minimize irregularities and saw cuttings. As shown in *Figure 17*, tubing cutters come in a variety of sizes to accommodate different tube sizes and for use in close quarters.

THIS DESIGN ALLOWS FOR QUICK ADJUSTMENT TO THE TUBE SIZE

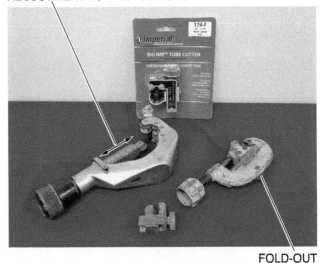

FOLD-OUT REAMER

Figure 17 Handheld tubing cutters.

Well-fitted gloves should be worn whenever working with copper tubing and the related tools, as well as safety glasses. To use the tube cutter, place it on the tube at the desired cutting point (*Figure 18*). Tighten the knob slightly, forcing the cutting wheel against the tube. It should not be tightened so tight as to prevent turning it around the tube. The cut is made by rotating the cutter around the tube under constant pressure. After one to three turns, the cutter will feel loose and can be tightened again, deepening the cut. Overtightening, especially when the blade is close to penetrating the tube, causes the sharp edge inside the lip of the tube to be more pronounced. Eventually, the two pieces will separate cleanly.

 WARNING!
The resulting edge inside the cut tube is extremely sharp! Avoid running the fingers across the cut edge, even when gloved.

Unrolling Soft Copper Tubing

To unroll soft copper tubing reasonably straight and minimize distortion of the tube, stand the roll up vertically and place the bottom on a flat surface. Hold the end of the tubing firmly to the surface with one hand, and roll the tubing away from your hand.

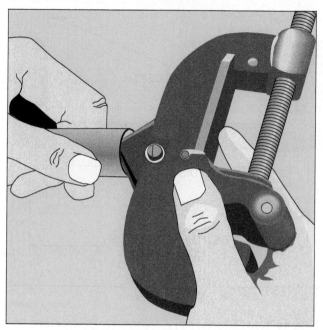

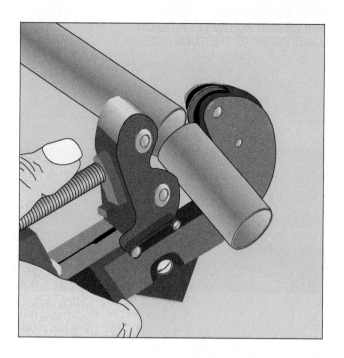

03103-13_F18.EPS

Figure 18 Cutting with a handheld tubing cutter.

Once the pieces separate, the tube ends must be deburred. Deburring can be done using a built-in deburring blade, or reamer, in the tubing cutter. It can also be done using a separate deburring tool (*Figure 19*). The tools are designed to carve out the sharp lip, or burr, left inside the tube edge. Not only is this sharp edge dangerous, it creates turbulence and resistance to flow. Further, over time, the surface can erode and transport small amounts of copper throughout the system. The soft copper molecules are then free to plate other metal surfaces, such as compressor piston rings, in the system.

Always point the tubing down whenever possible while reaming. This will prevent any burrs removed from falling down inside the tubing. To make sure any particles are removed, run a brush or cloth swab through the pipe after cutting and deburring operations are completed. If you are unsure all the particles can be removed, don't deburr the pipe. Although the sharp edge is not desirable, it may still be better than having large burrs and shavings remain in the piping system.

When using a tubing cutter is not an option, a hacksaw and sawing fixture such as a miter box may be used. The miter box helps keep the end square and allows for a more accurate cut. The blade of the hacksaw should have at least 32 teeth per inch (TPI). A saw blade that is too coarse will chatter and snag on the tubing walls as the blade reaches the mid-point of the tube. Avoid leaving saw cuttings inside the tubing. File the end and outer edge lightly with a single-cut file to produce

03103-13_F19.EPS

Figure 19 Reaming tool. This style reams the inside of the tube as well as the outside.

a smooth surface if necessary. Deburr the inside edge and carefully clean the inside of the tubing with a cloth. Turning the tube end down and lightly tapping it may help to dislodge any remaining particles.

2.1.3 Bending Tubing

Soft copper tubing can be bent around obstructions, as opposed to cutting and using fittings. Smaller-diameter tubing can easily be bent for large-radius turns by hand, but care must be taken to avoid flattening the tube with a bend that is too sharp. When bending is done properly, the tubing will not collapse or buckle. Following the bending process, the tube often becomes harder due to work hardening.

Table 3 provides the minimum bending radius, in inches, for various sizes of tubing. Although hard-drawn tubing can be bent, it is rarely done on the job site. However, manufacturers often bend hard-drawn tubing in the factory environment to construct HVAC equipment.

Tube-bending springs (*Figure 20*) are available for placement on the outside of the tube to prevent it from collapsing during hand bending. The spring slides over the tubing, and helps the tube maintain its shape. Slip the bending spring over the tube to the area where the bend is desired, and then form the tube to the desired shape. Make the bend radius as large as possible to avoid damaging the tube. Note that the spring may be more difficult to remove after the bend, as the tube shape does change some, applying more pressure to the inner walls of the spring. Twisting the spring helps relieve some of the friction. Remember to make the bend prior to flaring the end of the tube or attaching any fittings. Otherwise, the spring cannot be used.

Handheld tubing benders (*Figure 21*) are also available for making accurate and reliable bends in both soft and hard tubing. The tool provides the necessary leverage to make a smooth bend. Various sizes of benders and attachments are available to bend tubing of various diameters. For tubing greater than 1" OD, larger manually operated or powered equipment is necessary. To ensure bends are made without damage to the tubing and to the proper radius, follow the manufacturer's recommendations for the tool in use.

Using handheld or other benders instead of bending springs is recommended any time there are a number of bends to be done in the same area. They provide highly consistent bends with a smaller radius, and the bends are usually more visually appealing.

2.2.0 Joining Copper Tubing

Because the copper tubing used for refrigerant piping is too thin for threading, mechanical joining or heat-bonding methods are used. The mechanical joining methods include flared and compression fittings that can be taken apart repeatedly if required. Heat bonding is accomplished by soldering or brazing. Although heat-

Table 3 Minimum Bend Radius for Copper Tubing

Nominal, Standard Size, in	Tube Type	Temper	Minimum Bend Radius[1], in
¼	K, L	Annealed	¾
⅜	K, L	Annealed	1½
	K, L, M	Drawn	1¾
½	K, L	Annealed	2¼
	K, L, M	Drawn	2½
¾	K, L	Annealed	3
	K, L	Drawn	3
1	K, L	Annealed	4
1¼	K, L	Annealed	9

[1] The radii stated are the minimums for mechanical bending equipment only.

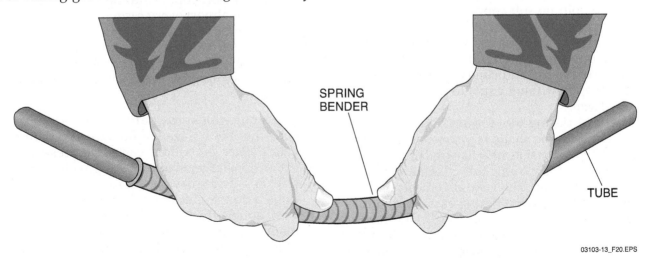

Figure 20 Using a bending spring.

03103-13_F20.EPS

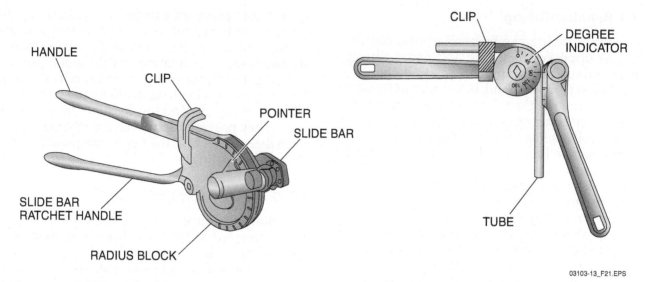

Figure 21 Handheld tubing bender.

bonding methods are considered permanent, the joints can be separated by reheating them. Some level of damage or hardened debris usually results when separating brazed joints. To make a new, reliable brazed joint, the end of the tubing sometimes must be cut off or a new section of tubing used. This section of the module focuses on mechanical joining methods, including both flare and compression joints.

Regardless of the type of joint being made, it is crucial that the end of the tubing be properly shaped, smooth, clean, and free of burrs inside and out. Burrs on the inside of the tubing do not affect the quality of a compression joint, but they do create other problems. However, burrs on the inside of cut tubing interfere with making a reliable and properly shaped flare.

2.2.1 Making Flared Connections

The flared connection (*Figure 22*) is a popular method of joining soft copper tubing. When it is done correctly, it will not leak and can be easily taken apart for service as necessary. Flared copper tubing joints are acceptable in some jurisdictions for fuel gas piping, and can typically be used to connect copper tubing equal to or less than a 1" nominal size.

A special flaring tool (*Figure 23*) expands the tube's end into the shape of a cone, or flare. The handheld tool consists of a flaring bar with several openings for various tube sizes, and a yoke with a flaring cone. The yoke also incorporates a clamp to attach it to the flaring bars. Flaring bars for larger sizes of tubing, such as 1" OD, may be usable with only one size of tubing to limit the size of the tool. The tool must match the size of the tubing in use. There are several different flar-

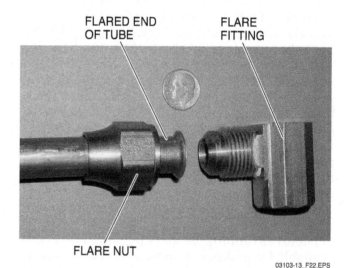

Figure 22 The components of a typical flared connection.

ing tool designs, some of which have accessories to also swage the end of a tube.

Annealed, soft copper tubing should be used for flaring in almost every case.

Flares can be made at different angles, and the flaring cone must match the desired angle. The required flare angle for most uses in the HVAC trade is 45 degrees. Two kinds of flare fittings are popular:

- The single-thickness flare forms a 45-degree cone that fits up against the face of a flare fitting (*Figure 24*). Under excessive pressure or expansion, larger sizes of tubing with a single-thickness flare may be weak. In these applications, a double-thickness flare is preferable.
- The double-thickness flaring tool and the method of forming it are shown in *Figure 25*. The single-thickness flare is formed in one operation, and then the lip is folded back onto

Figure 23 Flaring tool.

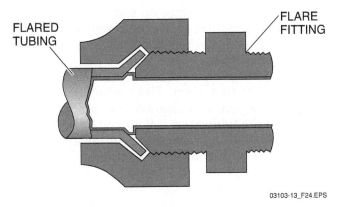

Figure 24 Cross-section of a single thickness flare joint.

FLARED TUBING

FLARE FITTING

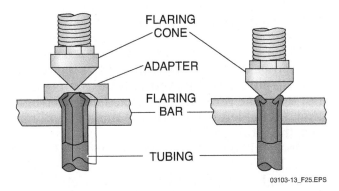

FLARING CONE

ADAPTER

FLARING BAR

TUBING

Figure 25 Forming a double-thickness flare.

itself and compressed again. Double-thickness flare connections can also be taken apart and reassembled more often without damage. Flaring kits with adapters can create both single- and double-thickness flares.

To make a flare, first measure and cut the tubing to the desired length. Once the tubing is cut, ream the cut end to remove burrs. Complete removal of the burr is important to the quality of the flare. It is the surface of the tubing in this area that becomes the sealing surface of the flare joint. Avoid gouging the tubing too deeply while reaming. Check to ensure the tube has a proper round shape. Remove oxidation on the outer surface using sand cloth or a nylon abrasive pad (*Figure 26*).

The sequence of flaring a piece of copper tubing is shown in *Figure 27*. Before constructing the flare, ream the cut tube and slide the flare nut onto the tube with the threaded portion facing the end of the tube. Then position the tube inside the flaring bar opening and tighten the bar screws to hold it in place. The end of the tube must protrude from the bar the correct distance to allow enough material to create the flare. Consult the tool manufacturer's literature for the proper position.

Place the yoke over the flaring bar with the cone centered over the tube. Tighten the yoke slowly and allow the cone to position itself in the center of the tube. Once the yoke is secure and centered, continue tightening it by turning clockwise. The cone will force its way down into the tube, spreading the edges to match the 45-degree angle face of the flaring bar.

Tighten the yoke until the cone is firmly bottomed out against the tube and flaring bar. Do not overtighten as it could damage or split the flare. When complete, loosen and remove the yoke by turning it counterclockwise. Remove the tube from the flaring bar. Examine the flare surface carefully, on both sides. The result should be a smooth, round flare of sufficient length to cover the mating surface of the flare nut. Splits or visible flaws are reasons to reject the flare and try again.

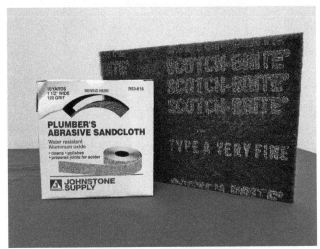

Figure 26 Nylon abrasive pads and sand cloth.

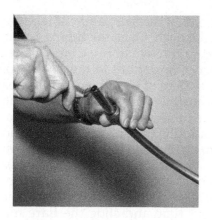

03103-13_F27.EPS

Figure 27 Flaring copper tubing.

If the flared surface is too long, it may interfere with the flare nut engaging the threads of the fitting. If it is too short, it will not cover the entire face of the fitting. The resulting length of the flare is dependent upon the initial position of the tube in the flaring bar.

The flare joint must be assembled using two wrenches. Do not use any sort of pipe thread sealant or similar materials on the face of the joint. Pipe thread sealant is also of no value on the flare nut threads. The primary sealing point of the joint is the front surface of the flare. The threads serve only to tighten the faces together with the flare

nut. Adding a drop of oil to the back of the flare, where it and the flare nut make contact, can help prevent galling.

2.2.2 Making Compression Connections

Like flare joints, compression fittings can be taken apart and reassembled. Compression connections are easier to make, since no special treatment of the tube is required. However, compression joints are not generally as reliable or as durable as flare joints under stress.

The tubing is prepared in the same manner as it is for a flare joint. The tubing must be cut square, reamed, and cleaned of oxidation. Make sure that the tubing is not flattened or bent and is free from any nicks or gouges in the area where the ferrule will land. If the compression nut and ring slip freely onto the tubing for a distance of an inch or so, the tubing is probably straight and round enough for a compression connection. If it binds and does not move freely, the tube is likely deformed.

Once the tube end is prepared, begin by sliding the compression nut onto the tube, with the open end toward the fitting. Then slide the ferrule onto the end of the tube. Insert the end of the tubing into the fitting socket, and pull the compression nut up to the fitting. Tighten the nut finger-tight first, all the while keeping the tubing firmly inside the fitting socket.

One of the keys to a successful joint is to keep the tubing tight against the bottom of the socket while tightening the nut. Since two wrenches are required—one to hold the nut and one to hold the fitting—this can be challenging. Once the nut has been tightened down roughly ½ turn, the tube will likely stay in position while the tightening is completed. One full turn after the fitting is finger-tight is usually sufficient to complete a leak-free joint. It is tempting and easy to overtighten compression joints, especially if the wrench is longer

In the Groove

The groove shown here in the roller of a tubing cutter allows a damaged flare to be cut off close to the end of the tube.

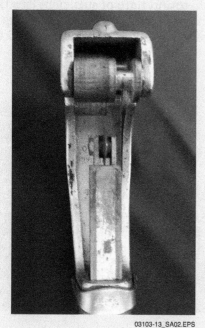

03103-13_SA02.EPS

than necessary. However, it is better for the joint to be slightly on the loose side. If it leaks, the fitting can still be tightened a little further. If the joint leaks due to overtightening, the tube end must be cut off and a new joint started. The ferrule cannot be removed once it is compressed onto the tube.

If a compression joint is disassembled, the end of the tubing may appear slightly bulged beyond the ferrule. This is normal, indicating that the ferrule has been compressed onto the tubing. Compression joints can easily be mastered with a little practice.

2.2.3 Swaging a Tube

Swaging a tube prepares it to accept the end of another tube for soldering and brazing. It is a method used to expand the inside diameter of a tube.

Hammer-driven swaging tools (*Figure 28*) are very simple devices. Stepped tools work with several different sizes of tubing. Other tools may be made to fit only a single size of tubing. The tool is simply inserted into the end of the tube, then hammered until the ring of the larger size meets the end of the tube (*Figure 29*). Swaging can also be done using a flaring tool set with the proper accessories (*Figure 30*). The swage is made by clamping the tube into the flaring bar and replacing the cone with the proper swage adapter. Turning the yoke handle clockwise pushes the swage adapter down into the tubing, expanding it as it goes.

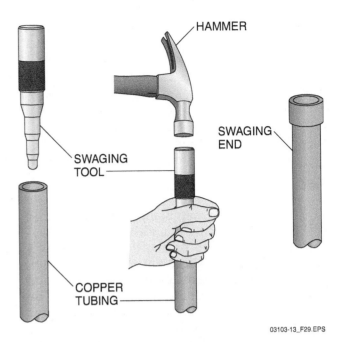

Figure 29 Swaging a tube with a hammer-driven tool.

03103-13_F30.EPS

Figure 30 Flaring tool set with swaging adapters.

2.2.4 Pressure Testing

When all joints are done and the installation is complete, it may be inspected by others to make sure that the work has been done in accordance with the job specifications and applicable codes. The system must also be leak-tested to be sure that all the joints are leak-free.

Testing under pressure is the best method of leak testing. Most of the testing in the HVAC trade is done using nitrogen. Nitrogen is inexpensive and, if there is a leak, the technician does not risk losing expensive refrigerant and violating laws regarding the release of refrigerants into the atmosphere. Nitrogen is also an inert and safe gas to use and handle. The test pressure will vary based on the purpose and design of the system, but it should never exceed the maximum pressure stated on the nameplate of any component. Although nitrogen is most commonly used, some manufacturers and codes permit the use of carbon dioxide (CO_2) for pressure testing. Sections of water systems that

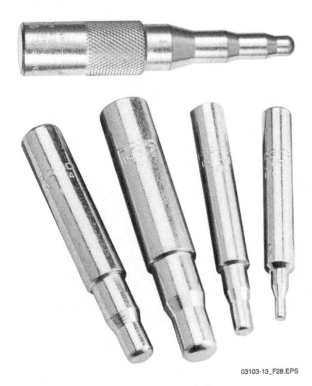

03103-13_F28.EPS

Figure 28 Hammer-driven swaging tool.

can be drained for repair are sometimes pressure tested by simply applying water pressure. A water leak is much easier to locate than escaping vapor.

Exercise caution when using nitrogen to build up pressure for testing at code specifications. A typical setup used to pressure test a system is shown in *Figure 31*. There must be a hand shut-off valve, pressure regulator, pressure gauge, and pressure-relief valve in the charging line. The relief valve should be adjusted to open at one or two psi above the test pressure.

A soap solution or commercial test liquid can be used to check for leaks when using nitrogen. This is done by wiping the outer surface of the joints with the solution and watching for any bubbles, which would indicate leaks. If no leaks are found, the nitrogen may be left in the system for 24 hours or longer to see if the pressure holds. Project specifications or codes may require a specific time period.

2.3.0 Hangers and Supports

Hangers and supports provide horizontal or vertical support of pipes and piping. Hangers generally describe a support that is anchored above the piping. Supports carry the piping load from below or from walls and other sturdy structural components. The principal purpose of hangers and supports is to keep the piping in alignment and to prevent it from bending or distorting.

Hangers that are in direct contact with copper usually must be copper-clad. It is best to keep copper and steel isolated from each other in piping systems. Under some conditions, corrosion can occur where there is contact between the two metals. Hangers of pure copper would be unnecessarily expensive and not as strong as steel.

To fasten a hanger assembly to beams and other metal structures, various styles of beam

Push-Connect Fittings

Push-connect fittings are another way that copper tubing can be joined. There is little that can be done to make a tubing joint any simpler—cut and prepare the tubing, then push it into the fitting with a twisting action. The joint is done. Fittings are available for use with copper tubing as well as several types of plastic tubing.

These fittings use a soft elastomeric sealing ring to seal the joint, and a stainless steel ring with teeth to grab and hold the tubing once it is inserted. While some fitting designs do not allow for easy removal, others can be disassembled quickly and easily. Most are rated to handle up to 200 psig and 250°F (121°C). Although many are still reluctant to consider them a permanent joint, their continued use in a variety of applications will eventually determine their long-term reliability and popularity.

03103-13_SA03.EPS

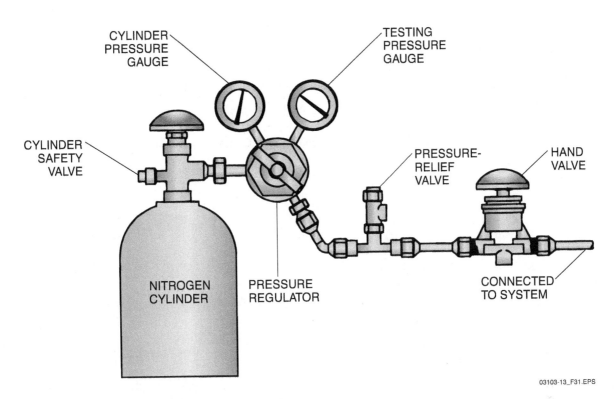

CYLINDER PRESSURE GAUGE

TESTING PRESSURE GAUGE

CYLINDER SAFETY VALVE

PRESSURE-RELIEF VALVE

HAND VALVE

NITROGEN CYLINDER

PRESSURE REGULATOR

CONNECTED TO SYSTEM

03103-13_F31.EPS

Figure 31 Nitrogen tank and accessories for pressure testing.

clamps (*Figure 32*) are a common choice. A piece of threaded rod or similar material is then connected to the clamp, and the pipe hanger is connected to the rod. *Figure 33* provides examples of two popular types of pipe hangers. Clevis hangers are very popular for all sizes of pipe and tubing. The split ring hanger can be used to suspend a pipe, but also to support it from below or horizontally.

A wide variety of straps are also available to mount tubing and pipe. Some simple examples of straps used for supports are shown in *Figure 34*.

Riser clamps (*Figure 35*) support pipe where the pipe penetrates floors vertically.

There are many different kinds of straps that are designed to hold tubing in place against a special type of channel (*Figure 36*). Copper or copper-clad straps can be used with the channel. Steel straps are also used with copper tubing, using a plastic insert to isolate the copper from the steel (*Figure 37*). The insert also provides a firmer grip around the tube and helps to isolate any vibration. In addition, the plastic strap provides some insulation at the location of the clamp. Any insulation needed for the tubing can often be butted against

Refrigerant Leak Detectors

Portable electronic leak detectors have been designed specifically for the location of refrigerant leaks. While some units are designed to detect a specific refrigerant compound or blend, others may be designed to detect a specific family of refrigerants. The units sound an alarm and/or register a meter reading when the compound is detected. Today's units are typically battery-powered. Many units have selectable and adjustable sensitivity ranges.

Some leak detectors are designed to detect the sound of leaking vapors. However, they can be unproductive in noisy environments.

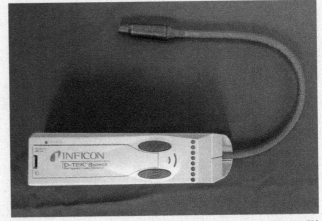

03103-13_SA04.EPS

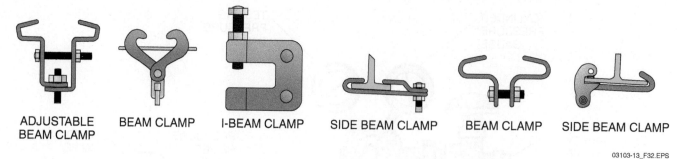

ADJUSTABLE BEAM CLAMP   BEAM CLAMP   I-BEAM CLAMP   SIDE BEAM CLAMP   BEAM CLAMP   SIDE BEAM CLAMP

03103-13_F32.EPS

Figure 32 Various beam clamps.

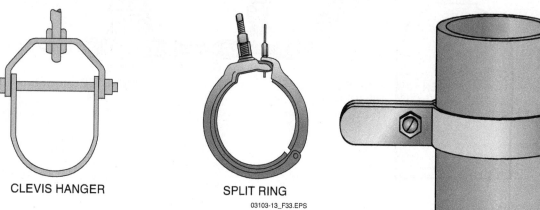

CLEVIS HANGER   SPLIT RING

03103-13_F33.EPS

Figure 33 Clevis and split ring hangers.

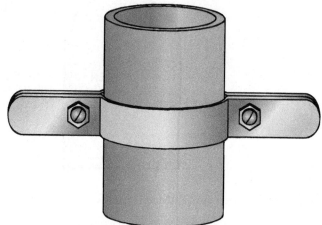

RISER CLAMP AROUND PIPE

RISER CLAMP WITHOUT PIPE

03103-13_F35.EPS

Figure 35 Riser clamp.

the clamp, rather than passed through it. With this type of clamp, an insulated insert that is rigid must be used. Otherwise, the clamp cannot grip the tube securely enough and would also crush the insulation. The channel material is available in a number of depths for different applications. The most popular depths are nominally ¾" and 1½". The channel can be mounted on walls, ceilings, or almost any other structurally reliable surface. When a number of copper tubes must be routed together, this method of support is very sound and provides a neat, consistent appearance.

Any time a pipe or tube must be insulated, the hanger system and hangers must be selected to

also accommodate the insulation thickness too. It is usually best, and often required by specifications, to ensure that the insulation is continu-

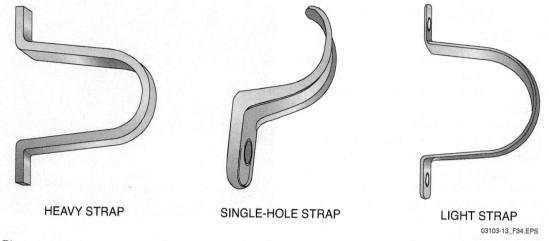

HEAVY STRAP   SINGLE-HOLE STRAP   LIGHT STRAP

03103-13_F34.EPS

Figure 34 Pipe straps.

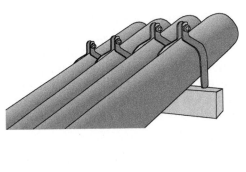

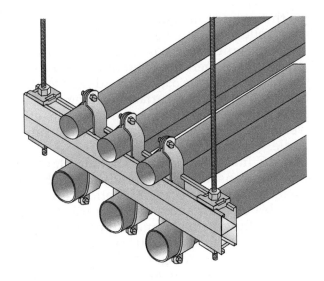

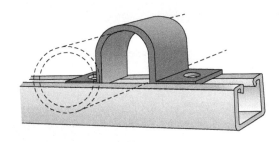

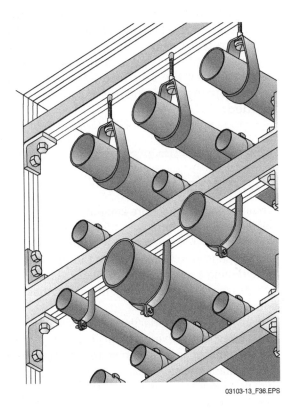

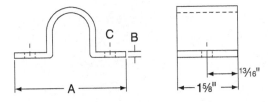

STANDARD PIPE STRAP

$1\frac{5}{8}''$

$\frac{13}{16}''$

03103-13_F36.EPS

Figure 36 Pipe straps and channel.

ous through the hanger or support rather than being spliced at each one. *Figure 38* shows some straps designed to accommodate pipe insulation. Clevis hangers are often fitted with a sheet metal saddle (*Figure 39*) to spread the weight of the load over a larger surface, keeping the insulation from being crushed and losing its effectiveness. The insulated line simply lies on the saddle.

HVAC installations often include plans and specifications that completely describe the hanger system. Project specifications are based on codes or ordinances and require strict adherence. Pipe hangers and supports are usually a part of the specifications. A specification for pipe hanger spacing, for example, may read as follows: "All piping shall be supported with hangers spaced not more than 10' apart (on center). Hangers shall be the malleable iron split-ring type and shall be as manufactured by XYZ Hangers, Inc. or other approved vendor." However, since the spacing of hangers and supports is a function of the type and size of piping used, a table to show the required hanger spacing for various piping and tubing materials is often provided.

Table 4 shows the recommended hanger spacing, as well as the hanger rod size, for copper tubing. Note that the spacing here is valid for copper tubes filled with liquid, so the added weight of the liquid has already been considered in the spacing. Hangers for steel piping can be spaced farther apart, due to the added strength and structural integrity of steel pipe.

03103-13_F37.EPS

Figure 37 Pipe straps with plastic inserts.

Pipe Insulation

Under certain temperature and humidity conditions, condensation will form on refrigerant and cold-water piping and may drip into equipment or occupied areas. Refrigeration piping can also pick up heat from the air, causing an air conditioning system to lose efficiency. Similarly, heat can escape from hot-water piping. To prevent these conditions, some piping is insulated.

Insulation is a material that slows down the transfer of heat. Cork, glass fibers, mineral wool, and polyurethane foams are examples of insulating materials. Insulation is ideally fire-resistant, moisture-resistant, and vermin-proof. It must also provide a vapor barrier to prevent moisture in the air from gaining access to the piping beneath. The most popular material for refrigerant lines is shown here. This is a closed-cell foam rubber material.

Type ACR soft copper tubing for refrigerant lines can be purchased with factory-installed insulation. If it is necessary to install the insulation at the job site, as much as possible should be added before the tubing is connected. That way, the insulation can be slid onto the tubing intact. The inside of the insulation is usually powdered to allow it to slip on easily.

If the insulation cannot be installed before the tubing is connected, it must be cut lengthwise to fit onto the pipe. The resulting seam must then be sealed with adhesive. Do not use tape.

Some pipes are always insulated, while others are insulated only under certain conditions. Local building codes and job specifications usually describe insulation requirements.

03103-13_SA05.EPS

Figure 38 Pipe straps designed to accommodate insulation.

Figure 39 A clevis hanger fitted with an insulation saddle.

Table 4 Recommended Hanger Spacing for Copper Tubing

Nominal Tubing Size		Maximum Span		Recommended Hanger Rod Size
In	mm	Ft	Meter	
½	(15)	5	(1.52)	⅜" – 16
¾	(20)	5	(1.52)	⅜" – 16
1	(25)	6	(1.83)	⅜" – 16
1¼	(32)	7	(2.13)	⅜" – 16
1½	(40)	8	(2.44)	⅜" – 16
2	(50)	8	(2.44)	⅜" – 16
2½	(65)	9	(2.74)	½" – 13
3	(80)	10	(3.05)	½" – 13
3½	(90)	11	(3.35)	½" – 13
4	(100)	12	(3.66)	½" – 13
5	(125)	13	(3.96)	½" – 13
6	(150)	14	(4.27)	⅝" – 11
8	(200)	16	(4.87)	¾" – 10

2.0.0 Section Review

1. Following the bending process, soft copper tubing may become softer.

   a. True
   b. False

2. What happens when a flare formed on a piece of tubing is too long?

   a. It will not cover the complete face of the fitting.
   b. The flare nut will slip over the flare at higher pressures.
   c. The flare nut may not be able to engage the fitting threads.
   d. The fitting and the flare face will not be at the same angle.

3. Nitrogen tanks used for pressure testing are routinely charged to a pressure in excess of _____.

   a. 1,500 psig
   b. 2,000 psig
   c. 3,000 psig
   d. 4,000 psig

4. The spacing between hangers and supports for copper tubing is based primarily on the _____.

   a. tubing size
   b. length of the tubing run
   c. use of the piping system
   d. tubing type, such Type L or Type M

SECTION THREE

3.0.0 PLASTIC PIPING

Objective

Recognize different types of plastic piping and show how it can be joined.

  a. Identify different types of plastic piping.
  b. Identify the tools and products needed and demonstrate how to join plastic piping.

Performance Tasks

3. Cut and join lengths of plastic piping.

Trade Terms

Chamfer: To create a symmetrical angled surface on an edge, breaking what would usually be a 90-degree angle. Chamfer is also used as a noun, as a name for the angled surface.

Witness mark: A mark made as a means of determining the proper positioning of two pieces of pipe or tubing when joining. A witness mark is typically in addition to a first mark, made in a location that will not be obscured by the joining process; the first mark is often obscured during the process.

Plastic pipe is used to drain condensate from furnaces and air conditioners, transport water in water-cooled HVAC systems, and even to vent combustion byproducts from some types of furnaces. Some plastic piping can be threaded, and threaded adapters are available to connect glued plastic piping to threaded-piping systems or components. There are several types of plastic pipe that are used in the HVAC industry, each with a different application. *Figure 40* shows how plastic pipe was used in a typical condensing furnace installation. The primary advantage that plastic piping offers over metal pipe is its resistance to corrosion or rot. Many types are highly resistant to chemical agents that would destroy metal piping systems in a relatively short period of time.

3.1.0 Types of Plastic Tubing and Pipe

There are many different types of plastic tubing and pipe in today's market. However, some types are encountered far more often than others are. You should become familiar with the types reviewed here.

3.1.1 ABS (Acrylonitrilebutadiene Styrene) Pipe

Acrylonitrilebutadiene styrene (ABS) pipe (*Figure 41*) was originally developed in the 1950s for use in the chemical industry and in oil fields. It is rigid, smooth, and has good impact strength at low temperatures. It is typically used for drain, waste, and vent (DWV) applications. It can perform reliably in the temperature range of –40°F to 180°F (–40°C to 82°C). Although it is generally used in applications that are at or near atmospheric pressure, ABS pipe is also available for pressurized systems in industrial environments. ABS pipe is sold in 10' and 20' lengths. Most ABS pipe is black in color, but other colors can be found.

Identifying markings on ABS pipe typically include the manufacturer and trademarks, the standard(s) that it was manufactured to, and the nominal size.

3.1.2 PE (Polyethylene) Tubing

Polyethylene (PE) tubing (*Figure 42*) is a thermoplastic material originating from ethylene. PE is available in rolled coils for sizes less than 6" diameter. Rigid, semi-rigid, and corrugated lengths up to 40' are available for the larger sizes. Diameters up to 63" are not uncommon. PE tubing and pipe is available in many different colors and wall thicknesses. In addition to DWV applications, it is also used for domestic water service and even fuel gases in some locations. One application in the HVAC industry is to provide water to water-source heat pumps. Since it is a highly inert material, it can be used in medical and food applications. Like ABS, PE has good impact strength at low temperatures. It is usually joined or connected with clamps. The required identifying markings on PE include the manufacturer and trademarks, the standard(s) that it was manufactured to, and the nominal size. It also must be marked for DWV service or with the pressure and temperature rating for all other uses. If it is to be used for potable water service, it must also bear a laboratory marking to prove it is acceptable.

3.1.3 PVC (Polyvinyl Chloride) Pipe

Polyvinyl chloride (PVC) is a rigid pipe with high impact strength, and it is used extensively in HVAC applications (*Figure 43*). It can be used in high-pressure systems at low temperatures. PVC pipe is usually joined with a solvent-cement. It comes in various wall thicknesses, some of which can be threaded. It is used in pressurized and non-pressurized applications, and is also used

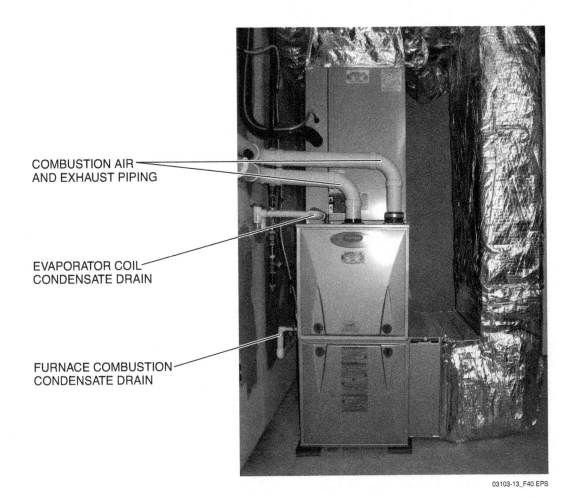

COMBUSTION AIR
AND EXHAUST PIPING

EVAPORATOR COIL
CONDENSATE DRAIN

FURNACE COMBUSTION
CONDENSATE DRAIN

03103-13_F40.EPS

Figure 40 Plastic piping used in a condensing furnace installation.

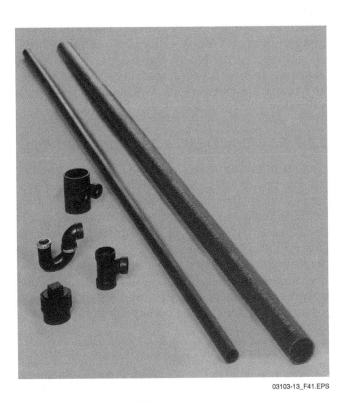

03103-13_F41.EPS

Figure 41 ABS pipe and fittings.

03103-13_F42.EPS

Figure 42 PE tubing.

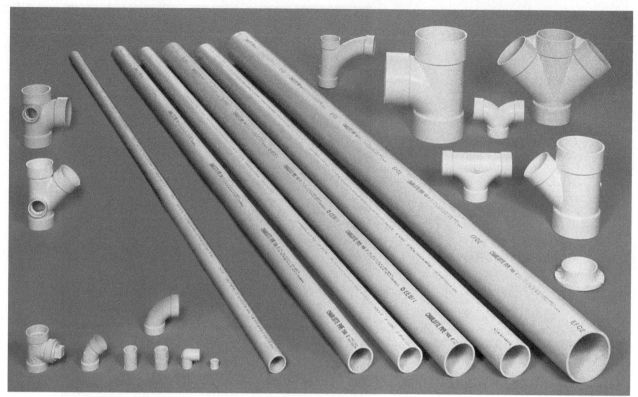

Figure 43 PVC pipe and fittings.

as an electrical conduit. PVC pipe can be thermo-welded, and it may be used for exhaust ducts where chemicals that attack metals are present. A somewhat informal color code does exist, but it is not universal:

- Green is generally used in sewage applications.
- Dark gray and blue, as well as white, are used for cold water piping.
- Dark gray is also used for industrial applications, and often for electrical conduit.
- White is used in many applications, and for both pressurized and non-pressurized service.

The identifying markings (*Figure 44*) are much the same as those required for PE piping. PVC is generally sold in 10' or 20' lengths.

3.1.4 CPVC (Chlorinated Polyvinyl Chloride) Pipe

The applications and joining methods for chlorinated polyvinyl chloride (CPVC) are very similar to those for PVC (*Figure 45*). However, CPVC can be used with fluid temperatures up to 180°F (82°C)

at a pressure of 100 psig. At a temperature of 73°F (23°C), it can withstand pressures up to 400 psig. Since CPVC pipe does not support combustion, it is suitable for use in fire sprinkler systems. This property also allows it to be used in areas where return air collects before flowing back to the system fan in an HVAC system. Combustible materials are not permitted in these areas. CPVC pipe is joined in much the same way as PVC, generally using solvent-cement. Unlike PVC though, it should not be threaded.

3.1.5 PEX (Cross-linked Polyethylene) Tubing

Cross-linked polyethylene (PEX) is a heat transfer tubing that is used in radiant floor heating systems, and to distribute hot and cold water (*Figure 46*). It is usually orange in color and is extremely resistant to chemicals. PEX tubing is rated for 200°F (93°C) at 80 psig, 180°F (82°C) at 100 psig, or 73°F (23°C) at 160 psig. A version of PEX tubing, identified as hePEX, is sealed with a special polymer barrier that prevents oxygen diffusion through the sides of the tubing. This type is spe-

¾" *Brandname* PVC-1120 SCH.40 PR 480 PSI @ 73°F ASTM D-1785 | NSF-pw

Figure 44 PVC pipe markings.

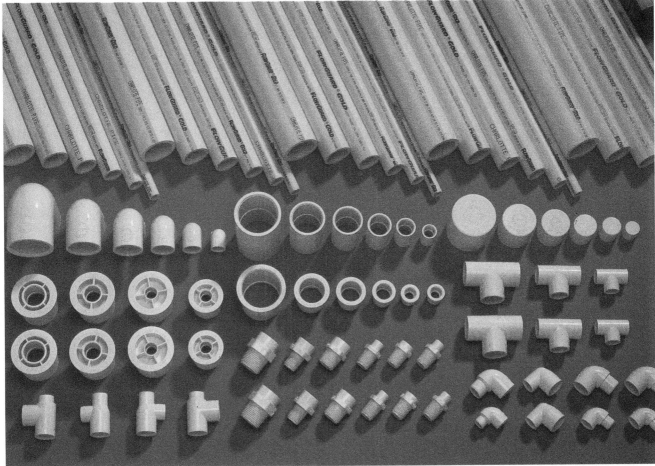

Figure 45 CPVC pipe and fittings.

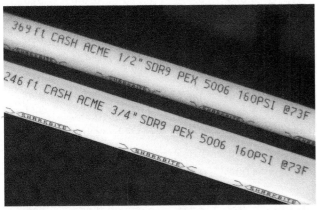

Figure 46 PEX tubing.

cifically designed for radiant heat applications. Today, PEX is the most widely used flexible pipe in the plumbing industry. It can be purchased in coils up to 1,000' in length, or as straight lengths of pipe. Unlike PVC and CPVC, PEX cannot be joined using solvent-cements. The required identifying markings are much the same as the others, including any pressure and temperature ratings.

3.1.6 Plastic Piping Schedules

Some plastic pipe, such as PVC and CPVC, is made to different schedules. Ferrous metal pipe also follows a schedule. The schedule of the pipe determines the wall thickness of the material. The schedule system for pipe can be compared to the system of assigning letters to copper tubing with different wall thicknesses.

As the schedule number increases, the wall thickness increases. It is important to note, though, that PVC and CPVC plastic pipe has the same outside diameter as the equivalent size of steel pipe. For example, the actual OD of 1" PVC pipe is 1.315", while 1" steel pipe has precisely the same OD. Like copper tubing, as the wall thickness of plastic (and steel) pipe increases, the inside diameter becomes smaller. The OD does not change. These dimensional standards are intentional, to make plastic and steel piping systems and the related components easier to work with.

The OD is not the only thing that PVC and CPVC pipe have in common with steel piping. The wall thickness of each schedule group is also the same. For example, 1" Schedule 40 PVC and 1" Schedule 40 steel pipe both have an OD of 1.315" and a wall thickness of 0.133". Obviously this results in PVC/CPVC pipe and steel pipe having the same ID as well, for a given schedule.

PVC and CPVC pipe are primarily manufactured in Schedules 40, 80, and 120. *Table 5* shows the dimensions of Schedule 40 pipe, and *Table 6* shows the same information for Schedule 80 pipe. Remember that these same dimensions apply to steel pipe of the same schedule. Steel piping, however, is available in more schedules than plastic.

Schedule 40 PVC and CPVC are the most popular types in the HVAC industry. They are typically solvent-welded to make a joint. Schedule 80 is used for applications where greater pressure will be applied, or where greater durability is needed. Schedule 80 PVC and CPVC plastic pipe can even be threaded like steel pipe, due to the increased wall thickness.

3.2.0 Joining Plastic Pipe

Plastic pipe may be cut with a tubing cutter, a hacksaw, or a special cutter known as a plastic tubing and pipe cutter (*Figure 47*). The cut must be square to ensure a good joint. The plastic tubing cutter provides a clean and smooth cut that is usually free of any significant burrs and is square. Many cutters, such as the one shown here, have a ratcheting action. Soft, flexible tubing may cut easily with a single stroke. Rigid plastic pipe like PVC however, will require multiple strokes of the cutter. However, these tools are generally limited to cutting pipe sizes up to roughly 2⅜" OD. Square cuts with saws on larger sizes may require a miter box or similar cutting guide. PVC hand saws (*Figure 48*) are also used to cut larger sizes of PVC and CPVC. The model shown here has a replaceable blade, with teeth designed specifically for rapid cutting of plastic pipe. When a great deal of cuts need to be made, a power tool, such as a chop saw or miter saw, is used to increase production.

A pipe wrap (*Figure 49*) provides a simple method of marking a straight line around any type of pipe

Table 5 Schedule 40 Pipe Dimensions

Nom. Pipe Size	OD	Avg ID	Min. Wall	Nom. Wt/Ft	Max. W.P. psi
⅛"	0.405	0.249	0.068	0.051	810
¼"	0.540	0.344	0.088	0.086	780
⅜"	0.675	0.473	0.091	0.115	620
½"	0.840	0.602	0.109	0.170	600
¾"	1.050	0.804	0.113	0.226	480
1"	1.315	1.029	0.133	0.333	450
1¼"	1.660	1.360	0.140	0.450	370
1½"	1.900	1.590	0.145	0.537	330
2"	2.375	2.047	0.154	0.720	280
2½"	2.875	2.445	0.203	1.136	300
3"	3.500	3.042	0.216	1.488	260
3½"	4.000	3.521	0.226	1.789	240
4"	4.500	3.998	0.237	2.118	220
5"	5.563	5.016	0.258	2.874	190
6"	6.625	6.031	0.280	3.733	180
8"	8.625	7.942	0.322	5.619	160
10"	10.750	9.976	0.365	7.966	140
12"	12.750	11.889	0.406	10.534	130
14"	14.000	13.073	0.437	12.462	130
16"	16.000	14.940	0.500	16.286	130
18"	18.000	16.809	0.562	20.587	130
20"	20.000	18.743	0.593	24.183	120
24"	24.000	22.544	0.687	33.652	120

Table 6 Schedule 80 Pipe Dimensions

Nom. Pipe Size	OD	Avg ID	Min. Wall	Nom. Wt/Ft	Max. W.P. psi
⅛"	0.405	0.195	0.095	0.063	1,230
¼"	0.540	0.282	0.119	0.105	1,130
⅜"	0.675	0.403	0.126	0.146	920
½"	0.840	0.526	0.147	0.213	850
¾"	1.050	0.722	0.154	0.289	690
1"	1.315	0.936	0.179	0.424	630
1¼"	1.660	1.255	0.191	0.586	520
1½"	1.900	1.476	0.200	0.711	470
2"	2.375	1.913	0.218	0.984	400
2½"	2.875	2.290	0.276	1.500	420
3"	3.500	2.864	0.300	2.010	370
3½"	4.000	3.326	0.318	2.452	350
4"	4.500	3.786	0.337	2.938	320
5"	5.563	4.768	0.375	4.078	290
6"	6.625	5.709	0.432	5.610	280
8"	8.625	7.565	0.500	8.522	250
10"	10.750	9.493	0.593	12.635	230
12"	12.750	11.294	0.687	17.384	230
14"	14.000	12.410	0.750	20.852	220
16"	16.000	14.213	0.843	26.810	220
18"	18.000	16.014	0.937	33.544	220
20"	20.000	17.814	1.031	41.047	220
24"	24.000	21.418	1.218	58.233	210

Figure 47 Plastic tubing and pipe cutter.

Figure 48 PVC hand saw.

Figure 49 Pipe wrap for measuring, marking straight cut lines, and laying out complex cuts.

to ensure a square cut. Wrap it around the pipe tightly so that the two ends meet squarely. Then use a pencil to mark the pipe completely around its circumference. Once it is cut, the pipe is de-burred inside and out using a sharp knife or file. A wrap-around designed for this purpose also has additional information and lines printed on it to help pipefitters with unusual cuts that are needed when welding pipe. However, for a simple cutting mark around a pipe, a piece of sand cloth long enough to wrap around the pipe works as well.

3.2.1 Solvent-Cementing Products

Although there are several ways to join rigid plastic pipe, solvent-cementing is the most commonly used. This technique is used on a daily basis by many HVAC craftworkers. PVC and CPVC pipe materials used in HVAC applications are most often joined this way.

A wide variety of products are available that are related to solvent-cementing PVC and CPVC pipe. They include cleaners, primers, and cements. Both PVC and CPVC pipe surfaces should first be thoroughly cleaned with an appropriate cleaner (*Figure 50*). The cleaner removes general debris and any oil or grease that may be on the pipe from manufacturing or handling.

Primers (*Figure 51*) contain an aggressive solvent that softens the surface of the material. This prepares the surface for the application of the cement, helping the cement penetrate deeper for a strong, leak-free joint. The use of a primer may be required by building and mechanical codes to ensure joints have the proper integrity. As a result, primers often have a color added that confirms a particular primer was used during the assembly. Inspectors can see from a distance that a primer was applied. The primer shown in *Figure 51* has had a purple colorant added, and also has been combined with a cleaner to eliminate the cleaning step.

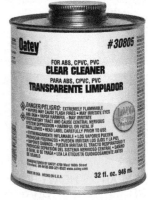

Figure 50 PVC/CVPC clear pipe cleaner.

Figure 51 PVC/CVPC primer/cleaner with a purple color.

Cements (*Figure 52*) are available in a much wider variety than cleaners and primers. Most cement products are not designed for use with both PVC and CPVC. The cleaners and primers can often be used on several different kinds of plastic piping. However, workers must ensure that the correct cement for the material is applied. Cements can be chosen for their body (thickness), color, curing time, ability to work in water, ability to perform better in hot environments, or for their safe use in potable water systems. On some projects, the characteristics of the cement to be used may be specified in the construction documents.

3.2.2 Solvent-Cementing Plastic Pipe

To solvent-cement PVC, CPVC, and ABS pipe, the pipe end is first deburred and examined to ensure it is square. If proper alignment of the pipe and fitting is required, clearly mark the pipe and fitting with a single line so that the two can be re-aligned as the fitting is slipped in place. Remember that cleaners and solvents may destroy or remove the line, so make it dark and bold or use a scratch awl to scratch the surface.

WARNING! PVC cleaners, primers, and cements are toxic and flammable. Do not use them near an open flame, and do not breathe the vapors. Do not use PVC cement in confined spaces without the appropriate ventilation or breathing apparatus. Refer to the applicable material safety data sheet (MSDS) for the product in use. Avoid getting it on the skin.

Use special care when solvent-cementing plastic pipe in low temperatures (below 40°F) or extremely high temperatures (above 100°F). In extremely hot environments, make sure both surfaces to be joined are still wet with cement when putting them together. Cements generally set up faster in higher temperatures. In cold temperatures, longer curing times are likely required. *Table 7* provides some general guidelines for handling and set-up times at different temperatures, based on the pipe size. Once the joint has been made, the cement must be allowed to cure before it can withstand pressure testing. *Table 8* provides the average cure times for both PVC and CPVC pipe. Note that the times shown in both tables can, and should, be extended 50 percent in wet or extremely humid environments. If this represents a problem, another cement can be selected that bonds and cures faster in humid conditions.

To join CPVC or PVC pipe and fittings with solvent cement, follow these steps. *Figure 53* shows a logical sequence of steps to fabricate a solvent-cemented joint. Note that these steps illustrate typical instructions. When joining plastic pipe, always follow the cement manufacturer's instructions for the environment and materials being joined.

Before beginning, put on the appropriate PPE. Safety glasses should be worn at all times when

   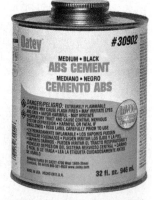

HEAVY-BODIED CLEAR PVC CEMENT

HEAVY-BODIED, FAST-DRYING GRAY PVC CEMENT

HEAVY-BODIED ORANGE CPVC CEMENT

MEDIUM-BODIED BLACK ABS CEMENT

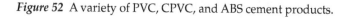

Figure 52 A variety of PVC, CPVC, and ABS cement products.

Table 7 Average Handling and Set-Up Times for PVC/CVPC Cements

Temperature During Assembly	Pipe Diameter					
	½" to 1¼"	1½" to 3"	4" to 5"	6" to 8"	10" to 16"	18"+
60°F–100°F	2 minutes	5 minutes	15 minutes	30 minutes	2 hours	4 hours
40°F–60°F	5 minutes	10 minutes	30 minutes	90 minutes	8 hours	16 hours
20°F–40°F	8 minutes	12 minutes	96 minutes	3 hours	12 hours	24 hours
0°F–20°F	10 minutes	15 minutes	2 hours	6 hours	24 hours	48 hours

Note: Joint cure time is the time required before pressure testing the system. In damp or humid weather, allow 50 percent additional cure time.

Table 8 Average Cure Times for PVC/CVPC Cements

RH 60% or Less Temperature During Assembly or Cure Period	Pipe Diameter									
	½" to 1¼"		1½" to 3"		4" to 5"		6" to 8"		10" to 16"	18"+
	Up to 180 psi	180 psi+	Up to 180 psi	180 psi+	Up to 180 psi	180 psi+	Up to 180 psi	180 psi+	Up to 100 psi	Up to 100 psi
60°F–100°F	1 hr	6 hrs	2 hrs	12 hrs	6 hrs	18 hrs	8 hrs	24 hrs	24 hrs	36 hrs
40°F–60°F	2 hrs	12 hrs	4 hrs	24 hrs	12 hrs	36 hrs	16 hrs	48 hrs	48 hrs	72 hrs
20°F–40°F	6 hrs	36 hrs	12 hrs	72 hrs	36 hrs	4 days	3 days	9 days	8 days	12 days
0°F–20°F	8 hrs	48 hrs	16 hrs	96 hrs	48 hrs	8 days	4 days	12 days	10 days	14 days

Note: Joint cure time is the time required before pressure testing the system. In damp or humid weather, allow 50 percent additional cure time.

working with pipe and solvent-cement products. In addition, rubber or latex gloves should also be worn during the joining process. Making a proper joint requires some timing; make sure that all the required materials are readily available and within easy reach.

Step 1 Cut the pipe to the required length (*Figure 53A*). Then ream the inside edge of the pipe with a reamer or sharp knife; chamfer the outside edge of the pipe with a suitable reamer, single-cut file, or sand cloth (*Figure 53B*). Examine the pipe end; if cracks are noted, the pipe should be cut off at least 2" away from the end of the crack.

Step 2 Test fit the pipe and fitting (*Figure 53C*). The pipe should fit more tightly as it reaches about two-thirds of the socket depth. Once the primer and cement are applied, the surfaces will be mildly dissolved and slippery, and the pipe and fitting should fit together fully.

Step 3 Mark the pipe and fitting with a felt-tipped pen or scratch awl to show the proper position for alignment, if necessary. Then mark the depth of the fitting socket on the pipe. Since the mark will be removed by cleaners and primers, make what is known as a witness mark another 2" away. Once the fitting is assembled, a measurement can quickly be made from the witness mark to the face of the fitting to make sure that the pipe bottomed-out in the socket. During the first few minutes of curing time, the pipe is sometimes pushed out of the socket somewhat by natural processes. The witness mark helps to ensure the joint stayed together during the cure.

Step 4 Wipe off the fitting and pipe to remove any loose dirt and moisture (*Figure 53D*). Clean the surfaces being joined by applying a proper cleaning product with the provided dauber. No additional action is required other than simply applying the cleaner and working it in a bit.

Step 5 Next, apply the primer to the surfaces being joined, if required. Remember that some types of plastic pipe, such as ABS, do not require a primer. Work the primer in with vigor; the dauber will begin to drag some when the surface starts to break down as desired. Apply primer to the fitting socket first, and then to the end of the pipe. Apply the primer to the pipe 1" to 1½" farther than the socket depth. Then, apply the primer to the socket of the fitting a second time.

Step 6 Work in a heavy, even coat of cement to the pipe end while the primer is still wet (*Figure 53E*). Apply the cement up to the depth of the socket only. Use the same applicator (without adding more cement) to apply a thin coat inside the fitting socket (*Figure 53F*). Do not allow excess cement to puddle in the fitting socket.

Step 7 Apply a second layer of cement to the pipe end.

Step 8 Immediately insert the pipe into the fitting socket, rotating the tube one-quarter to one-half turn while inserting (*Figure 53G*). This motion ensures an even distribution of cement in the joint. Properly align the fitting.

Step 9 Hold the assembly together firmly for about 30 seconds. This helps prevent the pipe from pushing out of the socket as the curing process begins. An even bead of cement should be visible around the joint. If this bead does not appear all the way around the socket edge, the cement was improperly applied. In this case, remake the joint to avoid the possibility of leaks. Wipe excess cement from the tubing and fitting surfaces; the bead should not be wiped away.

The procedure for joining ABS pipe with cement is the same as for PVC or CPVC, except that no primer is required on ABS pipe. Unless the pipe and joint is being used for DWV service, pressure test the piping system after the last joint assembled has reached the recommended curing time for the conditions.

3.2.3 Plastic Pipe Support Spacing

Since even rigid plastic piping products are somewhat flexible and are more sensitive to heat than metals, support must be provided at closer intervals. *Table 9* provides the recommended support spacing intervals for several schedules of PVC pipe, as well as ABS pipe. Note that most building codes require that support be provided at a maximum of 4' (1.2m) intervals regardless of the pipe size. Smaller pipe sizes require closer intervals. It is also important to note the operating temperature of the piping system in the table. As the operating temperature increases, the support interval becomes closer. Other tables are available for higher operating temperatures for products that can safely tolerate it. Local codes may require even closer support intervals for one or more products.

Confined Space Hazards

A confined space is any space that has limited or restricted openings for entry or exit. Rotating machinery, dangerous levels of toxic air contaminants, flammable or explosive air mixtures, excessive temperatures, lack of oxygen, or excess oxygen are some of the dangers in a confined space. Plastic solvents and cements also represent health hazards in these environments.

HVAC equipment, especially commercial or industrial systems, may be located in confined spaces. In many facilities, hazard reviews are conducted to identify non-permit required and permit-required confined spaces. The facility safety office must always be contacted to obtain permission to enter any confined space. OSHA regulations specify that a permit-required confined space must be tested for hazardous atmosphere with a calibrated instrument before a worker is allowed to enter the space. In addition, the atmosphere must be continuously monitored while work is in progress in order to detect atmospheric hazards, including oxygen deficiency. Continuous communication must be maintained between workers inside and outside the confined space.

HVAC workers entering any confined space should always observe proper precautions for ventilation, working in excessive temperatures, and working around operating machinery. No worker should enter any such space without appropriate rescue equipment and a means to communicate with a person stationed outside the space. The person stationed outside the space must be able to operate the necessary rescue equipment and summon help in the event of an emergency.

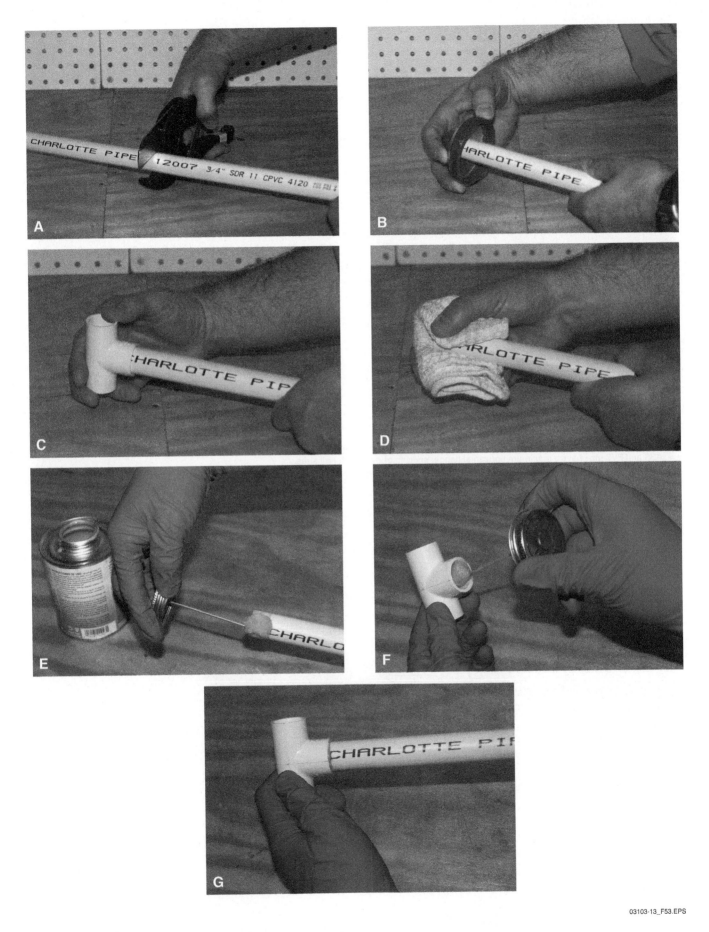

Figure 53 Fabricating a solvent-cemented joint.

03103-13_F53.EPS

Table 9 General Guidelines for Horizontal Support Spacing – Plastic Pipe

Nom. Pipe Size (in)	SDR 21 PR200 & SDR 26 PR160 Operating Temp (°F)					PVC Pipe Schedule 40 Operating Temp (°F)					Schedule 80 Operating Temp (°F)					ABS Pipe Schedule 40 Operating Temp (°F)				
	60	80	100	120	140	60	80	100	120	140	60	80	100	120	140	60	80	100	120	140
½	3½	3½	3	2		4½	4½	4	2½	2½	5	4½	4½	3	2½					
¾	4	3½	3	2		5	4½	4	2½	2½	5½	5	4½	3	2½					
1	4	4	3½	2		5½	5	4½	3	2½	6	5½	5	3½	3					
1¼	4	4	3½	2½		6	5½	5	3	3	6	6	5½	3½	3					
1½	4½	4	4	2½		6	5½	5	3½	3	6½	6	5½	3½	3½	6	6	5½	3½	3
2	4½	4	4	3		6	5½	5	3½	3	7	6½	6	4	3½	6	6	5½	3½	3
2½	5	5	4½	3		7	6½	6	4	3½	7½	7½	6½	4½	4					
3	5½	5½	4½	3		7	7	6	4	3½	8	7½	7	4½	4	7	7	7	4	3½
4	6	5½	5	3½		7½	7	6½	4½	4	9	8½	7½	5	4½	7½	7½	7	4½	4
6	6½	6½	5½	4		8½	8	7½	5	4½	10	9½	9	6	5	8½	8½	8	5	4½
8	7	6½	6	5		9	8½	8	5	4½	11	10½	9½	6½	5½					
10						10	9	8½	5½	5	12	11	10	7	6					
12						11½	10½	9½	6½	5½	13	12	10½	7½	6½					
14						12	11	10	7	6	13½	13	11	8	7					
16						12½	11½	10½	7½	6½	14	13½	11½	8½	7½					

Note:
Always follow local code requirements for hanger spacing. Most plumbing codes have the following hanger spacing requirements:
- ABS and PVC pipe have a maximum horizontal hanger spacing of every 4' for all sizes.
- CPVC pipe or tubing has a maximum horizontal hanger spacing of every 3' for 1" and under, and every 4' for sizes 1¼" and larger.

Additional Resources

Plastics Pipe Institute, Inc. **plasticpipe.org**

The Plastic Pipe and Fittings Association. **http://www.ppfahome.org**

3.0.0 Section Review

1. The schedule system for plastic pipe _____.
   a. is based on the materials of construction
   b. can be used to determine the outside diameter
   c. identifies the type of cement that should be used
   d. provides a way to compare its wall thickness to copper tubing

2. Cement is first applied to the fitting, then to the pipe, and then a second coat is applied to the fitting.
   a. True
   b. False

SUMMARY

An HVAC service technician must be able to work with several kinds of piping, including copper and plastic. Copper is the most common type used in the HVAC industry for refrigerant piping. As a general rule, only Type ACR copper tubing is used for refrigerant because it is factory cleaned, charged with inert nitrogen gas, and capped to maintain its condition.

Soft copper tubing can easily be cut, bent, and mechanically joined. It can also be soldered or brazed. Hard copper tubing is rarely bent or formed on site, and is usually soldered and brazed. Mechanical joints are relatively rare in hard copper piping.

Refrigerant tubing must be protected from exposure to dirt and moisture. Tube ends must remain capped, and a tubing cutter should be used for cutting whenever possible. If a hacksaw is used, the tubing requires more preparation before it can be joined.

Copper tubing can be joined with flare or compression fittings. However, it is common to solder or braze the connections. Runs of pipe must be properly supported and often insulated. The requirements for pipe hangers and insulation are usually determined from the job specifications or local building codes.

Once the piping installation is complete, it must be inspected and tested. A common method of leak testing involves pressurizing the system with nitrogen and then using a soap solution to check for leaks.

Plastic piping is also used extensively in the HVAC industry. Although heavier schedules of plastic piping can be threaded, it is most often solvent-cemented to make strong, durable joints. PVC and CPVC plastic piping shares common dimensions with steel piping, allowing for more consistency in their installation and related components.

1. The organization that establishes manufacturing standards for copper tubing is _____.
   a. SAE
   b. CDA
   c. ASTM
   d. ASHRAE

2. The working pressure rating of hard-drawn copper tubing is typically _____.
   a. 30 percent higher than soft copper tubing
   b. 40 percent higher than soft copper tubing
   c. 50 percent higher than soft copper tubing
   d. 60 percent higher than soft copper tubing

3. As the wall thickness of copper tubing increases, the inside diameter of the tube is decreased.
   a. True
   b. False

4. One of the main uses of flare fittings in HVAC work is to _____.
   a. connect various pressure controls
   b. connect runs of tubing greater that 1⅛" OD
   c. provide a tubing socket for heat bonded joints
   d. provide the most reliable type of joint wherever needed

5. One of the best reasons to use a tubing cutter instead of a hacksaw is that _____.
   a. stronger joints will result
   b. it does not create loose cuttings
   c. the tubing can be cut off at any angle
   d. it does not create a ridge inside the tube edge

6. Single-thickness flares may not be strong enough for _____.
   a. any water piping joint
   b. any refrigerant piping joint
   c. the larger sizes of copper tubing
   d. the smallest sizes of copper tubing

7. Swaging a copper tube prepares it for a sound, reliable fit with a compression ferrule.
   a. True
   b. False

8. The pressure-relief valve on a nitrogen pressure test set should be set to relieve pressure at _____.
   a. 1 or 2 psig above the required test pressure
   b. at a pressure 50 percent higher than the required test pressure
   c. the maximum working pressure of the nitrogen regulator
   d. the maximum safe working pressure of the tubing being tested

9. Which type of tubing or pipe would be expected to require the closest hanger spacing?
   a. ½" copper tubing
   b. 1½" copper tubing
   c. ¾" steel pipe
   d. 2" steel pipe

10. The type of plastic pipe that does not support combustion is _____.
    a. ABS
    b. PEX
    c. PVC
    d. CPVC

11. Which plastic product is the most widely used in the plumbing industry?
    a. ABS
    b. PVC
    c. PEX
    d. CPVC

12. Which type of plastic piping can be threaded?
    a. PEX
    b. Schedule 40 PVC
    c. Schedule 80 PVC
    d. Schedule 40 CPVC

13. The most common way to join plastic pipe is
    _____.
    a. solvent-cementing
    b. heat-bonding
    c. threading
    d. clamps

14. Plastic pipe cements take longer to cure in hot weather.
    a. True
    b. False

15. The handling and curing times in wet environments for solvent-cement joints in plastic piping should be extended _____.
    a. 20 percent
    b. 35 percent
    c. 50 percent
    d. 65 percent

Trade Terms Quiz

Fill in the blank with the correct trade term that you learned from your study of this module.

1. When a metal can be pressed or formed to some degree without cracking and breaking, it is said to be _____.

2. A(n) _____ component has been coated or covered with a thin layer of copper.

3. The compression ring that is placed around a tube, and then squeezed against the tube by the compression nut, is called a(n) _____.

4. A heat-bonding method of joining metals using an alloy with a melting point below 842°F (450°C) is called _____.

5. When one metal, sometimes precious such as gold, is going to be used to cover another metal, the intent is to _____ the second metal.

6. A(n) _____ metal has been heat-treated to soften it, making it formable.

7. When a symmetrical, angled surface is created on the edge of a tube or pipe, breaking what would otherwise be a 90-degree angle, the angled edge is called a(n) _____.

8. The process of _____ occurs when a metal is repetitively reformed, causing a permanent change in its crystalline structure, which results in an increase in its hardness.

9. The end of a tube can be expanded to create a socket, allowing it to be connected to another tube or fitting of equal size, through the process of _____.

10. When the process of _____ is used to heat-bond metals together, an alloy is used that has a melting point above 842°F (450°C).

11. Refrigerants that contain carbon as well as one or more halogens, such as chlorine, are referred to as _____.

12. When a tube is marked to indicate the depth of insertion for a pipe into a fitting, and then a second mark is made to ensure it will not be obscured by the joining process, the mark is called a(n) _____.

13. When one element combines with oxygen at a molecular level, the process is referred to as _____.

14. The characteristic of a material that describes the ease and rate in which energy passes through it is known as its _____.

15. Copper tubing that is pulled through a series of progressively smaller dies to reach the desired size and shape is commonly known as _____ tubing.

16. The reduction of pressure that occurs between two points in a piping system is known as _____.

17. The tool designed to remove the sharp lip from tubing or pipe that results from the cutting process is known as a(n) _____.

Trade Terms

Annealed
Brazing
Chamfer
Conductivity

Copper-clad
Ferrule
Halocarbon
  refrigerants

Hard-drawn
Malleable
Oxidation
Plate

Pressure drop
Reamer
Soldering
Swaging

Witness mark
Work hardening

Chris Sterrett

Program Director/Instructor
Fort Scott Community College

How did you get started in the construction industry (i.e., took classes in school, a summer job, etc.)?

I got a late start in the industry by enrolling in a certificate program at the age of 35, and then continuing those studies at Pittsburg State University.

Who or what inspired you to enter the industry (i.e., a family member, school counselor, etc.)? Why?

After getting laid off from an aircraft manufacturing assembly line job, I looked for a career in an industry that would both challenge me and allow some geographic mobility and variety in my day-to-day work environment. I found that in the HVACR trade.

What do you enjoy most about your career?

I love meeting people. As a technician, I also enjoy the satisfaction that comes from entering an often stressful and sometimes dire situation for a customer and solving it. Seeing the relief on their faces and being showered with their gratitude is very satisfying. As a teacher, I like watching a student progress from being very timid around electricity and the refrigerant circuit to confidently approaching a unit and getting down to business.

Why do you think training and education are important in construction?

As a service technician and later as a service manager, I would come across technicians that lacked a basic understanding of the fundamentals of the HVACR trade. They would just guess at a solution to a problem, make an incorrect diagnosis, change out a part needlessly, or condemn a unit. They would sometimes make multiple trips to the same job, costing both the contractor and the consumer time, money, and frustration. The solid understanding of the basics that a person gets from a good training program will help all the field experiences make sense. Thorough training programs connect the dots, so to speak.

Why do you think credentials are important in construction?

Very few programs have a national reputation with credentials that are portable. It may be a very good program, but 100 miles away or in another state, maybe no one will recognize it. Because of the increasing mobility of our society, having a portable credential allows a person to validate their proficiency in a trade to an employer that may be unfamiliar with the training program attended. That portability is one valuable aspect of the NCCER program.

How has training/construction impacted your life and career (i.e., advancement opportunities, better wages, etc.)?

My HVAC training allowed me the flexibility to move my family out of the city to a small town 150 miles away. When I decided to return to school and earn a teaching degree, I could work in the trade around my class schedule. When I got out of class, I would jump in a service van and start running service calls. This would not have been possible in most other career fields.

Would you recommend construction as a career to others? Why?

I would enthusiastically recommend a career in HVACR to anyone that likes the satisfaction and respect that comes from contributing to the health, comfort, and well being of their community's residents.

What does craftsmanship mean to you?

Craftsmanship to me is that attitude that drives a person to be one of the best at what they do. It is embodied in those who are not satisfied with just good, and not content with haphazardly making something work.

Trade Terms Introduced In This Module

Annealed: Metal that has been heat-treated to soften it, making it formable. Hard-drawn copper tubing is annealed to make it soft and formable.

Brazing: A heat-bonding method of joining metals using another alloy with a melting point lower than the metal(s) being joined. The alloy is used as a filler metal to bond and fill any gaps between the pieces by capillary action. Brazing uses filler metals that have a melting point above 842°F (450°C).

Chamfer: To create a symmetrical angled surface on an edge, breaking what would usually be a 90-degree angle. Chamfer is also used as a noun, as a name for the angled surface.

Conductivity: The ease and rate in which energy passes through a material. Thermal conductivity refers to the ease and rate at which heat passes through a material; electrical conductivity refers to the rate at which electricity passes.

Copper-clad: Components that have been coated or covered with a thin copper layer.

Ferrule: A ring or bushing placed around a tube that squeezes or bites into the tube beneath when compressed, forming a seal.

Halocarbon refrigerants: Any of a class of organic compounds containing carbon and one or more halogens, such as chlorine or fluorine. Refrigerants such as R-22 contain chlorine, which has been linked to the destruction of the ozone layer. As a result, refrigerants bearing chlorine are being phased out.

Hard-drawn: The process of heating copper and drawing it through dies to form its shape and size. Each die it is drawn through is progressively smaller until the desired size is reached.

Malleable: A characteristic of metal that allows it to be pressed or formed to some degree without breaking or cracking. Copper is malleable, while cast iron is not.

Oxidation: The process of combining with oxygen at a molecular level. Copper and oxygen join to form copper oxides, appearing as darkened deposits on the copper surface. Common rust is an iron oxide.

Plate: To apply a layer of a metal on the surface of another metal. Other, less-precious metals are often plated with gold to reduce the value or expense.

Pressure drop: The reduction in pressure between one point in a pipe or tube to another point. Pressure drop results from friction in piping systems, which robs energy from the flowing fluid or vapor. Pressure drops increase as flow velocity (speed) increases.

Reamer: A tool designed to remove the sharp lip and burrs left inside a pipe or tube after cutting.

Soldering: A heat-bonding method of joining metals using another alloy with a melting point lower than the metal(s) being joined. The alloy is used as a filler metal to bond and fill any gaps between the pieces by capillary action. Soldering uses filler metals that have a melting point below 842°F (450°C).

Swaging: The process of using a tool to shape metal. In context, it describes the process of forming a socket in the end of a copper tube that is the correct size to accept another piece of tubing or a component.

Witness mark: A mark made as a means of determining the proper positioning of two pieces of pipe or tubing when joining. A witness mark is typically in addition to a first mark, made in a location that will not be obscured by the joining process; the first mark is often obscured during the process.

Work hardening: The repetitive reforming (such as bending or flexing) of a material, causing a permanent change in its crystalline structure and an increase in its strength and hardness.

Additional Resources

This module is presented as a thorough resource for task training. The following resource material is suggested for further study.

The Copper Tube Handbook. A publication of the Copper Development Association.
http://www.copper.org/publications/pub_list/pdf/copper_tube_handbook.pdf

Plastics Pipe Institute, Inc., **plasticpipe.org**

The Plastic Pipe and Fittings Association. **http://www.ppfahome.org**

Figure Credits

Courtesy of Mueller Industries, Module opener, Figure 1

Courtesy of Copper Development Association Inc., Tables 2, 3, Figure 27

Viega, Figure 11

Courtesy of RIDGID®, Figures 12, 13, 23, 28 (bottom)

Courtesy of Imperial® Brand Tools - Stride Tool Inc., Figures 19, 28 (top)

Klein Tools, Inc., Figure 30

Cash Acme, SA03

Buckaroos, Inc., Figure 39

Courtesy of Cooper B-Line – USA, Table 4

Courtesy of American Heat and Hot Water www.americanheatandhotwater.com, Figure 40

Courtesy of Charlotte Pipe and Foundry, Figures 41, 43, 45, 53, Table 9

W.B. Noble/Mississippi Construction Education Foundation, Figure 42

Tables courtesy GF Harvel LLC, an entity under the Division Americas of Georg Fischer Piping Systems. ©2012 GF Harvel; reprinted with permission-All Rights Reserved, Tables 5, 6

Courtesy of Milwaukee Electric Tool Corporation, Figure 48

Courtesy of Flange Wizard®, Figure 49

Courtesy of Oatey, Figures 50 – 52, Tables 7, 8

Section Review Answer Key

Answer	Section Reference	Objective
Section One		
1. a	1.1.1; 1.1.3	1a
2. d	1.2.2	1b
Section Two		
1. b	2.1.3	2a
2. c	2.2.1	2b
3. b	2.2.4	2b
4. a	2.3.0; Table 4	2c
Section Three		
1. d	3.1.6	3a
2. b	3.2.2	3b

NCCER CURRICULA — USER UPDATE

NCCER makes every effort to keep its textbooks up-to-date and free of technical errors. We appreciate your help in this process. If you find an error, a typographical mistake, or an inaccuracy in NCCER's curricula, please fill out this form (or a photocopy), or complete the online form at **www.nccer.org/olf**. Be sure to include the exact module ID number, page number, a detailed description, and your recommended correction. Your input will be brought to the attention of the Authoring Team. Thank you for your assistance.

Instructors – If you have an idea for improving this textbook, or have found that additional materials were necessary to teach this module effectively, please let us know so that we may present your suggestions to the Authoring Team.

NCCER Product Development and Revision

13614 Progress Blvd., Alachua, FL 32615

Email: curriculum@nccer.org
Online: www.nccer.org/olf

❏ Trainee Guide   ❏ Lesson Plans   ❏ Exam   ❏ PowerPoints   Other _____

Craft / Level: _____   Copyright Date: _____

Module ID Number / Title: _____

Section Number(s): _____

Description: _____

_____

_____

_____

Recommended Correction: _____

_____

_____

_____

Your Name: _____

Address: _____

_____

Email: _____   Phone: _____

Soldering and Brazing

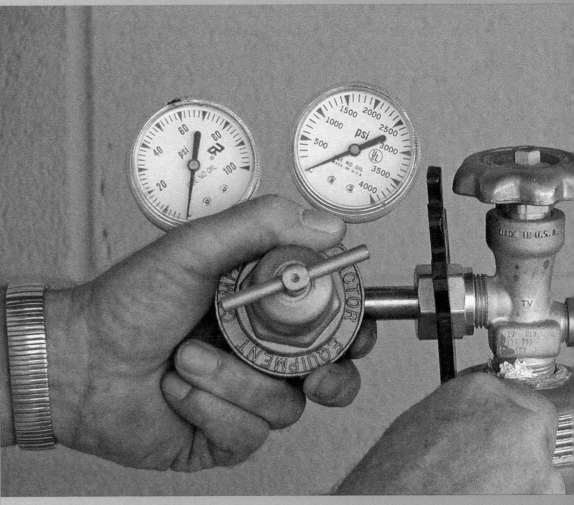

OVERVIEW

Soldering is used primarily to join copper water lines and condensate lines in the HVACR craft. When mechanically strong, pressure-resistant joints are needed for refrigerant lines, the process used is brazing. Both soldering and brazing demand careful attention to safety due to the hazards associated with flammable gases and open flames. With some practice, HVACR technicians can soon master soldering and brazing techniques.

Module 03104

Trainees with successful module completions may be eligible for credentialing through NCCER's Registry. To learn more, go to **www.nccer.org** or contact us at **1.888.622.3720**. Our website has information on the latest product releases and training, as well as online versions of our *Cornerstone* magazine and Pearson's product catalog.

Your feedback is welcome. You may email your comments to **curriculum@nccer.org**, send general comments and inquiries to **info@nccer.org**, or fill in the User Update form at the back of this module.

This information is general in nature and intended for training purposes only. Actual performance of activities described in this manual requires compliance with all applicable operating, service, maintenance, and safety procedures under the direction of qualified personnel. References in this manual to patented or proprietary devices do not constitute a recommendation of their use.

Objectives

When you have completed this module, you will be able to do the following:

1. Describe and demonstrate the safe process of soldering copper tubing.
   a. Describe and demonstrate the use of the PPE, tools, and materials needed to solder copper tubing.
   b. Describe and demonstrate the preparation required for soldering.
   c. Describe and demonstrate the soldering process.
2. Describe and demonstrate the safe process of brazing copper tubing.
   a. Describe and demonstrate the use of the PPE, tools, and materials needed to braze copper tubing.
   b. Describe and demonstrate the preparation used for brazing.
   c. Describe and demonstrate the brazing process.
   d. Describe and demonstrate the process of brazing copper tubing to dissimilar metals.

Performance Tasks

Under the supervision of the instructor, you should be able to do the following:

1. Properly set up and shut down oxyacetylene equipment.
2. Properly set up and shut down an acetylene single tank.
3. Properly prep and safely solder copper tubing in various planes, using various fittings.
4. Properly prep and safely braze copper tubing using various fittings.

Trade Terms

Acetone
Alloy
Capillary action
Cup depth
Fillet

Flashback arrestor
Flux
Nonferrous
Solder
Wetting

Industry Recognized Credentials

If you are training through an NCCER-accredited sponsor, you may be eligible for credentials from NCCER's Registry. The ID number for this module is 03104. Note that this module may have been used in other NCCER curricula and may apply to other level completions. Contact NCCER's Registry at 888.622.3720 or go to **www.nccer.org** for more information.

Contents

1.0.0 Soldering 1

  1.1.0 PPE, Tools, and Materials Used for Soldering 2

    1.1.1 Soldering PPE and Safety Guidelines 2

    1.1.2 Soldering Tools and Equipment 2

    1.1.3 Solders and Soldering Fluxes 4

  1.2.0 Preparing Tubing and Fittings for Soldering 6

  1.3.0 Soldering Joints 7

2.0.0 Brazing Copper Fittings and Tubing 13

  2.1.0 PPE, Tools, and Materials Used for Brazing 13

    2.1.1 Brazing Safety 13

    2.1.2 Brazing Equipment 14

    2.1.3 Filler Metals and Fluxes 18

  2.2.0 Preparing for Brazing 20

    2.2.1 Brazing Equipment Setup 20

    2.2.2 Lighting the Oxyacetylene Torch 24

    2.2.3 Air/Acetylene Equipment Setup 27

    2.2.4 Purging Refrigerant Lines 27

  2.3.0 Brazing Joints 29

  2.4.0 Brazing Dissimilar Metals 31

Figures and Tables

Figure 1   Capillary action..2
Figure 2   Handheld fuel gas cylinder...................................2
Figure 3   B and MC acetylene tanks 3
Figure 4   Self-igniting soldering torch...............................3
Figure 5   Air/acetylene torch 3
Figure 6   A cup-type striker 4
Figure 7   95-5 solder 5
Figure 8   Two types of flux and a flux brush.....................6
Figure 9   Face-to-face method of measuring pipe length.........6
Figure 10 Preparing copper for soldering 8
Figure 11 Acetylene B tank and related equipment 9
Figure 12 Applying heat to the fitting 10
Figure 13 Wiping away excess flux 11
Figure 14 Acetylene cylinder with a valve safety cap 14
Figure 15 Oxyacetylene brazing setup 15
Figure 16 Portable oxyacetylene equipment...................... 15
Figure 17 Oxygen and fuel gas regulators 16
Figure 18 Torch wrench.. 17
Figure 19 Medium- and light-duty brazing torch handles........... 17
Figure 20 Flashback arrestor 18
Figure 21 Various brazing tips... 18
Figure 22 Torch tip cleaners 18
Figure 23 BCuP-5 brazing rods comprised of 15 percent silver,
           80 percent copper, and 5 percent phosphorus 19
Figure 24 A high-performance brazing flux used in many applications for
both non-ferrous and ferrous metals 20
Figure 25 Typical empty cylinder marking 21
Figure 26 Inspecting the fittings... 22
Figure 27 Installing the regulator....................................... 23
Figure 28 Types of flames.. 26
Figure 29 Reduce the acetylene flow until the flame returns to the tip........ 26
Figure 30 Air/acetylene torch tips 28
Figure 31 Nitrogen purging setup....................................... 29
Figure 32 Cone valve plug.. 29
Figure 33 Multi-flame, or rosebud, heating tip.................. 30
Figure 34 Working in overlapping sectors 31
Figure 35 A 45 percent silver brazing alloy, in roll form (BAg-5) 31

Table 1   Air/Acetylene Torch Tip Performance Data........... 4
Table 2   Pressure-Temperature Ratings of Soldered Joints 5
Table 3   Manufacturer's Makeup Chart............................. 7
Table 4   Tip Sizes Used for Common Tubing Sizes 18
Table 5   Brazing Filler Metals... 19
Table 6   Brazing Tip Operational Data............................... 25

SECTION ONE

1.0.0 SOLDERING

Objective

Describe and demonstrate the safe process of soldering copper tubing.

  a. Describe and demonstrate the use of the PPE, tools, and materials needed to solder copper tubing.

  b. Describe and demonstrate the preparation required.

  c. Describe and demonstrate the soldering process.

Trade Terms

Acetone: A colorless organic solvent that is volatile and extremely flammable. In the HVACR trade, it is used as a carrier for acetylene gas in cylinders.

Alloy: Any substance made up of two or more metals.

Capillary action: The movement of a liquid along the surface of a solid in a kind of spreading action.

Cup depth: The distance that a tube inserts into a fitting, usually determined by a stop inside the fitting.

Fillet: A rounded internal corner or shoulder of filler metal often appearing at the meeting point of a piece of tubing and a fitting when the joint is soldered or brazed.

Flux: A chemical substance that prevents oxides from forming on the surface of metals as they are heated for soldering, brazing, or welding.

Nonferrous: A group of metals and metal alloys that contain no iron.

Solder: A fusible alloy used to join metals, with melting points below 842°F.

Wetting: A process that reduces the surface tension so that molten (liquid) solder flows evenly throughout the joint.

The two methods used for joining copper tubing and fittings are soldering and brazing. Both methods fasten the metals together using a nonferrous filler metal that adheres to the surfaces being joined. The filler metal is drawn into and distributed between the closely fitted surfaces by capillary action.

The difference between soldering and brazing is the temperature needed to melt the filler metal in order to make the joint. Soldering uses filler metals that melt at temperatures below 842°F (450°C), usually in the 375°F to 500°F range (191°C to 260°C). The filler metals used for brazing melt at temperatures above 842°F. This temperature dividing line between the two processes has been determined by the American Welding Society (AWS).

Soldering, also called sweating or soft soldering, is the common method of joining copper tubing and fittings for water lines and other low-pressure applications. Sweat fittings are sometimes used for low-pressure refrigerant lines, but brazing is generally preferred for all refrigerant line work. Soldered joints are used in piping systems that carry liquids at temperatures of 250°F (121°C) or below. Soldered joints are typically used in situations where specs call for copper condensate lines. They are also used in some cases to join piping in low-pressure hydronic heating systems.

Soldering involves joining two metal surfaces by using heat and a nonferrous filler metal. A nonferrous filler metal is a metal that contains no iron and is therefore nonmagnetic. The melting point of the filler metal must be lower than that of the two metals that are being joined.

Soldered joints depend on capillary action to pull and distribute the melted solder into the small gap between the fitting and the tubing. When the joint is filled, the solder will form a tiny bead around the joint called a fillet. Capillary action is the flow of liquid, in this case solder, into a small space between two surfaces. Capillary action is most effective when the space between the tubing and the fitting is between 0.002" and 0.005". To maintain the proper spacing, it is important to check the joint for correct alignment before soldering. *Figure 1* shows how capillary action works.

Did You Know?

To solder copper fittings, the fitting must be heated until the soldering paste starts to melt. The melting paste looks like beads of sweat on the pipe, which led to the name *sweating* to describe the soldering process.

1.1.0 PPE, Tools, and Materials Used for Soldering

Soldering does not require a lot of tools or sophisticated equipment. The equipment is relatively simple and easy to operate as long as safe work habits are maintained. The process does require some practice to master, however.

1.1.1 Soldering PPE and Safety Guidelines

Soldering involves the use of a flammable gas under pressure, open flame, and possible exposure to molten metal. In order to avoid injury, workers performing soldering must wear appropriate PPE and follow correct safety procedures. The following safety guidelines apply:

- Wear clothing made of nonflammable fabric such as cotton. Avoid synthetic fabrics such as nylon, which can melt and adhere to the skin.
- Wear a long-sleeve shirt to help prevent burns from sparks or molten metal. Pants should not have cuffs, as they can catch and hold hot materials.
- Wear eye and face protection.
- Wear flame-resistant gloves and high-top work boots.
- Make sure the gas cylinder is securely turned off when not in use. Gases used in soldering are flammable and explosive if exposed to a spark or flame.
- Keep a fire extinguisher handy while soldering in case material near the work ignites. When working close to flammable material such as wood, use a fire blanket or other flame/heat blocking material to protect it.
- Always solder in a well-ventilated area because fumes from the flux can irritate your eyes, nose, throat, and lungs. Wear respiratory equipment if working in an area with poor ventilation.
- Specific procedures are often in place for the use of an open flame. When working at field sites, be sure to acquire and follow all necessary guidelines and safety requirements related to the site.

1.1.2 Soldering Tools and Equipment

The heat used for soldering is provided by a torch that is attached to a cylinder of flammable gas. Flammable gases such as propane, MAP-Pro (propylene), and acetylene are commonly used to make sweat joints in copper tubing. Propane and MAP-Pro, which is also known as MAPP, are sold in small cylinders such as the one shown in *Figure 2* for handheld use. Such cylinders are fine for small projects such as occasionally joining fittings to water lines or condensate lines. When larger, more repetitive projects are involved, the use of acetylene

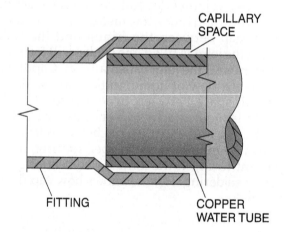

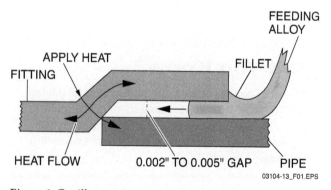

Figure 1 Capillary action.

03104-13_F02.EPS

Figure 2 Handheld fuel gas cylinder.

is far more common. It is generally sold by welding supply companies and trade-related distributors. When portability is important, acetylene can be carried in small, easily transportable cylinders (*Figure 3*). The size MC tank holds 10 ft^3, while the larger size B tank holds 40 ft^3. Larger acetylene cylinders are used when portability is not as important.

Acetylene cylinders are packed with a porous material of some type, such as wood fibers or charcoal, that is partially saturated with acetone to allow the safe storage of acetylene. The packing helps prevent high-pressure gas pockets from forming in the cylinder. Acetone is a colorless, flammable liquid. It is added to the cylinder until about 40 percent of the porous material becomes saturated. The acetone can hold large quantities of dissolved acetylene under pressure—as much as 25 times its own volume. Being a liquid, acetone can be drawn from an acetylene cylinder when it is not upright, or when the usage rate of the acetylene gas is excessive. The sudden appearance of flaming liquid at the end of the torch while working can be somewhat alarming.

In addition to the fuel, soldering requires a torch to provide the heat, such as the self-igniting soldering torch shown in *Figure 4*. A press of the yellow button ignites the air/gas mixture. This particular torch is designed to be connected directly to a propylene or propane gas cylinder. Propylene burns slightly hotter than propane, making it better suited for larger joints. An air/acetylene torch kit is shown in *Figure 5*. Note that a connecting hose and bottle fitting is used with acetylene. Since acetylene tanks are larger and heavier, the torch assembly with its fuel tank cannot be comfortably handheld.

03104-13_F04.EPS

Figure 4 Self-igniting soldering torch.

03104-13_F03.EPS

Figure 3 B and MC acetylene tanks.

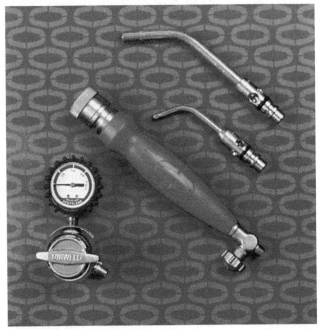

03104-13_F05.EPS

Figure 5 Air/acetylene torch.

Air/acetylene torches can usually accommodate different tips for a smaller or larger flame. The tips are numbered by the manufacturer, with larger numbers typically identifying larger tips. The tip specifications include the actual opening size, the fuel consumption at a given pressure (often rated at 14 psig), and the tubing sizes for which it is recommended. *Table 1* provides an example of the performance data for one manufacturer's series of air/acetylene torch tips.

Because soldering requires relatively low heat, torches that mix the flammable gas directly with air is all that is needed. Oxygen is required for combustion, but the air in the environment provides a sufficient amount of oxygen to produce the needed heat for soldering. Air/fuel torches are designed to draw in the correct amount of air for combustion from the atmosphere. Since air is roughly 80 percent nitrogen, the flame produced by an air/fuel torch burns at a lower temperature than the flame produced by an oxygen/fuel torch.

For torches that are not self-igniting, a cup-type striker (*Figure 6*) is used to light the torch. Other similar styles that produce a spark are also available. A handheld pocket lighter or common cigarette lighter should never be used for this purpose. Strikers keep the hands a safe distance from the flame and do not contain a flammable substance.

1.1.3 Solders and Soldering Fluxes

Solder is a nonferrous metal or metal alloy with a melting point below 842°F. An alloy is any substance made up of two or more metals. Solder selections are based mainly on the operating temperature and pressure of the piping system. If the system experiences dramatic temperature changes, then thermal stress may also be a selection factor. Another important point to consider is whether the system will carry potable water. Note that solders containing more than 0.2 percent lead were banned for use on potable water systems by a 1986 amendment to the Federal Safe Drinking Water Act.

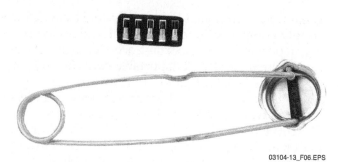

03104-13_F06.EPS

Figure 6 A cup-type striker.

The most common solder used on water lines and other low-pressure applications is an alloy made of 95 percent tin and 5 percent antimony. *Table 2* provides a comparison of pressure-temperature ratings for two different solder alloys. The 50-50 alloy of tin and lead provides very limited strength and is only suitable for low- or non-pressure applications. It also cannot be used on potable water lines. Solders composed of tin-antimony, tin-copper-silver, or tin-silver are usually recommended for applications requiring greater joint strength. The tin-antimony solder typically melts between 430°F and 480°F (221°C to 249°C) and solidifies rapidly. Generally, these solders, in the form of thick yet soft wire, are supplied on spools for easier use (*Figure 7*). The job specifications often identify the type of solder that is to be used for a specific project.

A flux (*Figure 8*) should be applied to all tube ends and fittings immediately after they are cleaned. However, since the flux is rather tacky, care must be taken to keep grit and debris off of the tube and fittings once it is applied. Choosing a proper flux is very important. Soldering flux performs many functions, and the wrong flux can ruin the soldered joint. Flux performs the following functions:

- It chemically cleans and protects the surfaces of the tubing and fitting from oxidation. Oxidation occurs when the oxygen in the air combines with the recently cleaned metal. Oxidation produces tarnish or rust in metal and prevents solder from adhering. The oxidizing

Table 1 Air/Acetylene Torch Tip Performance Data

Tip Model	Tip Size		Gas Flow (SCFH) @ 14 psig	Required Tube Size	
	in	mm		Up to 180 psi	180 psi+
T-2	⅛	4.8	2.1	⅛ – ½	⅛ – ¼
T-3	¹⁄₁₆	6.4	3.5	¼ – 1	⅛ – ½
T-4	¼	7.9	5.8	¾ – 1½	¼ – ¾
T-5	⁵⁄₁₆	9.5	8.2	1 – 2	½ – 1
T-7	⁷⁄₁₆	11.1	11.1	1½ – 3	⅞ – 1⅝
T-8	½	12.7	14.4	2 – 3½	1 – 2

Table 2 Pressure-Temperature Ratings of Soldered Joints

Joining Material[4]	Service Temperature (°F)	Fitting Type	Maximum Working Gauge Pressure (psi) for Standard Water Tube Sizes[1]				
			Nominal or Standard Size (inches)				
			⅛–1	1¼–2	2½–4	5–8	10–12
Alloy Sn50 50-50 Tin-Lead Solder[5]	100	Pressure[2]	200	175	150	135	100
		DWV[3]	–	95	80	70	–
	150	Pressure[2]	150	125	100	90	70
		DWV[3]	–	70	55	45	–
	200	Pressure[2]	100	90	75	70	50
		DWV[3]	–	50	40	35	–
	250	Pressure[2]	85	75	50	45	40
		DWV[3]	–	–	–	–	–
	Saturated Steam	Pressure	15	15	15	15	15
Alloy Sb5 95-5 Tin-Antimony Solder	100	Pressure[2]	1090	850	705	660	500
		DWV[3]	–	390	325	330	–
	150	Pressure[2]	625	485	405	375	285
		DWV[3]	–	225	185	190	–
	200	Pressure[2]	505	395	325	305	230
		DWV[3]	–	180	150	155	–
	250	Pressure[2]	270	210	175	165	125
		DWV[3]	–	95	80	80	–
	Saturated Steam	Pressure	15	15	15	15	15

For extremely low working temperatures in the 0°F to –200°F range, it is recommended that a joint material melting at or above 1,100°F be employed (see reference[6]).

[1] Standard water tube sizes per *ASTM B 88*.

[2] Ratings up to 8 inches in size are those given in *ASME B16.22 Wrought Copper and Copper Alloy Solder Joint Pressure Fittings* and *ASME B16.18 Cast Copper and Copper Alloy Solder Joint Fittings*. Rating for 10- to 12-inch sizes are those given in *ASME B16.18 Cast Copper and Copper Alloy Solder Joint Pressure Fittings*.

[3] Using *ASME B16.29 Wrought Copper and Wrought Copper Alloy Solder Joint Drainage Fittings – DWV*, and *ASME B16.23 Cast Copper Alloy Solder Joint Drainage Fittings – DWV*.

[4] Alloy designations are per *ASTM B 32*.

[5] The Safe Drinking Water Act Amendment of 1986 prohibits the use in potable water systems of any solder having a lead content in excess of 0.2%.

[6] These joining materials are defined as brazing alloys by the American Welding Society.

process is accelerated many times over when the tube and fitting is heated.

- It allows the soldering alloy or filler metal to flow more easily into the joint by promoting wetting of the metals.
- It floats out remaining oxides ahead of the molten filler metal.

Fluxes can be classified into three general groups: highly corrosive, less corrosive, and non-corrosive. The soldering process must render the flux inert; that is, lacking any chemical action. If not rendered inert, the flux gradually destroys the soldered joint over time. The fluxes used for joining copper tubing and copper fittings should be noncorrosive fluxes. Although corrosive types can do a better job of removing oxides and cleaning the copper, they represent a long-term risk to the joint. One noncorrosive type is composed of water and

03104-13_F07.EPS

Figure 7 95-5 solder.

03104-13_F08.EPS

Figure 8 Two types of flux and a flux brush.

white rosin dissolved in an organic or benzoic acid base. The best noncorrosive fluxes for joining copper pipe and fittings are compounds of mild concentrations of zinc and ammonium chloride with petroleum bases. Like solders, job specifications often identify the type of flux that is to be used on a specific project. Otherwise, it is best to make use of the manufacturer's recommendations.

An oxide film begins forming on copper immediately after it has been mechanically cleaned. Therefore, it is important to apply flux immediately after cleaning copper fittings and tubing. Flux should be applied to the clean metal with a brush or swab, never with your fingers. Not only is there a chance of causing infection if you have a cut, but there is also a chance that flux could be carried to the eyes or mouth. In addition, body contact with the cleaned fittings and tubing adds to unwanted contamination of the metal.

Flux must be stirred before each use. If a can of flux is not closed immediately after use, or if a can is not used for a considerable length of time, the chlorides separate from the petroleum base.

> **CAUTION**
>
> Brazing flux and soldering flux are not the same products. Do not allow these fluxes to become mixed or interchanged. Carelessness can ruin the soldering or brazing task.

1.2.0 Preparing Tubing and Fittings for Soldering

To prepare tubing and fittings for soldering, the tubing must be measured, cut, and reamed, and the tubing and fittings must be cleaned. It is criti-cal to use proper cleaning techniques in order to produce a solid, leakproof joint. Use the following procedure to prepare the tubing and fittings for soldering.

Step 1 If using the face-to-face method, measure the distance between the faces of the two fittings (*Figure 9*).

Step 2 Determine the cup depth engagement of each of the fittings. The cup depth engagement is the distance that the tubing penetrates the fitting. This distance can be found by measuring the fitting or by using a manufacturer's makeup chart. *Table 3* shows a manufacturer's makeup chart.

Step 3 Add the cup depth engagement of both fittings to the measurement found in Step 1 to find the length of tubing needed.

Step 4 Cut the copper tube to the correct length using a tubing cutter.

Step 5 Ream the inside of both ends of the copper tube using a reamer.

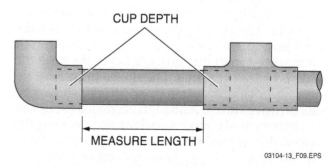

CUP DEPTH

MEASURE LENGTH

03104-13_F09.EPS

Figure 9 Face-to-face method of measuring pipe length.

Table 3 Manufacturer's Makeup Chart*

Pipe Size	Depth of Cup	Pipe Size	Depth of Cup
1/4	5/16	2	1 11/32
3/8	3/8	2 1/2	1 15/32
1/2	1/2	3	1 21/32
5/8	5/8	3 1/2	1 29/32
3/4	3/4	4	2 5/32
1	29/32	5	2 21/32
1 1/4	31/32	6	3 3/32
1 1/2	1 3/32	—	—

* All measurements are in inches.

> **NOTE**
>
> The pipe preparation instructions in Steps 6 through 9 are illustrated in *Figure 10*.

> **CAUTION**
>
> Take care when cleaning copper tubing and fittings to remove all of the abrasions on the copper without removing a large amount of metal. Abrasions can weaken or ruin a copper joint. Do not touch or brush away filings from the tube or fitting with your bare fingers because the oils in your fingers will contaminate the freshly cleaned metal.

Step 6 Clean the tubing and fitting to a bright finish using No. 00 steel wool, an abrasive pad, an emery cloth, or a tubing brush (*Figures 10A, B,* and *C*). Examine the fitting and tube to ensure they are well cleaned and no significant flaws exist in the finish. Flaws in the tubing can usually be seen more easily after cleaning, since oxidation may remain in a dent or gouge. The tube should be cleaned back from the end slightly farther than the cup depth of the fitting.

Step 7 Stir the flux before applying it. Using a brush or swab, apply flux to the copper tubing and to the inside of the copper fitting socket immediately after cleaning the parts (*Figures 10D* and *E*).

Step 8 Insert the tube into the socket, while turning the tube, until the tube touches the inside shoulder of the fitting.

Step 9 Wipe away any excess flux from the joint (*Figure 10F*).

Step 10 Check the tube and fitting for proper alignment before soldering.

1.3.0 Soldering Joints

With the tubing and fitting properly prepared, the soldering process can begin. Joints that have been prepared should be soldered within a few hours. Of course, it is best to have the soldering equipment prepared and ready so that the soldering can be done as soon as the joint has been prepared. Prepared joints should not be left for long periods of time or overnight, for example.

It is very important to ensure the tubing and fittings to be soldered are well supported. Once the heating process begins and the copper begins to expand, joints tend to sag. Remember that excess heat can damage certain components, such as valves with seals and other heat-sensitive materials. If a heat-sensitive component cannot be temporarily removed, then it should be protected by wrapping it with a very wet cloth. Heat absorbing pastes can also be used to protect components from excess heat.

> **WARNING!**
>
> Always solder in a well-ventilated area because fumes from the flux can irritate your eyes, nose, throat, and lungs. Wear a respirator if required.

Step 1 Prepare the work area and protect any components in the piping system that can be damaged by the soldering process. Remove any loose, flammable materials in the area and ensure there is adequate room for physical movement around the joint. Place an appropriate fire extinguisher within easy reach.

Acetylene

Acetylene has a lower heat content per cubic foot than any of the other fuel gases except methane or natural gas. Butane, propane, MAPP gases, and propylene all have twice as much heat content per cubic foot. This high-heat content allows much lower fuel consumption for the same amount of heating from the other gases. However, the major advantage of a properly adjusted air/acetylene torch is that higher combustion temperatures can be reached using acetylene than with any of the other gases.

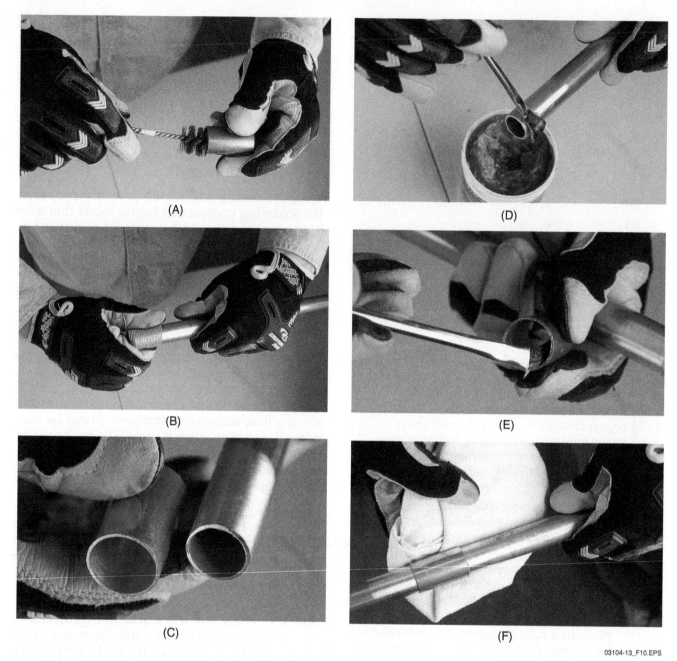

(A)

(B)

(C)

(D)

(E)

(F)

03104-13_F10.EPS

Figure 10 Preparing copper for soldering.

Step 2 Obtain either an acetylene tank and the related equipment (*Figure 11*) or a propane bottle and torch assembly. Also obtain the required solder. The instructions provided here are related to an air/acetylene torch, but the equipment manufacturer's instructions should always take precedence. For all fuel gases, follow the manufacturer's instructions carefully.

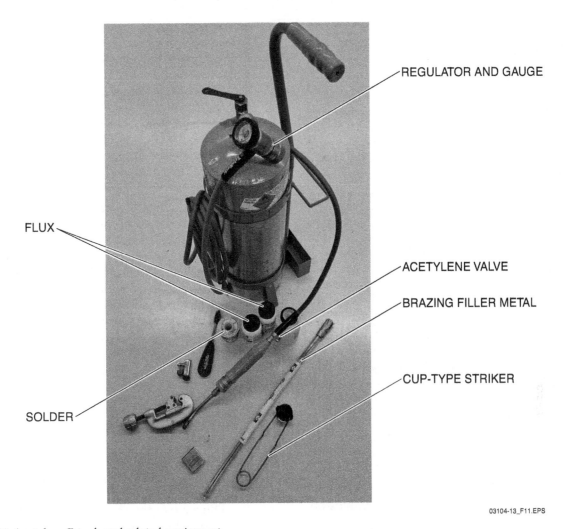

FLUX

SOLDER

REGULATOR AND GAUGE

ACETYLENE VALVE

BRAZING FILLER METAL

CUP-TYPE STRIKER

03104-13_F11.EPS

Figure 11 Acetylene B tank and related equipment.

Heat Sink Products

Overheating air conditioning and refrigeration piping components has always been an issue. Many components that must be soldered or brazed into the piping system, such as reversing valves and thermostatic expansion valves, are also sensitive to the extreme heat applied during the process. Some components can be disassembled before the installation, leaving only the metal body exposed to the heat. However, this is not always possible.

Heat sinks are materials that have a high capacity for absorbing heat. A variety of products are available to help absorb the heat that would otherwise be transferred to the component. Rather than bathe the component in the paste or gel, apply it in a ring around the tubing between the heat source and the component. If the component itself is being brazed, try to apply it on the component body away from the joint. The product should be applied with good contact and without leaving any voids where the heat can simply pass it by.

03104-13_SA01.EPS

Step 3 Purge (clear out) the acetylene hose.

- Crack the acetylene cylinder valve open slowly until a small amount of pressure registers on the acetylene high-pressure gauge. Then open it about ½ turn.
- Turn the acetylene regulator adjusting screw clockwise until a small amount of pressure shows on the acetylene working-pressure gauge. Allow a small volume of acetylene through to purge the hose, clearing it of any loose debris.
- Turn the acetylene regulator adjusting screw counterclockwise until it is loose. This shuts off the regulator output.

Step 4 Install the brazing torch handle on the ends of the hoses and close the valves on the torch.

Step 5 Check the equipment for leaks.

- Adjust the acetylene regulator adjusting screw for 10 psig on the working-pressure gauge.
- Close the acetylene cylinder valve and check for leaks. If the working-pressure gauge remains at 10 psig for several minutes, there should be no significant leaks in the system. If the reading drops, there is a leak. Use a soap solution or commercial leak detection fluid to check the acetylene connections for leaks. Never use hoses that are in questionable condition, dried or cracked, or appear to have a leak. Such hoses should be destroyed and replaced.

Step 6 Light the heating equipment according to the type of equipment you are using and the manufacturer's instructions.

- Open the torch handle valve and then set the acetylene regulator to the required pressure while the gas is flowing. Under no circumstances should the outlet pressure exceed 15 psig.
- Using an appropriate device for ignition, light the torch.
- Note that the torch handle valve is not generally used to control the size of the flame. It should be opened fully. The size of the flame is controlled by the torch size and the acetylene pressure regulator setting.

Step 7 Heat the tubing first, with the flame perpendicular (at a right angle) to the tube. Then move the flame onto the fitting cup (*Figure 12*). Make sure not to overheat the joint. The size of the tubing determines how long this step requires.

03104-13_F12.EPS

Figure 12 Applying heat to the fitting.

Step 8 Momentarily move the flame back to the tubing, and then to the back shoulder of the cup. As the flame is moved back to the cup, also back the flame away some to avoid overheating.

Step 9 Touch the end of the solder to the area between the fitting and the tube. The solder will be drawn into the joint by capillary action. The solder can be fed upward or downward into the joint. If the solder does not melt quickly on contact with the joint, move the solder away and heat the joint further. Do not melt the solder with the flame; the solder must be melted by the copper.

Step 10 Continue to feed the solder into the joint until a ring of solder appears around the joint, indicating that the joint is filled. On nominal ¾" diameter tubing and smaller, the solder can generally be fed into the joint from one point. On larger tubing, the solder should be fed quickly around the entire circumference of the joint.

Cup-Type Striker

When using a cup-type striker to ignite a torch, hold the cup of the striker slightly below and to the side of the tip, parallel with the fuel gas stream. This helps prevent the ignited gas from deflecting back toward you from the cup. It also reduces the amount of soot that accumulates in the cup. The flint in most cup-type strikers can be replaced when it is worn out.

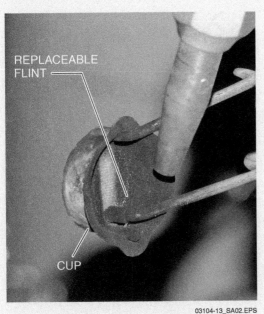

REPLACEABLE FLINT

CUP

03104-13_SA02.EPS

Step 11 Allow the joint to cool until the solder solidifies. While the joint is still warm but not hot, wipe it with a clean damp cloth to remove any excess flux (*Figure 13*). Shocking the joint with water while it is very hot may cause unnecessary stress and eventual failure.

As a general rule, the amount of solder needed is equal to the diameter of the tubing for sizes 1" and below. For example, with ¾" tubing, ¾" of solder wire should properly fill the joint. As the size increases above 1", more solder than this is necessary. For example, a 2" joint with an average joint clearance of 0.005" will require about 3½" of solder. However, in all cases, the joint clearance significantly affects the amount of solder that is needed.

If the joint is being done with the tubing horizontal, the top of the joint will likely heat up on its own and will not need preheating. Concentrate on the bottom and sides to avoid burning the flux at the top. Start feeding the solder near the bottom of the joint and work up towards the top. Then bring the solder back to the starting point and work up the other side toward the top. Remember that, although gravity does have an effect on the solder, it is capillary action rather than gravity that pulls the solder into the joint.

Soldering is an art that requires practice. Watching how the solder and flux behave during practice provides the necessary information to move and position the torch flame. With practice, soldering can be mastered by anyone.

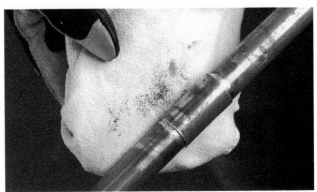

03104-13_F13.EPS

Figure 13 Wiping away excess flux.

Copper Fittings

Two kinds of solder fittings are available with copper tubing. The first is a wrought copper fitting, which is made from copper tubing that is shaped into different types of fittings. Wrought copper fittings are generally lightweight, smooth on the outside, and have thin walls. The second type is a cast fitting. This type of fitting is made using a mold. The first cast fittings had holes in the sockets to put solder in. Today, the heated copper is poured into the mold and allowed to cool. Cast copper fittings have a rougher exterior and come in a wide variety of shapes. They are typically heavier than wrought copper fittings. Wrought copper fittings are the type used in HVACR work.

Additional Resources

A Guide to Brazing and Soldering. The Harris Products Group.
Available at **www.harrisproductsgroup.com**.

Gas Welding, Cutting, Brazing, & Heating Torch Instruction Manual. The Harris Products Group.
Available at **www.harrisproductsgroup.com**.

Filler Metal Selection Guide. The Harris Products Group. Available at: **www.harrisproductsgroup.com**.

1.0.0 Section Review

1. Clothing made of synthetic fabrics should be worn whenever soldering or brazing.

    a. True
    b. False

2. When applying soldering flux, _____.

    a. brazing flux can be used if no soldering flux is available
    b. the flux is applied only to the tube end using a brush
    c. the flux is applied to the tube end and fitting socket using a brush
    d. the flux can be applied using the fingers or a brush

3. When soldering a joint, _____.

    a. heat is applied directly to the solder
    b. only the tubing is heated
    c. only the fitting is heated
    d. heat is applied to the tubing, then to the fitting

2.0.0 BRAZING COPPER FITTINGS AND TUBING

Objective

Describe and demonstrate the safe process of brazing copper tubing.

   a. Describe and demonstrate the use of the PPE, tools, and materials needed to braze copper tubing.
   b. Describe and demonstrate the preparation required.
   c. Describe and demonstrate the brazing process.
   d. Describe and demonstrate the process of brazing copper tubing to dissimilar metals.

Performance Tasks

1. Properly set up and shut down oxyacetylene equipment.
2. Properly set up and shut down an acetylene single tank.
3. Properly prep and safely solder copper tubing in various planes, using various fittings.
4. Properly prep and safely braze copper tubing using various fittings.

Trade Terms

Flashback arrestor: A valve that prevents a flame from traveling back from the tip and into the hoses.

Brazing, like soldering, uses nonferrous filler metals to join base metals that have a melting point above that of the filler metals. Brazing uses filler metals that melt at temperatures above 842°F. Brazed tubing and fittings are used in the following:

- Low-pressure steam lines
- High-pressure refrigeration lines
- Medical gas lines
- Compressed air lines
- Vacuum lines
- Fuel lines
- Other chemical lines that need extra corrosion resistance in the piping joints

Brazing, sometimes incorrectly referred to as hard soldering, produces mechanically strong, pressure-resistant joints. The strength of a brazed joint results from the ability of the filler metal to adhere to the base metal. However, reliable adhesion can occur only if the base metals are properly cleaned, the proper flux and filler metal are selected, and the clearance gap between the outside of the tubing and the inside of the fitting is appropriate. Ideally, the clearance is approximately 0.003" to 0.005".

2.1.0 PPE, Tools, and Materials Used for Brazing

To properly braze copper tubing and fittings, it is necessary to understand the following:

- Required PPE and safety practices
- Brazing equipment and accessories
- Filler metals and fluxes
- Tube and fitting preparations
- Setting up and lighting a torch
- Nitrogen purging equipment and use
- Brazing techniques

2.1.1 Brazing Safety

Brazing is often done with a combination of oxygen and acetylene. Although the hazards are similar to those encountered in soldering, they are compounded by the use of two flammable and explosive gases. In addition, when fuel gases are directly combined with oxygen instead of air, the temperature of the flame rises significantly. When brazing, the following safety guidelines apply:

- Wear clothing made of nonflammable fabric such as cotton. Avoid synthetic fabrics such as nylon, which can melt and adhere to the skin.
- Wear a long-sleeve shirt to avoid prevent burns from sparks or molten metal. Pants should not have cuffs.
- Wear eye and face protection.
- Wear flame-resistant gloves and high-top work boots.
- Make sure the gas bottle is securely turned off when not in use. Gases used in brazing are flammable and explosive if exposed to a spark or flame.
- Keep a fire extinguisher handy while brazing in case material near the work ignites. When working close to flammable material such as wood, use a fire blanket or other flame/heat-blocking material to protect it.
- Always braze in a well-ventilated area because fumes from the process can irritate your eyes, nose, throat, and lungs. Wear respiratory equipment if necessary.

There are specific safety concerns associated with the use of oxyacetylene equipment that must also be considered:

- Acetylene cylinders must be stored, transported, and used in the upright position. Otherwise, liquid acetone could be pulled out of the tank.
- Cylinders must be secured to a cart or structure using a safety chain or approved strap.
- When not in use, cylinders must be stored in the upright position with a safety cap installed (*Figure 14*). Note that some small acetylene cylinders, such as the MC tank, cannot be capped. When in transit, the regulator must be removed from all cylinders and the safety cap securely installed. In many states, it is against the law to transport cylinders with the regulators in place.
- Oxygen and fuel gas cylinders must be stored at least 20' (6.096m) apart or separated by a wall at least 5' high with a 30-minute minimum fire rating.
- Oxygen can cause ignition even when there is no obvious flame or spark around to set it off, especially when it comes in contact with oil or grease. Never handle oxygen cylinders with oily hands or gloves. Keep grease away from the cylinders and do not use oil or grease on cylinder attachments or valves. Never use an oxygen regulator for any other gas or try to use a regulator with oxygen that has been used for other service.
- A pressure-reducing regulator set for not more than 15 psig must be used with acetylene. Acetylene becomes unstable and volatile above 15 psig. An acetylene cylinder that has been laid down should be placed in the upright position for at least 20 minutes before using.
- Do not allow anyone to stand in front of the oxygen cylinder valve when opening because the oxygen is under high pressure (about 2,000 psig) and could cause severe injury when released.

03104-13_F14.EPS

Figure 14 Acetylene cylinder with a valve safety cap.

- Acetylene cylinder valves should only be opened ½ to 1 turn, and the wrench should never be removed from the valve during use. The valve on oxygen cylinders, however, should always be opened fully.

2.1.2 Brazing Equipment

Brazing is typically done using either an acetylene B-tank with an air/acetylene torch or an oxyacetylene setup (*Figure 15*). A complete B-tank setup was previously shown in *Figure 11*. The air/acetylene setup is generally used on smaller refrigerant lines, since air/acetylene produces less heat than the oxyacetylene equipment. For lines greater than 1" in diameter, installers usually select the oxyacetylene method, which provides both greater heat and better control over the movement of the filler metal. Oxyacetylene equipment mixes acetylene, as a fuel gas, with pure oxygen to produce a hotter flame.

The equipment shown in *Figure 15* is usually mounted on a hand truck and is very heavy. This type of equipment is typically used in a shop or on a job site when extensive brazing must be accomplished in a limited area. For small installations and service-related brazing, portable equipment that can be hand-carried by one person is generally chosen (*Figure 16*).

Oxygen and acetylene are compressed and shipped under medium to high pressures in cylinders. Oxygen is supplied in cylinders at pressures of about 2,000 psig. Acetylene cylinders are pressurized to about 250 psig. These cylinders must not be moved unless the protective caps are in place. Dropping a cylinder without the cap installed may result in breaking the valve off the cylinder. This allows the pressure inside to escape, propelling the cylinder like a rocket.

Transporting Fuel Gas Cylinders

Although OSHA has specific standards for the handling of fuel gas cylinders, found in *OSHA Standard 1910.253*, the Department of Transportation has primary jurisdiction over the cylinders themselves and their transportation requirements. These requirements are contained in the Code of Federal Regulations, *49 CFR, Parts 171 through 177*. One interesting provision is that drivers transporting hazardous fuel gases of a quantity that requires placards to be placed on the vehicle are prohibited from both texting and cell phone use while driving.

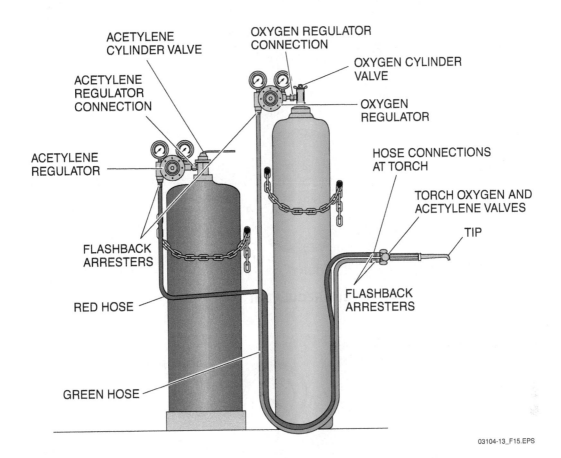

ACETYLENE CYLINDER VALVE

OXYGEN REGULATOR CONNECTION

ACETYLENE REGULATOR CONNECTION

OXYGEN CYLINDER VALVE

ACETYLENE REGULATOR

OXYGEN REGULATOR

HOSE CONNECTIONS AT TORCH

TORCH OXYGEN AND ACETYLENE VALVES

TIP

FLASHBACK ARRESTERS

RED HOSE

FLASHBACK ARRESTERS

GREEN HOSE

03104-13_F15.EPS

Figure 15 Oxyacetylene brazing setup.

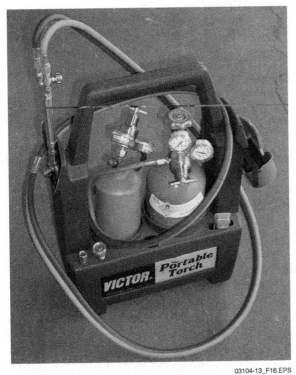

03104-13_F16.EPS

Figure 16 Portable oxyacetylene equipment.

The oxygen and acetylene tanks require regulators to be installed before they can be used. The regulator reduces the high pressure in the tank to the desired pressure for use. Regulators (*Figure 17*) must be selected for the correct gas. The connecting fittings are typically designed to prevent a regulator for one gas from being installed on the wrong gas tank. Different tank sizes of the same gas may also have different regulator connections. Regulators must be handled carefully; avoid dropping or otherwise shocking them.

> **WARNING!**
>
> A regulator should always be installed on a tank before the valve is opened. The high pressure inside the tank, if released suddenly, can cause the tank to become a dangerous projectile. Although it is a generally accepted practice to slightly open (crack) the valve of a cylinder to blow out any debris before attaching a regulator, it must be done with great care to ensure the cylinder does not tip over as a reaction to the release.

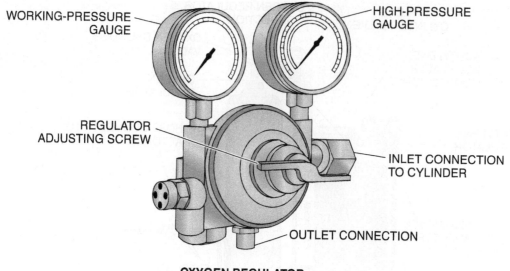

WORKING-PRESSURE GAUGE

HIGH-PRESSURE GAUGE

REGULATOR ADJUSTING SCREW

INLET CONNECTION TO CYLINDER

OUTLET CONNECTION

OXYGEN REGULATOR

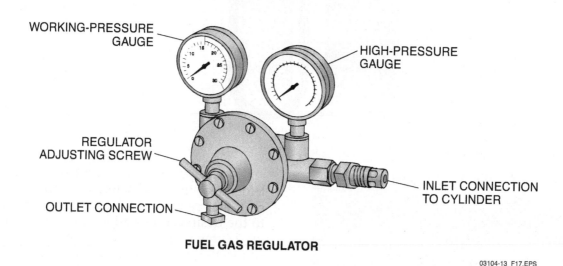

WORKING-PRESSURE GAUGE

HIGH-PRESSURE GAUGE

REGULATOR ADJUSTING SCREW

INLET CONNECTION TO CYLINDER

OUTLET CONNECTION

FUEL GAS REGULATOR

03104-13_F17.EPS

Figure 17 Oxygen and fuel gas regulators.

Many regulators have two stages of pressure regulation. The first stage reduces the tank pressure to an intermediate level, and has a fixed pressure setpoint. The second stage of the regulator reduces the intermediate pressure to the outlet pressure desired by the user, and is adjustable from 0 psig to the maximum outlet pressure specified by the manufacturer. The final adjustment of gas pressure at the nozzle is set using the valves on the torch handle.

A regulator usually has two pressure gauges. One shows the pressure in the cylinder, while the other shows the outlet pressure setpoint. The outlet gauge is monitored as the technician sets the desired outlet pressure. The regulator is adjusted by turning the adjusting screw clockwise (turning it in).

Some simpler or cheaper regulators, such as those used on small acetylene tanks, provide only one stage of regulation. A single-stage regulator needs adjustment more often, as the outlet pressure tends to fall as the pressure in the bottle changes. Two-stage regulators generally prevent this from happening. Small single-stage acetylene regulators often do not show the pressure in the cylinder, but instead have a less accurate gauge that shows the volume that remains.

Hoses designed specifically for brazing and welding work must be used to connect the regulator outlets to the torch handles. The hoses are color coded for visual identification. In the United States, the oxygen hose is green and the fuel gas hose is red. Other countries may have different standards. This is an important feature, as the hose fittings are designed so that they can only be connected to the proper gas regulator and torch handle port. Oxygen hoses have a right-handed thread, while fuel gas hoses use a left-handed thread. The female fittings with left-handed threads have a groove cut into the nut for further visual identification. New hoses should be purged with high-pressure air before use, as they often contain a white powder.

Figure 18 shows a universal torch wrench designed for regulators, hose connections, check valves, flashback arrestors, torches, and torch tips. The fittings for torch equipment are usually brass or bronze, and certain components are often fitted with soft, flexible, O-ring seals. The seal surfaces of the fittings or O-rings can be easily damaged by overtightening with standard mechanics wrenches. The length of a torch wrench is limited to reduce the chances of damage to fittings from application of excessive torque. In some cases, manufacturers specify only hand-tightening certain fitting connections of a torch set (tips or cutting/welding attachments). In any event, follow the manufacturer's specific instructions when connecting the components of a torch set. The torch wrench can be tethered to the torch outfit so that it is always available when needed and not easily misplaced.

Oxyacetylene brazing equipment offers a wide array of torch handles and tips from which to choose. The style and size of the tip is determined by the size of the tubing and, to a lesser degree, the type of joint to be made. Brazing torch handles are offered in heavy-, medium-, and light-duty models. For most HVACR brazing applications, medium- and light-duty handles are sufficient and generally preferred since they are more maneuverable (*Figure 19*). The handle typically provides a mixing chamber for the two fuel gases before they reach the nozzle.

Note that the two handles shown in *Figure 19* offer different configurations that can make a difference to technicians. The medium-duty handle places the fuel gas adjusting valves below the technician's hand. With the light-duty handle, the adjusting valves are placed above the hand. This affects the balance of the torch to some degree. Having the adjusting valves above the hand can be more convenient, but they can also get in the way in tight spots.

Many torch handles today have built-in flashback arrestors (*Figure 20*). Flashback arrestors are one-way valves that prevent a pressurized flame from traveling back up the tip and into the hoses and regulators. A flashback can sometimes happen if either the oxygen or the fuel gas flow rate is inadequate, or if both the oxygen and fuel gas are

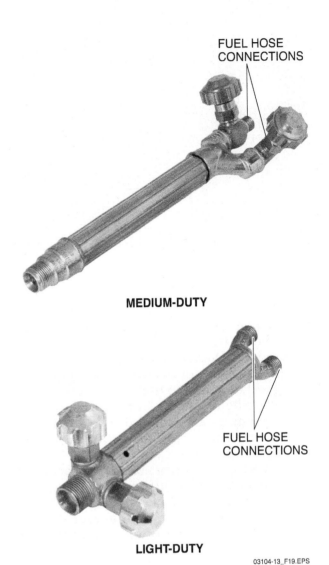

MEDIUM-DUTY

LIGHT-DUTY

FUEL HOSE CONNECTIONS

FUEL HOSE CONNECTIONS

03104-13_F19.EPS

Figure 19 Medium- and light-duty brazing torch handles.

turned on and then ignited at the brazing tip. If a torch handle is not already equipped with them, flashback arrestors should be installed where the hoses connect to the torch handle. Although regulators can be equipped with them at their outlet, torch-mounted models prevent a flashback from entering the hoses. To be effective, flashback arrestors must be installed with the flow arrow pointing toward the torch handle.

It is very important to note that the brazing tip must be designed to properly fit the handle. Most brazing tips from a given manufacturer are designed to fit only that manufacturer's handles. Further, tips do not fit every handle that a manufacturer offers. When selecting and purchasing tips, always be sure that the tips are compatible with the torch handle in use.

Several different brazing tips are shown in *Figure 21*. Brazing nozzles and tips are sized by a number corresponding to the size of the nozzle opening. Larger numbers indicate larger open-

03104-13_F18.EPS

Figure 18 Torch wrench.

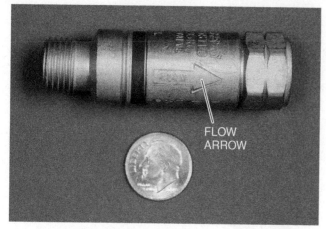

Figure 20 Flashback arrestor.

03104-13_F20.EPS

Table 4 Tip Sizes Used for Common Tubing Sizes

Tip Size (No.)	Rod Size (inches)	Tubing Diameter (inches)
4	³⁄₃₂	¼–³⁄₈
5	⅛	½–¾
6	³⁄₁₆	1–1¼
7	¼	1½–2
8		2–2½
9	³⁄₈	3–3½
10	⁷⁄₁₆	4–6

ings. *Table 4* provides a guide for the selection of tips based on the size of the tubing to be brazed. Much smaller tips are available but are not typically used for HVACR work.

The multi-flame, or rosebud, torch tip is used for larger tubing and to heat larger surfaces. Instead of a single opening in the nozzle, several smaller openings are provided. The multi-flame tip is an excellent choice for larger tubing and fittings, to broaden the area of the flame. The universal tip accepts different nozzles so that one tip can be used with many different nozzle sizes.

The brazing tip does occasionally get fouled with solids from the combustion process. This is especially true when the torch is not set up to burn cleanly. The relatively soft brass can also be damaged at the tip, causing the flame to be erratic. Torch tip cleaners (*Figure 22*) are used to clean the tip. Tip cleaners are like small, round files that clear the material out of the way. Each tip cleaner fits a specific nozzle size. As shown here, tip cleaners typically come as a set in folding packages. Tip drills are used for major cleaning and for holes that are plugged. Tip drills are tiny drill bits that are sized to match the diameters of the tip openings. A drill bit is mounted in a small handle for manual use.

2.1.3 Filler Metals and Fluxes

Filler metals used to join copper tubing are of two groups: alloys that contain 8 to 60 percent silver (the BAg series), and copper alloys that contain phosphorus (the BCuP series). *Table 5* lists brazing filler materials according to their American Welding Society (AWS) classification and principal elements. *Figure 23* shows a package of BCuP-5 brazing filler rods that contain 15 percent silver. They have been an industry standard for many years, although other alloys have dramatically increased in popularity as silver prices have increased. Unlike soldering filler metals, brazing filler metals are typically in the form of a stick.

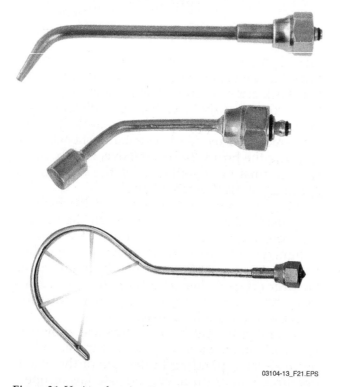

03104-13_F21.EPS

Figure 21 Various brazing tips.

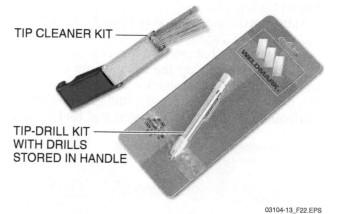

TIP CLEANER KIT

TIP-DRILL KIT WITH DRILLS STORED IN HANDLE

03104-13_F22.EPS

Figure 22 Torch tip cleaners.

Table 5 Brazing Filler Metals

| AWS Classification | Percent of Principal Element | | | | | Copper Classification |
	Silver	Phosphorus	Zinc	Cadmium	Tin	
BCuP-2	–	7–7.5	–	–	–	Balance
BCuP-3	4.75–5.25	5.75–6.25	–	–	–	Balance
BCuP-4	5.75–6.25	7–7.5	–	–	–	Balance
BCuP-5	14.5–15.5	4.75–5.25	–	–	–	Balance
BAg-1	44–46	–	14–18	23–25	–	14–16
BAg-2	34–36	–	19–23	17–19	–	25–27
BAg-5	44–46	–	23–27	–	–	29–31
BAg-7	55–57	–	15–19	–	4.5–5.5	21–23

> **WARNING!**
> BAg-1 and BAg-2 contain cadmium. Heating when brazing can produce highly toxic fumes. Use adequate ventilation and avoid breathing the fumes.

The two major groups of filler metals differ in their melting, fluxing, and flowing characteristics. These characteristics should be considered when selecting a filler metal. When joining copper tubing, any of the filler metals in *Table 4* can be used. However, the filler metals used most often for close tolerances and reasonable economy are BCuP-3 and BCuP-4. BCuP-5 is better where close tolerances cannot be held in the joint, but it is generally a more expensive product.

As with soldering, fluxes are also used in brazing. Brazing fluxes must have significantly differ-

ent properties due to the temperatures involved. However, a flux is not always needed in brazing. Alloys that contain phosphorus, such as BCuP-3 through BCuP-5, are self-fluxing when used in

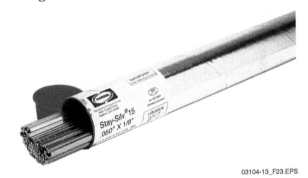

03104-13_F23.EPS

Figure 23 BCuP-5 brazing rods comprised of 15 percent silver, 80 percent copper, and 5 percent phosphorus.

Tiny Tips

Oxyacetylene torch tips are available with extremely tiny openings for jewelry work and other precision applications. Some nozzles are so small that the level of precision needed cannot be achieved by drilling the nozzle opening in the brass. Instead, a hole as small as 0.003" is drilled into a precious stone, such as a sapphire, and then stone is the inserted into the tip.

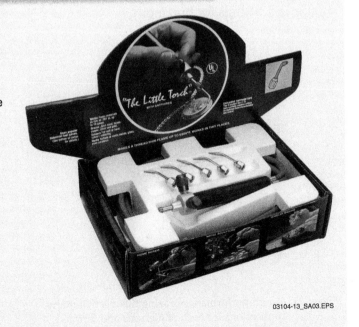

03104-13_SA03.EPS

clean copper-to-copper applications. When they are used with brass, bronze, or ferrous metals, a flux is required.

Brazing fluxes (*Figure 24*) are applied using the same methods and rules as soldering fluxes. However, high-temperature fluxes used in brazing are usually applied only to the end of the tube and not on the inside of the fitting. Brazing fluxes are more corrosive than soldering fluxes, so care must be taken never to confuse a soldering flux with a brazing flux. The two types should certainly never be mixed. For best results, use the flux recommended by the manufacturer of the brazing filler metals.

Since brazing temperatures are significantly higher than soldering temperatures, oxides form even more rapidly. Without the proper use of flux when needed, the brazed joint will not reach an acceptable level of quality or strength. Brazed joints often must be disassembled in the field for service purposes. Whenever a hermetic compressor is replaced, for example, a number of brazed joints usually must be taken apart. Flux should be applied to these joints as well, before they are heated. Apply the flux on top of the visible filler metal and on the adjacent areas of the tube and fitting.

2.2.0 Preparing for Brazing

To prepare tubing and fittings for brazing, you must follow the same procedures as you would to prepare tubing and fittings for soldering. It is critical that proper cutting and cleaning techniques be used in order to produce a solid, leakproof joint. Use the following procedure to prepare the tubing and fittings for brazing:

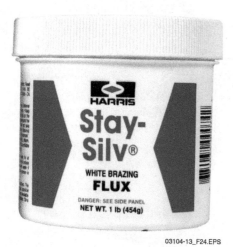

03104-13_F24.EPS

Figure 24 A high-performance brazing flux used in many applications for both nonferrous and ferrous metals.

Step 1 Measure the distance between the faces of the two fittings.

Step 2 Determine the cup depth engagement of each of the fittings.

Step 3 Add the cup depth engagement of both fittings to the measurement found in Step 1 to find the length of tubing needed.

Step 4 Cut the copper tubing to the correct length using a tubing cutter.

Step 5 Ream the inside and outside of both ends of the copper tubing using a reamer.

> **CAUTION**
>
> Take care when cleaning the tubing and fittings to remove all the abrasions on the copper without removing a large amount of metal. Abrasions can weaken or ruin a copper joint. Do not touch or brush away filings from the tube or fitting with your fingers because your fingers will also contaminate the freshly cleaned metal. Filings and burrs can also easily cause an injury.

Step 6 Clean the tubing and fitting using No. 00 steel wool, a piece of emery cloth, an abrasive pad, or a special copper-cleaning tool.

Step 7 Apply flux (if required) to the end of the copper tubing before inserting it into the fitting. Flux does not need to be applied to the fitting cup.

Step 8 Insert the tube into the fitting socket, and push and turn the tube into the socket until the tube touches the inside shoulder of the fitting.

Step 9 Wipe away any excess flux from the outside of the joint.

Step 10 Check the tube and fitting for proper alignment before brazing.

2.2.1 Brazing Equipment Setup

The brazing heating procedure differs from soldering in that different equipment may be required to raise the temperature of the metals to be joined well above 842°F (450°C). Actually, most brazed joints are made at temperatures between 1,200°F and 1,550°F (649°C to 843°C). This is the range of temperatures where the popular brazing alloys are fluid enough to flow, without becoming too fluid. The torch flame must be significantly higher than this to achieve good control over the heating process. Because of the higher temperatures needed, oxyacetylene equipment is typically used for brazing and is the primary approach covered here.

During use, transportation, and storage, oxygen and acetylene cylinders must be secured with a stout cable or chain in the upright position to prevent them from falling and injuring people or damaging equipment. When stored at the job site, oxygen and acetylene cylinders must be stored separately with at least 20' between them, or with a 5' high, half-hour minimum fire wall separating them. As a general rule, cylinders should be stored without a regulator attached, and the regulator should be removed when a cylinder is moved. With a regulator attached, the chances of shearing off the cylinder valve stem are increased, since it protrudes well beyond the body of the cylinder. Store empty cylinders away from partially full or full cylinders and make sure they are properly marked to clearly show that they are empty (*Figure 25*).

Follow this basic procedure to prepare oxyacetylene brazing equipment for use:

> **WARNING!**
>
> Do not handle acetylene and oxygen cylinders with oily hands or gloves. Keep grease away from the cylinders and do not use oil or grease on cylinder attachments or valves. The mixture of oil and oxygen could cause an explosion. Make sure that the protective caps are in place on the cylinders before transporting or storing the cylinders.

03104-13_F25.EPS

Figure 25 Typical empty cylinder marking.

Alternate High-Pressure Cylinder Valve Cap

High-pressure cylinders can also be equipped with a clamshell cap that can be closed to protect the cylinder valve with or without a regulator installed on the valve. This enables safe movement of the cylinder after the cylinder valve is closed. This type of cap is usually secured to the cylinder body cap threads when it is installed so that it cannot be removed. When the clamshell is closed, it can also be padlocked to prevent unauthorized operation of the cylinder valve.

CLAMSHELL OPEN TO ALLOW
CYLINDER VALVE OPERATION

LATCH PIN
(OR PADLOCK)

CLAMSHELL CLOSED FOR MOVEMENT OR
PADLOCKED TO PREVENT OPERATION
OF CYLINDER VALVE

CLAMSHELL CLOSED FOR TRANSPORT

Step 1 Install and securely fasten the oxygen and acetylene cylinders in a bottle cart or in an upright position against a wall or other substantial structure.

Step 2 Install the oxygen regulator on the oxygen cylinder.
- Remove the cylinder protective cap.
- Open (crack) the oxygen cylinder valve just long enough to allow a small amount of oxygen to pass through and blow out any debris, then close it.
- Carefully inspect the regulator attaching threads, cylinder threads, and mating surfaces for damage (*Figure 26*).
- Turn the adjusting screw on the oxygen regulator counterclockwise (out) until it is loose. This shuts off the regulator output and prevent accidental overpressurizing of the hose and torch during hookup.
- Using a torch wrench, install the oxygen regulator on the cylinder (*Figure 27*). Oxygen cylinders and regulators have right-hand threads. Tighten the nut snugly. Be careful not to overtighten the nut because this may strip the threads.

CHECK THAT FITTINGS ARE CLEAN

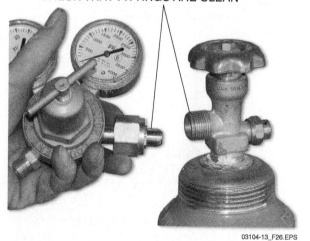

03104-13_F26.EPS

Figure 26 Inspecting the fittings.

Step 3 Install the acetylene regulator on the acetylene cylinder.
- Remove the cylinder's protective cap.
- Open (crack) the acetylene cylinder valve, using the valve wrench, just long enough to allow a small amount of acetylene to pass through the valve, then close it.

Alternate Acetylene Cylinder Safety Cap

Acetylene cylinders can be equipped with a ring guard cap that protects the cylinder valve with or without a regulator installed on the valve. This enables safe movement of the cylinder after the cylinder valve is closed. This type of cap is usually secured to the cylinder body cap threads when it is installed so that it cannot be removed.

03104-13_SA05.EPS

- Turn the adjusting screw on the acetylene regulator counterclockwise until it is loose. This shuts off the regulator output and prevent accidental overpressurizing of the hose and torch during hookup.
- Using a torch wrench, install the acetylene regulator on the cylinder. Acetylene cylinders and regulators have left-hand threads. Tighten the nut snugly. Be careful not to overtighten the nut because this may strip the threads.

Step 4 Connect the hoses and brazing torch.
- If they are not already in place or a part of the torch handle, install flashback arrestors on the oxygen and acetylene hoses at the torch handle inlets.
- Connect the green hose to the oxygen regulator and the red hose to the acetylene regulator. Tighten the hoses snugly. Be careful not to overtighten the fittings because this may strip the threads.

Step 5 Purge (clean) the oxygen hose.
- Crack the oxygen cylinder valve slowly until the pressure stops rising on the cylinder, or high-pressure, gauge. This now indicates the pressure in the cylinder. Then slowly open the valve completely.

TORCH WRENCH

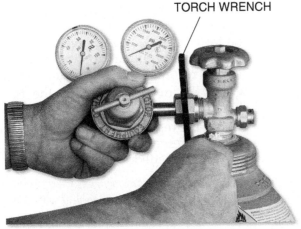

03104-13_F27.EPS

Figure 27 Installing the regulator.

- Turn the oxygen regulator adjusting screw clockwise (in) until a small amount of pressure (3 to 5 psig) shows on the oxygen working-pressure gauge. Allow a small amount of pressure to build up and purge the oxygen hose, clearing it of any loose debris.
- Turn the oxygen regulator adjusting screw counterclockwise (out) until it is again loose. This will shut off the regulator output.

Step 6 Purge (clean) the acetylene hose.
- Crack the acetylene cylinder valve open slowly until the cylinder pressure registers on the acetylene cylinder, or high-pressure, gauge. Then open it about ½ to ¾ of a turn.
- Turn the acetylene regulator adjusting screw clockwise (in) until a small amount of pressure shows on the acetylene working-pressure gauge. Allow a small amount of pressure to build up and purge the acetylene hose, clearing it of any loose debris.
- Turn the acetylene regulator adjusting screw counterclockwise (out) until it is again loose. This will shut off the regulator output.

Step 7 If not already in place, install a flashback arrestor on the acetylene inlet of the torch.

Step 8 Install the brazing torch on the ends of the hoses and close the valves on the torch. Remember that the oxygen hose has normal right-hand threads while the acetylene hose has left-hand threads.

Step 9 Check the oxyacetylene equipment for leaks.
- Adjust the acetylene regulator adjusting screw for 10 psig on the working-pressure gauge.
- Adjust the oxygen regulator adjusting screw for 20 psig on the working-pressure gauge.

- Close the oxygen and acetylene cylinder valves and check for leaks while the pressure remains in the hoses and torch handle. If the working-pressure gauges remain at 10 and 20 psig for several minutes, there should be no significant leaks in the system. If the readings drop, there is a leak. Use a soap solution or commercial leak detection fluid to check the oxygen or acetylene connections for leaks. Note that the leak could also be on the regulator assembly. If the regulator assembly is found to be leaking, stop using it immediately. Do not attempt to repair it in the field.

- Open both valves on the torch handle to release the pressure in the hoses. Watch the working-pressure gauges until they register zero, then close the valves on the torch handle.

- Turn the oxygen and acetylene regulator valves counterclockwise until they are loose. This releases the spring pressure on the regulator diaphragms and completely closes the regulators.

Step 10 Coil the hoses and hang them on the hose holder.

2.2.2 Lighting the Oxyacetylene Torch

After the oxyacetylene brazing equipment has been properly prepared, the torch can be lit and the flame adjusted for brazing. Before beginning however, it is important to know the correct oxygen and acetylene operating pressures for the tip in use. *Table 6* provides one example of a manufacturer's table with this data for a particular series of tips. It is important that the information for the actual tip in use be consulted. This table should not be considered a general guide for all tips. For hose lengths greater than 25 feet, the pressure at the regulator is generally increased 2 to 3 psig to compensate for pressure drop. However, remember that the acetylene pressure should never exceed 15 psig while the gas is flowing.

There are three types of flames: neutral, carburizing (reducing), and oxidizing. *Figure 28* shows the differences in the types of flames. The neutral flame burns equal amounts of oxygen and acetylene. The inner cone is bright blue in color, surrounded by a fainter blue outer flame envelope that results when the oxygen in the air combines with the superheated gases from the inner cone. A neutral flame is used for almost all fusion welding or heavy brazing applications.

A carburizing (reducing) flame has a white feather created by excess fuel. The length of the feather depends on the amount of excess fuel in the flame. The outer flame envelope is brighter than that of a neutral flame and is much lighter in color. The excess fuel in the carburizing flame produces large amounts of carbon. The carburizing flame is cooler than the neutral flame and is used for light brazing to prevent melting of the base metal.

An oxidizing flame has an excess amount of oxygen. Its inner cone is shorter, with a bright blue edge and a lighter center. The cone is also more pointed than the cone of a neutral flame. The outer flame envelope is very short and often fans out at the ends. The hottest flame, it is sometimes used for brazing cast iron or other metals.

Use the following procedure to light an oxyacetylene torch.

Step 1 Set up the oxyacetylene torch as discussed previously, heeding all warnings. Make sure the desired tip is installed before lighting the torch (refer back to *Table 4* for recommended tip sizes). Be aware of the needed gas pressure settings for the tip in use.

Step 2 Adjust the torch oxygen flow.
- Open the oxygen cylinder valve slightly until pressure registers on the oxygen cylinder high-pressure gauge and then stops rising; then open the valve fully.
- Turn the oxygen regulator adjusting screw clockwise until pressure shows on the oxygen working-pressure gauge. Adjust the regulator for the manufacturer's recommended pressure setting.
- Open the oxygen valve on the torch handle.
- Check the regulator and ensure that the pressure remains at the desired setting. Adjust as needed.

NOTE

Always fine-tune the pressure setting with the torch valve open. When it is closed, the pressure may register slightly higher. Regulators are incapable of making pressure adjustments without flow.

Table 6 Brazing Tip Operational Data

Part #	UPC #	Acetylene Consumption SCFH	Fuel Pressure psig	Oxygen Pressure psig	Welding Metal Thickness	Brazing Copper Tubing
Type 17-000	30250	1–2	5	5	1/32"	1/8"
Type 17-00	30251	2–3	5	5	3/64"	1/4"
Type 17-0	30252	2–4	5	5	5/64"	1/2"
Type 17-1	30253	3–6	5	5	3/32"	3/4"
Type 17-2	30254	5–10	5	5	1/8"	1"
Type 17-3	30255	8–18	6	7	3/16"	1½"
Type 17-4	30256	10–25	7	10	1/4"	2"
Type 17-5	30257	15–35	8	12	1/4"–1/2"	3"
Type 17-6	30258	25–45	9	14	1/4" – 3/4"	4"
Type 17-7	30259	30–60	10	16	3/4" – 1¼"	6"

- Close the oxygen valve on the torch handle.

Step 3 Adjust the torch acetylene flow.
- Open the acetylene cylinder valve slightly until the pressure registers and stabilizes on the acetylene high-pressure gauge; then open the valve about ½ turn.

> **CAUTION**
>
> Be sure to leave the valve wrench on the acetylene cylinder valve while the cylinder is in use so that the valve can be closed quickly in case of an emergency.

- Turn the acetylene regulator adjusting screw clockwise until the desired pressure shows on the acetylene working-pressure gauge.
- Open the acetylene valve on the torch handle.
- Check the regulator and ensure that the pressure remains at the desired setting. Adjust as needed.
- First close, and then open the torch acetylene valve about ½ turn to prepare for ignition.

> **WARNING!**
>
> When lighting the torch, be sure to wear flame-resistant gloves and goggles; use an appropriate device, such as a cup-type striker, to light the torch; and point the torch away from you. Always light the fuel gas first, and then open the oxygen valve on the torch handle.

Step 4 Light the torch.
- Hold the striker in one hand and the torch in your other hand. Strike a spark in front of the escaping acetylene gas.
- Open the acetylene valve on the torch until the flame jumps away from the tip about . Then slowly close the valve until the flame just returns to the tip (*Figure 29*). This sets the proper fuel gas flow rate for the size tip being used.

> **NOTE**
>
> It is not unusual for visible soot to form in strings and float away from the acetylene-only flame. However, they can fall on an otherwise clean surface and be somewhat messy. Take the surroundings into account as you light the torch and be aware that soot is likely to form from the initial flame.

- Open the oxygen valve on the torch slowly to add oxygen to the burning acetylene. Refer again to *Figures 28* and *29*. Observe the luminous cone at the tip of the nozzle and the long, greenish envelope around the flame, which is excess acetylene that represents a carburizing flame. As you continue to add oxygen, the envelope of acetylene should disappear. The inner cone will appear soft and luminous, and the torch will make a soft, even, blowing sound. This indicates a neutral flame, which is the ideal flame for welding. If too much oxygen is added, the flame becomes smaller, more pointed, and white in color, and the torch makes a sharp

CARBURIZING FLAME

INITIAL FLAME

NEUTRAL FLAME

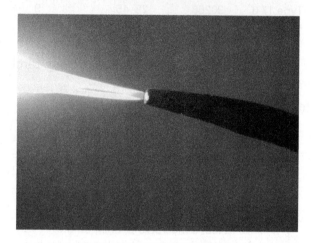

FINAL FLAME ADJUSTMENT

03104-13_F29.EPS

Figure 29 Reduce the acetylene flow until the flame returns to the tip.

Step 5 Shut off the torch when finished brazing.
- Shut off the oxygen valve on the torch handle first.
- Shut off the acetylene valve on the torch handle. Close it quickly to avoid excess carbon buildup.
- Shut off both the oxygen and acetylene cylinder valves completely.
- Open both valves on the torch handle to release the pressure in the hoses. Watch the gauges on both regulators. With the cylinder valves closed, the pressure should fall on all regulator gauges until they register zero. Then close the valves on the torch handle.
- Turn the oxygen and acetylene regulator adjusting knobs counterclockwise (out) until loose. This releases the spring pressure on the diaphragms in the regulators.

OXIDIZING FLAME

03104-13_F28.EPS

Figure 28 Types of flames.

snapping or whistling sound. For brazing thin materials, a cooler, carburizing flame can be used to help prevent melting the base metal accidentally.

Step 6 Properly stow the hoses while any final preparations to braze are made.

2.2.3 Air/Acetylene Equipment Setup

In many instances, the heat provided by an air/acetylene torch alone is suitable for brazing refrigerant lines. This is especially true for smaller lines. The procedure for setting up and using the air/acetylene torch is the same as that used in the soldering process.

Air/acetylene torch tips are available in different sizes (*Figure 30*). However, it should be noted that a larger tip does not increase the temperature of the flame. It does, however, provide more heat through additional coverage of the material. Follow the manufacturer's guidelines for the selection of torch tips.

2.2.4 Purging Refrigerant Lines

Oil inside the tubing or part being brazed can vaporize when the heat of the brazing torch is applied. Oil vapor mixed with air can catch fire or explode if a source of ignition is added. In addition, when copper is heated during brazing, it

Tank Identification Transducers and Safety Plugs

Many gas suppliers mount transducers on their tanks so that they can readily identify the tanks. The transducer is electronically scanned, and a coded number is matched against the supplier's records for identification. This aids in quickly determining the purchaser or user of the tank as well as the required retesting date. Acetylene cylinders are also equipped with safety plugs that will release if the temperature exceeds 220°F (104°C). The safety plugs release the gas in the event of a fire in order to prevent the cylinder from exploding.

CYLINDER TOP SAFETY PLUGS (1 OF 2)

VALVE HANDWHEEL

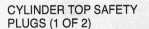

IF PRESENT, GAS SUPPLIER TRANSDUCER FOR CYLINDER IDENTIFICATION

CYLINDER BOTTOM SAFETY PLUGS

03104-13_SA06.EPS

03104-13_F30.EPS

Figure 30 Air/acetylene torch tips.

reacts rapidly with the oxygen in the air to form copper oxide. The oxidation occurs in such abundance, it forms thin, flaky deposits that separate from the tubing as they cool. If oxygen from the atmosphere is present in the tubing during brazing, these deposits form on the inside as well. Refrigerant flow later washes away the copper oxide particles, which can plug orifices, cause abrasion, and generally pollute the system. As a precaution, all the air should be removed from the tubing being brazed. This can best be done by purging the tubing with nitrogen.

> **WARNING!**
>
> Never use oxygen to purge tubing. An explosion could result when oil and oxygen are mixed.

The pressure in a full nitrogen cylinder is 2,000 psig or more. An accurate nitrogen pressure regulator and an adjustable pressure relief valve must always be used when purging. See *Figure 31*. The relief valve should be adjusted to open at a relatively low pressure, such as 7 to 10 psig.

Some careful thought about the process provides the information needed to properly purge a system for brazing. First, the goal is to simply push the air (and the oxygen it contains) out of

the tubing and keep it out while the brazing is being done. However, nitrogen pressure cannot be allowed to build in the tubing while brazing proceeds. There is a real risk that leaks will develop as nitrogen escapes through a joint while the filler metal is in its liquid state. For this reason, the tubing circuit must remain freely open to the atmosphere at one or more points, allowing the nitrogen to escape without a pressure build-up. Remember that there is no need to use any significant pressure. Once the air is initially pushed out, the goal is to simply create and maintain a nitrogen atmosphere inside the tubing. The nitrogen pressure inside the tubing, therefore, need only be slightly higher than the atmospheric pressure.

The best approach is to connect the nitrogen source to a refrigerant gauge manifold. The manifold valves provide much finer control over the nitrogen pressure applied than is possible with the regulator. The manifold hoses provide multiple outlets to connect the hoses to the system if necessary. When brazing independent tubing assemblies or circuits that have no gauge ports, ports can be installed or cone valve plugs (*Figure 32*) can be used. They are pushed into the end of the tubing or fitting to provide a point of connection. Note that these same valve plugs are very useful in clearing clogged condensate drains as well.

Once the connections are made and the regulator is installed, slowly open the nitrogen cylinder valve. With the refrigerant gauge manifold valves closed, set the regulator for roughly 3 to 5 psig. Then open the gauge manifold valve(s) and allow the nitrogen to begin circulating through the circuit. Once the air has been pushed out of the area, use the gauge manifold valve to reduce the flow to a very low volume—just enough to keep the air from being able to flow back in. As a general rule, the flow is sufficient when it can be felt gently in the palm of your hand. Allow the flow to continue at this pace throughout the brazing process. It should also continue to flow as the joint(s) cools.

If a refrigerant circuit is being closed and the final joints are being brazed, it is essential that a port remain open nearby the work to relieve any pressure. As the last few joints are completed, it is often best to turn off the nitrogen flow altogether. Heat from the process typically maintains enough of a draft leaving the circuit to prevent a significant amount of air from entering.

If it is necessary to braze tubing located in a confined space, make sure to attach a hose, pipe, etc., to the tubing being brazed so that the purge gas leaving the tubing is vented into the atmosphere outside. Otherwise, a hazardous atmosphere can

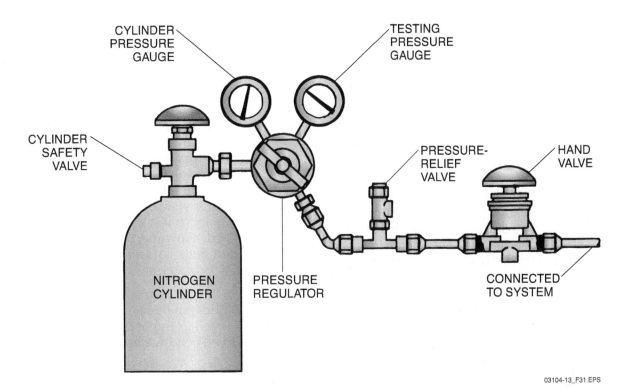

Figure 31 Nitrogen purging setup.

be created within the space as the oxygen is displaced, resulting in an oxygen deficiency. OSHA regulations require that the internal atmosphere of a permit-required confined space be tested for hazardous atmosphere with a calibrated direct-reading instrument before an employee is allowed to enter the space. In addition, the atmosphere must be continuously monitored while work is in progress, in order to detect any conditions indicating the presence of a hazardous atmosphere, including an oxygen-deficient atmosphere.

Nitrogen purging is a valuable aid in maintaining the cleanliness of refrigerant circuits and creating a safer working environment. Although it is all too often omitted in field work, its value cannot be overstated and it should be used whenever possible.

Figure 32 Cone valve plug.

2.3.0 Brazing Joints

Once the flame is properly adjusted, the torch is ready to braze joints. Use the following procedure to braze a joint.

Step 1 Prepare the work area and protect any components in the piping system that can be damaged by the heat of the brazing process. Remove any loose, flammable materials in the area and ensure there is adequate room for physical movement around the joint. Place an appropriate fire extinguisher within easy reach.

Step 2 Set up the nitrogen gas to purge the tubing following the guidelines and precautions described previously.

Step 3 Put on welding goggles with a No. 4 or 5 tint.

Step 4 Set up and light the oxyacetylene brazing equipment as described previously.
  • Place the chosen filler metal within easy reach and be certain that the material is freely accessible while the process is in progress.
  • Light and adjust the torch to produce a neutral flame.

Step 5 Apply the heat to the tubing first, in the area adjacent to the fitting. Watch the flux (if used). It will first bubble and turn white and then melt into a clear liquid, becoming calm. At this time, shift the flame to the fitting and hold it there until any flux on the fitting turns clear.

Step 6 Continue to move the heat back and forth over the tubing and the fitting. Allow the fitting to receive more heat than the tubing by briefly pausing at the fitting while continuing to move the flame back and forth. Pause at the back of the fitting cup. For and larger tubing, it may be difficult to bring the whole joint up to temperature at one time with a single-orifice tip. It will often be desirable to use a multi-flame (rosebud) heating tip such as the one shown in *Figure 33* to maintain a more uniform temperature over large areas. A mild preheating of the entire fitting and adjacent tubing is recommended for larger sizes, and the use of a second torch to retain a uniform preheating of the entire fitting assembly may be necessary with the largest diameters.

Step 7 Touch the filler metal rod to the joint. If the filler metal does not melt on contact, continue to heat and test the joint until the filler metal melts. Be careful to avoid melting the base metal.

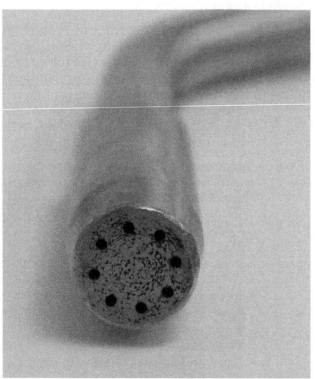

03104-13_F33.EPS

Figure 33 Multi-flame, or rosebud, heating tip.

Step 8 Hold the filler metal rod to the joint, and allow the filler metal to enter into the joint while holding the torch slightly ahead of the filler metal and directing most of the heat to the shoulder of the fitting. When the filler metal is applied to the top of the fitting, it should run down to the bottom and fill in the joint through capillary action. Applying heat at the bottom of the joint should help to make sure the filler metal penetrates. Make sure the filler metal is visible all around the shoulder of the joint. If it is not, apply additional filler metal. However, building an excessive fillet does not improve joint quality and wastes material.

Step 9 For larger joints, small sections of the joint can be heated and brazed. Be sure to overlap the previously brazed section as you continue around the fitting (*Figure 34*).

Step 10 After the filler metal has hardened and while it is still hot, wash the joint with a wet rag to clean any excess brazing flux from the joint. This also helps the copper oxides to crack and break away in flakes.

Step 11 Allow the joints to finish cooling naturally. Maintain the nitrogen flow until the joint(s) is cool.

The technique for brazing varies somewhat, depending on the position in which the brazing is being done. The method just described is good for horizontal runs of pipe, but some alterations are needed for vertical-up and vertical-down brazing.

- *Vertical-up joints* – Heat the tube first, and then transfer the heat to the fitting. Move the heat back and forth between the fitting and the tube, but be sure not to overheat the tube below the fitting. Doing so could cause the filler metal to flow out of the joint. When the brazing temperature is reached, apply the filler metal to the joint while applying heat to the wall of the fitting. This should cause the filler metal to run up into the fitting. Heat and capillary action must be used to defeat gravity.

- *Vertical-down joints* – Heat the tube before heating the fitting. When the brazing temperature is reached, apply more heat to the fitting while applying the filler metal to the joint. The filler metal should run down into the fitting.

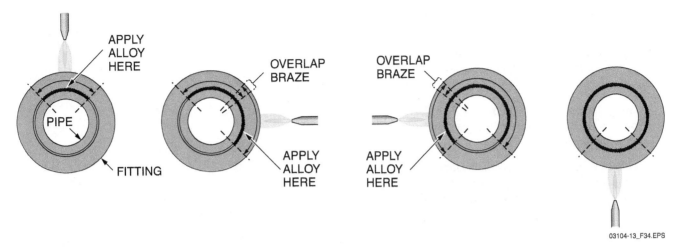

Figure 34 Working in overlapping sectors.

2.4.0 Brazing Dissimilar Metals

Some accessories used in HVACR equipment have fittings that are made of steel, copper-clad steel, or brass. Accumulators and receivers, for example often have steel fittings, while some filter-driers have copper-clad steel fittings. Sight glasses are sometimes made of brass. Special attention must be given to these situations. A 45-percent silver brazing alloy meeting BAg-5 specifications (*Figure 35*) is usually a better choice.

For example, when brazing dissimilar metals such as copper to brass or copper to steel, flux and a silver-bearing brazing alloy such as a BAg-5 should be used. A non-phosphorous alloy is generally required because the phosphorous could cause the joint to become brittle and fail. BAg-5

Figure 35 A 45-percent silver brazing alloy, in roll form (BAg-5).

products do not contain phosphorous. Remember that, for copper-to-copper joints, a phosphorous-bearing filler metal acts as a flux, so no additional flux is required. When a non-phosphorous filler metal is used, a flux must be used, even on copper. As always, the pieces being joined must be very clean and free from any oil or grease.

Aside from choosing the correct filler metal and flux, the other important aspect of brazing dissimilar metal is the application of heat. The fittings on accessories that use steel or copper-clad steel fittings reach their melting point faster than copper, so there is a potential to overheat and distort the fittings during brazing. In order to avoid overheating, it is necessary to move the heat rapidly and uniformly around the joint, and not let it linger too long in one place. Thin steel can turn cherry-red very quickly, since it transfers heat away much more slowly than copper, allowing its own temperature to rise quickly. When joining copper to steel, more heat needs to be applied to the copper, since is the better conductor and thus will dissipate heat more quickly than the steel.

Also keep in mind that metals expand at different rates. Copper and brass will expand more than steel when heated. This may cause the joint to tighten up so much that the filler metal will not flow into the joint. For that reason, a joint clearance of 0.010" may be more suitable than the usual 0.001" to 0.003" clearance when brazing copper or brass to steel. However, if the heat is likely to loosen the fit between the two metals, start with a tighter fit.

Additional Resources

www.brazingbook.com. Lucas-Milhaupt, Inc. – An interactive resource on the subject of brazing.

A Guide to Brazing and Soldering. The Harris Products Group. Available at **www.harrisproductsgroup.com**

Gas Welding, Cutting, Brazing, & Heating Torch Instruction Manual. The Harris Products Group. Available at **www.harrisproductsgroup.com**

2.0.0 Section Review

1. An acetylene tank must always be used in an upright position.
   a. True
   b. False

2. Oxygen tanks are typically pressurized to about _____.
   a. 15 psig
   b. 250 psig
   c. 500 psig
   d. 2000 psig

3. A rosebud tip is used for _____.
   a. soldering
   b. brazing larger tubing
   c. cutting
   d. brazing small tubing

4. Which of the following materials would be used for brazing dissimilar metals?
   a. A BCuP-3 filler metal
   b. A BAg-5 filler metal
   c. A BCuP-5 filler metal
   d. Any phosphorous-bearing alloy

SUMMARY

Soldering is the method commonly used to join water and condensate lines made of copper. It is also sometimes used to join small, low-pressure refrigerant tubing. Soldering, or sweating, is done using a filler metal that melts at temperatures below 842°F. The joint is heated to a high temperature, then the filler metal is applied. The heat draws the filler metal into the joint by capillary action, thus sealing the joint. Sweat joints are typically made using a gas such as propane.

In general, solder joints are not suitable for the pressures found in refrigeration systems, so brazing is the joining process most often used to join refrigerant lines. Brazing uses filler metals that melt at temperatures above 842°F. Acetylene, or acetylene combined with oxygen, is generally used for brazing. The air/acetylene method is commonly used for smaller-diameter tubing. The oxyacetylene method generates a much greater heat than the air/acetylene method. Joining is faster using that method and thus results in less oxidation of the joint. In order to avoid contamination during brazing, refrigerant lines must be purged with an inert gas such as nitrogen during the brazing process.

Soldering and brazing procedures require a step-by-step approach that includes:

- Cleaning the joints
- Applying flux
- Heating the parts to the correct temperature
- Bringing the nonferrous filler metal into contact with the tubing where it melts and flows into the joint by capillary action
- Cleaning and cooling the joint

The use of high heat, flame, and flammable gases creates unique hazards for workers performing brazing work. Acetylene is volatile and must be carefully handled. Oxygen is flammable and can explode if it comes into contact with oil or grease with a source of ignition. There are special safety procedures for handling, storing, and transporting oxygen and acetylene cylinders that must be followed.

1. The filler metals commonly used for soldering usually melt in a temperature range of _____.
   a. 200°F to 375°F
   b. 375°F to 500°F
   c. 750°F to 850°F
   d. 1,000°F to 1,200°F

2. The volume capacity of an acetylene B tank is _____.
   a. 10 ft³
   b. 20 ft³
   c. 30 ft³
   d. 40 ft³

3. Which of the following is an *incorrect* statement regarding the filler metal used for soldering?
   a. It is a nonferrous metal.
   b. It is a metal alloy.
   c. It has a melting point below 842°F.
   d. It always contains some percentage of lead.

4. The measured length of a pipe should include the distance between the fitting faces plus _____.
   a. the cup depth of the two fittings
   b. ½ the total length of the two fittings
   c. the distance to the centers of the two fittings
   d. approximately ½" on either end of the pipe

5. Acetylene becomes unstable when the pressure setting is above _____.
   a. 5 psig
   b. 7 psig
   c. 10 psig
   d. 15 psig

6. The desired joint clearance for a brazed copper to copper joint is _____.
   a. 0.001" to 0.003"
   b. 0.003" to 0.005"
   c. approximately 0.015"
   d. 5 percent of the pipe diameter

7. All brazing operations require the use of oxygen and acetylene.
   a. True
   b. False

8. The type of brazing filler metal that contains cadmium is _____.
   a. BCuP-1
   b. BCuP-2
   c. BAg-2
   d. BAg-5

9. The filler metals commonly used for brazing melt at a temperature range of _____.
   a. 375°F to 500°F
   b. 750°F to 850°F
   c. 850°F to 1,200°F
   d. 1,200°F to 1,550°F

10. The type of flame used for almost all heavy brazing applications is a(n) _____.
    a. oxidizing flame
    b. carburizing flame
    c. neutral flame
    d. feathering flame

Trade Terms Quiz

Fill in the blank with the correct trade term that you learned from your study of this module.

1. A process that reduces the surface tension so that molten (liquid) solder flows evenly throughout the joint is called _____.

2. Metals and metal alloys that contain no iron are referred to as being _____.

3. _____ is the movement of a liquid along the surface of a solid in a kind of spreading action.

4. A(n) _____ is any substance made up of two or more metals.

5. _____ is a chemical substance that prevents oxides from forming on the surface of metals as they are heated for soldering, brazing, or welding.

6. _____ is a fusible alloy that melts below 842°F and is used to join metals.

7. A(n) _____ is a valve that prevents a flame from traveling up the hoses.

8. The distance that a copper tube penetrates into a fitting is known as the _____.

9. The flammable solvent that is used as a carrier for acetylene gas is known as _____.

10. The rounded shoulder left after brazing where a fitting and a piece of tubing intersect is called the _____.

Trade Terms

Acetone
Alloy
Capillary action
Cup depth

Fillet
Flashback arrestor
Flux
Nonferrous

Solder
Wetting

Norman Sparks
Project Manager
Lee Company

How did you get started in the construction industry (i.e., took classes in school, a summer job, etc.)?

At the age of 20, while attending Nashville State Community College as a business major, I took a job as a parts delivery person with Lee Company. That exposure to the trade was the start of my career in the construction industry.

Who or what inspired you to enter the industry (i.e., a family member, school counselor, etc.)? Why?

My inspiration was really money at the time. A 30-year career in the trade really wasn't planned.

What do you enjoy most about your career?

I enjoy the technical aspect of the business and working with the new products and innovation. The HVAC industry is a rapidly changing industry and there is always something new to learn.

Why do you think training and education are important in construction?

Education and training are very important because technology is changing almost daily, and without continuous education it is impossible to provide an excellent service.

Why do you think credentials are important in construction?

Credentials provide the verification that employers need to document that education and training have been obtained, which is what management is looking for. When I am looking for individuals to place on jobs, I begin with their credentials and proceed from there.

How has training/construction impacted your life and career (i.e., advancement opportunities, better wages, etc.)?

Training has allowed me to advance within the industry and it has allowed me an opportunity to train others. I have been able to experience both professional and personal growth through the education and training opportunities presented to me. Even after 30 years, training and education remain a continuous part of my growth and personal success.

Would you recommend construction as a career to others? Why?

The construction industry is a tough industry, but if the individual enjoys a challenge, is willing to work hard, and is open to a continuous learning process, then it can be a great opportunity and I would highly recommend it.

What does craftsmanship mean to you?

Craftsmanship means taking pride in everything you do. It means making a profit for yourself or your employer while ensuring that the quality of service provided exceeds the customer's expectation and that service failures are avoided.

Acetone: A colorless organic solvent that is volatile and extremely flammable. In the HVACR trade, it is used as a carrier for acetylene gas in cylinders.

Alloy: Any substance made up of two or more metals.

Capillary action: The movement of a liquid along the surface of a solid in a kind of spreading action.

Cup depth: The distance that a tube inserts into a fitting, usually determined by a stop inside the fitting.

Fillet: A rounded internal corner or shoulder of filler metal often appearing at the meeting point of a piece of tubing and a fitting when the joint is soldered or brazed.

Flashback arrestor: A valve that prevents a flame from traveling back from the tip and into the hoses.

Flux: A chemical substance that prevents oxides from forming on the surface of metals as they are heated for soldering, brazing, or welding.

Nonferrous: A group of metals and metal alloys that contain no iron.

Solder: A fusible alloy used to join metals, with melting points below 842°F.

Wetting: A process that reduces the surface tension so that molten (liquid) solder flows evenly throughout the joint.

Additional Resources

This module presents thorough resources for task training. The following resource material is suggested for further study.

A Guide to Brazing and Soldering. The Harris Products Group. Available at **www.harrisproductsgroup.com**.

Gas Welding, Cutting, Brazing, & Heating Torch Instruction Manual. The Harris Products Group. Available at **www.harrisproductsgroup.com**.

Filler Metal Selection Guide. The Harris Products Group. Available at **www.harrisproductsgroup.com**.

www.brazingbook.com. Lucas-Milhaupt, Inc. – An interactive resource on the subject of brazing.

Figure Credits

Courtesy of BernzOmatic, Worthington Cylinders, Figures 2, 4, 6

Courtesy of Uniweld Products, Inc., Figures 5, 19, 21, 32, Tables 1 and 6

Courtesy of Copper Development Associations Inc., Table 2

Courtesy of Oatey, Figures 7, 8

Courtesy of Smith Equipment, SA03

Courtesy of Harris Products Group, Figures 23, 24, 35

Answer	Section Reference	Objective
Section One		
1. b	1.1.1	1a
2. c	1.2.0	1b
3. d	1.3.0	1c
Section Two		
1. a	2.1.1	2a
2. d	2.2.1	2b
3. b	2.3.0	2c
4. b	2.4.0	2d

NCCER CURRICULA — USER UPDATE

NCCER makes every effort to keep its textbooks up-to-date and free of technical errors. We appreciate your help in this process. If you find an error, a typographical mistake, or an inaccuracy in NCCER's curricula, please fill out this form (or a photocopy), or complete the online form at **www.nccer.org/olf**. Be sure to include the exact module ID number, page number, a detailed description, and your recommended correction. Your input will be brought to the attention of the Authoring Team. Thank you for your assistance.

Instructors – If you have an idea for improving this textbook, or have found that additional materials were necessary to teach this module effectively, please let us know so that we may present your suggestions to the Authoring Team.

NCCER Product Development and Revision
13614 Progress Blvd., Alachua, FL 32615

Email: curriculum@nccer.org
Online: www.nccer.org/olf

❏ Trainee Guide ❏ Lesson Plans ❏ Exam ❏ PowerPoints Other _____

Craft / Level: _____ Copyright Date: _____

Module ID Number / Title: _____

Section Number(s): _____

Description: _____

_____

_____

_____

_____

Recommended Correction: _____

_____

_____

_____

_____

Your Name: _____

Address: _____

_____

Email: _____ Phone: _____

Basic Carbon Steel Piping Practices

OVERVIEW

Iron and steel pipe are used in many HVAC systems. Steel pipe is used to carry water in hydronic systems, while black iron is used to supply gas to furnaces. The ability to properly cut, thread, and join steel pipe is an important skill for the HVAC installer.

Module 03105

Trainees with successful module completions may be eligible for credentialing through NCCER's Registry. To learn more, go to **www.nccer.org** or contact us at **1.888.622.3720**. Our website has information on the latest product releases and training, as well as online versions of our *Cornerstone* magazine and Pearson's product catalog.

Your feedback is welcome. You may email your comments to **curriculum@nccer.org**, send general comments and inquiries to **info@nccer.org**, or fill in the User Update form at the back of this module.

This information is general in nature and intended for training purposes only. Actual performance of activities described in this manual requires compliance with all applicable operating, service, maintenance, and safety procedures under the direction of qualified personnel. References in this manual to patented or proprietary devices do not constitute a recommendation of their use.

03105 V5

Objectives

When you have completed this module, you will be able to do the following:

1. Describe and identify the various types of steel pipe and fittings.
   a. Identify the characteristics and uses of steel pipe.
   b. Describe how pipe threads are classified and measured.
   c. Identify the various types of fittings used on steel pipe and describe how they are used.
   d. Describe how to properly measure lengths of steel pipe.
2. Describe the tools and methods used to cut and thread steel pipe.
   a. Identify pipe cutting and reaming tools and describe how they are used.
   b. Identify threading tools and describe how they are used.
3. Explain and demonstrate the methods of installing and mechanically joining steel pipe.
   a. Explain and demonstrate the methods and use of the tools to connect threaded pipe.
   b. Explain and demonstrate an understanding of pipe grooving methods.
   c. Describe how to assemble flanged steel pipe.
   d. Describe how to correctly install steel pipe.

Performance Tasks

Under the supervision of the instructor, you should be able to do the following:

1. Cut, ream, and thread steel pipe.
2. Join lengths of threaded pipe using selected fittings.

Trade Terms

Annealing
Black iron pipe
Chain vise
Chain wrench
Die
Fitting allowance
Flange
Galvanized
Grooved pipe
Makeup
Malleable iron

Nipple
Pipe dope
Pitch
Riser
Seismic
Stock
Strap wrench
Water hammer
Yoke vise

Industry Recognized Credentials

If you are training through an NCCER-accredited sponsor, you may be eligible for credentials from NCCER's Registry. The ID number for this module is 03105. Note that this module may have been used in other NCCER curricula and may apply to other level completions. Contact NCCER's Registry at 888.622.3720 or go to **www.nccer.org** for more information.

Contents:

1.0.0 Steel Pipe and Fittings 1
  1.1.0 Types and Characteristics of Steel Pipe 1
    1.1.1 Pipe Sizes and Wall Thickness 2
  1.2.0 Pipe Threads 3
  1.3.0 Steel Pipe Fittings 4
    1.3.1 Tees and Crosses 4
    1.3.2 Elbows 4
    1.3.3 Unions and Flanges 5
    1.3.4 Couplings 5
    1.3.5 Nipples 5
    1.3.6 Plugs, Caps, and Bushings 7
  1.4.0 Measuring Pipe 7
2.0.0 Tools and Methods Used to Cut and Thread Steel Pipe 11
  2.1.0 Cutting Pipe 11
    2.1.1 Reaming the Pipe 11
  2.2.0 Pipe Threading 12
    2.2.1 Using a Hand Threader and Vise 13
    2.2.2 Using a Bench or Tripod Threading Machine 14
3.0.0 Pipe Joining and Installation Procedures 16
  3.1.0 Assembling Threaded Pipe 16
    3.1.1 Wrenches 16
  3.2.0 Grooved Pipe 18
    3.2.1 Roll Grooving 19
    3.2.2 Cut Grooving 19
    3.2.3 Selecting Gaskets 20
    3.2.4 Installing Grooved Pipe Couplings 20
  3.3.0 Flanged Pipe 22
  3.4.0 Installing Steel Piping 23
    3.4.1 Pipe Hangers and Supports 24
Appendix Comparison of Imperial and Metric Pipe Sizes 34

Figures and Tables

Figure 1 Black iron gas piping 1
Figure 2 Galvanized and black iron pipe 2
Figure 3 Steel pipe diameters 2
Figure 4 American National Standard taper (NPT) pipe threads 3
Figure 5 Tees 5
Figure 6 Elbows 6
Figure 7 Unions 6
Figure 8 Couplings 7
Figure 9 Nipples 7
Figure 10 Plugs, caps, and bushings 7
Figure 11 Makeup and fitting allowance 8
Figure 12 Methods of measuring pipe 9
Figure 13 Pipe cutters 11
Figure 14 Pipe vises and portable tripod 12
Figure 15 Pipe reamer 12
Figure 16 Hand threader 12
Figure 17 Threading power drive 13
Figure 18 Pipe threading machines 13
Figure 19 Pipe mounted in a threading machine 14
Figure 20 Starting the die on the end of the pipe 15
Figure 21 Pipe threads at completion 15
Figure 22 PTFE tape and pipe dope applied to pipe 16
Figure 23 Applying PTFE tape 17
Figure 24 Pipe wrenches 17
Figure 25 Tightening pipe using two pipe wrenches 18
Figure 26 Two views of a grooved pipe coupling 18
Figure 27 Details of pipe grooves 19
Figure 28 Power roll-grooving machine 20
Figure 29 Joining grooved pipe 21–22
Figure 30 Completed grooved joint 22
Figure 31 Flanged valve 23
Figure 32 Flange bolt tightening pattern 23
Figure 33 Example of seismic protection for a pipe 25
Figure 34 Individual pipe supports 25
Figure 35 Surface mounting methods 26
Figure 36 Trapeze hangers 26
Figure 37 Rollers and spring hangers 27

Table 1 American National Standard Taper Pipe Thread
(NPT) Dimensions 4
Table 2 Suggested Pipe Wrench-to-Pipe Sizes 17
Table 3 Recommended Maximum Hanger Spacing Intervals
for Carbon Steel Pipe 24

1.0.0 STEEL PIPE AND FITTINGS

Objective

Describe and identify the various types of steel pipe and fittings.

a. Identify the characteristics and uses of steel pipe.
b. Describe how pipe threads are classified and measured.
c. Identify the various types of fittings used on steel pipe and describe how they are used.
d. Describe how to measure lengths of steel pipe.

Trade Terms

Annealing: A process in which a material is heated, then cooled to strengthen it.

Black iron pipe: Carbon steel pipe that gets its black coloring from the carbon in the steel.

Die: A tool insert used to cut external threads by hand or machine.

Flange: A flat plate attached to a pipe or fitting and used as a means of attaching pipe, fittings, or valves to the piping system.

Fitting allowance: The distance from the end of the pipe to the center of the fitting. Also called *takeoff*.

Galvanized: Describes steel with a zinc-based coating to prevent rust.

Grooved pipe: A method for connecting piping systems using a groove installed in the pipe. The use of grooved piping eliminates the need for threading, flanging, or welding when making connections. Connections are made with gaskets and couplings installed using a wrench and lubricant.

Makeup: The amount by which the end of the pipe penetrates a fitting.

Malleable iron: Iron that has been toughened by gradual heating or slow cooling.

Nipple: A short length of pipe that is used to join fittings. It is usually less than 12" long and has male threads on both ends.

Pitch: The number of threads per inch on threaded pipe and other objects.

Stock: A tool used to hold and turn dies when cutting external threads.

Carbon steel pipe is part of a family of ferrous metals. The term *ferrous* simply means that it contains iron. Therefore, ferrous metal pipe is pipe that either contains, or is made from iron. Steel is made by adding carbon and other material such as manganese to iron. The ferrous metal pipe used in HVAC work is actually steel pipe. Iron pipe is used in some plumbing applications such as drains and sewer lines, but is not used in HVAC work. However, black iron pipe is a term often heard in the HVACR trade. When it is used by trade workers in reference to piping, they are actually referring to steel pipe that has not been galvanized. Technically, black iron is iron that has been wrought and tempered, and the name differentiates it from cast iron.

Many large commercial systems use water to transfer heat from the indoors to the outdoors. Hydronic heating systems use hot water to deliver heat to the conditioned space. In these applications, the water may be carried in galvanized or black iron pipe. Black iron pipe is typically used to supply natural or LP gas to a gas-fired furnace. *Figure 1* shows black iron pipe feeding gas to a furnace.

For many applications, steel pipe has a number of advantages over other piping methods, including the following:

- It is very durable.
- It has exceptional structural strength.
- Its cost is low in comparison to copper.
- It holds heat well.
- It does not expand and contract too much when exposed to heat and cold.

1.1.0 Types and Characteristics of Steel Pipe

The two most common types of carbon steel pipe are black iron pipe and galvanized pipe. Black iron pipe is manufactured exactly like galvanized pipe. The only difference between the two is that black iron pipe is not coated with zinc. *Figure 2* shows a comparison of the two types. Carbon gives black iron pipe its color. Black iron pipe is most often used in the HVACR trade for gas piping, chilled and hot water, steam, and air-pressure applications. It is used where corrosion will not affect its uncoated surfaces, or where internal chemical treatment helps prevent corrosion.

Galvanized pipe is steel pipe that has been dipped in molten zinc. The zinc protects the surfaces from abrasive or corrosive materials. It gives the pipe a mottled, silvery color when it is new, and a dull, grayish color after it has aged. Galvanized pipe is most often used when specifications require it. The outer coating is particularly valuable in out-

Figure 1 Black iron gas piping.

03105-13_F02.EPS

Figure 2 Galvanized and black iron pipe.

door applications. However, the galvanizing material on the inside causes concern in some situations.

Both galvanized steel and black iron pipe are joined using threaded, grooved, or flange fittings. Flanges can be welded or threaded onto the pipe. The fittings for threaded or grooved pipe are ready for use, but the pipe itself must be cut to size and threaded or grooved on the job. For that reason, the ability to measure, cut, thread, groove, and join galvanized and black iron pipe is an important skill for anyone installing HVACR systems.

1.1.1 Pipe Sizes and Wall Thickness

In the United States, pipe is sized in inches by its nominal size. For pipe sizes up to and including 12", the nominal size is an approximation of the inside diameter (ID). From 14" on, the nominal size reflects the outside diameter (OD) of the pipe. There are times when the nominal size of a pipe and its actual inside or outside diameter differ greatly, but nominal size, not actual size, is always used to select and discuss piping. *Figure 3* shows the actual inside and outside diameters for a 1" nominal-sized steel pipe.

There are two ways to describe the wall thickness of a pipe. The first is by schedule. As schedule numbers get larger, pipe walls get thicker and stronger. The schedule numbers used for pipe are 5, 10, 20, 30, 40, 60, 80, 100, 120, 140, and 160. It is important to remember that a schedule number only describes the wall thickness of a pipe for a given nominal size. Thus, ¾" Schedule 40 does not have the same wall thickness as 1" Schedule 40.

The second way to describe pipe wall thickness is by manufactured weight. There are three classifications in common use. In ascending order of wall thickness, they are:

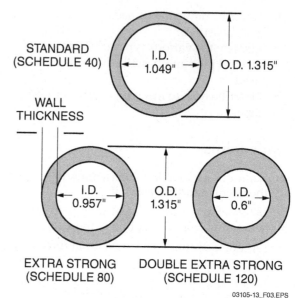

03105-13_F03.EPS

Figure 3 Steel pipe diameters.

- *STD* – Standard
- *XS* – Extra strong
- *XXS* – Double extra strong

The wall thickness and the inside diameter differ with each weight. The thicker the wall, the smaller the inside diameter, but the more pressure the pipe will withstand. Standard weight will prove adequate in most piping situations; however, the two stronger weights are needed for higher-pressure systems such as high-pressure steam systems. XS and XXS pipe may also be used to contain highly toxic or hazardous materials, even at low pressures.

All schedules or weights for a specific pipe size have the same outside diameter, so the same threading dies fit all of them. Threads are always formed on the outside of the pipe. Specifications, drawings, or data sheets sometimes specify pipe sizes in metric dimensions. In such cases, it is necessary to convert these dimensions to imperial dimensions. The table in the *Appendix* gives the dimensions of commonly used steel pipe in both imperial and metric sizes.

1.2.0 Pipe Threads

Lengths of steel pipe are commonly joined by threading the end of the pipe and using threaded fittings. Other joining methods include welding and the use of grooved fittings that mate with a groove cut into the pipe or piping component. The portion of a threaded pipe end that is screwed into a fitting is referred to as the thread engagement, or makeup.

American National Standard pipe threads can be either straight or tapered. Tapered threads are identified with the acronym NPT, while straight threads are identified by NPS. Only tapered pipe threads (NPT) are used for HVACR work because they produce leak-tight and pressure-tight connections. When tight, they also produce a mechanically rigid piping system. Tapered threads can be cut by hand with a die and stock or with an electric pipe threading machine.

The tapered thread used on pipe (*Figure 4*) is V-shaped with a thread angle of 60 degrees, very slightly rounded at the top. The taper is 1/16" per inch of length (1/32" per inch from each wall). There are about seven to 12 perfect threads and several imperfect threads for each joint. Imperfect threads are those that are not cut to their full depth as a result of the taper, as opposed to perfect threads. The actual number of perfect threads used depends on the size of the pipe being threaded. As shown in *Figure 4*, the threads closest to the end of the pipe are perfect threads; they are sharp at

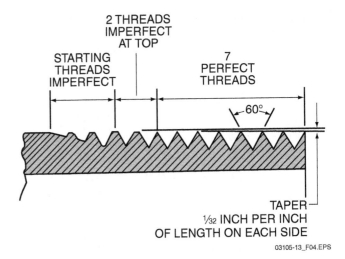

Figure 4 American National Standard taper (NPT) pipe threads.

the top and bottom. The remaining threads are non-perfect because they are not completely cut, resulting in rounded or imperfect edges. The primary sealing power of the threads occurs as the female threads begin to meet the tapered imperfect threads. If the perfect threads are marred or broken, they also lose their sealing power.

Threads are designated by specifying in sequence the nominal pipe size, the number of threads per inch, and the thread series symbols. The number of threads per inch is referred to as the pitch of the thread. For example, the thread specification 3/4 – 14 NPT means:

- 3/4 = 3/4" nominal pipe size, ID
- 14 = 14 threads per inch (pitch)
- NPT = American National Standard taper pipe thread

Taper pipe threads are engaged or made up in two phases: hand-tight engagement and wrench makeup. *Table 1* shows the dimensions for hand-tight engagement as well as other NPT specifications for commonly used pipe sizes. In practice, about three turns are done by hand, followed by several turns with a wrench. When a pipe is threaded and properly tightened, about three threads (most or all imperfect) should remain showing.

It should be noted that pipe threads are also referred to with several other acronyms, both in print and in conversation. The acronyms include the following:

- MIP (Male Iron Pipe)
- FIP (Female Iron Pipe)
- IPT (Iron Pipe Thread)
- FPT (Female Pipe Thread)
- MPT (Male Pipe Thread)

How Long Have Iron and Steel Been Around?

The first use of iron has been traced back to a time nearly 5,000 years ago. A lot of early iron was extracted from meteorites that landed on Earth. The extensive use of iron for weapons and other purposes dates back to the so-called Iron Age, which began around 3,200 years ago, or 1200 BC. The Iron Age was preceded by the Bronze Age, a period of about 800 years, which was in turn preceded by the Stone Age.

Bronze is a blend of copper and tin. The tin was added to help overcome the brittle nature of copper. Steel is an iron-based alloy, made by adding material such as carbon, which acts as a hardening agent and gives the steel its exceptional strength. Steel has been around for thousands of years, but the manufacture of steel as we know it today dates back about 400 years to the invention of the Bessemer furnace.

Table 1 American National Standard Taper Pipe Thread (NPT) Dimensions

Nominal Pipe Size (in inches)	Threads per Inch	No. of Usable Threads	Hand-Tight Engagement	Total Thread Makeup	Total Thread Length
1/8	27	7	3/16	1/4	3/8
1/4	18	7	1/4	3/8	5/8
3/8	18	7	1/4	3/8	5/8
1/2	14	7	5/16	7/16	3/4
3/4	14	8	5/16	1/2	3/4
1	11 1/2	8	3/8	9/16	7/8
1 1/4	11 1/2	8	7/16	9/16	1
1 1/2	11 1/2	8	7/16	9/16	1
2	11 1/2	9	7/16	5/8	1
2 1/2	8	9	11/16	7/8	1 1/2
3	8	10	3/4	1	1 1/2
3 1/2	8	10	13/16	1 1/16	1 5/8
4	8	10	13/16	1 1/16	1 5/8
5	8	11	15/16	1 3/16	1 3/4
6	8	12	15/16	1 3/16	1 3/4

The MIP, MPT, FIP, and FPT acronyms are particularly useful when selecting pipe fittings and piping system components. They designate whether the component is male or female so that installers know what gender the mating portion must be. Note that these same acronyms can be used for straight pipe threads as well as tapered. However, straight pipe threads are rarely encountered in the HVACR environment.

1.3.0 Steel Pipe Fittings

Pipe fittings for steel pipe are generally made of cast iron, malleable iron, or galvanized (zinc-coated) iron. Malleable iron is produced by prolonged annealing of ordinary cast iron. This process makes the iron tough. Also, it can be bent or pounded to some extent without breaking. Malleable iron fittings are typically used for gas piping. Cast-iron fittings are used for steam and hydronic system piping. Galvanized fittings should be used with galvanized pipe. Fittings can be used to join lengths in a straight line (unions and couplings); join three lengths at the same location (tees); change direction (elbows); and close or reduce the size of openings in pipe (plugs, caps, and bushings). Variations of these fittings are used to change the pipe size when necessary.

1.3.1 Tees and Crosses

Tees can be purchased in a great number of sizes and patterns. They are used to make a branch at a right angle to the main pipe. If all three outlets are the same size, the fitting is called a regular tee (*Figure 5*). If outlet sizes vary, the fitting is called a reducing tee.

Tees are specified by giving the straight-through, or run, dimensions first, then the branch dimensions. For example, a tee with one run outlet of 2", a second run outlet of 1", and a branch outlet of 3/4" is known as a 2 × 1 × 3/4 tee (always state the larger run size first, the small run size next, and the branch size last). It should be noted that a tee with three different pipe sizes is rather rare and very difficult to find. The reality is that bushings or reducing couplings are usually required to change the outlet size(s) of a tee. Tees are also available with male threads on a run or branch outlet but again, these are also quite rare. Unusual pipe fittings that are seldom requested are not only difficult to find, but are also significantly more expensive due to the laws of supply and demand.

Keep Piping Fittings to a Minimum

Pipe fittings add resistance and reduce flow in piping systems. All piping systems should be planned in advance of the installation in order to minimize the number of fittings. In some large industrial or commercial installations, bending pipes in curves instead of using elbows may be preferable in order to reduce flow resistance.

A cross is a four-way distribution device. Instead of three connections as a tee has, a cross has four connections. They are rarely used in the HVAC industry.

1.3.2 Elbows

Elbows (*Figure 6*), often called ells, are used to change the direction of pipe. The most common ells are the 90-degree ell; the 45-degree ell; the street ell, which has a male thread on one end; and the reducing ell, which has outlets of different sizes. Ells are also available to make 11¼-, 22½-, and 60-degree bends.

1.3.3 Unions and Flanges

Unions (*Figure 7*) make it possible to disassemble a threaded piping system. After disconnecting the union, the length of pipe on either end of the union may then be turned. The two most common types of unions are the ground joint and the flange.

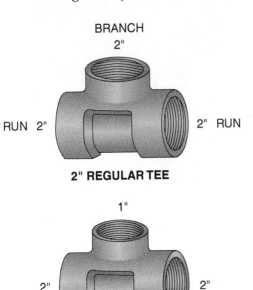

BRANCH
2"

RUN 2"   2" RUN

2" REGULAR TEE

1"

2"   2"

**2" × 2" × 1"
REDUCING TEE**

¾"

2"   1"

**2" × 1" × ¾"
REDUCING TEE**

03105-13_F05.EPS

Figure 5 Tees.

The ground joint union connects two pipes by screwing the thread and shoulder pieces onto the pipes. Then, both the shoulder and thread parts are drawn together by the collar. This union creates a gas-tight and water-tight joint that can be disassembled rather quickly.

The flange union also connects two separate pipes. The flanges screw to the pipes to be joined and are then pulled together with nuts and bolts. A gasket between the flanges makes this connection gas-tight and water-tight. Flange unions are most often used for larger sizes of pipe than are ground joint unions.

1.3.4 Couplings

Couplings (*Figure 8*) are short fittings with female threads in both openings. They are used to connect two lengths of pipe when making straight runs. The pipes can be of the same size or different sizes. Reducing couplings are used to join pipes of different sizes. Most are concentric, but reducing couplings that are eccentric can be found. An eccentric reducing coupling allows the bottom of the two pipes to maintain the same level. Couplings cannot be used in place of unions because they cannot be disassembled.

1.3.5 Nipples

Nipples (*Figure 9*) are pieces of pipe 12" or less in length and threaded on both ends. They are used to make extensions from a fitting or to join two fittings. Nipples are manufactured in many sizes beginning with the close, or all-thread, nipple. Close nipples have little or no unthreaded surface area.

Pipe Nipples

Pipe nipples are available from plumbing and heating suppliers in a variety of short lengths. Sizes up to 6" in length are very common, but longer nipples are also available. It generally makes more sense to buy commonly used nipples in bulk, especially the close and shoulder nipples, rather than to cut and thread them in the field. Most field-threading equipment cannot be used to make close or shoulder nipples. However, longer nipples can be fabricated using pieces of scrap piping gathered from field installations. This type of task can be accomplished in the shop during slow periods.

90-DEGREE ELL

45-DEGREE ELL

REDUCING ELL

STREET ELL

03105-13_F06.EPS

Figure 6 Elbows.

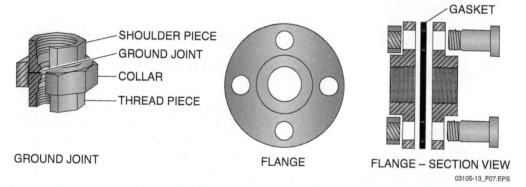

GASKET

SHOULDER PIECE
GROUND JOINT
COLLAR
THREAD PIECE

GROUND JOINT

FLANGE

FLANGE – SECTION VIEW

03105-13_F07.EPS

Figure 7 Unions.

ORDINARY STANDARD, FULL,
AND HALF COUPLINGS

CAP   BUSHING

CONCENTRIC
REDUCING
COUPLING

REDUCING COUPLING

SQUARE AND HEXAGONAL SOCKET
COUNTERSUNK PIPE PLUGS

03109-13_F08.EPS

Figure 8 Couplings.

1.3.6 Plugs, Caps, and Bushings

Plugs are male threaded fittings used to close female openings in other fittings. There are a variety of heads (square, slotted, and hexagon) found on plugs, as shown in *Figure 10*.

A cap is a fitting with a female thread. It is used for the same purpose as a plug, except that the cap fits on the male end of a pipe or nipple.

Bushings are fittings with a male thread on the outside and a female thread on the inside. They are usually used to connect the male end of a pipe to a fitting of a larger size. The ordinary bushing has a hexagon nut at the female end. Bushings can be used in place of a reducing fitting to accommodate a smaller pipe.

SQUARE SOCKET PIPE PLUG

03105-13_F10.EPS

Figure 10 Plugs, caps, and bushings.

1.4.0 Measuring Pipe

In order to accurately measure the length of a section of pipe, it is important to understand the relationship between the pipe and the fittings. That in turn requires an understanding of two key terms. *Makeup*, also known as thread engagement, is the amount that the end of the pipe penetrates the fitting. Refer back to *Table 1* to see the required makeup for various sizes of pipe, along

BLACK AND GALVANIZED WELDED PIPE NIPPLES

LONG NIPPLE

03105-13_F09.EPS

Figure 9 Nipples.

with other common characteristics. **Fitting allowance**, or *takeoff*, is the distance from the end of the pipe to the center of the fitting. The combination of makeup and fitting allowance equals the distance from the outer shoulder of the fitting to its center (*Figure 11*).

There are a number of methods used to state the length of a threaded pipe, including end-to-end, end-to-center, face-to-end, center-to-center, and face-to-face. *Figure 12* shows the different measuring techniques.

An end-to-end measurement is accomplished by measuring the full length of the pipe, including the threads at both ends. An end-to-center measurement is used for a piece of pipe with a fitting screwed onto one end only. The pipe length is equal to the total end-to-center measurement, minus the center-to-face dimension of the fitting, plus the length of the thread engagement.

A center-to-center measurement is used to measure pipe with fittings screwed onto both ends. The pipe length is equal to the total center-to-center measurement, minus the sum of the two center-to-face dimensions of the fittings, plus two times the length of the thread engagement.

A face-to-face measurement can be used under the same conditions as the center-to-center method. It is figured by measuring the length of

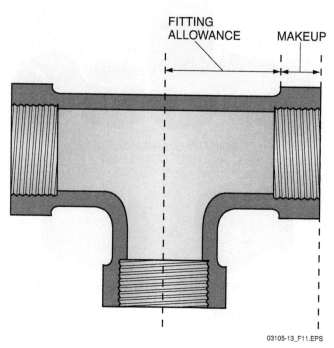

03105-13_F11.EPS

Figure 11 Makeup and fitting allowance.

pipe plus two times the length of the thread engagement.

Although these different methods may seem confusing at first, they become clear very quickly once a pipefitting job is in progress.

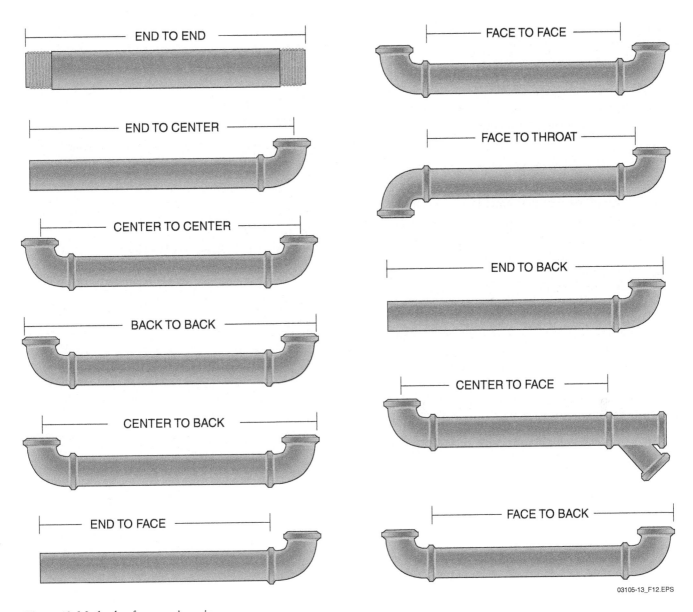

Figure 12 Methods of measuring pipe.

Additional Resources

Pipefitter's Handbook. Forrest R. Lindsey New York, NY: Industrial Press, Inc.

The Pipefitter's Blue Book. W.V. Graves. Webster, TX: Grave Publishing Company.

1.0.0 Section Review

1. Which of these measurements remains the same for all schedules of a pipe of a given nominal size?

   a. inside diameter
   b. outside diameter
   c. wall thickness
   d. length

2. The number of perfect threads required for a threaded joint is _____.

   a. at least 10 but not more than 12
   b. dependent upon the size of the pipe
   c. dependent upon the weight of the pipe
   d. dependent upon the type of fitting to be used

3. A close nipple is one that _____.

   a. is shorter than 2 inches
   b. is threaded on one end
   c. has no threads at all
   d. has little or no unthreaded area

4. A pipe length measurement that is made by measuring the length of pipe and adding twice the length of the thread engagement is called a(n) _____.

   a. face-to-end measurement
   b. end-to-end measurement
   c. center-to-center measurement
   d. face-to-face measurement

2.0.0 TOOLS AND METHODS USED TO CUT AND THREAD STEEL PIPE

Objective

Describe the tools and methods used to cut and thread steel pipe.

   a. Identify pipe cutting and reaming tools and describe how they are used.

   b. Identify threading tools and describe how they are used.

Performance Tasks

1. Cut, ream, and thread steel pipe.

Trade Terms

Chain vise: A device used to clamp pipe and other round metal objects. It has one stationary metal jaw and a chain that fits over the pipe and is clamped to secure the pipe.

Yoke vise: A holding device used to hold pipe and other round objects. It has one movable jaw that is adjusted with a threaded rod.

Before threaded pipe is installed, the pipe must be cut to the proper length and threaded. Cutting and threading of pipe must be done correctly and precisely in order to avoid leaks in the piping. This section describes the tools and methods used for that purpose.

2.1.0 Cutting Pipe

Pipe cutters (*Figure 13*) may have from one to four cutting wheels. The cutter must be rotated around the pipe, so the more cutting wheels a cutter has, the less space it requires to cut the pipe. In other words, a four-wheel cutter does not have to make a complete revolution around the pipe to make a cut. However, a single-wheel, or conventional, cutter must make a complete revolution. This may be a disadvantage if there is limited clearance around the pipe. The pipe cutter is made of a cutting wheel, adjusting screw, and, depending on the type, either guide rollers or additional cutting wheels.

If the tube or pipe becomes mashed or flattened, or excessive force must be used when cutting, replace the cutting wheel or wheels. Lubricating oil must be applied periodically to all movable parts on the pipe cutter to ensure smooth operation. Applying cutting oil to the pipe helps when cutting steel pipe.

The pipe cutter is rotated around the pipe. The adjusting screw is tightened ¼ revolution with each turn, pushing the wheel tighter against the pipe. Avoid overtightening the cutting wheel, as this can cause a larger burr to form inside the pipe and dull the cutting wheel quickly. It may also result in a shattered cutting wheel. Make sure to save any usable scrap pieces of pipe for nipples or other fit-up requirements on the job.

A pipe vise (*Figure 14*) can be used to hold the pipe while it is being prepared. The yoke vise is the most commonly used vise for smaller pipe sizes. Its jaws hold the pipe firmly and prevent it from turning while the pipe is threaded. This vise can typically handle pipe from ⅛" to 3½" in diameter, depending on the yoke size used.

The chain vise is used in the same way as the yoke vise; however, the chain vise can hold much larger pieces of pipe than the yoke vise. The chain must be kept oiled or it will become stiff. A stiff chain makes the chain vise operate poorly.

Either type of vise can be bench-mounted or mounted on a portable folding stand, known as a tripod, for field use, as shown in *Figure 14*.

2.1.1 Reaming the Pipe

Once the pipe is cut, a reamer (*Figure 15*) is used to remove the burr that forms on the inside of the pipe. If the burr is left unattended, it collects de-

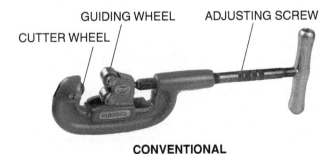

CONVENTIONAL

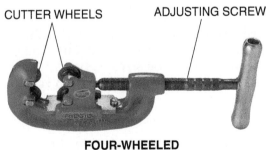

FOUR-WHEELED

03105-13_F13.EPS

Figure 13 Pipe cutters.

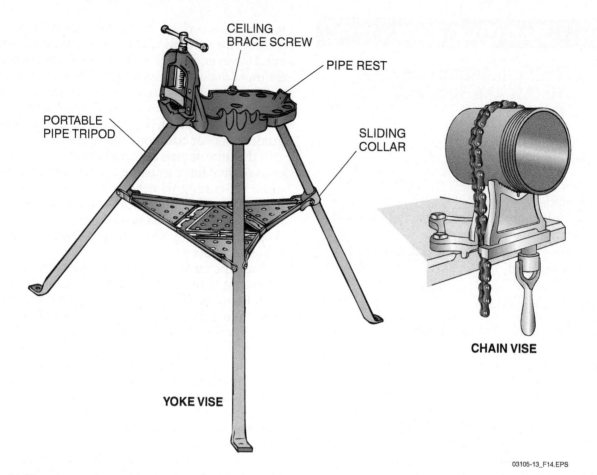

CEILING
BRACE SCREW

PIPE REST

PORTABLE
PIPE TRIPOD

SLIDING
COLLAR

CHAIN VISE

YOKE VISE

03105-13_F14.EPS

Figure 14 Pipe vises and portable tripod.

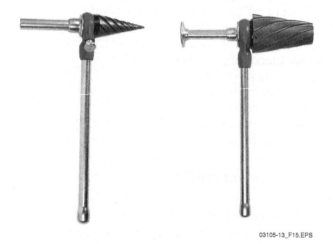

03105-13_F15.EPS

Figure 15 Pipe reamer.

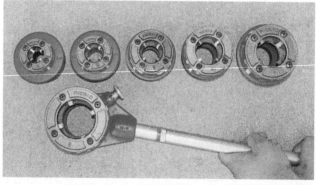

03105-13_F16.EPS

Figure 16 Hand threader.

posits and slows the flow of liquid within the pipe. Because reamers are tapered, one reamer can be used on many pipe sizes. A typical tapered pipe reamer can handle pipe sizes from ¼" through 2".

2.2.0 Pipe Threading

There are two types of pipe threaders: hand threaders and power threaders. Hand threaders (*Figure 16*) are made up of two major parts: the die

and the stock (handle). Dies are used to cut the threads, and the stock is the device that holds the die and provides turning leverage. The pipe die consists of the holder and cutters. Although they are mostly used manually, a hand threader can be used with a power drive (*Figure 17*).

A pipe threading machine is used when large quantities of pipe must be threaded. These are multipurpose machines equipped to rotate, cut, thread, and ream pipe. The reamer and pipe cutter are nearly identical to those used independently. They are permanently mounted on most

Figure 17 Threading power drive.

power threaders for quick access and use. The pipe is mounted through the machine's speed chuck. The speed chuck has a wheel that spins easily with the hand to quickly lock the pipe into the teeth of the vise. Power threaders mounted on tripods, wheels, or benches are operated using a foot pedal switch. Like pipe vises, threading machines (*Figure 18*) can be mounted on a tripod or a shop bench.

> CAUTION
>
> Do not use the pipe threading machine to tighten fittings onto a threaded pipe, as this could cause the fitting to be overtightened.

2.2.1 Using a Hand Threader and Vise

Pipe is often threaded by hand on small jobs where only a few joints need to be done. It is also not an effective way to thread larger pipe. Although hand threaders and dies are available to thread up to 2" pipe, threading pipe this size requires a tremendous amount of power. An average-sized human is not likely to succeed. Most manual threading is reserved for pipe sizes 1¼" or less due to the physical effort and time required.

When using a hand threader and vise, the pipe is first placed into the vise and secured. Prepare by donning the appropriate PPE, including safety glasses and gloves. Threading produces extremely sharp and stiff metal debris and sharp edges, so the gloves should be chosen accordingly. To cut threads using a hand threader and vise, proceed as follows:

Step 1 Select the correct size die for the pipe being threaded.

Step 2 Inspect the die to make sure the cutters are free of nicks, excessive wear, and debris.

Step 3 Lock the pipe securely in a vise.

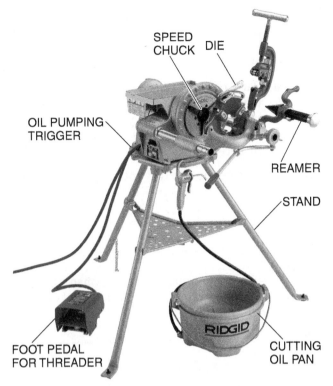

A. TRIPOD-MOUNTED

B. BENCH-MOUNTED

Figure 18 Pipe threading machines.

Step 4 Slide the die over the end of the pipe, pipe guide end first. The guide end is the side of the threader with the round opening, with no teeth.

Step 5 Push the die against the pipe with the heel of one hand. Take three or four short, slow, clockwise turns. Most threader stocks have a ratcheting function to allow it to turn backwards freely. Be careful to keep the die pressed firmly against the pipe. When enough thread is cut to hold the die firmly on the pipe, apply some thread-cutting oil. This oil prevents the pipe from overheating due to friction and lubricates the die. Oil the threading die every two or three downward strokes.

Step 6 Back off ¼ turn after each full turn forward to clear out the metal chips. Continue until the pipe projects two threads from the die end of the stock. Too few threads is just as bad as too many threads.

Step 7 To remove the die, rotate it counterclockwise.

Step 8 Wipe off excess oil and any chips.

2.2.2 Using a Bench or Tripod Threading Machine

Each threading machine is slightly different. Become familiar with the manufacturer's operating procedures before attempting to operate any threading machine. Also, become thoroughly familiar with the maintenance and safety instructions for the machine. A poorly maintained

Handheld Power Pipe Threaders

A handheld power pipe threader can be used in conjunction with a portable stand when a number of pipes must be threaded in the field. The rotating power head of this threader turns at about 30 rpm and uses the same dies as an equivalent manual threader. They are typically limited in capacity to 2" pipe. When the larger sizes are being threaded, a support arm (not shown) is clamped to the pipe and the threader is rested against it to counteract the torque of the threader.

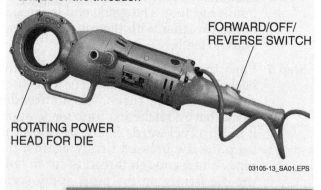

FORWARD/OFF/
REVERSE SWITCH

ROTATING POWER
HEAD FOR DIE

03105-13_SA01.EPS

machine is a safety hazard. To thread pipe using a power threading machine, don the appropriate PPE and follow these basic steps:

Step 1 Select the pipe stock, and inspect/clean the correct size die before installing it in the threader.

Step 2 Mount the pipe stock into the chuck (*Figure 19*). Long stock must have additional support from a pipe roller.

Step 3 Check the pipe and die alignment.

Step 4 With the power switch in the forward position and the machine running, move the tool carriage and start the die on the end of the pipe (*Figure 20*). Apply cutting oil during the threading operation.

Step 5 Cut threads until two threads appear at the other end of the die (*Figure 21*). Stop threading.

Step 6 Reverse the machine and back off the die until it is clear of the pipe.

Step 7 Remove the pipe from the machine chuck. Be careful not to mar the threads.

Step 8 Using a rag, wipe the pipe clean of oil and metal chips.

SPEED
CHUCK

PIPE CLAMPED
IN SPEED CHUCK

03105-13_F19.EPS

Figure 19 Pipe mounted in a threading machine.

MOVABLE TOOL CARRIAGE

03105-13_F20.EPS

Figure 20 Starting the die on the end of the pipe.

03105-13_F21.EPS

Figure 21 Pipe threads at completion.

Additional Resources

Pipefitter's Handbook. Forrest R. Lindsey New York, NY: Industrial Press, Inc.

The Pipefitter's Blue Book. W.V. Graves. Webster, TX: Grave Publishing Company.

2.0.0 Section Review

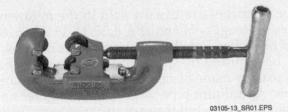

03105-13_SR01.EPS

1. The tool shown here is used to _____.
   a. cut pipe
   b. thread pipe
   c. ream pipe
   d. deburr pipe

2. A threading operation stops when _____.
   a. five threads have been cut
   b. the threader stops automatically
   c. two threads are visible beyond the die
   d. the threads extend 1' from the end of the pipe

SECTION THREE

3.0.0 PIPE JOINING AND INSTALLATION PROCEDURES

Objectives

Explain and demonstrate the methods of installing and mechanically joining steel pipe.

   a. Explain and demonstrate the methods and use of the tools to connect threaded pipe.
   b. Explain and demonstrate an understanding of pipe grooving methods.
   c. Describe how to assemble flanged steel pipe.
   d. Describe how to correctly install steel pipe.

Performance Tasks

2. Join lengths of threaded pipe using selected fittings.

Trade Terms

Chain wrench: An adjustable tool for holding and turning large pipe up to 4" in diameter. A flexible chain replaces the usual wrench jaws.

Pipe dope: A putty-like pipe joint material used for sealing threaded pipe joints.

Riser: A pipe that travels vertically. It must be supported at each level of the structure.

Seismic: Related to the vibration and movement associated with earthquakes.

Strap wrench: A tool for gripping pipe. The strap is made of nylon web.

Water hammer: The noise and concussion that occurs when a volume of water or other liquid moving in a pipe suddenly stops or loses momentum.

Once the pipe has been threaded, it can be assembled using the various fittings or connected directly to components. In most cases, unions are used to install important components into the piping system. This allows for the final assembly to be done. Since the pipe can only be tightened in one direction, unions provide a point where the final joints can be assembled. They also provide a means of easy component removal for service and replacement.

3.1.0 Assembling Threaded Pipe

Apply a joint compound or polytetrafluoroethylene (PTFE) tape to the male threads before assembling a pipe connection (*Figure 22*). PTFE tape made specifically as a pipe joint sealer may often be used with the proper application. Joint compound, commonly called pipe dope, is a putty-like material applied to seal a joint and provide lubrication for assembly. The dope or tape should be applied only to the male threads. The tape must not hang over the threaded end of the pipe. Pipe dope should be kept out of the inside of the pipe and fittings. Specific types of pipe dope are required for various applications depending on the material to be contained in the piping. Some project specifications may prohibit the use of PTFE tape. Always check to make sure the compound is compatible with the piped material.

When using PTFE tape, no more than three wraps should be made. A special yellow PTFE tape is available for use on natural gas piping. When using PTFE tape, apply the tape in a clockwise direction looking from the end of the pipe, the same direction as the fitting turns (*Figure 23*). Apply two to three layers, starting at least one thread from the end of the pipe. Remember to check the local codes to see if the use of tape is permitted.

3.1.1 Wrenches

Pipe wrenches are usually sold in the following sizes (by length of handle): 6, 8, 10, 14, 18, 24, 36, 48, and 60 inches. Smaller and larger sizes can possibly be acquired by special order. Pipe wrenches are usually made of either cast iron or cast aluminum. Aluminum pipe wrenches are lighter and easier to handle, but they are also more expensive.

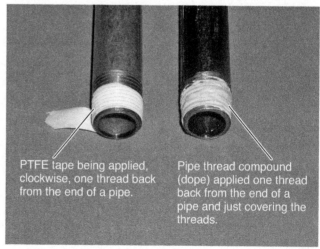

PTFE tape being applied, clockwise, one thread back from the end of a pipe.

Pipe thread compound (dope) applied one thread back from the end of a pipe and just covering the threads.

03105-13_F22.EPS

Figure 22 PTFE tape and pipe dope applied to pipe.

NCCER – *HVAC Level One* 03105

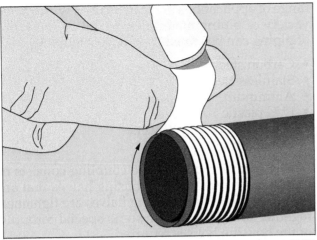

03105-13_F23.EPS

Figure 23 Applying PTFE tape.

Teeth and jaw kits can be purchased to repair worn wrenches that have not been seriously damaged.

Straight and offset pipe wrenches (*Figure 24*) are used to grip and turn round stock. They have teeth that are set at an angle. This angle allows the teeth of the wrench to grip in one direction only. The chain wrench is used on pipe that is over 2" in diameter. The chain must be oiled often to prevent it from becoming stiff or rusty. The strap wrench is used to hold chrome-plated or other types of finished or polished pipe. The strap wrench does not

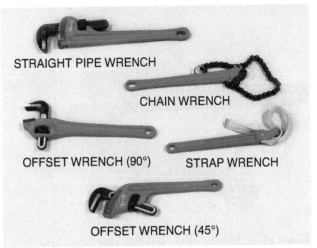

STRAIGHT PIPE WRENCH

CHAIN WRENCH

OFFSET WRENCH (90°)   STRAP WRENCH

OFFSET WRENCH (45°)

03105-13_F24.EPS

Figure 24 Pipe wrenches.

leave jaw marks or scratches on the pipe. Resin applied to the strap adds to the holding power of the wrench by reducing slippage.

Selecting the correct wrench size is important to ensure that the joint is properly tightened. A wrench that is too small does not provide sufficient leverage. A wrench that is too large allows even a small, lightweight worker to overtighten a joint with ease. Refer to *Table 2* to assist in selecting the proper wrench for a given pipe size. Using an extension on the wrench handle, such as a length of pipe often referred to as a cheater bar, is not recommended. The extension can slip off or the handle of the wrench can fail, causing an injury. Never use a pipe wrench that has a bent handle. Wrenches that are bent should be immediately taken out of service.

The pipe wrench must also be adjusted to fit the pipe properly. Maintain a gap of approximately ½" between the pipe wall and the shank of the hook jaw on the wrench. See *Figure 25*. The adjusting nut should never be at the bottom of its travel.

PTFE Tape

PTFE tape has long been referred to as Teflon® tape among trade workers. Teflon® is one brand name for the synthetic material known chemically as polytetrafluoroethylene. However, the owner of the Teflon® brand, DuPont Co., does not manufacture thread tape, nor do they license the brand name for its manufacture.

PTFE was discovered by accident in 1938. A chemical engineer named Roy Plunkett, while working for Kinetic Chemicals to develop a new chlorofluorocarbon refrigerant, discovered that a bottle of tested material in gaseous form lacked pressure before the weight had fallen to expected levels. His curiosity led him to cut the bottle open to look for the source of the non-gaseous weight. He found the inside coated with a white, waxy material that was especially slippery. The material was patented in 1941 and the Teflon® trademark was registered in 1945. DuPont eventually became the owner of the patent and trademark, since Kinetic Chemicals was created under a joint venture with General Motors.

Table 2 Suggested Pipe Wrench-to-Pipe Sizes

Wrench Size	Pipe Nominal Size
6"	⅛" – ½"
8"	¼" – ¾"
10"	¼" – 1"
12"	½" – 1½"
14"	½" – 1½"
18"	1" – 2"
24"	1½" – 2½"
36"	2" – 3½"
48"	3" – 5"
60"	3" – 8"

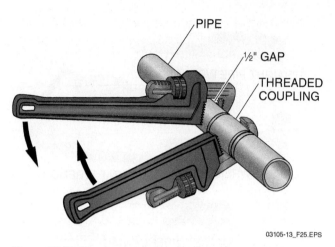

Figure 25 Tightening pipe using two pipe wrenches.

Start the fitting onto the threaded pipe by hand. Turn the fitting clockwise. Finish tightening the fitting using pipe wrenches. Note the position and rotation of the two wrenches in *Figure 25*. The use of two wrenches may be necessary even when the pipe is firmly held by a pipe vise. Refer to *Table 1* for the total thread makeup expected for various sizes of threaded pipe. Remember that the dimensions shown in the table are approximate. However, they do provide some guidance in determining how far to tighten a threaded fitting. Overtightening a fitting may create stress that later results in a leak or failure, while under-tightening typically results in immediate leaks.

3.2.0 Grooved Pipe

Grooved pipe is so called because grooves are used to join the pieces together instead of welded, flanged, or threaded joints. Each joint in a grooved piping system serves as a union, allowing easy access to any part of the piping system for cleaning or servicing. Grooved piping systems have a wide range of applications and can be used with a wide variety of piping materials. The following types of piping can be joined by grooved couplings:

- Carbon steel
- Stainless steel
- Aluminum
- PVC plastic
- High-density polyethylene
- Ductile iron

A standard grooved pipe coupling consists of a rubber gasket and two housing halves that are bolted together. The housing halves are tightened together until they touch, so no special torquing of the housing bolts is required.

The grooved piping system offers varied mechanical benefits, including the option of rigid or flexible couplings. Rigid couplings (*Figure 26*) create a rigid joint useful for risers, mechanical rooms, and other areas where positive clamping with no flexibility within the joints is desired. Flexible couplings provide allowance for controlled pipe movement that occurs with expansion, contraction, and deflection. There is very little difference in the two types visually. The primary difference is in the way that the fittings fit the grooves. Rigid fittings grip and fit the grooves very tightly. Flexible couplings offer some freedom of movement in the pipe grooves. Flexible couplings may eliminate the need for expansion joints, cold springing, or expansion loops, and will provide a virtually stress-free piping system.

Grooved pipe can be delivered to the job site precut to length and grooved, or it can be cut and grooved on the job. Full lengths of pipe often arrive at the site with grooves. Pieces that must be cut to fit at the site are then grooved after the cut is made. Use of the grooved piping method is based on the proper preparation of a groove in

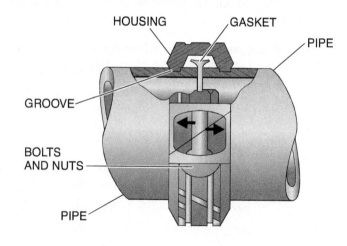
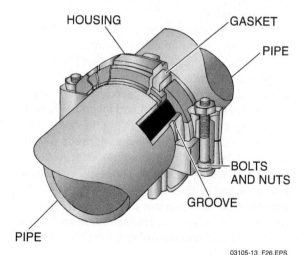

Figure 26 Two views of a grooved pipe coupling.

---

the pipe end to receive the coupling housing key. The groove has enough depth to secure the coupling housing, but leaves enough wall thickness for a full pressure rating. Many types of tools are available to properly groove pipe in the shop or in the field. Groove preparation varies with different pipe materials and wall thicknesses. The two methods of forming a groove in pipe are roll grooving and cut grooving.

Figure 27 shows the details of three different pipe grooves. Note that the cut groove removes material, while the roll groove does not. The dimensions indicated by the letters in the figure must comply with specifications.

- The A dimension is the distance from the pipe end to the first groove wall. This area is where the gasket seals against the wall of the pipe; it must be free from indentations, projections, or roll marks to provide a leakproof surface for the gasket.
- The B dimension is the groove width; it controls expansion and angular deflection based on its distance from the pipe end and its width in relation to the width of the coupling housing key.
- The C dimension is the proper inside diameter tolerance and is concentric with the outside diameter of the pipe.
- The D dimension must be changed if necessary to keep the C dimension within the stated tolerances.
- The F dimension is used with the standard roll groove only and gives the maximum allowable pipe end flare. As the groove is rolled in, the end of the pipe tends to flare out from the stress. If it flares too much, the joint cannot be used.

- The T dimension is the lightest grade or minimum thickness of pipe suitable for roll or cut grooving.
- The R dimension is the radius necessary at the bottom of the groove to eliminate a point of stress concentration for cast-iron and PVC plastic pipe.

3.2.1 Roll Grooving

Power roll-grooving machines (*Figure 28*) are used to roll grooves at the ends of pipe to prepare the piping for groove-type fittings and couplings. Power grooving machines are available to groove 2" to 16" standard and lightweight steel, aluminum, stainless steel, and even PVC plastic pipe.

Roll grooving removes no metal from the pipe. The process forms a groove by displacing the metal, pushing it into the pipe. Since the groove is cold-formed, it has rounded edges that reduce the pipe movement after the joint is made up.

3.2.2 Cut Grooving

Cut grooving differs from roll grooving in that a groove is actually cut into the pipe, removing material in the process. Cut grooving is basically intended for standard weight or heavier pipe. Pipe weights less than Schedule 40 are generally not thick enough for cut grooving. The cut removes slightly less than one-half of the pipe wall.

Cut-grooving machines are designed to be driven around a stationary pipe. This creates a groove that is of uniform depth and is concentric with the outside diameter of the pipe. Grooves can be cut with power equipment or a manually operated unit.

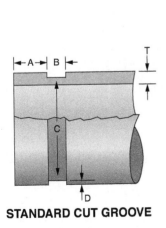
STANDARD CUT GROOVE

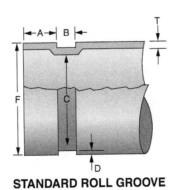
STANDARD ROLL GROOVE

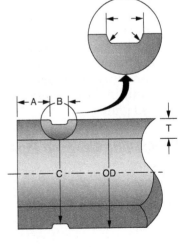
RADIUS CUT GROOVE

03105-13_F27.EPS

Figure 27 Details of pipe grooves.

Figure 28 Power roll-grooving machine.

03105-13_F28.EPS

3.2.3 Selecting Gaskets

There are many types of synthetic rubber gaskets available to provide the option of selecting grooved piping products for the widest range of applications. In order to provide maximum life for the service intended, the proper gasket selection and specification is essential.

Several factors must be considered in determining the best gasket to use for a specific service. The first consideration is the temperature of the product flowing through the pipe. Temperatures beyond the recommended limits decrease gasket life. Also, the concentration of the product, duration of service, and continuity of service must be considered because there is a direct relationship between these factors and gasket life. It should also be noted that there are services for which specific gaskets are not recommended. Always refer to the manufacturer's recommendations for selecting and installing gaskets. Job specifications may also outline the specific gaskets to be used for grooved joints.

3.2.4 Installing Grooved Pipe Couplings

Figure 29 shows the joining sequence for grooved pipe. The procedure used to join grooved pipe is as follows:

Step 1 Check the pipe ends. To make a leakproof seal, the ends must be free from indentations, projections, or roll marks.

Step 2 Check to make sure that the gasket is suitable for the intended use. Some manufacturers color-code their gaskets. Apply a thin coat of lubricant to the lips and the outside of the gasket. Lubricants are typically silicone-based and available from the gasket and/or fitting manufacturer.

Step 3 Install the gasket over the pipe end. Be sure the gasket lip does not hang over the pipe end.

Step 4 Align and bring the two pipe ends together. Slide the gasket into position and center it between the grooves on each pipe. Be sure that no part of the gasket extends into the groove on either pipe.

In-Air Groover

Another tool that is used to roll grooves is the portable in-air (in place) groover. The portable in-air groover is a manually powered machine used to roll-groove piping that is already installed. Portable in-air grooving machines are available to groove a wide variety of pipe, up to Schedule 40 in most materials. In-air groovers require a different roll groove and drive roll to maintain the correct gasket seat and groove width dimension when grooving different types and sizes of pipe. Note the ratcheting drive handle.

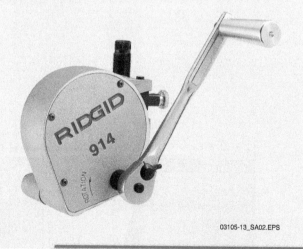

03105-13_SA02.EPS

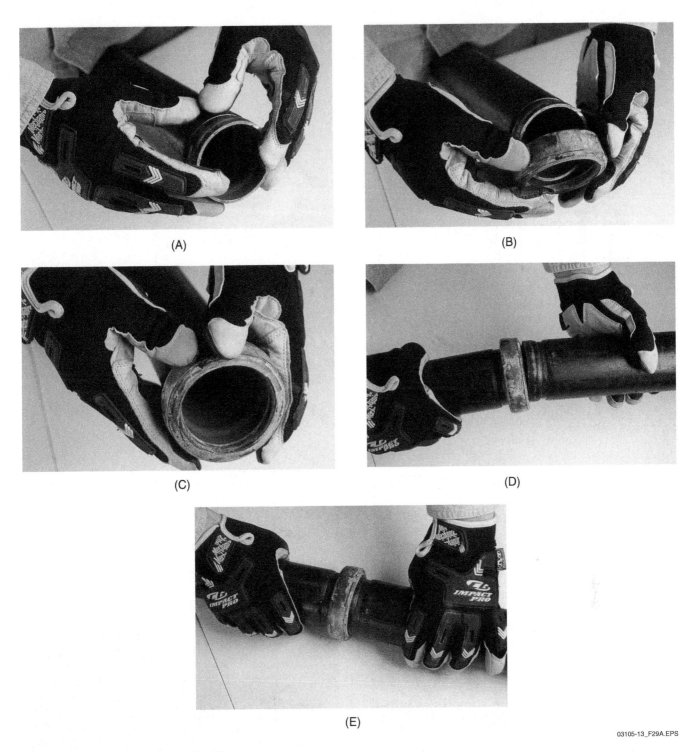

(A)

(B)

(C)

(D)

(E)

03105-13_F29A.EPS

Figure 29 Joining grooved pipe (1 of 2).

Step 5 Assemble the housing segments loosely, leaving one nut and bolt off to allow the housing to swing over the joint.

Step 6 Install the housing, swinging it over the gasket and into position in the grooves on both pipes.

Step 7 Insert the remaining bolt and nut. Be sure that the bolt track head engages into the recess in the coupling housing (*Figure 30*).

Step 8 Tighten the nuts alternately and equally to achieve metal-to-metal contact at the angle bolt pads.

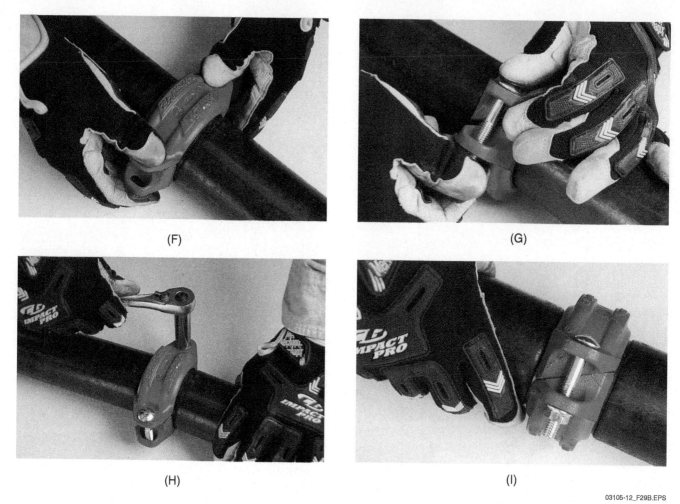

(F)

(G)

(H)

(I)

03105-12_F29B.EPS

Figure 29 Joining grooved pipe (2 of 2).

Note in *Figure 30* that the two coupling halves meet at an angle. This is a characteristic of one type of rigid (non-flexible) grooved coupling. The angular mating surfaces causes the coupling to pull harder against the walls of the groove, making it more rigid and resistant to movement that tries to pull the ends of the two pipes apart.

3.3.0 Flanged Pipe

Larger pipes and valves are sometimes joined with flange fittings (*Figure 31*). Flange fittings are assembled using gaskets and bolts. As previously discussed under grooved fittings, the proper selection of gaskets is important in establishing and maintaining a tight seal. As is the case with grooved pipe, the gasket material must be compatible with the transported material and its temperature range. Flange pipe systems can use the same types of fittings as other systems, including tees, elbows, crosses, and reducers. A flange can be welded or threaded onto a pipe.

ANGULAR MATING SURFACES

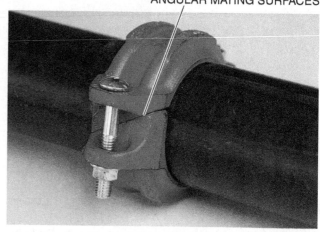

03105-13_F30.EPS

Figure 30 Completed grooved joint.

Piping systems are rarely assembled entirely with flanges. Most of the time, flanged fittings are used to connect valves and other components, or to provide access points in the piping. Most of the

Figure 31 Flanged valve.

other pipe joints are then threaded, grooved, or welded.

Note in *Figure 32* that a portion of the flange face, nearest the opening, is raised. Flanges are available in two basic styles: raised-face and full-face. With a raised-face flange, the gasket is much smaller in diameter, since it only makes contact with the raised portion. A gasket for a full-face flange is roughly the same size as the entire flange face, with openings for the bolts to pass through. A raised-face and a full-face flange surface should never be mated together.

A torque wrench should be used to tighten flange bolts. The bolts must be tightened in a specific pattern, as shown in *Figure 32*. Rather than tightening the bolts to the full torque the first time around, torque is applied in increments to prevent the joint from becoming distorted.

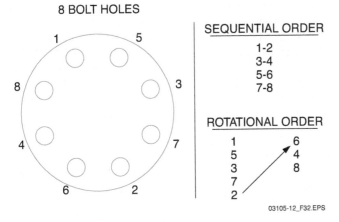

Figure 32 Flange bolt tightening pattern.

3.4.0 Installing Steel Piping

When installing steel piping, one important thing to remember is that the method used to install a piping system is generally defined by the builder's specifications. Air conditioning and heating installations planned by architects and engineering consultants include plans and specifications that describe the proposed installation in detail. After the job has been awarded to a contractor, the engineering drawings are supplemented by the working drawings of the installation contractor.

Although detailed piping drawings should be available for every piping system that is installed, this is not always the case. Therefore, the HVACR installer often has to select the best route and install the piping. This is known as field-fabricating or field-routing a pipe run. One significant challenge is working around the electrical conduit, ductwork, and other piping system installations.

Some important points to consider when installing steel piping systems are as follows:

- Many specifications require that a pipe hanger or support be provided within 1 foot of any change in direction. For example, when an elbow is installed at the top of a riser to change it to a horizontal run, a hanger should be placed within 1 foot of the elbow. Another common requirement is to install a hanger or support where the piping is connected to primary equipment. The piping system must support itself and not rely on the equipment for significant support.
- To the extent possible, keep piping runs straight and avoid the excessive use of elbows, tees, and other fittings.
- In hydronic and steam piping systems, make sure piping is installed in accordance with the specifications in order to reduce noise and avoid damage from water hammer, thermal expansion, and vibration.
- Install valves in the piping system so selected portions of the piping can be isolated for repairs. Heavy, serviceable components in a piping system should always have hangers installed adjacent to them, on each side.
- Install gas piping systems in accordance with the latest editions of the *National Fuel Gas Code (ANSI Z223.1)*, local codes, and the related equipment manufacturer's installation instructions. Similarly, install oil piping systems in accordance with the National Board of Fire Underwriters standards and local regulations.

- If piping drawings and specifications are supplied for the job, they must be followed when installing the piping runs.
- If the equipment is not yet in place, rough-in the pipe to the approximate location of the unit, and then plug or cap the pipe. The final piping can be accomplished when the unit is in place. If a material specification sheet exists for the job, check the sheet for the unit rough-in location information.

> **WARNING!**
>
> Always make sure that any pipe dope or PTFE tape used on piping is compatible with the substance being carried in the piping. The solvents in certain substances may dissolve a noncompatible pipe dope, resulting in dangerous leaks. A certain type of pipe dope must be used with natural gas and a different pipe dope must be used for LP (liquefied petroleum) gas. The same is true for PTFE tape. A special, yellow PTFE tape must be used for natural gas. When using PTFE tape to seal piping, use care to prevent any small pieces from breaking off during the application or assembly of the fittings. These small pieces may lodge in the orifice of a metering or control device and cause equipment to malfunction.

3.4.1 Pipe Hangers and Supports

Pipe must be properly supported in order to maintain the integrity of joints. Hangers must be correctly and securely installed and spaced in accordance with the job specifications. Steel pipe uses many of the same hanger and support styles used for copper and plastic. However, steel pipe is generally larger and much heavier, so anchoring and spacing requirements are much more stringent. Hangers and supports provide horizontal or vertical support of pipes and piping. Hangers are generally supports that are anchored above the piping. Supports carry the piping load from below or from walls and other structural components. The principal purpose of hangers and supports is to keep the piping in alignment and to prevent it from bending or distorting.

The installer has to determine the placement of the pipe hangers based on the size, weight, and type of pipe being run. In order for a hanger system to do its job, it must support the pipe at regular intervals. Evenly spacing the hangers prevents any individual hanger from being overloaded. Proper piping support also relieves components such as pumps from carrying any significant piping load. *Table 3* lists the recommended maximum hanger spacing intervals for carbon steel pipe installed horizontally. Risers should always be supported using a riser clamp at every floor level at a minimum.

Table 3 is intended to serve only as a guide. Always check the job-site specifications when determining pipe hanger spacing. Note that larger pipe requires fewer hangers, since the pipe is stronger. If the pipe sags, add more hangers. Piping systems must run straight without sagging. It also should be noted that some piping systems must be installed with a pitch in one direction or another. In these cases, hanger rods suspended from a level surface must increase or decrease in length at every hanger to provide the proper pitch.

In areas with the potential for earthquakes, special care is required in selecting and installing pipe supports. Since pipes are attached directly or indirectly to the building structure, seismic action that applies stresses to the building structure will also apply stresses to pipe joints. In an earthquake, a great deal of the casualties and property damage that occur result from broken pipes that release flammable gas, water, or high-pressure steam. Special seismic restraints and methods are used to counteract this effect. The purpose of the restraints is to ensure that the pipe is securely fastened to the structure in the event of excessive vibration. For example, some codes require hangers and supports to be used at closer intervals than in nonseismically active areas. In addition, they may require that you leave extra spacing for pipe where it meets walls and floors to allow for anticipated movement.

Seismic action against a structure and its pipes can be lateral (up and down), as well as longitudinal (side to side). Therefore, seismic protection must account for both conditions. There are numerous methods and devices used to protect pipes from seismic action. *Figure 33* provides one example. The two cables limit the side-to-side movement of the hanger. The hanger itself is equipped with a spring to isolate vibration. Springs and other vibration isolation components are primarily selected based on the weight of the object being suspended or supported. Remember that local codes, as well as the project drawings

Table 3 Recommended Maximum Hanger Spacing Intervals for Carbon Steel Pipe

Pipe Size	Rod Diameter	Maximum Spacing
Up to 1¼"	⅜"	8'
1½" and 2"	⅜"	10'
2½" to 3½"	½"	12'
4" and 5"	⅝"	15'
6"	¾"	17'
8" to 12"	⅞"	22'

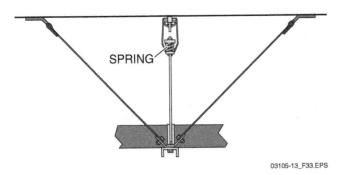

Figure 33 Example of seismic protection for a pipe.

and specifications, specify the type of seismic protection to be used. Failure to follow these requirements will create liability for the contractor and could lead to personal injury.

There are many ways to support pipe runs. A number of these are covered in NCCER Module 03103, *Basic Copper and Plastic Piping Practices*. This section shows methods that are commonly used to support steel piping. Single pipes can be supported using a clevis hanger or pipe clamp as shown in *Figure 34*. *Figure 35* shows examples of methods used to secure piping runs to walls and ceilings. Piping runs are often supported on trapeze hangers suspended from a ceiling or beam. *Figure 36* shows several examples of this approach.

The thermal expansion of pipes must be considered in hot water and steam heating systems. In order to compensate for thermal expansion, rollers and spring hangers such as those shown in *Figure 37* are used.

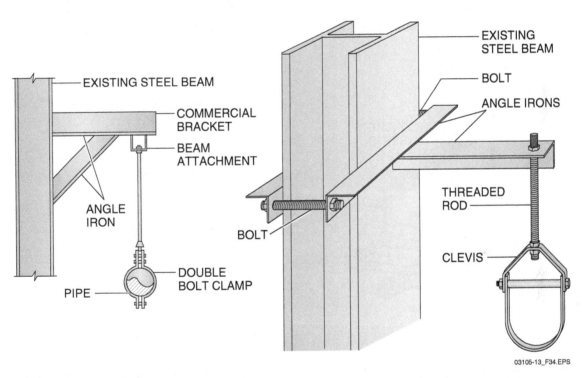

Figure 34 Individual pipe supports.

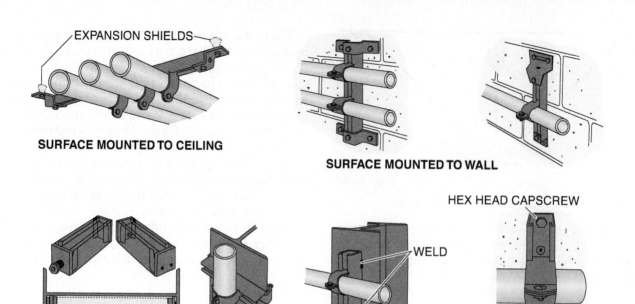

EXPANSION SHIELDS

SURFACE MOUNTED TO CEILING

SURFACE MOUNTED TO WALL

HEX HEAD CAPSCREW

WELD

SURFACE MOUNTED TO STEEL COLUMNS

03105-13_F35.EPS

Figure 35 Surface mounting methods.

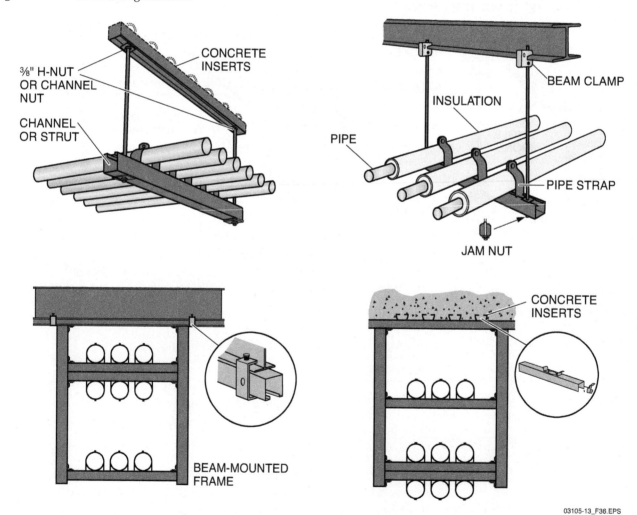

³⁄₈" H-NUT
OR CHANNEL
NUT

CONCRETE
INSERTS

CHANNEL
OR STRUT

BEAM CLAMP

INSULATION

PIPE

PIPE STRAP

JAM NUT

BEAM-MOUNTED
FRAME

CONCRETE
INSERTS

03105-13_F36.EPS

Figure 36 Trapeze hangers.

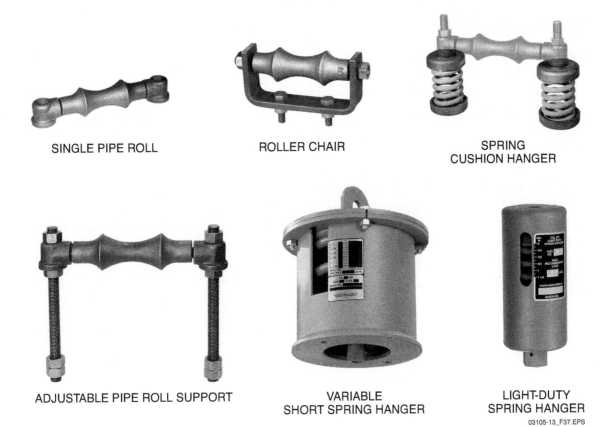

SINGLE PIPE ROLL   ROLLER CHAIR   SPRING
CUSHION HANGER

ADJUSTABLE PIPE ROLL SUPPORT   VARIABLE
SHORT SPRING HANGER   LIGHT-DUTY
SPRING HANGER

03105-13_F37.EPS

Figure 37 Rollers and spring hangers.

Additional Resources

Pipefitter's Handbook. Forrest R. Lindsey. New York, NY: Industrial Press, Inc.

The Pipefitter's Blue Book. W.V. Graves. Webster, TX: Grave Publishing Company.

3.0.0 Section Review

1. The type of wrench that is best used on polished stainless steel or chrome-plated pipe is the _____.
   a. chain wrench
   b. offset wrench
   c. straight wrench
   d. strap wrench

2. The two main categories of grooved fittings are _____.
   a. large and small
   b. rigid and flexible
   c. flanged and standard
   d. left-handed and right-handed

3. Flanged fittings should be tightened with a torque wrench.
   a. True
   b. False

4. PTFE tape typically approved for use on natural gas piping is colored _____.
   a. white
   b. yellow
   c. green
   d. blue

Summary

Carbon steel pipe is used in many HVAC applications. Black iron pipe is used most often for gas and air-pressure applications. Galvanized steel pipe is electroplated with zinc to protect the pipe from abrasive or corrosive materials inside and out.

Steel pipe is categorized by both its size (inside or outside diameter) and strength. Schedule 40 is a standard-weight pipe, Schedule 80 is an extra strong weight, and Schedule 120 is a double extra strong weight.

Steel pipe can be threaded or grooved. Threads used on carbon steel pipe are standardized. This means that thread sizes on all manufactured pipe products are the same. Threaded pipe can be measured in several ways, depending on whether there are fittings on one or on both ends. Incorrect measuring results in a pipe that is too long or too short for the installation. A variety of tools are available for working with and cutting and threading carbon steel pipe. Familiarity with tool use and maintenance ensures that your workmanship is of the best quality.

All pipe requires adequate support during installation to keep it from sagging, bending, and losing proper alignment. Codes and building specifications indicate which hangers, fasteners, and connectors are appropriate for different applications. Seismically active areas require special support systems.

Review Questions

1. The type of pipe used for natural gas supply to a furnace is _____.
   a. galvanized
   b. wrought iron
   c. cast iron
   d. black iron

2. The inside diameter of a given nominal size of Schedule 80 pipe is _____.
   a. greater than that of Schedule 40
   b. less than that of Schedule 40
   c. less than that of Schedule 120
   d. the same as that of Schedule 40 and 120

3. Of the following types of pipe, the strongest is _____.
   a. Schedule 5
   b. Schedule 20
   c. Schedule 40
   d. Schedule 120

4. All weights of steel pipe have the same _____.
   a. inside diameter
   b. outside diameter
   c. thread diameter
   d. nominal size

5. The standard thread pitch for a ½" pipe size is _____.
   a. 11½ threads per inch
   b. 14 threads per inch
   c. 17 threads per inch
   d. 18 threads per inch

6. A tee fitting described as 2 × 1½ × 1 has a _____.
   a. large run size of 2", small run size of 1½", and branch size of 1"
   b. large run size of 1", small run size of 1½", and branch size of 2"
   c. large run size of 1½", small run size of 2", and branch size of 1"
   d. large run size of 2", small run size of 1", and branch size of ½"

7. Pipe nipples are used to _____.
   a. close openings in other fittings
   b. connect two lengths of pipe when making straight runs
   c. change the direction of the pipe run
   d. make extensions from a fitting or join two fittings

8. A measurement of the full length of pipe including the threads is a(n) _____.
   a. end-to-center measurement
   b. end-to-end measurement
   c. face-to-end measurement
   d. face-to-face measurement

9. What should be done to the wheels of the cutting tool if the pipe end is flattened while it is being cut?
   a. Sharpen them
   b. Replace them
   c. Polish them
   d. Adjust them

10. Overtightening the cutting wheel on a pipe cutter _____.
    a. can cause a large burr to form on the inside of the pipe
    b. will make the threading process go faster
    c. is better for cutting through galvanized pipe
    d. will make it difficult to thread the pipe

11. Before using a threading machine to thread pipe, you should always _____ .
    a. apply a coat of heavy grease
    b. become familiar with its operating and safety instructions
    c. wipe the unthreaded end of the pipe with cutting oil
    d. apply pipe dope to the unthreaded end of the pipe

12. PTFE tape is installed by wrapping it counterclockwise.
    a. True
    b. False

13. The pipe joining method that uses two housing halves bolted together is _____.

    a. flanged
    b. threaded
    c. grooved
    d. brazed

14. Generally, 2" steel pipe should be supported with metal hangers at intervals of _____.

    a. 4'
    b. 6'
    c. 8'
    d. 10'

15. A clevis hanger would be used to _____.

    a. secure a pipe to a wall
    b. support multiple pipes suspended from a ceiling
    c. support an individual pipe suspended from a ceiling
    d. provide seismic protection for piping runs

Trade Terms Quiz

Fill in the blank with the correct trade term that you learned from your study of this module.

1. The tool insert used to cut external threads in a pipe is a(n) _____.

2. _____ pipe is carbon steel pipe that has been coated with zinc.

3. A(n) _____ is a short length of pipe with male threads on both ends.

4. The tool used to hold a die when cutting external threads on a pipe is called a(n) _____.

5. _____ is joined to another piece using special hardware that can be rigid or somewhat flexible.

6. A device with a chain used to clamp and hold pipe and other round objects for working is a(n) _____.

7. The type of pipe that gets its coloring from the carbon in the steel is _____.

8. The _____ of a thread refers to the number of threads per inch.

9. A(n) _____ is a device with one movable jaw that is used to secure pipe while working on it.

10. A special type of wrench used to turn large pipe is a(n) _____.

11. The putty-like substance used to seal threaded pipe joints is referred to as _____.

12. A process in which a material is heated then cooled to give it strength is called _____.

13. A tool that uses a nylon web to grip a pipe is called a(n) _____.

14. A(n) _____ is a plate attached to a pipe or fitting and is used with a gasket and hardware to create a joint.

15. Vibrations and movement related to an earthquake is referred to as _____.

16. The amount that the threaded end of a pipe penetrates into a fitting is known as _____.

17. A pipe that is installed vertically is a(n) _____.

18. A loud noise in a pipe caused by a change in motion is known as _____.

19. Iron that has been toughened by gradual heating or slow cooling is _____.

20. The distance from the end of the pipe to the center of the fitting is known as the _____.

Trade Terms

Annealing	Die	Grooved pipe	Pipe dope	Stock
Black iron pipe	Fitting allowance	Makeup	Pitch	Strap wrench
Chain vise	Flange	Malleable iron	Riser	Water hammer
Chain wrench	Galvanized	Nipple	Seismic	Yoke vise

Trade Terms Introduced in This Module

Annealing: A process in which a material is heated, then cooled to strengthen it.

Black iron pipe: Carbon steel pipe that gets its black coloring from the carbon in the steel.

Chain vise: A device used to clamp pipe and other round metal objects. It has one stationary metal jaw and a chain that fits over the pipe and is clamped to secure the pipe.

Chain wrench: An adjustable tool for holding and turning large pipe up to 4" in diameter. A flexible chain replaces the usual wrench jaws.

Die: A tool insert used to cut external threads by hand or machine.

Fitting allowance: The distance from the end of the pipe to the center of the fitting. Also called *takeoff*.

Flange: A flat plate attached to a pipe or fitting and used as a means of attaching pipe, fittings, or valves to the piping system.

Galvanized: Describes steel with a zinc-based coating to prevent rust.

Grooved pipe: A method for connecting piping systems using a groove installed in the pipe. The use of grooved piping eliminates the need for threading, flanging, or welding when making connections. Connections are made with gaskets and couplings installed using a wrench and lubricant.

Makeup: The amount by which the end of the pipe penetrates a fitting.

Malleable iron: Iron that has been toughened by gradual heating or slow cooling.

Nipple: A short length of pipe that is used to join fittings. It is usually less than 12" long and has male threads on both ends.

Nominal size: The approximate dimension(s) by which standard material is identified.

Pipe dope: A putty-like pipe joint material used for sealing threaded pipe joints.

Pitch: The number of threads per inch on threaded pipe and other objects.

Riser: A pipe that travels vertically. It must be supported at each level of the structure.

Seismic: Related to the vibration and movement associated with earthquakes.

Stock: A tool used to hold and turn dies when cutting external threads.

Strap wrench: A tool for gripping pipe. The strap is made of nylon web.

Water hammer: The noise and concussion that occurs when a volume of water or other liquid moving in a pipe suddenly stops or loses momentum.

Yoke vise: A holding device used to hold pipe and other round objects. It has one movable jaw that is adjusted with a threaded rod.

Appendix

COMPARISON OF IMPERIAL AND METRIC PIPE SIZES

Nominal Pipe Size (in)	Nominal Pipe Size (mm)	OD (in)	OD (mm)	Schedule Designations (ANSI/ASME)	Wall Thickness (in)	Wall Thickness (mm)	Lbs/Ft	Kg/M
⅛	6	0.405	10.30	10/10S	0.049	1.24	0.1863	0.28
⅛	6	0.405	10.30	STD/40/40S	0.068	1.73	0.2447	0.36
⅛	6	0.405	10.30	XS/80/80S	0.095	2.41	0.3145	0.47
¼	8	0.540	13.70	10/10S	0.065	1.65	0.3297	0.49
¼	8	0.540	13.70	STD/40/40S	0.088	2.24	0.4248	0.63
¼	8	0.540	13.70	XS/80/80S	0.119	3.02	0.5351	0.80
⅜	10	0.675	17.10	10/10S	0.065	1.65	0.4235	0.63
⅜	10	0.675	17.10	STD 40/40S	0.091	2.31	0.5676	0.84
⅜	10	0.675	17.10	XS 80/80S	0.126	3.20	0.7388	1.10
½	15	0.840	21.30	5/5S	0.065	1.65	0.5383	0.80
½	15	0.840	21.30	10/10S	0.083	2.11	0.671	1.00
½	15	0.840	21.30	STD 40/40S	0.109	2.77	0.851	1.27
½	15	0.840	21.30	XS 80/80S	0.147	3.73	1.088	1.62
½	15	0.840	21.30	160	0.188	4.78	1.309	1.95
½	15	0.840	21.30	XX	0.294	7.47	1.714	2.55
¾	20	1.050	26.7	5/5S	0.065	1.65	0.6838	1.02
¾	20	1.050	26.7	10/10S	0.083	2.11	0.8572	1.28
¾	20	1.050	26.7	STD/40/40S	0.113	2.87	1.131	1.68
¾	20	1.050	26.7	XS/80/80S	0.154	3.91	1.474	2.19
¾	20	1.050	26.7	160	0.219	5.56	1.944	2.89
¾	20	1.050	26.7	XX	0.308	7.82	2.441	3.63
1	25	1.315	33.4	5/5S	0.065	1.65	0.8678	1.29
1	25	1.315	33.4	10/10S	0.109	2.77	1.404	2.09
1	25	1.315	33.4	STD/40/40S	0.133	3.38	1.679	2.50
1	25	1.315	33.4	XS/80/80S	0.179	4.55	2.172	3.23
1	25	1.315	33.4	160	0.25	6.35	2.844	4.23
1	25	1.315	33.4	XX	0.358	9.09	3.659	5.45
1¼	32	1.660	42.2	5/5S	0.065	1.65	1.107	1.65
1¼	32	1.660	42.2	10/10S	0.109	2.77	1.806	2.69
1¼	32	1.660	42.2	STD/40/40S	0.140	3.56	2.273	3.38
1¼	32	1.660	42.2	XS/80/80S	0.191	4.85	2.997	4.46
1¼	32	1.660	42.2	160	0.250	6.35	3.765	5.60
1¼	32	1.660	42.2	XX	0.382	9.70	5.214	7.76

Nominal Pipe Size (in)	Nominal Pipe Size (mm)	OD (in)	OD (mm)	Schedule Designations (ANSI/ASME)	Wall Thickness (in)	Wall Thickness (mm)	Lbs/Ft	Kg/M
1½	40	1.900	48.3	5/5S	0.065	1.65	1.274	1.90
1½	40	1.900	48.3	10/10S	0.109	2.77	2.085	3.10
1½	40	1.900	48.3	STD/40/40S	0.145	3.68	2.718	4.05
1½	40	1.900	48.3	XS/80/80S	0.200	5.08	3.631	5.40
1½	40	1.900	48.3	160	0.281	7.14	4.859	7.23
1½	40	1.900	48.3	XX	0.400	10.16	6.408	9.54
2	50	2.375	60.3	5/5S	0.065	1.65	1.604	2.39
2	50	2.375	60.3	10/10S	0.109	2.77	2.638	3.93
2	50	2.375	60.3	STD/40/40S	0.154	3.91	3.653	5.44
2	50	2.375	60.3	XS/80/80S	0.218	5.54	5.022	7.47
2	50	2.375	60.3	160	0.344	8.74	7.462	11.11
2	50	2.375	60.3	XX	0.436	11.07	9.029	13.44
2½	65	2.875	73	5/5S	0.083	2.11	2.475	3.68
2½	65	2.875	73	10/10S	0.120	3.05	3.531	5.26
2½	65	2.875	73	STD/40/40S	0.203	5.16	5.793	8.62
2½	65	2.875	73	XS/80/80S	0.276	7.01	7.661	11.4
2½	65	2.875	73	160	0.375	9.53	10.01	14.9
2½	65	2.875	73	XX	0.552	14.02	13.69	20.37
3	80	3.500	88.9	5/5S	0.083	2.11	3.029	4.51
3	80	3.500	88.9	10/10S	0.120	3.05	4.332	6.45
3	80	3.500	88.9	STD/40/40S	0.216	5.49	7.576	11.27
3	80	3.500	88.9	XS/80/80S	0.300	7.62	10.25	15.25
3	80	3.500	88.9	160	0.438	11.13	14.32	21.31
3	80	3.500	88.9	XX	0.600	15.24	18.58	27.65
3½	90	4.000	101.6	5/5S	0.083	2.11	3.472	5.17
3½	90	4.000	101.6	10/10S	0.120	3.05	4.973	7.40
3½	90	4.000	101.6	STD 40/40S	0.226	5.74	9.109	13.56
3½	90	4.000	101.6	XS 80/80S	0.318	8.08	12.500	18.6
3½	90	4.000	101.6	XX	0.636	16.15	22.850	34.01

Additional Resources

This module presents thorough resources for task training. The following resource material is suggested for further study.

Pipefitter's Handbook. Forrest R. Lindsey New York, NY: Industrial Press, Inc.
The Pipefitter's Blue Book. W.V. Graves. Webster, TX: Grave Publishing Company.

Figure Credits

Answer	Section Reference	Objective
Section One		
1. b	1.1.1	1a
2. b	1.2.0	1b
3. c	1.3.5	1c
4. d	1.4.0	1d
Section Two		
1. a	2.1.0; *Figure 13*	2a
2. c	2.2.1	2b
Section Three		
1. d	3.1.1	3a
2. b	3.2.0	3b
3. a	3.3.0	3c
4. b	3.4.0	3d

NCCER CURRICULA — USER UPDATE

NCCER makes every effort to keep its textbooks up-to-date and free of technical errors. We appreciate your help in this process. If you find an error, a typographical mistake, or an inaccuracy in NCCER's curricula, please fill out this form (or a photocopy), or complete the online form at **www.nccer.org/olf**. Be sure to include the exact module ID number, page number, a detailed description, and your recommended correction. Your input will be brought to the attention of the Authoring Team. Thank you for your assistance.

Instructors – If you have an idea for improving this textbook, or have found that additional materials were necessary to teach this module effectively, please let us know so that we may present your suggestions to the Authoring Team.

NCCER Product Development and Revision

13614 Progress Blvd., Alachua, FL 32615

Email: curriculum@nccer.org
Online: www.nccer.org/olf

❏ Trainee Guide   ❏ Lesson Plans   ❏ Exam   ❏ PowerPoints   Other _____

Craft / Level: _____ Copyright Date: _____

Module ID Number / Title: _____

Section Number(s): _____

Description: _____

_____

_____

_____

_____

Recommended Correction: _____

_____

_____

_____

_____

Your Name: _____

Address: _____

_____

Email: _____ Phone: _____

Glossary

Absolute pressure: Positive pressure measurements that start at zero (no atmospheric pressure). Gauge pressure plus the pressure of the atmosphere (14.7 psi at sea level at 70°F) equals absolute pressure. Absolute pressure is expressed in pounds per square inch absolute (psia). Absolute pressure = gauge pressure + atmospheric pressure; the total pressure that exists in a system. Absolute pressure is expressed in pounds per square inch absolute (psia)

Acetone: A colorless organic solvent that is volatile and extremely flammable. In the HVACR trade, it is used as a carrier for acetylene gas in cylinders.

AK factor: The area factor of registers, grilles, and diffusers that reflects the free area for airflow relative to a square foot.

Alloy: Any substance made up of two or more metals.

Alternating current (AC): An electrical current that changes direction on a cyclical basis.

Ammeter: A test instrument used to measure current flow.

Ampere (A): The basic unit of measurement for electrical current, represented by the letter A.

Analog meter: A meter that uses a needle to indicate a value on a scale.

Annealed: Metal that has been heat-treated to soften it, making it formable. Hard-drawn copper tubing is annealed to make it soft and formable.

Annealing: A process in which a material is heated, then cooled to strengthen it.

Annual Fuel Utilization Efficiency (AFUE): HVAC industry standard for defining furnace efficiency, expressed as a percentage of the total heat available from a fuel. The AFUE goes beyond the specific thermal efficiency of a unit in that it accounts for both the peak measurement of fuel-to-heat conversion efficiency and the losses of efficiency that occur during startup, shutdown, etc.

Arc: A visible flash of light and the release of heat that occurs when an electrical current crosses an air gap.

Area: The amount of surface in a given plane or two-dimensional shape.

Atmospheric pressure: The pressure exerted on all things on Earth's surface due to the weight of the atmosphere. It is roughly 14.7 psi at sea level at 70°F; the standard pressure exerted on Earth's surface. Atmospheric pressure may also be expressed as 29.92 inches of mercury.

Atomized: Broken into tiny pieces or fragments, such as a liquid being broken into tiny droplets to create a fine spray

Balance point: An outdoor temperature value that represents an exact match of a structure's heat loss with the capacity of a heat pump to produce heat. At the heating balance point, a heat pump would need to run continuously to maintain the indoor setpoint, without falling below or rising above it.

Barometer: An instrument used to measure atmospheric pressure, typically in units of inches of mercury (in. Hg).

Barometric pressure: The actual atmospheric pressure at a given place and time.

Black iron pipe: Carbon steel pipe that gets its black coloring from the carbon in the steel.

Blower door: An assembly containing a fan used to depressurize a building by drawing air out, to support duct leak testing. It is usually installed temporarily in place of an outside door.

Bond: Short for surety bond. It is a funded guarantee that a contractor will perform as agreed.

Boots: Sheet metal fittings designed to transition from the branch duct to the receptacle for the grille, register, or diffuser to be installed.

Brazing: A heat-bonding method of joining metals using another alloy with a melting point lower than the metal(s) being joined. The alloy is used as a filler metal to bond and fill any gaps between the pieces by capillary action. Brazing uses filler metals that have a melting point above 842°F (450°C).

Brine: Water that is saturated with, or contains, large amounts of salt. In the HVACR industry, the term typically describes any water-based mixture that contains substances such as salt or glycols that lower its freezing point.

British thermal unit (Btu): The amount of heat needed to raise the temperature of one pound of water one degree Fahrenheit.

Capillary action: The movement of a liquid along the surface of a solid in a kind of spreading action.

Carbon monoxide (CO): A common by-product of combustion processes, CO is a colorless, tasteless, and odorless gas that is lighter than air and quite toxic. CO reacts with blood hemoglobin to form a substance that significantly reduces the flow of oxygen to all parts of the body. In some countries, CO is responsible for the majority of fatal air poisoning events.

Chain vise: A device used to clamp pipe and other round metal objects. It has one stationary metal jaw and a chain that fits over the pipe and is clamped to secure the pipe.

Chain wrench: An adjustable tool for holding and turning large pipe up to 4" in diameter. A flexible chain replaces the usual wrench jaws.

Chamfer: To create a symmetrical angled surface on an edge, breaking what would usually be a 90-degree angle. Chamfer is also used as a noun, as a name for the angled surface.

Chiller: A high-volume hydronic cooling unit.

Chlorofluorocarbon (CFC) refrigerant: A class of refrigerants that contains chlorine, fluorine, and carbon. CFC refrigerants have a very adverse effect on the environment.

Clamp-on ammeter: An ammeter with operable jaws that are placed around a conductor to sense the magnitude of the current flow.

Coefficient: A multiplier (e.g., the numeral 2 as in the expression 2b).

Combustion: The process by which a fuel is ignited in the presence of oxygen.

Comfort cooling: Cooling related to the comfort of humans in buildings and residences. The temperature range for comfort cooling is typically considered to be 60°F (15.5°C) to 80°F (26.6°C).

Compound: As related to refrigerants, a substance formed by a union of two or more elements in definite proportions by weight; only one molecule is present.

Compressor: In a refrigeration system, the mechanical device that converts low-pressure, low-temperature refrigerant gas into high-temperature, high-pressure refrigerant gas.

Condenser: A heat exchanger that transfers heat from the refrigerant flowing inside it to the air or water flowing over it.

Condensing furnace: A furnace that contains a secondary heat exchanger that extracts latent heat by condensing exhaust (flue) gases.

Conduction: In the context of heat transfer, the process through which heat is directly transmitted from one substance to another when there is a difference of temperature between regions in contact.

Conductivity: The ease and rate in which energy passes through a material. Thermal conductivity refers to the ease and rate at which heat passes through a material; electrical conductivity refers to the rate at which electricity passes.

Conductor: A material through which it is relatively easy to maintain an electrical current.

Conductors: Relevant to heat transfer, materials that readily transfer heat by conduction.

Constant: An element in an equation with a fixed value.

Contactor: A control device consisting of a coil and one or more sets of contacts used as a switching device in high-voltage circuits.

Continuity: A continuous current path. The absence of continuity indicates an open circuit.

Convection: The movement caused within a fluid (air or water, for example) by the tendency of hotter fluid to rise and colder, denser material to fall, resulting in heat being transferred.

Copper-clad: Components that have been coated or covered with a thin copper layer.

Cubic feet per minute (cfm): A unit for the volume of air flowing past a point in one minute. Cubic feet per minute can be calculated by multiplying the velocity of air, in feet per minute (fpm), times the area it is moving through, in square feet (cfm = fpm × area). The metric value is cubic meters per hour (m^3/h).

Cup depth: The distance that a tube inserts into a fitting, usually determined by a stop inside the fitting.

Current: The rate or volume at which electrons flow in a circuit. Current (I) is measured in amperes.

Desiccant: A material or substance, such as silica gel or calcium chloride, that seeks to absorb and hold water from any adjacent source, including the atmosphere.

Dew point: The temperature at which air becomes saturated with water vapor, and the water starts to condense into droplets; a state of 100% relative humidity.

Die: A tool insert used to cut external threads by hand or machine.

Digital meter: A meter that provides a direct numerical reading of the value measured.

Direct current (DC): An electric current that flows in one direction. A battery is a common source of DC voltage.

Dry-bulb temperature: The temperature measured using a standard thermometer. It represents the measure of sensible heat present.

Dual-fuel system: A heating system typically comprised of a heat pump and a fossil-fuel furnace. Heat pumps supplemented by electric heat are not considered dual-fuel.

Easement: A portion of a property that is set aside for public utilities or municipalities.

Electromagnet: A coil of wire wrapped around a soft iron core. When current flows through the coil, magnetism is created.

Electronically commutated motor (ECM): A DC-powered motor able to operate at variable speeds based on the programming of its electronic control module, which is AC-powered and converts AC to three-phase DC. ECMs also operate more efficiently than standard motors.

Enthalpy: The total heat content (sensible and latent) of a refrigerant or other substance.

Evaporator: A heat exchanger that transfers heat from the air flowing over it to the cooler refrigerant flowing through it.

Expansion device: Also known as the *liquid metering device* or *metering device*. Provides a pressure drop that converts the high-temperature, high-pressure liquid refrigerant from the condenser into the low-temperature, low-pressure liquid refrigerant entering the evaporator.

Exponent: A small figure or symbol placed above and to the right of another figure or symbol to show how many times the latter is to be multiplied by itself (e.g., $b^3 = b \times b \times b$).

External static pressure (ESP): The total resistance of all objects and ductwork in the air distribution system beyond the blower assembly itself.

Ferrule: A ring or bushing placed around a tube that squeezes or bites into the tube beneath when compressed, forming a seal.

Fillet: A rounded internal corner or shoulder of filler metal often appearing at the meeting point of a piece of tubing and a fitting when the joint is soldered or brazed.

Fitting allowance: The distance from the end of the pipe to the center of the fitting. Also called *takeoff*.

Flame rectification: A method of proving the existence of a pilot flame by applying an AC current to a flame rod, which is then rectified to a DC current as it flows back to a ground source. Monitoring the DC current flow provides the means of proving a pilot flame has been established.

Flange: A flat plate attached to a pipe or fitting and used as a means of attaching pipe, fittings, or valves to the piping system.

Flashback arrestor: A valve that prevents a flame from traveling back from the tip and into the hoses.

Fluorocarbon: A compound formed by replacing one or more hydrogen atoms in a hydrocarbon with fluorine atoms.

Flux: A chemical substance that prevents oxides from forming on the surface of metals as they are heated for soldering, brazing, or welding.

Force: A push or pull on a surface. In this module, force is considered to be the weight of an object or fluid. This is a common approximation.

Free air delivery: The condition that exists when there are no effective restrictions to airflow (no external static pressure) at the inlet or outlet of an air-moving device.

Fusible link: An electrical safety device that melts to open a circuit but does not respond to current like a common fuse.

Galvanized: Describes steel with a zinc-based coating to prevent rust.

Gauge pressure: The pressure measured on a gauge, expressed as pounds per square inch gauge (psig) or inches of mercury vacuum (in. Hg vac.). Pressure measurements that are made without including atmospheric pressure.

Geothermal heat pump: A heat pump that transfers heat to or from the ground using the earth as a source of heat in the winter and a heat sink in the summer. The system makes use of water as the heat-transfer medium between the earth and the refrigerant circuit.

Grooved pipe: A method for connecting piping systems using a groove installed in the pipe. The use of grooved piping eliminates the need for threading, flanging, or welding when making connections. Connections are made with gaskets and couplings installed using a wrench and lubricant.

Ground fault: Accidental contact between an energized conductor and a connection to ground, such as an equipment frame. The fault current passes through the grounding system, as well as a person or other conductive surface in the path.

Halocarbon: Hydrocarbons, like methane and ethane, that have most or all their hydrogen atoms replaced with the elements fluorine, chlorine, bromine, astatine, or iodine.

Halocarbon refrigerants: Any of a class of organic compounds containing carbon and one or more halogens, such as chlorine or fluorine. Refrigerants such as R-22 contain chlorine, which has been linked to the destruction of the ozone layer. As a result, refrigerants bearing chlorine are being phased out.

Halogens: Substances containing chlorine, fluorine, bromine, astatine, or iodine.

Hard-drawn: The process of heating copper and drawing it through dies to form its shape and size. Each die it is drawn through is progressively smaller until the desired size is reached.

Heat exchanger: A device that is used to transfer heat from a warm surface or substance to a cooler surface or substance.

Heat pump: A comfort air conditioner that is able to produce heat by reversing the mechanical refrigeration cycle.

Heat transfer: The transfer of heat from a warmer substance to a cooler substance.

Hot surface igniter (HSI): A ceramic device that glows when an electrical current flows through it. Used to ignite the fuel/air mixture in a gas furnace.

Hydrocarbons: Compounds containing only hydrogen and carbon atoms in various combinations.

Hydrochlorofluorocarbon (HCFC) refrigerant: A class of refrigerants that contains hydrogen, chlorine, fluorine, and carbon.

Hydronic: A heating or air conditioning system that uses water as a heat transfer medium.

Hygroscopic: Describes a material that readily, and sometimes aggressively, absorbs water from the atmosphere or other adjacent material.

In-line ammeter: A current-reading meter that is connected in series with the circuit under test.

Induced-draft furnace: A furnace in which a motor-driven fan draws air from the surrounding area or from outdoors to support combustion and create a draft in the heat exchanger.

Inductive: Describes an operation where a conductor becomes electrically charged from being near another electrically charged body, or becomes magnetized by being within an existing magnetic field. The process itself is called *induction*.

Infiltration: Air that unintentionally and naturally enters a building through doors, windows, and cracks in the structure.

Inrush current: A significant rise in electrical current associated with energizing inductive loads such as motors.

Insulator: A device or substance that inhibits the flow of current; the opposite of conductor.

Insulators: Materials that resist heat transfer by conduction.

International Building Code (IBC): A series of model construction codes. These codes set standards that apply across the country. This is an ongoing process led by the International Code Council (ICC).

Ladder diagram: A simplified schematic diagram in which the load lines are arranged like the rungs of a ladder between vertical lines representing the voltage source.

Latent heat: The heat energy absorbed or rejected when a substance is in the process of changing state (solid to liquid, liquid to gas, or the reverse of either change).

Latent heat of condensation: The heat given up or removed from a gas in changing back to a liquid state (steam to water).

Latent heat of fusion: The heat gained or lost in changing to or from a solid (ice to water or water to ice).

Latent heat of vaporization: The heat gained in changing from a liquid to a gas (water to steam).

Line duty: A protective device connected in series with the supply voltage.

Loads: Devices that convert electrical energy into another form of energy (heat, mechanical motion, light, etc.). Motors are the most common significant loads in HVACR systems.

Makeup: The amount by which the end of the pipe penetrates a fitting.

Malleable: A characteristic of metal that allows it to be pressed or formed to some degree without breaking or cracking. Copper is malleable, while cast iron is not.

Malleable iron: Iron that has been toughened by gradual heating or slow cooling.

Manometer: An instrument that measures air or gas pressure by the displacement of a column of liquid.

Mass: The quantity of matter present.

Mechanical refrigeration: The use of machinery to provide cooling.

Mechanical refrigeration cycle: The process by which a circulating refrigerant absorbs heat from one location and transfers it to another location.

Mixture: As related to refrigerants, a blend of two or more components that do not have a fixed proportion to one another and that, however thoroughly blended, are conceived of as retaining a separate existence; more than one type of molecule remains present.

Motor starter: A magnetic switching device used to control heavy-duty motors.

Multimeter: A test instrument capable of reading voltage, current, and resistance. Also known as a *volt-ohm-milliammeter* (VOM).

Multipoise furnace: A furnace that can be configured for upflow, counterflow, or horizontal installation.

Natural-draft furnace: A furnace in which the natural flow of air from around the furnace provides the air to support combustion and venting of combustion byproducts.

Newton (N): The amount of force required to accelerate one kilogram at a rate of one meter per second.

Nipple: A short length of pipe that is used to join fittings. It is usually less than 12" long and has male threads on both ends.

Nominal size: The approximate dimension(s) by which standard material is identified.

Nonferrous: A group of metals and metal alloys that contain no iron.

Noxious: Harmful to health.

Ohm (Ω): The basic unit of measurement for electrical resistance, represented by the symbol Ω.

On the job learning (OJL): Learning obtained while working under the supervision of a journeyman.

Orifices: Precisely drilled holes that control the flow of gas to the burners.

Oxidation: The process of combining with oxygen at a molecular level. Copper and oxygen join to form copper oxides, appearing as darkened deposits on the copper surface. Common rust is an iron oxide.

Pilot duty: A protective device that opens the motor control circuit, which then shuts off the motor.

Pipe dope: A putty-like pipe joint material used for sealing threaded pipe joints.

Pitch: The number of threads per inch on threaded pipe and other objects.

Pitot tube: A tool used to capture pressure measurements in a moving air stream.

Plate: To apply a layer of a metal on the surface of another metal. Other, less-precious metals are often plated with gold to reduce the value or expense.

Plenum: A chamber at the inlet or outlet of an air handler. The air distribution system attaches to the plenum.

Polygon: A shape formed when three or more straight lines are joined in a regular pattern.

Power: The rate of doing work, or the rate at which energy is dissipated. Electrical power is measured in watts.

Pressure drop: The reduction in pressure between one point in a pipe or tube to another point. Pressure drop results from friction in piping systems, which robs energy from the flowing fluid or vapor. Pressure drops increase as flow velocity (speed) increases.

Pressure: The force exerted by a substance against its area of containment; mathematically defined as force per unit of area.

Primary air: Air that is pulled or propelled into the initial combustion process along with the fuel.

Psychrometric chart: A graphic method of showing the relationship of various air properties.

Pump-down control: A control scheme that includes a liquid refrigerant solenoid valve that closes to initiate the system off cycle. Once the valve closes, the compressor continues to operate, pumping most or all of the remaining refrigerant out of the evaporator coil. Pump-down control eliminates the possibility of excessive liquid refrigerant forming in the evaporator coil during the off cycle, and is primarily used in refrigeration applications. A thermostat typically controls the liquid solenoid valve, while a low-side pressure switch and/or a timer controls the compressor.

R-value: A number, such as R-19, that is used to indicate the ability of insulation to resist the flow of heat. The higher the R-value, the better the insulating ability.

Radiation: In the context of heat transfer, the direct transmission of heat through electromagnetic waves through space or another medium. No contact between the two substances is required.

Radiation: The movement of heat in the form of invisible rays or waves, similar to light.

Reamer: A tool designed to remove the sharp lip and burrs left inside a pipe or tube after cutting.

Reclaimed: Used refrigerant that has been re-manufactured to bring it up to the standards required of new refrigerant.

Recovery: The removal and temporary storage of refrigerant in containers approved for that purpose.

Rectifier: A device that converts AC voltage to DC voltage.

Recycling: Circulating recovered refrigerant through filtering devices that remove moisture, acid, and other contaminants.

Redundant gas valve: A gas control containing two gas valves in series. If one fails, the other is available to shut off the gas when needed.

Refrigerant floodback: A significant amount of liquid refrigerant that returns to the compressor through the suction line during operation. Refrigerant floodback can have several causes, such as a metering device that overfeeds refrigerant or a failed evaporator fan motor.

Relative humidity (RH): The ratio of the amount of moisture present in a given sample of air to the amount it can hold at saturation. Relative humidity is expressed as a percentage.

Relay: A magnetically operated device consisting of a coil and one or more sets of contacts.

Resistance: An electrical property that opposes the flow of current through a circuit. Resistance (R) is measured in ohms.

Revolutions per minute (rpm): The number of rotations made by a spinning object over the course of one minute.

Riser: A pipe that travels vertically. It must be supported at each level of the structure.

Rupture disk: A pressure-relief device that protects a vessel or other container from damage if pressures exceed a safe level. A rupture disk typically consists of a specific material at a precise thickness that will break or fracture when the pressure limit is reached, creating a controlled weakness, thereby protecting the rest of the container from damage.

Saturation temperature: The boiling temperature of a substance at a given pressure. In the saturated condition, both liquid and vapor is likely present in the same space.

Secondary air: Air that is added to the mix of fuel and primary air during combustion.

Sectional boiler: A boiler consisting of two or more similar sections that contain water, with each section usually having an equal internal volume and surface area. Sectional boilers are often shipped in pieces and assembled at the installation site.

Seismic: Related to the vibration and movement associated with earthquakes.

Sensible heat: Heat energy that can be measured by a thermometer or sensed by touch.

Short circuit: A situation in which a conductor bypasses the load, causing a very high current flow.

Slug: Traditionally refers to a significant volume of oil returning to the compressor at once, primarily at startup. For example, a trap in a suction line may fill with oil during the off cycle, and then leave the trap as a slug at start-up. However, slugging is also a term incorrectly used by some to describe refrigerant floodback.

Solder: A fusible alloy used to join metals, with melting points below 842°F.

Soldering: A heat-bonding method of joining metals using another alloy with a melting point lower than the metal(s) being joined. The alloy is used as a filler metal to bond and fill any gaps between the pieces by capillary action. Soldering uses filler metals that have a melting point below 842°F (450°C).

Solenoid coil: An electromagnetic coil used to control a mechanical device such as a valve or relay contacts.

Specific heat: The amount of heat required to raise the temperature of one pound of a substance one degree Fahrenheit. Expressed as Btu/lb/°F. The specific heat of water (H_2O) is 1.0, thus providing the basis for defining one Btu.

Spud: A threaded metal device that screws into the gas manifold. It contains the orifice that meters gas to the burners.

Standing pilot: A gas pilot that remains lit continuously.

Static pressure (s.p.): The pressure exerted uniformly in all directions within a duct system, usually measured in inches of water column (in. w.c.) or centimeters of water column (cm H_2O).

Stock: A tool used to hold and turn dies when cutting external threads.

Strap wrench: A tool for gripping pipe. The strap is made of nylon web.

Stratify: To form or arrange into layers. In air distribution, layers of air at different temperatures will tend to stratify unless an outside influence forces them to move and mix. Warmer air will stratify on top of cooler air.

Subcooling: The reduction in temperature of a refrigerant liquid after it has completed the change in state from a vapor. Only after the phase change is complete can the liquid begin to decrease in temperature as heat is removed.

Superheat: The additional, increase in temperature of a refrigerant vapor after it has completed the change of state from a liquid to a vapor. Only after the phase change is complete can the vapor begin to increase in temperature as additional heat is applied.

Surge chamber: A vessel or container designed to hold both liquid and vapor refrigerant. The liquid is generally fed out of the bottom into an evaporator, while vapor is drawn from the top of the container by the compressor to maintain refrigerant temperature through pressure.

Sustainable construction: Construction that involves minimum impact on land, natural resources, raw materials, and energy over the building's life cycle. It uses material in a way that preserves natural resources and minimizes pollution.

Swaging: The process of using a tool to shape metal. In context, it describes the process of forming a socket in the end of a copper tube that is the correct size to accept another piece of tubing or a component.

Takeoffs: Connection points installed on a trunk duct that allow the connection of a branch duct.

Thermistor: A semiconductor device that changes resistance with a change in temperature.

Thermocouple: A device comprised of two different metals that generates electricity when there is a difference in temperature from one end to the other.

Time-delay fuses: Fuses with a built-in time delay to accommodate the inrush current of inductive loads.

Ton of refrigeration: Large unit for measuring the rate of heat transfer. One ton is defined as 12,000 Btus per hour, or 12,000 Btuh.

Total heat: Sensible heat plus latent heat.

Total pressure: The sum of the static pressure and the velocity pressure in an air duct.

Toxic: Poisonous.

Transfer grille: A grille usually installed in walls or doors, with a grille of the same size mounted on each side, that allows air to pass freely in or out of an enclosed space.

Transformer: Two or more coils of wire wrapped around a common core. Used to raise and lower voltages.

Unit: A definite standard of measure of dimension or quantity.

Vacuum: Any pressure that is less than the prevailing atmospheric pressure.

Vapor barrier: A barrier placed over insulation to stop water vapor from passing through the insulation and condensing on a cold surface.

Variable: An element of an equation that may change in value.

Velocity: The speed at which air is moving. The rate of airflow usually measured in feet per minute.

Velocity pressure: The pressure in a duct due to the linear movement of the air. It is the difference between the total pressure and the static pressure.

Venturi: A ring or panel surrounding the blades on a propeller fan to improve fan performance.

Volt (V): The unit of measurement for voltage, represented by the letter V. One volt is equivalent to the force required to produce a current of one ampere through a resistance of one ohm.

Volt-ohm-milliammeter (VOM): See multimeter.

Voltage: The driving force that makes current flow in a circuit. Voltage, often represented by the letter E, is measured in volts. Also known as *electromotive force (emf)*, *difference of potential*, or *electrical pressure*.

Volume: The amount of space contained in a given three-dimensional shape.

Water hammer: The noise and concussion that occurs when a volume of water or other liquid moving in a pipe suddenly stops or loses momentum.

Watts (W): The unit of measure for power consumed by a load.

Wet-bulb temperature: Temperature taken with a thermometer that has a wick wrapped around its sensing bulb, saturated with distilled water before taking a reading. The reading from a wet-bulb thermometer, through evaporation of the water, takes into account the moisture content of the air. It reflects the total heat content (sensible and latent) of the air.

Wetting: A process that reduces the surface tension so that molten (liquid) solder flows evenly throughout the joint.

Witness mark: A mark made as a means of determining the proper positioning of two pieces of pipe or tubing when joining. A witness mark is typically in addition to a first mark, made in a location that will not be obscured by the joining process; the first mark is often obscured during the process.

Work hardening: The repetitive reforming (such as bending or flexing) of a material, causing a permanent change in its crystalline structure and an increase in its strength and hardness.

Yoke vise: A holding device used to hold pipe and other round objects. It has one movable jaw that is adjusted with a threaded rod.

Index

A

A. *See* Ampere (A)
ABS. *See* Acrylonitrilebutadiene styrene (ABS) plastic piping
Absenteeism, (03101):19–20
Absolute pressure (psia)
  calculating, (03102):10, 19, 40
  defined, (03102):1, 35, (03107):1, 80
  gauge pressure (psig) vs., (03109):6
Absolute pressure (psia) scale, (03107):10, (03109):6
Absolute vacuum pressure, (03102):11
Absolute zero, (03101):6, (03102):14, (03107):2
Absorption heat pumps, (03108):40
AC. *See* Alternating current (AC)
ACCA. *See* Air Conditioning Contractors' Association (ACCA)
Accumulator, (03108):39
Acetone, (03104):1, 3, 37
Acetylene-B tank, (03104):9, 14
Acetylene cylinders, (03104):3, 9, 21, 27
Acetylene gas, (03104):2–3, 7, 10, 14
Acrylonitrilebutadiene styrene (ABS) plastic piping, (03103):26, 27, 32
Acute angle, (03102):23, 24
Acute triangle, (03102):26
Adjacent angles, (03102):23, 24
Adjustable refrigerant metering devices, (03107):51–53
AFUE. *See* Annual Fuel Utilization Efficiency (AFUE)
AHU. *See* Air handling unit (AHU)
Air/acetylene equipment, (03104):21
Air/acetylene torch, (03104):4, 14, 17, 27
Air cleaners, (03101):4–5
Air Conditioning Contractors' Association (ACCA), (03101):16
Air conditioning systems
  basic principle, (03101):5–6
  compressors, (03101):8
  HVACR principles, (03101):5–7
  modern, invention of, (03101):21
  refrigeration cycle, (03107):15–17
  reverse-cycle, (03101):7
  split-system, (03103):6
  temperature-pressure relationship, (03102):6
  tubing packages, (03103):6
Air-cooled condensers, (03107):40
Air distribution systems
  air handling unit (AHU), (03109):1–2
  determining, (03109):35
  energy efficiency
    duct sealing, (03109):43–46
    insulation for, (03109):41–42
    requirements for, (03109):41
    vapor barriers for, (03109):41–42
  friction in, (03109):3–4
  function, (03109):1
  furred-in, (03109):38, 39–40
  perimeter-loop, (03109):36, 39
  pressure values, (03109):4
  radial, (03109):36
  reducing-trunk, (03109):36–37
  room airflow, (03109):39–40

typical residential system, (03109):7–9
Air filters, (03108):18–20, 38–39
Air filtration, (03101):4
Airflow, (03109):39–40
Air flow volume change, (03102):40
Air-handling systems, (03101):4
Air handling unit (AHU), (03109):1–2
Air pressure, (03102):7, (03109):2, 4
Air pressure measurement devices
  differential pressure gauges, (03109):12
  gauge manifolds, (03109):22–24
  manometer, (03109):4–5, 10–12, 24
  pitot tube, (03109):12–13
  static pressure tips, (03109):12–13
Air quality, (03101):4–5
Air velocity, (03109):3
Air velocity measurement devices
  anemometers, (03109):3, 13–14
  differential pressure gauges, (03109):12
  velometers, (03109):3, 13–14
Air volume, (03109):3
A_K factor, (03109):16, 32, 53
Aldehyde, (03108):4
Algebra
  defined, (03102):16
  rules of, (03102):19–20
  sequence of operations, (03102):18
Algebraic equations, solving
  rules of algebra, (03102):19–20
  sequence of operations, (03102):18
  simplification in, (03102):18
Algebraic expressions, (03102):18
Algebraic terms
  coefficient, (03102):1, 16, 17, 35
  constant, (03102):16
  constants, (03102):16
  equations, (03102):16
  exponent, (03102):16, 17, 35
  mathematical operators, (03102):16
  operators, (03102):16
  powers, (03102):17–18
  powers and roots, (03102):17–18
  roots, (03102):17–18
  variable, (03102):16, 35
  variables, (03102):16
Alloy, (03104):1, 4, 37
Alternating current (AC), (03106):1, 3, 48
Alternative energy sources, (03106):4
Aluminum tubing, (03103):1
American Society for Testing and Materials (ASTM), (03103):2, 6
American Society of Heating, Refrigerating, and Air-Conditioning Engineers (ASHRAE), (03107):25–26, (03108):19, (03109):28
American Society of Refrigeration and Air Conditioning Engineers (ASHRAE), (03101):16
American Welding Society, (03104):1, 18
Ammeter, (03106):19, 21–22, 48
Ammonia (NH_3) refrigerants, (03107):27–28

Ampere (A), (03106):1, 7, 12, 48
Analog meter, (03106):19, 48
Anemometers, (03109):3, 13–14
Angles
  acute, (03102):23, 24
  adjacent, (03102):23, 24
  classification of, (03102):26
  complementary, (03102):23, 24
  degrees, (03102):22–23
  identifying, (03102):23
  interior, of polygons, (03102):23, 26
  labeling in right triangles, (03102):26
  obtuse, (03102):23, 24
  rays, (03102):22
  straight, (03102):23, 24
  sum of, in triangles, (03102):26
  supplementary, (03102):23, 24
  types of, (03102):23, 24
Anhydrous ammonia (R-717), (03107):27
Annealed, (03103):1, 2, 42
Annealed copper tubing, (03103):2
Annealing, (03105):1, 4, 33
Annual Fuel Utilization Efficiency (AFUE), (03108):1, 5–6, 47
Apprenticeship programs, (03101):21–23
Apps, (03101):17
Aquastat, (03108):34
Arc, (03106):11, 12, 48
Area
  of circles, (03102):4, 40
  common units of, (03102):3
  conversions, (03102):36, 38–39
  defined, (03102):1, 35
  highlighted in text, (03102):2
  of squares and rectangles, (03102):3–4, 5, 40
  of triangles, (03102):40
  unit conversions, (03102):3–5
Area formula, (03102):3, 4, 5, 16, 40
ASHRAE. See American Society of Heating, Refrigerating, and Air-Conditioning Engineers (ASHRAE); American Society of Refrigeration and Air Conditioning Engineers (ASHRAE)
ASTM. See American Society for Testing and Materials (ASTM)
Atmospheric pressure
  boiling point and, (03107):2
  defined, (03102):1, 35, (03107):1, 80
  measurement of, (03109):6
  measuring, (03107):9–10
  pounds per square inch (psi), (03102):8, 10
  units of measure, (03102):9
Atomized, (03108):1, 4, 47
Automatic gas valves, (03108):20–21

B

b. See Bar (b) unit of measure
Backward-inclined centrifugal blowers, (03109):17–18
Balance point, (03108):31, 40, 47
Balancing damper, (03109):32
Bar (b) unit of measure, (03102):9
Bare-tube evaporators, (03107):47
Barometer, (03107):1, 9, 13, 80, (03109):6
Barometric pressure, (03102):1, 9, 10, 35, (03107):13
Baseboard diffusers, (03109):30, 31
Baseboard heating, (03101):1
BCuP series flux, (03104):18, 19
Beam clamps, (03103):21, 22
Bellows valve, (03107):69
Belt-drive blowers, (03109):16

Belt-drive fans, (03109):16
Bending copper tubing, (03103):15
Bimetal thermostat, (03107):65
Bimetal valve operator, (03108):21
Biomass, (03106):4
Black iron pipe, (03105):1–2, 33
Blower door, (03109):1, 43, 53
Blowers
  belt-drive, (03109):16
  centrifugal, (03109):16–18
  direct-drive, (03109):16
  duct system, (03109):2–4
  heating system, (03108):18
  multi-speed, (03108):38
  performance, rules governing, (03109):19–21
  pressure measurement, (03102):10–11
  radial, (03109):17–18
  rotational speed measurement, (03109):14–15
Boilers, (03101):1–2, 4, (03108):31, 33–34
Bond, (03101):10, 13, 32
Boots, (03109):16, 27, 53
Brazing
  copper tubing and fittings
    joints, (03104):29–30
      lighting the oxyacetylene torch, (03104):24–27
      preparations, (03104):20
      purging refrigerant lines, (03104):27–29
  defined, (03103):1, 42
  dissimilar metals, (03104):31
  oxidation when, (03103):2
  process, (03104):1
  safety, (03104):13–14, 22
  soldering vs., (03104):1
  temperature, (03104):20–21
Brazing equipment
  acetylene-B tank, (03104):9, 14
  air/acetylene equipment, (03104):27
  air/acetylene torch, (03104):14, 17, 27
  fittings, (03104):18
  flashback arrestors, (03104):18
  fuel gas regulators, (03104):16–17
  hoses, (03104):17–18
  oxyacetylene equipment, (03104):14–15, 17, 24–27
  oxyacetylene torch, (03104):14–15, 17, 24–27
  setup
    air/acetylene equipment, (03104):27
    oxyacetylene equipment, (03104):14–15, 17, 24–27
    procedure, (03104):20–24
  torch handles and tips, (03104):17–18, 19, 25, 28
Brine, (03107):33, 48, 80
British thermal unit (Btu), (03107):1, 2, 80, (03108):4
Bromine, (03107):25
Bronze, (03105):3
Btu. See British thermal unit (Btu)
Btuh. See Btus per hour (Btuh)
Btus per hour (Btuh), (03107):8, (03108):4
Building codes, (03101):15–16
Bulb-type valve operators, (03108):21
Bushings, (03103):8–9, (03105):7
Butane gas, (03108):7

C

Calculators for squares and square roots, (03102):17–18, 27
Capillary action, (03104):1, 2, 37
Capillary tubes, (03107):50–51
Caps, (03105):7
Carbon dioxide (CO_2), (03108):4
Carbon dioxide (CO_2) pressure testing, (03103):19

Carbon monoxide (CO), (03101):10, 11, 13, 32, (03108):4, 5, 22, 25
Career opportunities, HVACR installers and technicians
    commercial/industrial, (03101):20–21
    in manufacturing, (03101):21, 24
    residential/light commercial, (03101):20, 21
Carrier, Willis, (03101):21
Carrier Corporation, (03101):21
Cast fittings, (03104):12
Casting, investment, (03101):25
Cast-iron steel pipe fittings, (03105):4
Celsius scale, (03101):6, (03102):12, 14, (03107):2, 76–78, (03108):4
Center-to-center measurement, (03105):8–9
Centigrade, (03102):14, (03107):2
Central heating systems
    electric, (03101):1, 4
    forced-air, (03101):1–2
    fuel oil, (03101):1
    hydronic, (03101):1
    natural gas, (03101):1
Centrifugal blowers, (03109):16–18
Centrifugal compressors, (03107):38–39
Centrifugal fans, (03108):18
Certification, HVACR installers and technicians, (03101):12–13
CFC. See Chlorofluorocarbon (CFC); Chlorofluorocarbon (CFC) refrigerant
cfm. See Cubic feet per minute (cfm)
Chain vise, (03105):11, 33
Chain wrench, (03105):16, 17, 33
Chamfer, (03103):26, 33, 42
Check valves, (03107):69
Child Labor Provisions, Fair Labor Standards Act, (03101):24
Chilled water systems, (03107):48
Chillers, (03101):18, 21, 22, 32, (03107):48
Chlorinated polyvinyl chloride (CPVC) plastic piping, (03103):28, 29, 31–32
Chlorine, (03107):25, 26
Chlorofluorocarbon (CFC), (03101):10, 13, 32
Chlorofluorocarbon (CFC) refrigerant, (03107):25–27
Circle
    area formula, (03102):4, 40
    defined, (03102):35
    dimensions of a, (03102):4
    properties of, (03102):22
    radius of a, (03102):4, 6
Circuit breakers, (03106):14, 30–31
Circuit diagrams, (03106):26
Circuits
    basic, (03106):14
    electronic, (03106):35–36
    parallel, (03106):15–16
    series, (03106):15
    series-parallel, (03106):16
    testing live, (03106):24
Circulating pump, (03108):32, 34
Circumference, (03102):22
Clamp-on ammeter, (03106):19, 21, 48
Clean Air Act, (03101):15, (03107):21, 26
Clevis hangers, (03103):21, 22, 24, 25
Climate, heat pumps and, (03101):5
Climate change, (03107):26–27
CO. See Carbon monoxide (CO)
CO_2. See Carbon dioxide (CO_2); Carbon dioxide (CO_2) pressure testing
Coal, heating with, (03101):4

Coal-fired power plants, (03106):1, 2
Coefficient, (03102):1, 16, 17, 35
Coils, (03106):46, 47
Combination circuit, (03106):16
Combustion
    byproducts, (03108):4, 25
    complete, (03108):4
    defined, (03108):1, 4, 47
    efficiency, (03108):5–6
    flames, (03108):6–7
    fuels, (03108):6–7
    incomplete, (03108):4–5
    requirements for, (03108):4
Combustion air, (03108):27
Comfort cooling, (03101):7–8, (03107):1, 15, 80
Commercial/industrial, career opportunities, (03101):20–21
Common-vented furnace and water heater, (03108):26
Complementary angles, (03102):23, 24
Complete combustion, (03108):4
Compounds, (03107):25, 27, 80
Compression adapters, (03103):7
Compression connections, (03103):18–19
Compression fittings, (03103):7–8, 12–13
Compression ring, (03103):7
Compressor muffler, (03107):63
Compressors
    air conditioning systems, (03101):8
    defined, (03101):1, 6, 32
    function, (03107):33
    motors, (03106):30, (03107):33–34
    refrigeration cycle function, (03107):12, 14
    schematic symbols, (03106):47
    testing, (03106):23–24
    types of
        centrifugal, (03107):38–39
        hermetic, (03101):8, (03107):34, 36
        open-drive, (03107):33–34
        reciprocating, (03107):35, 36
        rotary, (03107):35–36
        rotary vane, (03107):36
        screw, (03107):37–38
        scroll, (03107):37
        semi-hermetic (serviceable), (03101):8, (03107):34
Condensation, (03107):1, 11
Condensers
    defined, (03101):1, 5, 32
    as evaporators, (03101):7
    function, (03107):12, 14, 39
    refrigeration cycle, (03107):12, 14
    types of
        air-cooled, (03107):40
        cooling towers, (03107):43–44
        evaporative, (03107):43–44
        fin-and-tube, (03107):40
        water-cooled, (03107):41–43, 69
Condenser water valves, (03107):69
Condensing boiler, (03108):33
Condensing coil, (03107):43, 46
Condensing furnace
    defined, (03108):1, 47
    efficiency, (03108):6
    heat exchanger, (03108):16–17
    multipoise applications, (03108):17
    primary components, (03108):15
    venting, (03108):25–26
Conduction, (03107):1, 7, 80, (03108):1–2, 47
Conductivity, (03103):1, 2, 42, (03108):1

Conductors
    defined, (03106):1, 14, 48, (03107):1, 8
    heat transfer, (03107):8
    overloaded, (03106):14
    schematic symbols, (03106):46
    shock and, (03106):3
    types of, (03106):14
Cone, (03102):40
Connection diagram, (03106):37–38
Constant, (03102):16, 35
Construction
    codes, (03101):15–16
    energy-efficient, (03101):4
    formulas used in, (03102):36
    permits, (03101):15–16
    sustainable, (03101):10, 12, 32
Construction trade members, ethical principals for,
    (03101):19
Contactor, (03106):26, 31–33, 48
Contacts, (03106):46
Contact tachometer, (03109):14
Continuity, (03106):11, 15, 48
Continuity tester, (03106):22–23
Contractors, mechanical, (03101):13
Control devices
    circuit breakers, (03106):14, 30–31
    contactors, (03106):26, 31–33
    electronic controls, (03106):35–36
    fuses, (03106):14, 30–31
    motor starter, (03106):31–33
    overload protection devices, (03106):34–35
    relays, (03106):11, 13, 31–33
    solenoid coil, (03106):31
    switches, (03106):28–29
    thermistor, (03106):11, 12, 34–35
    transformer, (03106):1, 7, 33–34
Controllers, programmable, (03106):36–37
Convection, (03107):1, 7, 8, 80, (03108):1, 47
Conversions
    area, (03102):36, 38–39
    inch-pound system
        common, (03102):2
        length, (03102):4, 27–28
        pressure, (03102):9
        volume, (03102):7, 8
        weight, (03102):6
        wet to dry measures, (03102):8
    length, (03102):36, 37, 38–39
    liquids, (03102):38–39
    metric system
        area, (03102):38–39
        length, (03102):4, 27–28, 37, 38–39
        liquids, (03102):38–39
        pressure, (03102):9, (03107):79
        volume, (03102):7, 8, 37, 38–39
        weight, (03102):6, 37, 38–39
        wet to dry measures, (03102):6, 8
    pressure, (03107):10, 79, (03109):6
    temperature, (03102):12–14, 38–39, 40, (03107):2, 76–78
    volume
        box-shaped objects, (03102):5, 7
        cylindrical objects, (03102):8
        to inch-pound system, (03102):8
        metric system, (03102):37
        metric to US, (03102):38–39
        US system, (03102):36
        wet to dry measures, (03102):7
    weight, (03102):36, 37, 38–39

Coolers, (03101):7–8
Cooling coils, (03101):9
Cooling system components
    compressors
        centrifugal, (03107):38–39
        function, (03107):33
        hermetic compressor, (03107):34, 36
        open-drive, (03107):33–34
        reciprocating, (03107):35, 36
        rotary, (03107):35–36
        screw, (03107):37–38
        scroll, (03107):37
        semi-hermetic (serviceable) compressor, (03101):8,
            (03107):34
    condensers
        air-cooled, (03107):40
        cooling towers, (03107):43–44
        evaporative, (03107):43–44
        fin-and-tube, (03107):40
        function, (03107):39
        water-cooled, (03107):41–43, 69
    evaporators
        bare-tube, (03107):47
        chilled water systems, (03107):48
        construction of, (03107):47
        direct-expansion (DX), (03107):45–46
        finned-tube, (03107):47
        flooded, (03107):46
        function, (03107):44–45
        natural-draft, (03107):48
        plate, (03107):47
        shell-and-coil, (03107):48
        shell-and-tube, (03107):48
    refrigerant circuit accessories
        compressor muffler, (03107):63
        crankcase heater, (03107):59–60
        filter-drier, (03107):57, 59
        heat exchanger, (03107):61
        moisture indicator, (03107):57–58
        oil separator, (03107):60–61
        receiver, (03107):62
        service valves, (03107):62–63
        sight glass, (03107):57–58
        suction line accumulator, (03107):59, 61
    refrigerant metering devices
        adjustable, (03107):51–53
        fixed, (03107):50
        function, (03107):49
    refrigerant piping
        basic principles, (03107):53
        function, (03107):53
        hot gas line, (03107):12, 55–56
        insulation, (03107):57
        layout requirements, (03107):53
        liquid line, (03107):12, 56–57
        suction line, (03107):12, 53–55
Cooling system controls
    primary
        function, (03107):65
        humidistats, (03107):68
        pressure switch, (03107):67–68
        thermostats, (03107):64–67
        time clock, (03107):68
    secondary
        check valves, (03107):69
        condenser water valves, (03107):69
        evaporator pressure regulator, (03107):69
        flow switches, (03107):71

function, (03107):68
   oil pressure safety switches, (03107):70
   operating controls, (03107):68
   pressure-relief devices, (03107):70
   safety controls, (03107):68
Cooling thermostat, (03107):64
Cooling tower condensers, (03107):43–44
Cooling towers, (03107):43
Cooperate, willingness to, (03101):19
Copper
  ammonia and, (03107):28
  cost of, (03103):2
  fittings
    cast, (03104):12
    wrought, (03104):12
  historical uses of, (03103):4
  recycling, (03103):2
Copper-clad, (03103):12, 20, 42
Copper mining, (03103):4
Copper ribbon circuits, (03106):36
Copper tubing
  annealed, (03103):2
  bending, (03103):15
  categories, (03103):2
  characteristics, (03103):1–2
  cutting, (03103):13–14
  deburring, (03103):14
  diameter, inside and outside, (03103):2–5, 4
  fittings, types of
    compression fittings, (03103):7–8, 12–13
    flare fittings, (03103):6–7
    pressed fittings, (03103):9, 10
    push-connect fittings, (03103):20
    swaged joints, (03103):9, 11
    sweat fittings, (03103):8–9, 12–13
    tee fittings, (03103):8
  hangers and supports, (03103):20–25
  imperial, (03103):5–6
  joining
    compression connections, (03103):18–19
    flared connections, (03103):16–18
    methods of, (03103):15–16
    pressure testing following, (03103):19–20
    push-connect fittings, (03103):20
    swaging, (03103):19
  markings, (03103):4
  measuring, (03103):12
  Medical Gas Type K, (03103):4
  Medical Gas Type L, (03103):4
  metric standards, (03103):5–6
  refrigeration systems, (03103):6
  sizing, (03103):4–6
  specifications, (03103):2
  Type ACR, (03103):3–5, 6
  Type DWV, (03103):3, 4
  Type K, (03103):3–5
  Type L, (03103):2–5, 6
  Type M, (03103):2–5
  unrolling, (03103):13
Copper tubing and fittings, joining
  by brazing
    filler metals and fluxes, (03104):18–19
    joints, (03104):29–30
    lighting the oxyacetylene torch, (03104):24–27
    preparations, (03104):20
    purging refrigerant lines, (03104):27–29
    safety, (03104):13–14
  by soldering

    joints, (03104):7–8, 10–11
    preparations, (03104):6–8
    safety, (03104):2
    solder and soldering fluxes, (03104):4–6
    sweating, (03104):1
    tools and equipment, (03104):2–3
  pipe preparation, (03104):6–7, 8
Counterflow gas furnace, (03108):13
Coupling nut, (03103):7
Couplings, (03103):8–9, (03105):5, 7
CPVC. See Chlorinated polyvinyl chloride (CPVC) plastic
    piping
Crankcase heater, (03107):59–60
Crosses, (03105):5
Cross-linked polyethylene (PEX) plastic tubing, (03103):28–
    29
Cubic feet per minute (cfm), (03109):1, 3, 53
Cup depth, (03104):1, 6, 37
Cup-type striker, (03104):4, 11
Current
  calculating, (03106):12
  defined, (03106):1, 48
  effects on the human body, (03106):7, 8
  expressions of, (03106):12
  in HVACR systems, (03106):11–12
  measuring, (03106):21–22
Current surge, (03106):30
Cut grooving, (03105):19
Cylinder, (03102):40
Cylinders
  acetylene, (03104):3, 9, 21, 27
  disposable, (03107):29
  oxygen cylinders, (03104):21
  recovery, (03107):30
  returnable, (03107):29
  transporting and storing, (03104):21
Cylinder valve caps, (03104):21
Cylindrical heat exchanger, (03108):16

D

Dampers, (03109):32–33
DC. See Direct current (DC)
Deaths
  electrocution, (03106):7
  power tools, (03106):7
Defrosting evaporator coils, (03101):8
Degrees
  of angles, (03102):22–23
  of a circle, (03102):22
Department of Energy (DOE), (03108):6
Department of Labor
  Child Labor Bulletin No. 101, (03101):24
  Office of Apprenticeship, (03101):21
Department of Transportation (DOT), (03107):30
Dessicant, (03107):33, 57, 80
Dew point, (03109):1, 10, 53
Diameter
  of circles, (03102):22
  pipe and tubing, inside and outside, (03102):24
Diameter, pipe and tubing, inside and outside, (03103):2–3
Diaphragm-operated valve, (03108):21
Die, (03105):1, 2, 33
Differential pressure gauges, (03109):12
Diffusers, (03109):30–32, 40
Digital meter, (03106):19, 48
Digital multimeters (DMMs), (03109):9
Direct current (DC), (03106):1, 3, 48
Direct-drive blowers, (03109):16

Direct-drive fans, (03109):16
Direct-expansion (DX) evaporators, (03107):45–46
Direct-expansion chiller, (03107):48
Discharge line, (03107):12
Discharge pressure, (03107):15, 21
Discharge valve
   reciprocating compressor, (03107):36
   scroll compressors, (03107):35
Disconnect switch, (03106):28, (03108):39
Distance-to-spot ratio (D:S), (03107):19
DMM. *See* Digital multimeters (DMMs)
DOE. *See* Department of Energy (DOE)
DOT. *See* Department of Transportation (DOT)
Dotson, Corky, (03102):33–34
Double-pole, double-throw (DPDT) switch, (03106):28
Double-pole, single-throw (DPST) switch, (03106):28
Downflow gas furnace, (03108):13
DPDT. *See* Double-pole, double-throw (DPDT) switch
DPST. *See* Double-pole, single-throw (DPST) switch
Dry-bulb temperature, (03109):1, 10, 53
Dry volume measurements, (03102):6
D:S. *See* Distance-to-spot ratio (D:S)
Dual-fuel system, (03108):31, 40, 47
Dual-purpose systems, (03101):9
Ducts, area of, (03102):3, 4
Duct systems
   airflow and pressure, (03109):2–4
   aspect ratio, (03109):23
   assembling
      closure systems, (03109):25
      fiberglass ductboard, (03109):25, 27
      metal duct, (03109):23, 25, 26–27
   cold climate, (03109):30
   components, (03109):21
      dampers, (03109):32–33
      diffusers, (03109):30–32
      fittings and transitions, (03109):25–27, 27–29
      grilles, (03109):1, 2, 4, 7, 9, 30–32
      registers, (03109):1, 30–31
   in concrete slabs, (03109):36
   design factors, (03109):2–4, 30
   extended plenum system, (03109):36
   fans, (03109):19
   furred-in, (03109):38, 39–40
   insulation, (03109):25
   materials
      fabric, (03109):22
      fiberglass, (03109):23, 25
      flexible duct, (03109):25, 27
      galvanized steel, (03109):22–23
      selecting, (03109):21–22
      sheet metal, (03109):22–23
   perimeter-loop, (03109):36, 39
   pressure ranges, (03109):6
   pressure values
      static pressure (s.p.), (03109):4
      total pressure, (03109):4
      velocity pressure, (03109):4
   radial system, (03109):36
   reducing-trunk system, (03109):36–37
   return air, (03109):2–4, 30
   sealing, (03109):43–46
   supply air, (03109):2–4, 30
   supports, (03109):23
Duct tape, (03109):45
DX. *See* Direct-expansion (DX) evaporators

E
Easement, (03101):10, 15, 32
ECM. *See* Electronically commutated motor (ECM)
Efficiency ratings, (03101):4
Elbows (ells), (03103):8, (03105):5
Electrical components in HVAC/R systems
   control devices
      circuit breakers, (03106):14, 30–31
      contactors, (03106):26, 31–33
      electronic controls, (03106):35–36
      fuses, (03106):14, 30–31
      motor starters, (03106):31–33
      overload protection devices, (03106):34–35
      relays, (03106):11, 13, 31–33
      solenoid coil, (03106):31
      switches, (03106):28–29
      thermistor, (03106):11, 12, 34–35
      transformer, (03106):1, 7, 33–34
   electrical diagrams, (03106):36–38
   load devices
      function, (03106):26
      heaters, (03106):27
      motors, (03106):14, 26–27
Electrical load, (03106):11–12
Electrical measuring instruments
   current, (03106):21–22
   historically, (03106):23
   resistance, (03106):22–24
   test meters, (03106):19
   voltage, (03106):20
Electrical power
   AC voltage, (03106):3, 5–6
   DC voltage, (03106):3, 5
   distribution, (03106):1, 3–4
   generation, (03106):1–2
   safety, (03106):7–8
Electrical safety, (03106):24
Electrical theory
   current, (03106):11–12
   Ohm's law, (03106):12–13, 14
   power, (03106):13–14
   resistance, (03106):11–12
   voltage, (03106):11–12
Electric furnaces, (03108):37–39
Electric heat, (03101):1, 4, (03108):8, 31
Electric motors, (03106):14
Electric spark ignition (re-ignition pilot), (03108):23
Electrocution, (03106):7
Electromagnets, (03106):1, 3, 5, 48
Electromechanical humidistats, (03107):68
Electron flow, (03106):11–12
Electronic air cleaners, (03101):4, 5
Electronically commutated motor (ECM), (03108):10, 17, 47
Electronic control devices, (03106):35–36
Electronic expansion valves (EXVs), (03107):52
Electronic humidistats, (03107):68
Electronic thermostats, (03107):67
End-to-center measurement, (03105):8–9
End-to-end measurement, (03105):8–9
Energy-efficient buildings, (03101):4
Energy sources, alternative, (03106):4
ENERGY STAR® Program, (03101):11, (03108):6
Enthalpy, (03107):1, 5, 80
Environmental Protection Agency (EPA)
   certification, (03101):14–15
   Clean Air Act, (03101):14
   ENERGY STAR® program, (03101):11, (03108):6
   refrigerant controls, (03107):26, 27, 30

EPA. *See* Environmental Protection Agency (EPA)
EPR. *See* Evaporator pressure regulator (EPR)
Equations, (03102):16
Equidistant points, (03102):22
Equilateral triangle, (03102):26
ESP. *See* External static pressure (ESP)
Ethane, (03107):25
Ethical principals, HVACR installers and technicians, (03101):19
Evaporative condensers, (03107):43–44
Evaporator coils, (03101):8, (03107):44, 46–47
Evaporator pressure regulator (EPR), (03107):69
Evaporators
   chilled water systems, (03107):48
   as condensers, (03101):7
   construction of, (03107):47
   defined, (03101):1, 5, 32
   function, (03107):44–45
   refrigeration cycle function, (03107):12, 14
   types of
      bare-tube, (03107):47
      direct-expansion (DX), (03107):45–46
      finned-tube, (03107):47
      flooded, (03107):46
      natural-draft, (03107):48
      plate, (03107):47
      shell-and-coil, (03107):48
      shell-and-tube, (03107):48
Excellence, commitment to, (03101):19
Expansion device, (03101):1, 5–6, 32, (03107):12
Expansion tank, (03108):36
Exponent, (03102):16, 17, 35
Extended plenum duct system, (03109):36
External static pressure (ESP), (03109):1, 7, 53
EXV. *See* Electronic expansion valves (EXVs)

F

Fabric duct, (03109):22
Face-to-end measurement, (03105):8–9
Face-to-face measurement, (03105):8–9
Fahrenheit scale, (03101):6, (03102):12, 14, (03107):2, 76–78, (03108):4
Fair Labor Standards Act, Child Labor Provisions, (03101):24
Fairness characteristic, (03101):19
Fan airflow vs. speed, (03102):20
Fan-assisted furnaces, (03108):27
Fan coil unit (FC), (03109):1–2
Fan curve charts, (03109):21
Fan Laws, (03109):19–21
Fan pressure measurement, (03102):10–11
Fan rules, (03102):20
Fans
   air-cooled condensers, (03107):40
   forced-draft cooling towers, (03107):43
   function, (03109):18
   gas furnaces, (03108):17–18
   performance, rules governing, (03102):20, (03109):19–21
   pressure measurement, (03102):10–11
   rotational speed measurement, (03109):14–15
   types of
      belt-drive, (03109):16
      direct-drive, (03109):16
      duct, (03109):19
      propeller, (03109):18–19
FC. *See* Fan coil unit (FC)
Federal Safe Drinking Water Act, (03104):4
Ferrous metal pipe, (03105):1

Ferrule, (03103):1, 7, 42
Fiberglass ductboard, (03109):23, 25, 27
Field-fabricating a pipe run, (03105):23
Field-routing a pipe run, (03105):23
Filler metals for brazing, (03104):18–20
Fillet, (03104):1, 37
Filter-drier, (03107):57, 59
Fin-and-tube condenser, (03107):40
Finned-tube evaporators, (03107):47
Fire dampers, (03109):32–33
Fitting allowance, (03105):1, 8, 33
Fittings. *See also* Copper tubing and fittings, joining
   copper
      cast fittings, (03104):12
      wrought copper fittings, (03104):12
   copper tubing
      compression fittings, (03103):7–8, 12–13
      flare fittings, (03103):6–7, 12
      pressed fittings, (03103):9, 10
      push-connect fittings, (03103):20
      swaged joints, (03103):9, 11
      sweat fittings, (03103):8–9, 12–13
      tee fittings, (03103):6, 7, 8
   galvanized pipe, (03105):4
   grooved pipe, (03105):2
   plastic tubing, (03103):7
   steel pipe
      bushings, (03105):7
      caps, (03105):7
      cast-iron, (03105):4
      couplings, (03105):5, 7
      crosses, (03105):4–5
      elbows (ells), (03105):5
      flanges, (03105):2, 5, 22–23
      galvanized iron, (03105):4
      grooved, (03105):2, 3
      malleable iron, (03105):4
      nipples, (03105):5, 7
      plugs, (03105):7
      tees, (03105):4–5
      threaded, (03105):3
      unions, (03105):5, 6
Fixed refrigerant metering devices, (03107):50
Fixed restrictors, (03107):52
Flame rectification, (03108):10, 23, 47
Flame rollout switches, (03108):24–25
Flames, (03108):6–7
Flange, (03105):1, 2, 33
Flange fittings, (03105):2, 5, 22–23
Flare adapter, (03103):7
Flared connections, making, (03103):16–18
Flare fittings, (03103):6–7, 12
Flare nuts, (03103):6, 16
Flare nut wrench, (03103):7
Flare-to-pipe adapters, (03103):7
Flaring tools, (03103):16, 17, 19
Flashback arrestor, (03104):13, 17, 37
Flexible couplings, (03105):18
Flexible duct, (03109):25, 27
Flooded evaporators, (03107):46
Flooded-type chiller, (03107):49
Floor registers, (03109):30, 39
Flow of heat, (03107):7
Flow switches, (03107):71
Flue gas venting, (03108):25–27, 28
Fluorine, (03107):25
Fluorocarbon, (03107):25, 80
Fluorocarbon refrigerants, (03107):25–27

Flux
  brazing, (03104):18–20
  defined, (03104):1, 37
  safety when working with, (03104):2
  soldering, (03104):4–6
Food processing plants, (03101):7
Force, (03102):1, 6, 35
Forced-air furnaces, (03101):1–2, 4–5, (03108):10
Forced-draft cooling towers, (03107):43, 45
Forward-curved centrifugal blowers, (03109):17
Free air delivery, (03109):16, 18, 53
Freezers, (03101):7–8
Fresh air ventilation, (03101):4–5
Friction in air distribution systems, (03109):3–4
Fuel gases, (03103):1, (03108):7
Fuel gas hoses, (03104):17
Fuel oil, (03108):4, 8
Fuel oil furnaces, (03101):1, (03108):7
Furnaces. *See also* Gas furnaces
  Btu ratings, (03108):4
  common-vented furnace and water heater, (03108):26
  condensing
    defined, (03108):1, 47
    efficiency, (03108):6
    heat exchanger, (03108):16–17
    multipoise applications, (03108):17
    primary components, (03108):15
    venting, (03108):25–26
  dual-purpose systems, (03101):9
  electric, (03108):37–39
  energy efficiency, (03108):4, 7, 20, 26
  fan-assisted, (03108):27
  forced-air, (03101):1–2, 4–5, (03108):10
  fuel oil, (03101):1, (03108):7
  high-efficiency, (03101):4
  humidifiers in, (03101):4
  induced-draft
    combustion air, (03108):27
    defined, (03108):1, 47
    efficiency, (03108):6
    gas burners, (03108):27
    manifold pressure, (03108):22–23
    venting, (03108):25–26
  multipoise, (03101):4, (03108):10, 13, 14, 47
  natural-draft
    combustion air, (03108):27
    defined, (03108):1, 47
    efficiency, (03108):5–6
    safety devices, (03108):24–25
    venting, (03108):25–26
  pilot-type, (03108):20
  temperature rise measurements, (03108):29
  twinned, (03108):25
  venting, (03108):25–27, 28, 36
  working with safely, (03101):11
Furred-in duct systems, (03109):38, 39–40
Fuses, (03106):14, 30–31, 46
Fusible link, (03108):31, 38, 47
Fusible plugs, (03107):70

G
Galvanized, (03105):1, 33
Galvanized pipe, (03105):1–2
Galvanized pipe fittings, (03105):4
Galvanized steel ductwork, (03109):22–23
GAMA. *See* Gas Appliance Manufacturers Association
  (GAMA)

Gas
  acetylene, (03104):2–3, 7, 10, 14
  butane, (03108):7
  liquified petroleum (LP), (03108):7, 8
  manufactured, (03108):7
  natural, (03101):1, (03105):1, (03108):4, 7, 8
  pressure regulation, (03108):20
  propane, (03104):2–3, (03108):7, 22
  propylene, (03104):2–3
Gas Appliance Manufacturers Association (GAMA),
  (03108):26
Gas burner flame, (03108):27–28
Gas burners, (03108):20, 22–23, 27–28
Gases, pressure exerted by, (03102):7, (03107):10
Gas-fired furnace, (03105):1, 2
Gas-fired unit heaters, (03108):18
Gas furnaces. *See also* Furnaces
  components
    air filters, (03108):18–20
    fans, (03108):17–18
    gas burner, (03108):22–23
    gas valve assemblies, (03108):20–22
    heat exchanger, (03108):16–17
    ignition devices, (03108):23–24
    manifold, (03108):22
    motors, (03108):17–18
    orifices, (03108):22
  installation
    clearances, (03108):25
    code requirements, (03108):25
    combustion air, (03108):25, 27
    location factor, (03108):25
    venting, (03108):25–27
  maintenance, (03108):27–29
  safety, (03101):11, (03108):24–25
  types of
    counterflow, (03108):13
    downflow, (03108):13
    forced-air, (03108):10
    horizontal, (03108):10, 12
    low-boy, (03108):10, 13
    multipoise, (03108):13, 14
    upflow, (03108):10, 11
Gas ignition
  re-ignition pilot (electric spark ignition), (03108):23
  standing pilot, (03108):23
Gaskets, (03105):20
Gas train, (03108):20
Gas valves
  assemblies, (03108):20–22, 28–29
  types of
    automatic, (03108):21
    bimetal, (03108):21
    bulb-type, (03108):21
    diaphragm-operated, (03108):21
    modulating, (03108):20–21
    redundant, (03108):20
    solenoid-operated, (03108):21
    two-stage, (03108):21
Gauge manifold set, (03102):12, (03107):20–22
Gauge pressure (psig)
  absolute pressure (psia) vs., (03109):6
  calculating, (03102):19
  defined, (03102):10, (03107):1, 80
  formula, (03102):40
Gauge pressure (psig) scale, (03107):10, (03109):6
Geometric objects
  angles, (03102):22–23, 24

circles, (03102):4, 6, 22, 35
  formulas for, (03102):40
  polygon, (03102):23, 25–26, 35
  triangles, (03102):26–28
Geothermal heat pump, (03108):31, 40, 47
GFCI. *See* Ground fault circuit interrupter (GFCI)
GHS. *See* Globally Harmonized System of Classification and
    Labeling of Chemicals (GHS)
Globally Harmonized System of Classification and Labeling
    of Chemicals (GHS), (03107):31
Global warming, (03101):13, (03107):26–27, 28
Global warming potential (GWP), (03101):14
Greenhouse effect, (03101):13
Grilles, (03109):1, 2, 4, 7, 9, 30–32, 39
Grooved pipe
  assemblage
    cut grooving, (03105):19
    gaskets, (03105):20
    installing couplings, (03105):20–22
    roll grooving, (03105):19, 20
  couplings, (03105):18, 20–22
  defined, (03105):1, 33
  fittings, (03105):2
Ground fault, (03106):26, 31, 48
Ground fault circuit interrupter (GFCI), (03106):8
GWP. *See* Global warming potential (GWP)

H

H₂O. *See* Water vapor (H$_2$O)
Hacksaws, (03103):13–14
Halocarbon, (03107):25, 80
Halocarbon refrigerants, (03103):1, 42
Halogens, (03107):25, 80
Hammer-driven swaging tools, (03103):19
Hand threaders, (03105):12
Hand threading, (03105):12–14
Hangers
  copper tubing, (03103):20–25
  steel pipe, (03105):23–26
Hard copper, (03103):2
Hard-drawn, (03103):1, 2, 42
Hard-drawn tubing, (03103):2
HCFC. *See* Hydrochlorofluorocarbon (HCFC);
    Hydrochlorofluorocarbon (HCFC) refrigerant
Head pressure, (03107):15
Heat
  absence of, (03101):6
  characteristics of
    heat content, (03107):2
    latent heat, (03107):4–6
    sensible heat, (03107):4–6
    specific heat capacity, (03107):7
    temperature, (03107):2
  flow of, (03107):7
  measurement, (03107):2, (03108):4
Heat content, (03107):2
Heaters, (03106):27
Heat exchangers
  boiler, (03108):33
  condensing furnace, (03108):16–17
  defined, (03108):1, 47
  function, (03101):1
  gas furnaces, (03108):16–17
  illustrated, (03101):2
  liquid-to-suction, (03107):61
  maintenance, (03108):27–28
  refrigerant water preheater, (03107):61

Heating
  central, (03101):1–2, 4
  electric, (03101):1, 4
  forced air, (03101):1, 2, 4–5
  fuel oil, (03101):1
  fuels for, (03101):1
  hydronic, (03101):1, 3
  natural gas, (03101):1
  principles of, (03101):1
  propane gas, (03101):1
  steam, (03101):1
Heating fundamentals
  combustion
    complete, (03108):4
    defined, (03108):4
    efficiency, (03108):5–6
    flames, (03108):6–7
    fuels, (03108):6–7
    incomplete, (03108):4
    requirements for, (03108):4
  heat transfer
    conduction, (03108):1–2
    convection, (03108):2
    heat measurement, (03108):4
    humidity and, (03108):3
    radiation, (03108):2–3
    temperature and, (03108):4
Heating systems
  heat pumps
    dual-fuel system, (03108):40
    geothermal, (03108):31, 40, 47
    operating principle, (03108):39–40
  hydronic
    advantages, (03108):31, 36
    air vents, (03108):36
    boiler components, (03108):31, 33–34
    boiler controls, (03108):34
    expansion tank, (03108):36
    make-up water valve, (03108):36
    pressure-relief valve, (03108):36–37
    zone controls, (03108):36
  radiant, (03108):2
Heating thermostat, (03107):65
Heat pumps
  augmented, (03101):4, (03106):27
  climate for, (03101):1, 4, 5
  defined, (03101):1, 32
  dual-fuel system, (03108):40
  efficiency, (03101):5
  geothermal, (03108):31, 40, 47
  operating costs, (03106):27
  operating principle, (03101):7, (03108):39–40
  suction line accumulator, (03107):59
  suction line accumulators, (03107):61
Heat pumps, tubing sets, (03103):6
Heat recovery ventilator (HRV), (03101):11
Heat sink products, (03104):9
Heat transfer
  anhydrous ammonia (R-717), (03107):27
  condensers, (03107):39, 40
  conduction, (03107):7, 8, (03108):1–2
  conductors, (03107):8
  convection, (03107):7, (03108):2
  defined, (03101):1, 32, (03107):7
  experiments, (03108):3
  finned-tube evaporators, (03107):47
  heat measurement, (03108):4
  humidity and, (03108):3

Heat transfer (*continued*)
  insulators, (03107):8
  radiation, (03107):7, 8, (03108):2–3
  rate of, (03107):8
  temperature and, (03108):4
Heights, working at, (03101):10
Hermetic compressor, (03101):8, (03107):34, 36
HFC. *See* Hydrofluorocarbon (HFC) refrigerant
High-limit switches, (03108):22
High-pressure side, (03107):15
High-side pressure measurement, (03107):20–21
High-sidewall registers, (03109):30, 39
High-temperature limit switch, (03108):24
Honesty characteristic, (03101):18, 19
Hopscotch troubleshooting, (03106):37
Horizontal gas furnace, (03108):10
Horsepower, (03106):14
Hoses, gauge manifold set, (03107):21
Hot gas line, (03107):12, 14, 55–56
Hot surface igniter (HSI), (03108):1, 4, 23–24, 47
Hot wire anemometer, (03109):13
HRV. *See* Heat recovery ventilator (HRV)
HSI. *See* Hot surface igniter (HSI)
Humidifiers, (03101):4, 5, (03108):3
Humidistats, (03106):29, (03107):68
Humidity
  heat transfer and, (03108):3
  relative (RH), (03108):3, (03109):1, 10, 53
Humidity measurement instruments, (03109):10
HVACR installers and technicians
  career opportunities
    commercial/industrial, (03101):20–21
    in manufacturing, (03101):21, 24
    residential/light commercial, (03101):20, 21
  EPA certification, (03101):14
  licensing, (03101):12–13
  responsibilities and characteristics
    absenteeism, (03101):19–20
    commitment to excellence, (03101):19
    ethical principals, (03101):19
    fairness, (03101):19
    honesty, (03101):18, 19
    integrity, (03101):19
    leadership, (03101):19
    loyalty, (03101):18–19
    obedience, (03101):19
    professionalism, (03101):18
    respect for others, (03101):19
    tardiness, (03101):19–20
    willingness to cooperate, (03101):19
    willingness to learn, (03101):19
    willingness to take responsibility, (03101):19
  safety, (03101):10–11, (03106):8
  training, (03101):21–23
HVACR systems
  critical environment, (03101):25
  inspection items for, (03101):16
Hydrocarbons, (03107):25, 80
Hydrochlorofluorocarbon (HCFC), (03101):10, 13, 32
Hydrochlorofluorocarbon (HCFC) refrigerant, (03107):25–27
Hydroelectric power plants, (03106):1–2
Hydrofluorocarbon (HFC) refrigerant, (03107):25–27
Hydronic, (03101):1, 32
Hydronic heating systems
  advantages, (03108):31, 36
  air vents, (03108):36
  boiler components, (03108):31, 33–34
  boiler controls, (03108):34
    defined, (03101):1
    expansion tank, (03108):36
    make-up water valve, (03108):36
    piping, (03105):1
    pressure-relief valve, (03108):36–37
    zone controls, (03108):36
Hygroscopic, (03107):65, 68, 80
Hypotenuse, (03102):26, 27

I

IAQ. *See* Indoor air quality (IAQ)
IBC. *See* International Building Code (IBC)
Ice storage systems, (03101):11–12
Igniters
  hot surface (HSI), (03108):1, 4, 23–24, 47
  intermittent, (03108):23
Ignition devices, gas furnaces, (03108):23–24
Imperial copper tubing, (03103):5–6
Imperial pipe sizes, (03105):34–35
In-air groover, (03105):20
Inches of mercury (in. Hg), (03107):79
Inch-pound system
  area, (03102):3–4
  common values, (03102):2
  conversions
    common, (03102):2
    length, (03102):4, 27–28
    pressure, (03102):9
    volume, (03102):7, 8
    weight, (03102):6
    wet to dry measures, (03102):8
  units of measure
    common, (03102):3
    pressure, (03102):9
    weight, (03102):7, 9
  volume, (03102):4–6
  wet measures, (03102):6–7
Inclined manometer, (03109):12
Inclined-vertical manometer, (03109):12
Incomplete combustion, (03108):4–5
Indoor air quality (IAQ), (03101):4–5, (03108):19
Induced-draft fans, (03108):17–18, 23
Induced-draft furnace
  combustion air, (03108):27
  defined, (03108):1, 47
  efficiency, (03108):6
  gas burners, (03108):27
  manifold pressure, (03108):22–23
  venting, (03108):25–26
Induction, (03106):33
Inductive, (03106):11, 13, 48
Infiltration, (03108):10, 27, 47
in. Hg. *See* Inches of mercury (in. Hg)
In-line ammeter, (03106):19, 21–22, 48
Inrush current, (03106):26, 30, 48
Insulated pipe, straps and hangers, (03103):24–25
Insulation, (03107):57, (03109):41–42
Insulators
  defined, (03106):11, 48, (03107):1, 8, 80
  function, (03106):14
  heat transfer, (03107):8
Integrity characteristic, (03101):19
Intermittent igniter, (03108):23
International Building Code (IBC), (03101):10, 15–16, 32
Iron, historical uses for, (03105):3
Iron pipe, (03105):1
Isosceles triangle, (03102):26

J

Joint compound, (03105):16
Joule, (03107):2

K

Kathabar Dehumidification Systems, Inc., (03101):25
Kelvin scale, (03101):6, (03102):12, 14
Kerosene heat, (03108):8
Kilohms (KΩ), (03106):12
Kilopascal (kPa), (03107):9
KΩ. *See* Kilohms (KΩ)
kPa. *See* Kilopascal (kPa)

L

Ladder diagram, (03106):26, 37–38, 48
Latent heat
    of condensation, (03107):1, 6, 80
    defined, (03107):1, 5, 80
    of fusion, (03107):1, 6, 80
    transforming ice to steam, (03107):5
    of vaporization, (03107):1, 6, 80
Leadership, (03101):19
Leadership in Energy and Environmental Design (LEED),
    (03101):12
Learn, willingness to, (03101):19
LEED. *See* Leadership in Energy and Environmental Design
    (LEED)
Length
    conversions, (03102):4, 36, 37, 38–39
    defined, (03102):3
Licensing, HVACR installers and technicians, (03101):12–13
Lightning, (03106):11
Light-sensitive switches, (03106):29
Line duty, (03106):26, 34, 48
Lines, parallel, (03102):22
Liquid line
    refrigerant pipe, (03107):56–57
    refrigeration cycle function, (03107):12, 14
Liquids
    conversions, (03102):38–39
    pressure exerted by, (03102):7
    volume measurements, (03102):6
Liquid-to-suction heat exchangers, (03107):61
Liquified petroleum (LP) gas, (03105):1, (03108):7, 8
Load, electrical, (03106):1, 11–12, 48
Load devices
    heaters, (03106):27
    motors, (03106):14, 26–27
Low-boy gas furnace, (03108):10, 13
Low-pressure side, (03107):15
Low-side pressure, (03107):20–21
Low-sidewall registers, (03109):30
Loyalty characteristic, (03101):18–19
LP. *See* Liquified petroleum (LP) gas

M

Magnahelic® gauge, (03109):12
Magnetism, (03106):3, 5
Makeup, (03105):1, 3, 7, 33
Make-up water valve, (03108):36
Malleable, (03103):1, 2, 42
Malleable iron, (03105):1, 4, 33
Malleable iron steel pipe fittings, (03105):4
Manifold, (03108):22
Manifold pressure, (03108):28
Manmade refrigerants, (03107):25
Manometer, (03107):22, (03108):10, 28, 47, (03109):4–5, 10–12
Manufactured gas, (03108):7

Manufacturing, career opportunities in, (03101):21, 24
MAPP, (03104):2
MAP-Pro, (03104):2
Markings
    ABS pipe, (03103):27
    brazing hoses, (03104):17
    copper tubing, (03103):4
    PVC pipe, (03103):28
    refrigerant containers, (03107):29, 31
Mass, (03102):1, 6, 35
Mathematical symbols, (03102):16
Measurement, (03102):1–2
Measurement units, (03102):1–2
Mechanical contractors, (03101):13
Mechanical refrigeration
    basic principle, (03101):5
    for comfort cooling, (03101):7–8
    defined, (03101):1, 32
    heat pumps, (03101):4
    system components, (03101):5
    temperature-pressure relationship, (03101):7
Medical Gas Type K copper tubing, (03103):4
Medical Gas Type L copper tubing, (03103):4
Megohmmeters, (03106):23–24
Megohms (MΩ), (03106):12
Mercury disposal, (03107):65
Mercury thermostat, (03107):65
MERV. *See* Minimum efficiency rating value (MERV)
Metal duct, (03109):22–23
Metallic electricity, (03106):12
Meter, defined, (03102):4
Metering device, refrigeration cycle function, (03107):12, 14
Methane, (03106):4, (03107):25
Metric system
    advantages, (03102):2
    area measures, (03102):3–4
    common values, (03102):2
    conversions
        area, (03102):38–39
        length, (03102):4, 27–28, 37, 38–39
        liquids, (03102):38–39
        pressure, (03102):9, (03107):79
        volume, (03102):7, 8, 37, 38–39
        weight, (03102):6, 37, 38–39
        wet to dry measures, (03102):6, 8
    heat content measures, (03107):2
    length measures, (03102):3, 4
    pipe sizes, (03105):34–35
    prefixes, (03102):2, 3, 9
    pressure measurement, (03107):9
    units of measure, (03102):1–2, 9, 14
    use of globally, (03102):1
    volume, (03102):4–6, 9
    weight units, (03102):6, 9
    wet measures, (03102):6–7
Microamps, (03106):12
Micron gauges, (03107):21
Microprocessor-controlled systems, (03106):35–36
Milliamp, (03106):12
Millibar, (03102):9
Minimum efficiency rating value (MERV), (03108):19
Mixture, (03107):25, 27, 80
Modulating gas valves, (03108):20–21
MΩ. *See* Megohms (MΩ)
Moisture indicator, (03107):57–58
Moisture-sensitive switches, (03106):29

Motors
  electric, (03106):14, 26–27, 30
  gas furnaces, (03108):17–18
  multi-speed, (03108):18
  rotational speed measurement, (03109):14–15
  schematic symbols, (03106):47
  tap-wound, (03108):18
  variable-speed, electronically commutated, (03108):18
  wiring schematics, (03106):27
Motor starter, (03106):26, 31–33, 48
Motor winding testing, (03106):23
Multi-flame torch tip, (03104):18
Multimeter, (03106):19–20, 20, 48
Multipoise furnace, (03101):4, (03108):10, 13, 14, 47
Multi-port burners, (03108):23
Multi-speed blowers, (03108):38

N

N. *See* Newton (N)
National Appliance Energy Conservation Act, (03108):5
National Electrical Code® (NEC®), (03101):16, (03106):8, 28, (03108):39
National Fire Protection Association (NFPA), (03101):16, (03109):32
National Fuel Gas Code, (03101):16
Natural-draft evaporators, (03107):48
Natural-draft furnace
  combustion air, (03108):27
  defined, (03108):1, 47
  efficiency, (03108):5–6
  safety devices, (03108):24–25
  venting, (03108):25–26
Natural-draft gas burners, (03108):22
Natural-draft tower, (03107):45
Natural gas, (03101):1, (03105):1, (03108):4, 7, 8
Natural-gas induced-draft furnace, (03108):22
NC. *See* Normally closed (NC) contacts
NCCER curriculum, (03101):22
NCCER National Registry, (03101):22
NEC®. See National Electrical Code® (NEC®)
Newton (N), (03102):1, 9, 35
Newtons per square meter (N/m²), (03102):9
NFPA. *See* National Fire Protection Association (NFPA)
NH₃. *See* Ammonia (NH₃) refrigerants
Nipples, (03105):1, 5, 7, 33
Nitrogen pressure testing, (03103):19–20, 21
N/m². *See* Newtons per square meter (N/m²)
Nominal size, (03105):33
Noncontact tachometer, (03109):14
Nonferrous, (03104):1, 37
Nonferrous filler metal, (03104):1
Normally closed (NC) contacts, (03106):32
Normally open (NO) contacts, (03106):32
Noxious, (03101):1, 4, 32
Noxious fumes, eliminating, (03101):4–5
NPS. *See* Straight pipe threads (NPS)
NPT. *See* Tapered pipe threads (NPT)
Nuclear power plants, (03106):1, 2

O

Obedience characteristic, (03101):19
Obtuse angle, (03102):23, 24
Obtuse triangle, (03102):26
Occupational Safety and Health Administration (OSHA) standards
  ammonia exposure, (03107):27
  employee responsibility for safety, (03106):8

hazardous chemical labeling, (03107):31
Lockout and Tagging of Circuits, (03106):9
ODP. *See* Ozone depletion potential (ODP)
Offset pipe wrenches, (03105):17
Ohm (Ω), (03106):11, 12, 49
Ohm, Georg, (03106):12
Ohmmeter, (03106):22, 23
Ohm's law, (03106):12–13, 14
Oil-fired gas furnace, (03108):10, 16
Oil furnaces, safety, (03101):11
Oil heat, (03101):1
Oil pressure safety switches, (03107):70
Oils, working with, (03101):10–11
Oil separator, (03107):60–61
OJL. *See* On the job learning (OJL)
Olive, (03103):7
On the job learning (OJL), (03101):18, 21, 32
Open-drive compressors, (03107):33–34
Orifice, (03108):1, 7, 22, 47
OSHA. *See* Occupational Safety and Health Administration (OSHA) standards
Overcurrent protection devices, (03106):14
Overload protection devices, (03106):34–35
Oxidation, (03103):1, 2, 42
Oxyacetylene brazing equipment, (03104):14–15, 17, 21–22, 24–27
Oxygen cylinders, (03104):21
Oxygen hoses, (03104):17
Ozone depletion potential (ODP), (03101):14
Ozone layer, (03107):26–27, 28

P

Pa. *See* Pascal (Pa)
Parallel circuits, (03106):15–16
Parallel lines, (03102):22
Parallel resistance, (03106):15–16
Pascal (Pa), (03102):9
Pascal, Blaise, (03102):9
PE. *See* Polyethylene (PE) plastic tubing
Perfect square, (03102):17
Perimeter-loop duct systems, (03109):36, 39
Perimeter of polygons, (03102):23
Personal protective equipment
  brazing, (03104):13–14, 25
  cutting tubing, (03103):13
  refrigerants, (03107):31
  soldering, (03104):2, 10
  solvent-cement products, (03103):32–33
PEX. *See* Cross-linked polyethylene (PEX) plastic tubing
Physical quantity, (03102):1
Pietrzak, Joseph, (03101):30–31
Pilot assembly, (03108):23
Pilot duty, (03106):26, 34, 49
Pilots
  flame rectification, (03108):23
  maintenance, (03108):28
  re-ignition pilot (electric spark ignition), (03108):23
  standing pilot, (03108):23
Pilot-type furnaces, (03108):20
Pipe
  black iron, (03105):1–2, 33
  galvanized, (03105):1–2, 4
  insulation, (03103):24
  measuring, (03105):7–9
  Schedule 40, (03103):30
  Schedule 80, (03103):30
  sizes, imperial and metric, (03105):34–35
  steel, (03103):1

threading
    bench or tripod threading machine, (03105):14–15
    equipment, (03105):12–14
    hand threading, (03105):12
  tubing vs., (03103):3
Pipe, grooved
  assemblage
    cut grooving, (03105):19
    gaskets, (03105):20
    installing couplings, (03105):20–22
    roll grooving, (03105):19, 20
  couplings, (03105):18, 20–22
  defined, (03105):1, 33
  fittings, (03105):2
Pipe, plastic
  advantages, (03103):26
  cutting, (03103):30–31
  deburring, (03103):32
  diameter, inside and outside, (03103):29–30
  joining
    cleaning step, (03103):31
    solvent-cementing process, (03103):32–34, 35
    solvent-cementing products, (03103):31–32
  schedules, (03103):29–30
  support spacing, (03103):34
  threaded, (03103):26, 30
  types of
    acrylonitrilebutadiene styrene (ABS), (03103):26, 27
    chlorinated polyvinyl chloride (CPVC), (03103):28, 29
    polyvinyl chloride (PVC), (03103):26, 28
  uses for, (03103):26
Pipe, steel
  advantages, (03105):1
  assembling
    flanged pipe, (03105):22–23
    grooved pipe, (03105):18–22
    joint compound for, (03105):16
    pipe dope for, (03105):16
    pipe wrenches, (03105):16–18
    polytetrafluoroethylene (PTFE) tape for, (03105):16, 17
    threaded pipe, (03105):14–16
  cutting, (03105):11–12
  dimensions
    diameter, inside and outside, (03105):2
    wall thicknesses, (03105):2
  fittings
    bushings, (03105):7
    caps, (03105):7
    cast-iron, (03105):4
    couplings, (03105):5, 7
    crosses, (03105):4–5
    elbows (ells), (03105):5
    flanges, (03105):2, 5, 22–23
    galvanized iron, (03105):4
    grooved, (03105):2, 3
    malleable iron, (03105):4
    nipples, (03105):5, 7
    plugs, (03105):7
    tees, (03105):4–5
    threaded, (03105):3
    unions, (03105):5, 6
  hangers and supports, (03105):24–27
  installing, (03105):23–24
  joining, (03105):2
  measuring, (03105):7–9
  reaming, (03105):11–12
  sizes, (03105):2

threading
    bench or tripod threading machine, (03105):14–15
    equipment, (03105):12–13
    hand threader and vise, (03105):13–14
  types of, (03105):1–2
Pipe cutters, (03105):11
Pipe dope, (03105):16, 33
Pipe reamer, (03105):11–12
Pipe straps, (03103):21–25
Pipe threads, (03105):3–4
Pipe vise, (03105):11, 12
Pipe wrap, (03103):30, 31
Pipe wrenches, (03105):16–18
Pitch, (03105):1, 3, 33
Pitot tube, (03109):1, 4, 12–13, 53
Plane geometry, (03102):22
Plastic pipe
  advantages, (03103):26
  cutting, (03103):30–31
  deburring, (03103):32
  diameter, inside and outside, (03103):29–30
  joining
    cleaning step, (03103):31
    solvent-cementing process, (03103):32–34, 35
    solvent-cementing products, (03103):31–32
  schedules, (03103):29–30
  support spacing, (03103):34
  threaded, (03103):26, 30
  types of
    acrylonitrilebutadiene styrene (ABS), (03103):26, 27
    chlorinated polyvinyl chloride (CPVC), (03103):28, 29
    polyvinyl chloride (PVC), (03103):26, 28
  uses for, (03103):26
Plastic tubing
  cross-linked polyethylene (PEX), (03103):28–29
  cutting, (03103):31
  fittings, (03103):7
  polyethylene (PE), (03103):26, 27
Plate, (03103):12, 14, 42
Plate-and-frame condensers, (03107):42–43
Plate evaporators, (03107):47
Plenum, (03109):16, 32, 53
Plenum systems, (03109):36
Plugs, (03105):7
Pole transformer, (03106):7
Polyethylene (PE) plastic tubing, (03103):26, 27
Polygon, (03102):22, 23, 25–26, 35
Polytetrafluoroethylene (PTFE) tape, (03105):16, 17
Polyvinyl chloride (PVC) plastic pipe, (03103):26, 28, 31–32,
    (03108):26
Pounds per square inch (psi), (03102):8, 9, 10, (03107):9
Power, electrical, (03106):1, 13–14, 49
Powers, mathematical, (03102):17–18
Power threaders, (03105):12–13, 14
Pressed fittings, (03103):9, 10
Pressing tools, (03103):9, 11
Pressure
  absolute (psia)
    calculating, (03102):10, 19
    defined, (03102):1, 35
    scale, (03107):10
  absolute vacuum, (03102):11
  atmospheric
    defined, (03102):1, 35, (03107):9
    pounds per square inch (psi), (03102):8, 10
    at sea level, (03102):40
    units of measure, (03102):9
  barometric, (03102):1, 9, 10, 35

Pressure (*continued*)
  boiling point and, (03107):3
  conversions, (03107):10, 79, (03109):6
  defined, (03107):1, 80
  exerted by
    gases, (03102):7, (03107):10
    liquids, (03102):7, (03107):10
    refrigerants, (03107):29
    solids, (03107):10
  formulas, (03102):40
  gauge (psig), (03102):10, 19, 40, (03107):10
  highlighted in text, (03107):8
  measurement devices
    differential pressure gauges, (03109):12
    gauge manifolds, (03107):20–22
    gauges, (03107):21
    manometer, (03107):22, (03109):4–5, 10–12
    pitot tube, (03109):12–13
    static pressure tips, (03109):12–13
  measuring
    absolute pressure, (03102):10, 40, (03107):10
    gauge pressure, (03107):10
    static head pressure, (03102):10–11
    vacuum, (03102):13
  refrigeration cycle fundamentals, (03107):9–10
  static head, (03102):10–11
  tire, (03102):9, 11
  units of measure, (03102):9
  vacuum, (03102):11, 13
Pressure drop, (03103):1, 6, 42
Pressure-relief valve, (03107):70, (03108):34, 36–37
Pressure-sensing switch, (03106):29
Pressure-sensitive thermostat, (03107):65
Pressure switch, (03107):67–68, (03108):25
Pressure-temperature relationship, (03102):6–8
Pressure testing copper tube joins, (03103):19–20
Pressure values
  static pressure (s.p.), (03109):4
  total pressure, (03109):4
  velocity pressure, (03109):4
Primary air, (03108):6, 22, 48
Professionalism, (03101):18
Propane furnace, (03108):22
Propane gas, (03104):2–3, (03108):7, 22
Propane gas heating systems, (03101):1
Propeller fans, (03108):18, (03109):18–19
Propylene gas, (03104):2–3
psi. *See* Pounds per square inch (psi)
psia. *See* Absolute pressure (psia); Absolute pressure (psia) scale
psig. *See* Gauge pressure (psig); Gauge pressure (psig) scale
Psychrometer, (03109):10
Psychrometric chart, (03109):1, 10, 53
PTFE. *See* Polytetrafluoroethylene (PTFE) tape
Pump-down control, (03107):33, 55, 81
Pushbuttons, (03106):46
PVC. *See* Polyvinyl chloride (PVC) plastic pipe
Pyramid, (03102):40
Pythagorean theorem, (03102):27, 28, 40

R
R-12 refrigerant, (03101):13
R-22 refrigerant, (03101):13
R-717. *See* Anhydrous ammonia (R-717)
Radial blowers, (03109):17–18
Radial duct system, (03109):36
Radiant heat, (03108):2

Radiation
  basics, (03108):2–3
  defined, (03107):1, 81, (03108):1, 48
  harmful, ozone layer and, (03107):26–27
  heat transfer, (03107):7, 8
Radius, (03102):4, 6, 22
Radon, (03109):38
Rankine scale, (03101):6, (03102):12, 14
Rays of angles, (03102):22
Reamer, (03103):12, 14, 42
Receiver, (03107):62
Reciprocating compressors, (03107):35, 36
Reclaimed, (03101):10, 13, 32
Recovery, (03101):10, 15, 32
Recovery cylinders, (03107):30
Rectangles, area of, (03102):3–4, 40
Rectifier, (03106):1, 3, 47, 49
Recycling
  copper, (03103):2
  defined, (03101):10, 32
  refrigerant containers, (03107):29
  refrigerants, (03101):15
Reducing couplings and bushings, (03103):8–9
Reducing-trunk duct system, (03109):36–37
Redundant gas valve, (03108):10, 20, 48
Refrigerant circuit accessories
  compressor muffler, (03107):63
  crankcase heater, (03107):59–60
  filter-drier, (03107):57, 59
  heat exchanger, (03107):61
  moisture indicator, (03107):57–58
  oil separator, (03107):60–61
  receiver, (03107):62
  service valves, (03107):62–63
  sight glass, (03107):57–58
  suction line accumulator, (03107):59, 61
Refrigerant cycle, (03107):15–17
Refrigerant floodback, (03107):33, 45, 81
Refrigerant lines
  electronic expansion valves (EXVs), (03107):52
  joining, (03104):1, 9
  thermostatic expansion valve (TXV or TEV), (03107):52
Refrigerant piping
  basic principles, (03107):53
  ells, (03103):8
  fittings, (03103):8, 9
  function, (03107):53
  hot gas line, (03107):12, 55–56
  insulation, (03107):57
  layout requirements, (03107):53
  liquid line, (03107):12, 56–57
  suction line, (03107):12, 53–55
Refrigerants
  blending, (03107):27
  boiling point, (03107):3
  boiling points, (03107):13
  chlorine content, (03101):13
  chlorofluorocarbon (CFC), (03101):10, 13, 32
  compound numbering, (03107):27
  containers
    disposable cylinders, (03107):29
    drums, (03107):30
    handling, (03107):29
    illustrated, (03107):30
    markings, (03107):29, 31
    recycling, (03107):29
    returnable cylinders, (03107):29–30
    storage, (03107):29

environmental damages, (03101):13–15
function, (03107):2, 25
halocarbon, (03103):1, 33, 42
hydrochlorofluorocarbon (HCFC), (03101):10, 13, 32
labeling, (03101):15
leak detectors, (03103):21
legislation affecting, (03107):26
metering devices
   adjustable, (03107):51–53
   fixed, (03107):50
   function, (03107):49
mixtures, (03107):27
naming, (03107):25–26
packaging, (03101):14, 15
reclaimed, (03101):13
recovery, (03101):15
recycling, (03101):15
safety precautions, (03107):31
states of matter, (03107):4–6
storage, (03107):62
temperature-pressure relationship, (03102):7–8,
   (03107):11–12, 15–16
transporting, (03103):1
types of
   ammonia (NH$_3$), (03107):27–28
   chlorofluorocarbon (CFC), (03107):25–27
   fluorocarbon, (03107):25–27
   hydrochlorofluorocarbon (HCFC), (03107):25–27
   hydrofluorocarbon (HFC), (03107):25–27
   manmade (synthetic), (03107):25
Refrigerant water preheater, (03107):61
Refrigeration, (03107):2
Refrigeration, mechanical
  basic principle, (03101):5
  for comfort cooling, (03101):7–8
  defined, (03101):1, 32
  heat pumps, (03101):4
  system components, (03101):5
  temperature-pressure relationship, (03101):7
Refrigeration circuit, (03107):15
Refrigeration cycle
  component functions, (03107):14
  principles, (03107):13
  typical air conditioning system, (03107):15–17
Refrigeration cycle fundamentals
  heat, characteristics of
   heat content, (03107):2
   latent heat, (03107):4–6
   sensible heat, (03107):4–6
   specific heat capacity, (03107):7
   temperature, (03107):2
  heat transfer
   condensers, (03107):39, 41
   conduction, (03107):7
   conductors, (03107):8
   convection, (03107):7
   defined, (03107):7
   finned-tube evaporators, (03107):47
   insulators, (03107):8
   radiation, (03107):7
   rate of, (03107):8
  measurement devices
   gauge manifolds (pressure), (03107):20–22
   thermometers, (03107):18–19
  pressure
   absolute pressure (psia), (03107):9–10
   defined, (03107):9
   gauge pressure (psig), (03107):9–10

   measuring, (03107):9–10
   temperature-pressure relationship, (03107):11–12
Refrigeration systems
  components
   compressor, (03107):12, 14
   condenser, (03107):12, 14
   evaporator, (03107):12
   hot gas line, (03107):12, 14, 55–56
   liquid line, (03107):12, 14, 56–57
   metering, or expansion device, (03107):12, 14
   piping, (03107):12–13, 14
   suction line, (03107):12, 14, 53–55
  copper tubing, (03103):6
  flare fittings, (03103):6
  operating temperatures, (03102):14
  temperature-pressure relationship, (03102):6–8
  tubing packages, (03103):6
Registers, (03109):1, 30–31, 39
Re-ignition pilot (electric spark ignition), (03108):23
Relative humidity (RH), (03108):3, (03109):1, 10, 53
Relay, (03106):11, 13, 31–33, 49
Relief valve, (03107):70
Remote bulb thermostat, (03107):65–66
Residential/light commercial, career opportunities,
  (03101):20, 21
Resistance
  calculating, (03106):12
  current and, (03106):7
  defined, (03106):1, 11, 49
  expressions of, (03106):11–12
  measuring, (03106):22–24
  parallel circuits, (03106):15–16
  series circuits, (03106):15
Resistance heaters, (03106):27
Resistors, (03106):12, 47
Respect characteristic, (03101):19
Responsibility, willingness to take, (03101):19
Returnable cylinders, (03107):29
Return air duct system, (03109):2–4
Reverse-cycle air conditioners, (03101):7
Reversing valve, (03101):7
Revolutions per minute (rpm), (03109):1, 14, 53
RH. *See* Relative humidity (RH)
RIDGID® press tool and jaws, (03103):10
Right angle, (03102):23, 24, 28
Right triangle
  angles and sides of, (03102):26–27
  calculations using the Pythagorean theorem, (03102):27,
   28, 40
Rigid couplings, (03105):18
Riser, (03105):16, 18, 33
Riser clamps, (03103):21, 23
Roll grooving, (03105):19, 20
Room airflow, (03109):39–40
Roots, (03102):17–18
Rosebud torch tip, (03104):18
Rotary compressors, (03107):35–36
Rotary vane compressors, (03107):36
Rotating vane anemometer, (03109):13, 14
Rotational speed measurement, (03109):14–15
rpm. *See* Revolutions per minute (rpm)
Rupture disk, (03107):25, 29, 30, 70, 81
R-value, (03109):35, 42, 53

S
Safety
  ammonia exposure, (03107):27, 28
  brazing, (03104):13–14, 22

Safety (*continued*)
electrical, (03106):7–8, 24
employee responsibility for, (03106):8
gas furnace controls for, (03108):24–25
gas furnaces, (03101):11, (03108):24–25
HVACR installers and technicians, (03101):10–11, (03106):8
lockout/tagout, (03106):9–10
oil furnaces, (03101):11
oils, working with, (03101):10–11
refrigerants, (03107):31
refrigerants, working with, (03101):10
soldering, (03104):2, 10
working at heights, (03101):10
Safety Data Sheet (SDS), (03107):31
Saturation temperature, (03107):1, 6, 81
Scalene triangle, (03102):26
Schedule 40 pipe, (03103):30
Schedule 80 pipe, (03103):30
Schematic symbols, (03106):46–47
Schrader valves, (03107):63
Screw compressors, (03107):37–38
Scroll compressors, (03107):37
SDS. *See* Safety Data Sheet (SDS)
Sealed compressor, (03107):34
Secondary air, (03108):6, 22, 48
Sectional boiler, (03108):31, 33, 48
Seismic, (03105):16, 24, 33
Semi-hermetic (serviceable) compressor, (03101):8, (03107):34
Sensible heat, (03107):1, 4–6, 81
Sequence of operations, (03102):18, 40
Series circuits, (03106):15
Series-parallel circuits, (03106):16
Series resistance, (03106):15
Service valves, (03107):62–63
Sheet Metal and Air Conditioning Contractors' National Association (SMACNA), (03101):16
Sheet metal duct, (03109):22–23
Shell-and-coil condensers, (03107):42–43
Shell-and-coil evaporators, (03107):48
Shell-and-tube condensers, (03107):41–42
Shell-and-tube evaporators, (03107):48
Shock, electrical, (03106):7–8
Short circuit, (03106):26, 30, 49
Sick building, (03101):4–5
Sight glass, (03107):57–58
Single-pole, single-throw (SPST) switch, (03106):28
Single-port burners, (03108):22
Sizing copper tubing, (03103):4–6
Sling psychrometer, (03109):10
Slow-blow fuse, (03106):30
Slug, (03107):33, 81
SMACNA. *See* Sheet Metal and Air Conditioning Contractors' National Association (SMACNA)
Smoke dampers, (03109):32–33
Snap-action thermostat, (03107):65
Soft copper, (03103):2, 13
Soft soldering, (03104):1
Solar energy, (03106):4
Solder
defined, (03104):1, 4, 37
tin-antimony, (03104):4
tin-copper-silver, (03104):4
tin-silver, (03104):4
Soldering
brazing vs., (03104):1
copper tubing and fittings

joints, (03104):7–8, 10–11
preparations, (03104):6–7
solder and soldering fluxes, (03104):4–6
sweating, (03104):1
defined, (03103):1, 42
oxidation when, (03103):2
pressure-temperature ratings of soldered joints, (03104):5
procedure, (03104):1
safety guidelines, (03104):2
soft, (03104):1
tools and equipment, (03104):2–3
Soldering flux, (03104):4–6
Soldering torch, (03104):3–4
Solenoid coil, (03106):26, 31, 49
Solenoid-operated gas valve, (03108):21
Solid geometry, (03102):22
Solvent-cementing plastic pipe, (03103):31–34, 35
Soot, (03108):4
s.p. *See* Static pressure (s.p.)
Sparks, Norman, (03104):36
Specifications
dimensions, (03105):34–35
steel pipe sizes, (03105):2
Specific heat, (03107):2, 7, 81
Specific heat capacity, (03107):7
Specific heat values, common substances, (03107):7
Sphere, (03102):40
Spider systems, (03109):36
SPST. *See* Single-pole, single-throw (SPST) switch
Spud, (03108):10, 48
Square, formula of a, (03102):40
Squares and square roots, (03102):17, 27
Standards vs. codes, (03101):16
Standing pilot, (03108):10, 23, 48
Static head pressure, (03102):10–11
Static pressure (s.p.), (03109):1, 4, 53
Static pressure tips, (03109):12–13
Static-rated dampers, (03109):32
Steam heating systems, (03101):1
Steel, historical uses of, (03105):3
Steel pipe
advantages, (03105):1
assembling
flanged pipe, (03105):22–23
grooved pipe, (03105):18–22
joint compound for, (03105):16
pipe dope for, (03105):16
pipe wrenches, (03105):16–18
polytetrafluoroethylene (PTFE) tape for, (03105):16, 17
threaded pipe, (03105):3–4, 14–16
cutting, (03105):11–12
dimensions
diameter, inside and outside, (03105):2
wall thicknesses, (03105):2
fittings
bushings, (03105):7
caps, (03105):7
cast-iron, (03105):4
couplings, (03105):5, 7
crosses, (03105):4–5
elbows (ells), (03105):5
flanges, (03105):2, 5, 22–23
galvanized iron, (03105):4
grooved, (03105):2, 3
malleable iron, (03105):4
nipples, (03105):5, 7
plugs, (03105):7
tees, (03105):4–5

threaded, (03105):3
  unions, (03105):5, 6
hangers and supports, (03105):24–27
installing, (03105):23–24
joining, (03105):2
measuring, (03105):7–9
reaming, (03105):11–12
refrigerant line, (03103):1
sizes, (03105):2
threading
  bench or tripod threading machine, (03105):14–15
  equipment, (03105):12–13
  hand threader and vise, (03105):13–14
types of, (03105):1–2
Step-up/step-down transformer, (03106):33
Sterrett, Chris, (03103):41
Stock, (03105):1, 3, 33
Straight angle, (03102):23, 24
Straight pipe threads (NPS), (03105):3
Straight pipe wrenches, (03105):17
Straps, (03103):21–25
Strap wrench, (03105):16, 17, 33
Stratify, (03109):16, 30, 53
Stronkowski, John, (03106):45
Subcooling, (03107):2, 6, 81
Suction line, (03107):12, 14, 53–55
Suction line accumulator, (03107):59, 61, (03108):39
Suction pressure, (03107):15, 21
Suction valve, (03107):35, 36
Superheat, (03107):2, 6, 81
Supplementary angles, (03102):23, 24
Supply air duct system, (03109):2–4
Supports
  copper tubing, (03103):20–25
  duct systems, (03109):23
  plastic pipe, (03103):34, 36
  steel pipe, (03105):23–26
Surge chamber, (03107):33, 46, 81
Sustainable construction, (03101):10, 12, 32
Swaged joints, (03103):9, 11
Swaging, (03103):1, 9, 19, 42
Swaging tools, (03103):10, 11, 19
Sweat adapters, (03103):8
Sweat fittings, (03103):8–9, 12–13, (03104):1
Sweating, (03104):1
Switches
  functions, (03106):28–29
  gas furnace safety controls, (03108):24–25
  schematic symbols, (03106):46
  wiring schematics, (03106):28–29
Synthetic refrigerants, (03107):25

T

Tachometer, (03109):14–15
Takeoff, (03105):8, (03109):16, 27, 53
Tapered pipe threads (NPT), (03105):3–4
Tardiness, (03101):19–20
Tee fittings, (03103):6, 7, 8, (03105):4–5
Temperature
  conversions, (03102):12–14, 38–39, 40, (03107):2, 76–78
  defined, (03102):12, (03108):4
  function, (03107):2
  heat pump operations and, (03101):5
  heat transfer and, (03108):4
  measurement scales, (03101):6, (03102):12, (03107):2, (03108):4
  sensible and latent heat, (03107):4–6
Temperature-controlled switch, (03106):29

Temperature limit switches, (03108):24, 34
Temperature measurement devices
  digital multimeters (DMMs), (03109):9
  electronic thermometers, (03109):9–10
  psychrometers, (03109):10
  thermometers
    accuracy, (03107):18
    calibrating, (03107):19–20
    dial-type, (03107):18–19
    dial-type pocket-style, (03107):18–19
    digital, (03102):15
    electronic, (03107):18, (03109):9–10
    infrared, (03107):18
    non-contact, (03107):18
    pocket-style, (03107):18–19
Temperature-pressure relationship
  in HVACR systems, (03101):7, (03102):6–8
  refrigerant, (03107):15–16
  refrigerants, (03107):11–12
  water, (03107):11
Temperature rise measurements, (03108):29
TEV. See Thermostatic expansion valve (TXV or TEV)
Thales, (03102):27
Thermal anemometer, (03109):13
Thermistor, (03106):11, 12, 34–35, 49, (03107):2, 18, 81
Thermistor probe, (03107):18, (03109):9
Thermocouple, (03107):2, 18, 81, (03108):10, 23, 48, (03109):9
Thermocouple probes, (03107):18
Thermometers
  accuracy, (03107):18
  calibrating, (03107):19–20
  dial-type, (03107):18–19
  dial-type pocket-style, (03107):18–19
  digital, (03102):15
  electronic, (03107):18, (03109):9–10
  infrared, (03107):18
  non-contact, (03107):18
  pocket-style, (03107):18–19
Thermopile, (03108):23
Thermostatic expansion valve (TXV or TEV), (03107):52
Thermostatic switch, (03106):28–29
Thermostats
  bimetal, (03107):65
  check filter warnings, (03108):20
  cooling, (03107):64
  electronic, (03107):67
  function, (03107):65
  heating, (03107):65
  heat pump, (03108):39
  hydronic systems, (03108):36
  mercury, (03107):65
  pressure-sensitive, (03107):65
  remote bulb, (03107):65–66
  snap-action, (03107):65
  types of, (03106):35–36
Threaded plastic pipe, (03103):26, 30
Thread engagement, (03105):7
3-4-5 rule, (03102):28
Three-dimensional objects
  solid geometry, (03102):22
  volume measurements, (03102):4–6, 7–8
Time clock, (03107):68
Time-delay fuses, (03106):26, 30, 49
Time-delay relays, (03106):32
Tin-antimony solder, (03104):4
Tin-copper-silver solder, (03104):4
Tin-silver solder, (03104):4
Tire pressure, (03102):9, 11

Ton of refrigeration, (03107):2, 8, 81
Torch handles and tips, (03104):17–18, 19, 25
Torch wrench, (03104):17
Total heat, (03107):2, 5, 81
Total pressure, (03109):1, 4, 53
Toxic, (03101):1, 4, 32
Toxic particles, eliminating, (03101):4–5
Training, HVACR, (03101):21–23
Transducers, (03104):27
Transfer grille, (03109):35, 39, 53
Transformers, (03106):1, 7, 33–34, 49
Triangles
    acute, (03102):26
    equilateral, (03102):26
    formulas for, (03102):40
    isosceles, (03102):26
    obtuse, (03102):26
    right
        angles and sides of, (03102):26–27
        calculations using the Pythagorean theorem,
            (03102):27, 28, 40
    scalene, (03102):26
    sum of angles of, (03102):26
    types of, (03102):26
Trigonometry, (03102):27
Tube-axial fans, (03109):19, 20
Tube-bending springs, (03103):15
Tube-in-tube condensers, (03107):41–42
Tubing. See also Copper tubing
    air conditioning systems, (03103):6
    aluminum, (03103):1
    bending, (03103):15, 16
    cutting, (03103):13, 31
    hard-drawn, (03103):2
    pipe vs., (03103):3
    plastic
        cross-linked polyethylene (PEX), (03103):28–29
        cutting, (03103):31
        fittings, (03103):7
        polyethylene (PE), (03103):26, 27
Tubing benders, (03103):15, 16
Tubing cutters, (03103):13
Tubing sets/packages, (03103):6
Tubular gas-fired radiant heater, (03108):3
Turbine-generators, (03106):1, 6
Twinned furnaces, (03108):25
Two-dimensional objects
    angles, (03102):22–23, 24
    area measurements, (03102):3–4, 5
    circles, (03102):22
    polygon, (03102):23, 25–26, 35
    triangles, (03102):26–28
Two-stage gas valves, (03108):21
TXV. See Thermostatic expansion valve (TXV or TEV)
Type ACR copper tubing, (03103):3–5, 6
Type DWV copper tubing, (03103):3, 4
Type K copper tubing, (03103):3–5, 5
Type L copper tubing, (03103):2–5, 6
Type M copper tubing, (03103):2–5

U
U-inclined manometer, (03109):12
UL Standard
    Leakage Rated Dampers for Use in Smoke Control Systems,
        (03109):25
Ultraviolet air cleaners, (03101):4
Unions, (03105):5, 6
Unit, (03102):1, 35

Unrolling copper tubing, (03103):13
Upflow gas furnace, (03108):10, 11
USGBC. See US Green Building Council (USGBC)
US Green Building Council (USGBC), (03101):12
U-tube manometer, (03109):12

V
V. See Volt (V)
VA. See Volt-amp (VA) rating
Vacuum, (03102):1, 11, 13, 35
Vacuum levels, measuring, (03107):21
Vane-axial fans, (03109):19, 20
Vapor barrier, (03109):16, 25, 53
Vapor barriers, (03109):16, 41–42
Variable, (03102):16, 35
Vazquez, Tony, (03109):51–52
Velocity, (03109):1, 3, 53
Velocity pressure, (03109):1, 4, 53
Velometers, (03109):3, 13–14
Vent damper, (03108):6
Ventilation, (03101):4–5
Venting
    electric furnace, (03108):36
    furnace, (03108):25–27, 28
    hydronic systems, (03108):36
    water heaters, common-vented, (03108):26
Vent pipe, (03108):26
Venturi, (03109):16, 18, 53
Vertex
    of angles, (03102):22
    of polygons, (03102):23
Vertical-inclined manometer, (03109):12
Viega LLC, (03103):8–9
Volt (V), (03106):1, 12, 49
Volta, Alessandro, (03106):12
Voltage
    alternating current (AC), (03106):3, 5–6
    calculating, (03106):12
    death from, (03106):7
    defined, (03106):1, 11, 49
    direct current (DC), (03106):3, 5
    expressions of, (03106):12
    historically, (03106):12
    in HVACR units, (03106):11–12
    measuring, (03106):20
    transformers, (03106):1, 7, 33–34
Voltage testers, (03106):20
Volt-amp (VA) rating, (03106):33
Voltmeter, (03106):20, 37
Volt-ohm-milliammeter (VOM), (03106):19, 49
Volume
    box-shaped objects, (03102):6, 7
    common units of, (03102):4
    conversions
        box-shaped objects, (03102):5, 7
        cylindrical objects, (03102):8
        to inch-pound system, (03102):8
        metric system, (03102):37
        metric to US, (03102):38–39
        US system, (03102):36
        wet to dry measures, (03102):7
    cylindrical objects, (03102):6, 8
    defined, (03102):1, 4, 35
    measuring
        dry volume, (03102):6
        formulas, (03102):5, 16, 40
        practical uses, (03102):4–5
        wet volume, (03102):6

metric system prefixes, (03102):9
units of measure, (03102):1
VOM. *See* Volt-ohm-milliammeter (VOM)

W

W. *See* Watts (W)
Water
  boiling point, (03102):14, (03107):3, 11
  electricity from, (03106):1–2
  freezing point, (03102):12
  heating with, (03101):1, 3
  heat needed to raise by one degree, (03107):2
  pressure exerted by, (03107):10
  specific heat of, (03107):7
  states of matter, (03107):4
  temperature-pressure relationship, (03107):11
Water-cooled condensers, (03107):41–43, 69
Water hammer, (03105):16, 23, 33
Water heaters, common-vented, (03108):26
Water pressure, (03102):10–11
Water systems pressure testing, (03103):19–20
Water vapor (H_2O), (03108):4, 25
Watts (W), (03106):11
Wax molds, (03101):25
Weather predictions, (03102):9, (03107):13
Weight
  conversions, (03102):36, 37, 38–39
  defined, (03102):7
  equivalents, (03102):9

Welded compressor, (03107):34
Wet-bulb temperature, (03109):1, 10, 53
Wetting, (03104):1, 5, 37
Wet volume measurements, (03102):6
Wind energy, (03106):4
Wiring schematics
  circuit breakers, (03106):31
  connection diagram, (03106):37–38
  fuses, (03106):30
  ladder diagram, (03106):37–38
  motors, (03106):27
  overload devices, (03106):35
  relay, (03106):32
  resistance heater, (03106):27
  solenoid coil, (03106):31
  switches, (03106):28–29
  symbols, (03106):46–47
  transformer, (03106):34
Witness mark, (03103):26, 33, 42
Wood, (03101):4
Work hardening, (03103):12, 15, 42
Working pressure rating, hard and soft copper, (03103):2
Wrought copper fittings, (03104):12

Y

Yoke vise, (03105):11, 33
Youth Apprenticeship Program, (03101):23